Sustainable Remediation of Contaminated Soil and Groundwater

Sustainable Remediation of Contaminated Soil and Groundwater

Materials, Processes, and Assessment

Edited by

Deyi Hou

School of Environment
Tsinghua University
Beijing

ELSEVIER

Butterworth-Heinemann
An imprint of Elsevier

Butterworth-Heinemann is an imprint of Elsevier
The Boulevard, Langford Lane, Kidlington, Oxford OX5 1GB, United Kingdom
50 Hampshire Street, 5th Floor, Cambridge, MA 02139, United States

Library of Congress Cataloging-in-Publication Data
A catalog record for this book is available from the Library of Congress

British Library Cataloguing-in-Publication Data
A catalogue record for this book is available from the British Library

ISBN: 978-0-12-817982-6

Publisher: Matthew Deans
Acquisition Editor: Matthew Deans
Editorial Project Manager: Fernanda A Oliveira
Production Project Manager: Selvaraj Raviraj
Cover Designer: Mark Rogers

Typeset by TNQ Technologies

Contents

11. Remedial process optimization and sustainability benefits

Jim Leu and Deyi Hou

12. Landscape architecture and sustainable remediation

Xiaodi Zheng and Niall G. Kirkwood

13. Phytoremediation value chains and modeling

Thomas Upcraft and Miao Guo

Contributors

Paul Bardos, Environmental Technology Ltd., Reading, and University of Brighton, Brighton, United Kingdom

Brian Bone, Bone Environmental Consultant Ltd. and University of Coventry, Coventry, United Kingdom

Mr Liang Chen, Department of Civil and Environmental Engineering, Hong Kong Polytechnic University, Kowloon, Hong Kong, China

Jingqi Dong, China University of Geosciences, Beijing, China; Chinese Academy for Environmental Planning, Beijing, China

Steve Edgar, FLI Global, Bristol, United Kingdom

Jason I. Gerhard, Department of Civil and Environmental Engineering, The University of Western Ontario, London, ON, Canada

Gavin P. Grant, Savron, Toronto, ON, Canada

Daniel R. Griffiths, Parsons Corporation, Denver, CO, United States

Miao Guo, Department of Chemical Engineering, Imperial College London, London, United Kingdom

Deyi Hou, School of Environment, Tsinghua University, Beijing, China

Fei Jin, School of Engineering, University of Glasgow, Glasgow, United Kingdom

Chih-Ming Kao, Institute of Environmental Engineering, National Sun Yat-Sen University, Kaohsiung, Taiwan

Niall G. Kirkwood, School of Architecture, Tsinghua University, Beijing, China; Department of Landscape Architecture, Harvard Graduate School of Design, Harvard University, Cambridge, United States

Solvita Klapare, The World Bank Group, Beijing, China

Petr Kvapil, Photon Water Technology s.r.o, Prague, Czech Republic

Jim Leu, Parsons Corporation, Walnut Creek, CA, United States

Wei-Han Lin, Institute of Environmental Engineering, National Sun Yat-Sen University, Kaohsiung, Taiwan

Yongming Luo, Institute of Soil Science, Chinese Academy of Sciences, Nanjing, China

Oliver McMillan, Department of Engineering, University of Cambridge, Cambridge, United Kingdom

David O'Connor, School of Environment, Tsinghua University, Beijing, China

Jiun-Hau Ou, Institute of Environmental Engineering, National Sun Yat-Sen University, Kaohsiung, Taiwan

Jason H. Prior, Institute for Sustainable Futures, University of Technology Sydney, Sydney, NSW, Australia

Zhengtao Shen, Department of Earth and Atmospheric Sciences, University of Alberta, Edmonton, AB, Canada

Yih-Terng Sheu, Institute of Environmental Engineering, National Sun Yat-Sen University, Kaohsiung, Taiwan

John A. Simon, Nathan Associates, Inc., Remediation Journal, John Wiley & Sons, Arlington, VA, United States

José L. Torero, Department of Civil, Environmental and Geomatic Engineering, University College London, London, United Kingdom

Dr Daniel C.W. Tsang, Department of Civil and Environmental Engineering, Hong Kong Polytechnic University, Kowloon, Hong Kong, China

Thomas Upcraft, Department of Chemical Engineering, Imperial College London, London, United Kingdom

Dr Lei Wang, Department of Civil and Environmental Engineering, Hong Kong Polytechnic University, Kowloon, Hong Kong, China

Jinnan Wang, Chinese Academy for Environmental Planning, Beijing, China

Hongzhen Zhang, Chinese Academy for Environmental Planning, Beijing, China

Yunhui Zhang, Department of Engineering, University of Cambridge, Cambridge, United Kingdom

Xiaodi Zheng, School of Architecture, Tsinghua University, Beijing, China; Department of Landscape Architecture, Harvard Graduate School of Design, Harvard University, Cambridge, United States

Chapter 1

Green and sustainable remediation: concepts, principles, and pertaining research

Deyi Hou, David O'Connor

School of Environment, Tsinghua University, Beijing, China

1. Background

The international community is eagerly seeking new scientific knowledge, policy instruments, and other endeavors to meet the United Nations' Sustainable Development Goals (SDGs), which balance social, economic, and environmental needs. While nearly all industries are seeking sustainability in their operation, one industry has been particularly active, namely, the remediation industry. Literature on sustainable remediation (SR) and green remediation (GR) has grown exponentially over the last decade (see Fig. 1.1), and sustainability is now viewed as an "imperative" in many developed countries such as the US and UK (Hou and Al-Tabbaa, 2014; Hou et al., 2014b).

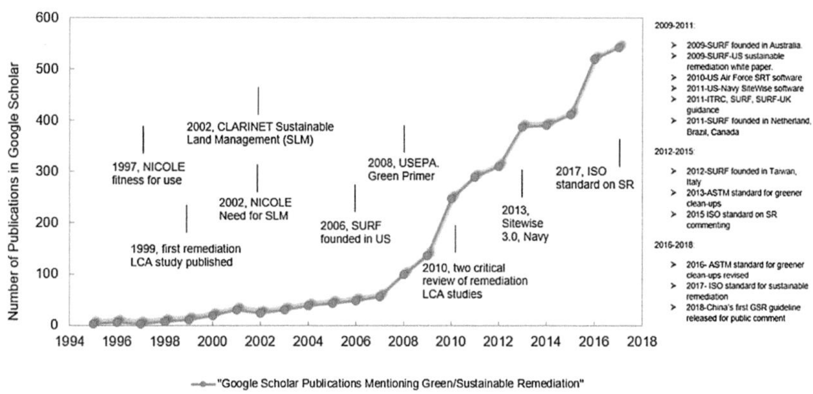

FIGURE 1.1 Growing numbers of publications on green and sustainable remediation.

Sustainable Remediation of Contaminated Soil and Groundwater. https://doi.org/10.1016/B978-0-12-817982-6.00001-X

Green and sustainable remediation (GSR) also renders benefits in emerging remediation markets (note: the subtle differences between SR, GR, and GSR are further discussed in Section 2.2). While traditional remediation focuses on the contaminated site itself and health risk to site users, GSR takes a more holistic approach by examining life-cycle impacts and encompassing wider socio-economic effects (Hou et al., 2014a; Song et al., 2018). Such a holistic approach leads to more efficient usage of limited resources. A strategic benefit for developing countries dealing with large numbers of contaminated sites. Such benefits have been well studied and demonstrated in developed countries but are probably more needed in developing countries (Hou et al., 2016b).

Technical standards will assist the successful implementation of GSR and help bring convergence. The American Society for Testing and Materials (ASTM) published its first Green Remediation standard in 2013 (revised in 2016): ASTM E2893—16e1 Standard Guide for Greener Cleanups. The International Organization for Standardization ISO published a Sustainable Remediation standard in 2017: ISO 18504:2017 Soil quality—Sustainable remediation. A task force, led by Tsinghua University, is currently working on China's first GSR technical standard, which is expected to be published in 2019. Such technical standards will undoubtedly enhance the capability of practitioners who implement GSR.

2. Concepts

2.1 Definition of sustainable remediation

Sustainable remediation (SR) is defined as a holistic approach where the environmental, social, and economic benefits of remediation are maximized for all stakeholders, inside and outside of the site boundary, in both current and future generations.

SR should meet the following five criteria:

1) All viable remediation alternatives are evaluated by an evidence-based sustainability assessment of environmental, social, and economic impacts.
2) The sustainability benefits of the chosen remedial alternative exceed the local and wider detrimental impacts on a life-cycle basis.
3) Relevant and up-to-date best management practice is applied to minimize secondary emissions, waste, energy and resource use, and ecological impacts.
4) The social impacts to workers and local communities are considered and addressed by stakeholder engagement.

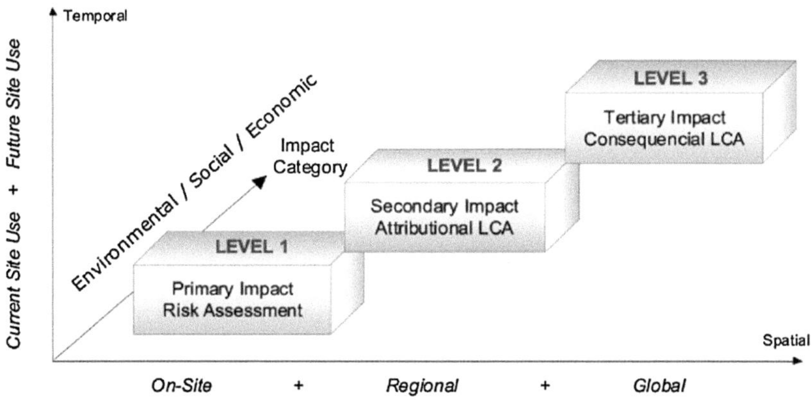

FIGURE 1.2 A framework of sustainability assessment for defining sustainable remediation.

5) The remediation minimizes life-cycle project costs and maximizes gains in the wider economy.

Fig. 1.2 depicts a framework for sustainability assessment at different levels. The new SR paradigm goes beyond the traditional boundary of traditional risk-based remediation in three aspects: temporally, spatially, and in the number of impact categories considered.

2.2 Pertaining concepts

2.2.1 Green remediation

In the United States Environmental Protection Agency (USEPA) definition, the practice of "green remediation" uses strategies such as the use of natural resources and energy efficiently, reduction of negative impacts on the environment, minimization or elimination of pollution at its source, and reduction of waste to the greatest extent possible, in order to take into account all environmental effects of remedy implementation for contaminated sites and to incorporate options that maximize the net environmental benefit of cleanup actions (USEPA, 2008).

2.2.2 Green and sustainable remediation

In the context of this book, green and sustainable remediation (GSR) is synonymous with sustainable remediation. It should be noted that GSR has been defined by the US-based Interstate Technology and Regulatory Council (ITRC) as the site-specific use of products, processes, technologies, and procedures that mitigate contaminant risk to receptors while balancing community goals, economic impacts, and net environmental effects.

2.2.3 Green materials

In the context of this book, "green materials", or more specifically, "green remediation materials", refer to materials that differ from their traditional alternatives in the following ways: (1) their design meets green chemistry principles; or (2) their manufacturing renders lower life-cycle environmental impacts compared with traditional remediation materials based on the common functional unit.

2.2.4 Primary impacts

Primary impacts refer to impacts associated with the state of contaminated sites and site contaminants. Examples of primary impacts include greenhouse gas (GHG) emission from contaminated land, health impacts from site contamination, landscape degradation (Hou et al., 2014c).

2.2.5 Secondary impacts

Secondary impacts refer to impacts associated with remedial activity. Secondary impacts often arise from the use of energy (e.g., fuel and electricity) and materials (e.g., the manufacture of zerovalent iron (ZVI) for permeable reactive barriers or substrates for enhanced in situ bioremediation [EIB]). Post-remediation monitoring can also cause significant secondary impacts, especially monitoring of remediation that has low certainty of meeting remedial objectives, which requires more frequent or longer-term monitoring (Hou et al., 2014d).

2.2.6 Tertiary impacts

Tertiary impacts refer to impacts associated with the holistic postremediation site use, often involving site redevelopment. For example, the redevelopment of brownfield sites in downtown urban areas is an alternative to greenfield development in suburban areas, resulting in less demand for new infrastructure such as road and utilities, as well as shorter commute distances, greater use of public transport, and less energy consumption (Hou et al., 2018b).

3. General principles

3.1 Going beyond the site boundary

Traditional decision-making in contaminated site management focuses on the site itself. Many secondary impacts, such as environmental emission associated with energy and material acquisition, air pollution due to transport, and the environmental risk of landfilling waste, are not typically considered and rarely quantified. In the new paradigm of SR, the impact assessment goes beyond the site boundary, sometimes encompassing the entire planet, e.g., in a life-cycle assessment (LCA). Breaking through the traditional spatial boundary enables more holistic decision-making and greater consideration of societal needs.

3.2 Looking beyond the contemporary time horizon

In SR, decision-making considers not only the possible contemporary impacts but also impacts on future generations. This principle is essential to address the issue of intergenerational inequality. A common mistake in traditional decision-making is to use too high a discount rate in cost—benefit analysis (CBA), which leads to contamination legacies being left for future generations to deal with. For example, passive containment technologies associated with low capital cost and high maintenance cost will appear superior to active remediation technologies where the magnitude of these costs is reversed. In such scenarios, the welfare future generations are seriously "discounted."

3.3 Expanding to social and economic sustainability

Sustainable remediation must address the multifaceted nature of sustainability (Hou et al., 2018a). Most existing sustainability appraisal studies have focused on life-cycle environmental impact (Hou et al., 2016a) but lack quantitative social and economic impact assessments. Among remediation practitioners, this information is often viewed as difficult to ascertain (especially quantitative data), but it is needed for holistic decision-making in remediation planning and design, balancing project costs with social well-being, and wider economic benefits. Social and economic sustainability appraisals can be simplified by combining qualitative and quantitative methods in multicriteria analysis (MCA). The selection of social and economic indicators for this purpose should be well acknowledged among remediation practitioners and measurable (quantitatively or qualitatively) in the context of a specific project.

3.4 Fostering resilience to environmental and social changes

Sustainable remediation renders optimum results not only for the present but also for the future. Given that the natural environment is dynamic, it is important that sustainable remediation strategies are resilient to environmental changes, such as sea level rise, groundwater depletion, and increased precipitation levels. The socioeconomic and regulatory environment may also change. SR needs to take these changes into consideration and should render the following characteristics (Hou and Al-Tabbaa, 2014): (1) capability to meet evolving human health and environmental standards; (2) adaptability to a variety of future site development choices; and (3) resistance to changing geophysical conditions (e.g., climate change).

According to complexity theories, the sustainability of coupled human—nature systems requires both change and persistence (Holling, 2001). Persistence in the remediation field can be considered as the effectiveness of a remediation program, commitment to cleanup, and the effectiveness of remediation systems. Change in the remediation field can be the adaptability of a remediation program, reception of new concepts by remediation practitioners, and flexibility of remediation systems to incorporate new technologies.

3.5 Embracing nature-based solutions

Nature-based solutions (NBSs) are actions inspired by, supported by, or copied from nature (van den Bosch and Sang, 2017). The adoption of NBS to deal with contaminated sites and brownfield redevelopment offers a number of environmental, social, and economic benefits (Song et al., 2019), which range from less energy use and greater material efficiency to fostering resilience in the face of ever-greater global environmental change (Chi, Zuo and Liu, 2017; Liang and Wang, 2017). Some of the NBS alternatives that have been selected at contaminated sites include: (1) revegetation and integration of landscape architecture and urban planning, (2) conversion of brownfields into greenspace, (3) development of green industrial heritage parks, (4) coupling ground source heating with groundwater remediation, and (5) creating nature reserves.

4. Pertaining research on green and sustainable remediation

4.1 Sustainability assessment and sustainable behavior

The path toward sustainability consists of numerous decisions made by individuals, groups, and organizations. A decision can be described as a conscious choice between two or more alternatives. Sustainability assessment is a tool to guide such decision-making. It is conducted after alternative options have been developed but before decisions have been made. Decision-making and carrying out decisions are considered an aspects of human behavior. Sustainable behavior reflects this in meeting sustainability objectives. The behavior spectrum is being increasingly studied to help explain social phenomena and guide decision-making. Irrationality has been identified as a basic feature of both individual behavior (Tversky et al., 1981) and organizational behavior (Brunsson, 1982). The path toward a sustainable future requires not only rational-based and micro-level sustainability assessment but also an understanding of sustainable behavior, which provides a broader and more comprehensive perspective. The latter issue also addresses "carrying out of actions" rather than simply "choosing actions" (Brunsson, 1982).

4.1.1 Life-cycle assessment and sustainability assessment

Sustainability assessments can be broadly defined as processes that "direct the planning and decision-making process toward achieving sustainable development" (Hacking and Guthrie, 2008). Alternatively, sustainability assessment may be defined as processes integrating natural and societal systems, addressing both local and global dimensions, and covering both short-term and long-term perspectives (Ness et al., 2007), or processes determining whether an initiative is sustainable or not (Pope et al., 2004), or an evaluation against a set of sustainability principles (George, 1999; Gibson, 2006). Sustainability has received increasing attention among the scientific research community in the past two decades (see Chapter 2 for an introduction of the history and outlook for future development) (Hou and O'Connor, 2019). GSR has received much attention from governments, industry, and academia. Many sustainability studies have been conducted in the GSR field. These have involved various approaches, such as LCA, MCA, and environmental footprint analysis.

In Chapter 3, important concepts involved in sustainability assessments are outlined, and various tools to aid sustainability assessment are discussed, including LCA and MCA (O'Connor and Hou, 2019). Of these, LCA, as standardized by the ISO 14040 series, is generally considered to be the most comprehensive tool available and is increasingly used by environmental consultants, public authorities, and researchers to support remediation decision-making. However, LCAs can be challenging to perform, particularly for new users. To help clarify its use, a systematic application and interpretation approach is presented. Advanced sustainability assessment approaches, such as economic input—output (IO) LCA and hybrid LCA, are also discussed. Finally, an outlook for the progression of quantitative sustainability assessment is provided.

Chapter 14 presents a sustainability assessment comparing nanoremediation with other in-situ remediation technologies (Bone et al., 2019). Public perceptions were garnered by two public dialogue exercises. More quantitative tiers of assessment and/or engaging the opinions of wider stakeholders are called for. Based on two case studies, the sustainability assessment reveals nanoremediation to be sustainably favorable. However, the authors point out that nanoremediation does not have a long track record, thus posing uncertainty in comparison with other more established methods such as in-situ chemical oxidation.

4.1.2 Best management practices (BMPs)

Best management practice (BMP) is a concept borrowed from civil engineering and adapted to the development of sustainable remediation. It has been adopted by an ASTM workgroup in developing the Standard Guide for Greener Cleanups, defining it as "activities that, if applicable to the situation, typically will reduce the environmental footprint of a cleanup activity". BMPs serve as a basic approach for GR. In Chapter 4, Simon provides a stepwise approach for selecting and implementing sustainable remediation BMPs (Simon, 2019). A summary of the Pharmacia and Upjohn Company LLC remediation in Connecticut, USA, illustrates the benefits of implementing this SR concept.

4.1.3 Norm, rules, and motivational values

Based on institutional theories, norms and rules are "shared linguistic constraint or opportunity that prescribe, permit, or advise actions or outcomes for actors (both individual and corporate)" (Crawford and Ostrom, 1995). The implementation of SR is transforming the norms and rules that guide remediation. Despite this transformation, few studies offer systematic insights into the diverse norms and rules behind SR. In Chapter 15, Bone et al., present a study involving a systematic analysis of the positioning, aggregation, and scoping rules and norms embedded in SR processes in Australia (Bone et al., 2019). The study reports strong interdependence between the norms and rules (sanctions) within the remediation processes and the normative principles operating within the broader domain of environmental management and planning.

4.1.4 Socioeconomic benefits

Assessing the socioeconomic benefits of remediation is an important factor for prioritizing contaminated sites. In Chapter 16, Dong et al. assess three contaminated sites in Chongqing, China, and then provide a preliminary social CBA for contaminated site remediation in China (Dong et al., 2019). The results suggest that under the estimated soil pollution prevention funding (150−200 billion RMB in 2016−20), 1000−1300 sites could be remediated if the strictest remediation Scenario 1 is used, 3300−4400 sites in case of Scenario 2, and 8300−11,000 sites using Scenario 3. As such, using phased risk-based approaches could provide the ability to remediate up to eight times the number of sites compared with business as usual. Analysis of the Chinese soil remediation industry indicates that the release of Soil Pollution Prevention and Control Action will promote a booming Chinese environmental remediation market. By applying a risk-based phased approach, China will form an innovation initiative mechanism and practical situation-based GSR supervision, technology promotion, and public involvement system.

4.2 Green remediation materials

With growing interest in GSR (Hou et al., 2018b; O'Connor and Hou, 2018; O'Connor et al., 2018a), conventional manufacturing of remediation reagents that employ hazardous reagents, including organic solvents, toxic chemicals, and nonbiodegradable stabilizing agents, is becoming a cause of concern (Kalaiselvi et al., 2015; Yang et al., 2016). For example, conventional bottom-up methods for nanoparticle (NP) synthesis often utilize toxic sodium borohydride. Moreover, gaseous hydrogen is formed during the process, which is a safety concern. Intuitively, it is not sensible to risk harm to human health and the environment during the production of NPs that are used to remediate the environment. Recently, various "green synthesis" procedures have been put forward. These involve either: (i) the use of green materials as synthesis reagents, or (ii) employing production methods that consume less energy or natural resources (Wang et al., 2019b, 2019c; Zhang et al., 2017, 2018). Researchers have explored a variety of microbial and phytosynthesis methods, which can be cost-effective, biocompatible, nontoxic, and eco-friendly (Kozma et al., 2016; Luo et al., 2014). Microbial synthesis refers to the use of microorganisms such as bacteria, fungi, yeasts, and viruses, whereas phytosynthesis utilizes plant extracts as reducing agents (Cheng Keong et al., 2017). In addition to green synthesis, researchers have also explored the conversion of biological waste into useable remediation materials, including biochar produced from agricultural waste - a particularly promising material (O'Connor et al., 2018b; Shen et al., 2018; Zhao et al., 2018). Moreover, a wide range of modification methods have been developed to render high-performance modified biochar (Shen et al., 2019a).

4.2.1 Waste-based stabilization materials

Cement-based solidification/stabilization (S/S) can be a relatively low-cost, time-efficient, and versatile way to treat contaminated soils. Portland cement (PC) is extensively used for this purposes. However, PC production is associated with abundant CO_2 emission and, therefore, high carbon footprints are a major limitation on S/S sustainability. By using industrial wastes, such as slags or ashes, as additives to PC, the environmental footprint can be significantly reduced. In recent years, researchers have paid increasing attention to the use of agricultural/industrial by-products to synthesize stabilization materials.

In Chapter 5, Wang et al. introduce a number of waste-based stabilization materials, including pulverized fly ash (PFA), silica fumes (SF), blast furnace slag (BFS) and granulated blast furnace slag (GGBS), incinerated sewage sludge ash (ISSA), calcium carbide residue (CCR), and metakaolin (MK) (Wang et al., 2019a). The state-of-art mechanisms, benefits, and challenges of these cements are systematically discussed. Accelerated CO_2 curing is also discussed where pressurized CO_2 gas reacts with CH to generate stable carbonation products in a short period of time (2–24 hours). This rapid reaction mitigates the negative effects of calcium–organic complexation and pollutants. The utilization of CO_2 curing also enables CO_2 sequestration and decreases the life-cycle carbon footprint. Overall, waste-based stabilization materials are an effective, economical, and environmentally friendly GR materials that fill a niche in the market.

4.2.2 Biochar for soil remediation

Biochar is the solid, carbon-rich product of the pyrolysis of biomass, which can be used to reduce contaminant bioavailability in soils. It has been studied particularly as a means to reduce Cd enrichment of rice crops. Dozens of biochar field trials at contaminated sites have been performed. Overall, the field effectiveness depends on a myriad of factors including the application time, site-specific factors (e.g., climate, biochar dosage rate, and mixing depth), biochar feedstock type, and biochar properties.

In Chapter 6, Shen et al. point out that the use of biochar in soil remediation offers a number of environmental benefits that align with GSR (Shen et al., 2019b). For example, (1) biochar can be produced using agricultural, household, or industrial wastes (e.g., crop residues, food scraps, manure, and sludges); (2) coproducts formed during biochar pyrolysis, such as syngas and bio-oil, can be used as green energy sources; (3) biochar can help increase soil fertility by adding nutrients or by improving soil structure or pH; and (4) biochar production converts labile carbon into a recalcitrant form and, therefore, can store carbon for hundreds to thousands of years.

4.2.3 Slow-release materials for groundwater remediation

There are various challenges for the effective long-term remediation of recalcitrant organic pollutants in groundwater, such as back diffusion, tailing, and rebound. Traditional remediation techniques, such as pump and treat (P&T), often require long periods of operation, resulting in high energy inputs and large environmental footprints. In recent years, a variety of slow-release materials (SRMs) have been put forward to counteract such issues. SRMs offer sustainability benefits such as reducing nonselective consumption of oxidants and the overall dosage of treatment reagents and prevent the generation of toxic oxidation by-products. SRMs typically consist of binding agents plus reactive reagents. The reactive materials can be broadly classified into three groups: (1) carbon sources (i.e., electron donors for anaerobic microbial stimulation), (2) oxygen sources (i.e., electron acceptors for aerobic microbial stimulation), and (3) oxidants (i.e., that participate in redox reactions). The unreactive binding agent provides a basis for the slow release of reactive constituents.

In Chapter 7, Kao et al. introduce the application of SRMs in passive permeable reactive barriers (PRBs), bio-barrier systems, and liquid substrate injection systems. SRMs that release chemical oxidants, dissolved oxygen, carbon substrates are described (Kao et al., 2019). Based on recent literature, a variety of contaminants in groundwater have been treated by SRMs. For example, TCE, PCE, BTEX, and 1,4-Dioxane have been treated via in situ chemical oxidation; TCE and 1,2-DCA have been treated by anaerobic dechlorination stimulated by SRM substrates; and BTEX, 1,4-Dioxane, and ammonia have been treated by aerobic biodegradation stimulated by oxygen-releasing SRMs. These SRMs can achieve contaminant removal efficiencies of up to 100%. The reaction kinetics depend on dosage ratio and release rate. Overall, the use of SRMs renders multiple benefits, including reduced secondary effects and sustained treatment of pollutants. However, a lack of field studies hinders the uptake of this technology in the remediation industry.

4.3 Sustainable remediation processes

Sustainable remediation is not limited to the appraisal of remediation alternatives and decision-making. An important aspect is the development and deployment of greener and cleaner processes. In general, the choice of either ex- or in-situ remediation has a strong bearing on the overall sustainability. For instance, in-situ groundwater remediation using permeable reactive barriers (PRBs) is associated with far less life-cycle impacts than ex-situ groundwater remediation with P&T (Higgins and Olson, 2009). Similarly, in-situ bioremediation (ISB) and in-situ thermal desorption are associated with lower life-cycle impacts than excavation with off-site treatment and disposal (Lemming et al., 2010). Studies have also shown that in-situ containment is more sustainable than ex-situ treatment or disposal, based on LCA (Blanc et al., 2004; Harbottle et al., 2008). Both quantitative assessment methods and innovative remediation technologies are needed to minimize life-cycle primary, secondary, and tertiary environmental impacts and maximize sustainability benefits.

4.3.1 Phytoremediation

Phytoremediation is a GR strategy in which plants are used to remove or reduce the risks associated with contaminated land. The umbrella term, phytoremediation, incorporates a range of specific processes including phytodegradation, phytovolatilization, phytoextraction, phytostabilization, and rhizofiltration. In Chapter 13, Guo and Upcraft present the concept of the phytoremediation value chain, covering metal removal, biomass cultivation and supply, refinery of bioproducts, storage, distribution, and demand. The authors discuss characteristics and design challenges, rendering an overview of phytoremediation processes and mechanisms, including empirical research in soil metal remediation by terrestrial plant species (Upcraft and Guo, 2019). A state-of-the-art summary highlights emerging research gaps and identified future research frontiers to unlock the complexity of phytoremediation value chains. The authors also discuss a phytoremediation modeling framework, underpinned by process systems engineering, to inform multicriteria multi-echelon decision-making.

4.3.2 In-situ bioremediation

In-situ bioremediation (ISB) technology offers remediation engineers a balance between cost and effectiveness while maintaining site function/availability. However, there are many potential pitfalls that may occur through either lack of sufficient site characterization, or inadequate application design, or improper installation. ISB can be applied in oxic and anoxic geochemical conditions, depending on the contaminants to be degraded and the degradation mechanism. Anaerobic ISB is a well-proven remedial option for addressing chlorinated solvents in groundwater with minimal disturbance at the ground surface. However, under certain conditions, it can lead to potentially dangerous consequences. In Chapter 8, Griffiths provides an overview of common problems encountered during anaerobic ISB (Griffiths, 2019), including impacts on groundwater quality (e.g., color, odor, dissolved iron, manganese, turbidity, total dissolved solids (TDS), sulfides, suppressed pH, biofouling). Elevated levels of total organic carbon (TOC), lowered oxidation–reduction potential (ORP), and elevated levels of metabolic by-products (e.g., CO_2, ferrous iron, methane) can migrate out of the treatment area with groundwater flow, impacting down-gradient groundwater quality. Secondary impacts vary widely in terms of risk (e.g., methane may represent a higher risk than ferrous iron) and impact importance (e.g., methane intruding into a building basement is more hazardous than dissolved methane discharging to a stream). Thus, responses to secondary water quality impacts can also vary widely and should be applied on a case-by-case basis.

4.3.3 Self-sustaining treatment for active remediation

Self-sustaining Treatment for Active Remediation (STAR) is an innovative thermal technology for treating soils contaminated by organic pollutants including coal tar, crude oil sludge, and other difficult-to-treat materials as well as some emerging contaminants. With this technology, air is applied to propagate smoldering—a low-temperature, flameless form of combustion. The self-sustaining reaction travels from an ignition location through contaminated soil, destroying contamination, while a small fraction is recovered as a vapor phase. In Chapter 9, Gerhard et al. discuss theories and application principles for STAR (Gerhard et al., 2019). Both in-situ and ex-situ processes are described, with case studies to illustrate the technology. Moreover, the authors discuss the sustainability implications of STAR. Based on a case study in the United States, STAR was found to render an order-of-magnitude lower environmental footprint compared with thermal desorption and dig and haul technologies.

4.3.4 In-situ solidification/stabilization

S/S treatments limit the release of harmful chemicals from hazardous solid wastes. The solidification aspect refers to encapsulation in a monolithic solid of adequate structural integrity, and stabilization refers to chemical reaction that converts contaminants into less mobile or less toxic forms. Stabilization reactions include adsorption, precipitation, isomorphous substitution, and lattice incorporation. The long-term effectiveness of S/S has been questioned because of various environmental stresses that potentially degrade S/S materials. In Chapter 10, Jin reviews the immobilization and leaching mechanisms of S/S treated soils and how their evolution over time is affected by numerous factors (Jin, 2019). Long-term field-based research is summarized to assess real-world long-term effectiveness. It is concluded that, generally, in-situ S/S treatments can be effective under natural conditions for many years. Finally, insights are provided on ways to improve the long-term performance and sustainability of in-situ S/S.

4.3.5 Remedial process optimization

Remedial process optimization (RPO) is a systematic approach for evaluating the performance and effectiveness of existing site remediation systems and identifying ways to move sites toward closeout more quickly or more cost-effectively. It applies the principles of value engineering to the implementation components of remediation. In Chapter 11, Leu and Hou summarize the principles of a phased approach to RPO and illustrate the application of RPO in five case studies (Leu and Hou, 2019). The sustainability benefits of RPO practices are further evaluated. It is suggested that the remediation industry should consider sustainability aspects when developing RPO recommendations to maximize all environmental, economic, and social benefits as well as expediting site closure. Overall, RPO is an effective way to reduce the life-cycle environmental footprints of remedial operations, making remediation greener and more sustainable.

4.3.6 Landscape architecture

Brownfield is the most widely distributed type of contaminated sites. In brownfield remediation and redevelopment, landscape architecture can play an important role. It can address the complicated issues involved in the process of brownfields regeneration from all three aspects of ecological process, social humanities, and physical environment. In Chapter 12, Zheng and Kirkwood explain the key aspects of brownfields regeneration tackled by the discipline of landscape architecture (Zheng and Kirkwood, 2019). They elaborated the concept of "brown earth-work" and used this concept to facilitate the integration of landscape and environmental engineering. The authors further proposed landscape architecture strategies to achieve sustainable remediation goals and provided prospects for future development.

References

Blanc, A., Métivier-Pignon, H., Gourdon, R., Rousseaux, P., 2004. Life cycle assessment as a tool for controlling the development of technical activities: application to the remediation of a site contaminated by sulfur. Advances in Environmental Research 8, 613–627.

Bone, B., Bardos, P., Edgar, S., Kvapil, P., 2019. The sustainability of nanoremediation – two initial case studies from Europe. In: Hou, D. (Ed.), Sustainable Remediation of Contaminated Soil and Groundwater: Materials, Processes, and Assessment. Elsevier.

van den Bosch, M., Sang, A.O., 2017. Urban natural environments as nature-based solutions for improved public health - A systematic review of reviews. Environmental Research 158, 373–384.

Brunsson, N., 1982. The irrationality of action and action rationality: decisions, ideologies and organizational actions. Journal of Management Studies 19, 29–44.

Chi, T., Zuo, J., Liu, F., 2017. Performance and mechanism for cadmium and lead adsorption from water and soil by corn straw biochar. Frontiers of Environmental Science & Engineering 11 (2).

Cheng Keong, C., Sunitha Vivek, Y., Salamatinia, B., Amini Horri, B., 2017. Green synthesis of ZnO nanoparticles by an alginate mediated ion-exchange process and a case study for photocatalysis of methylene blue dye. Journal of Physics: Conference Series 829, 012014.

Crawford, S.E., Ostrom, E., 1995. A grammar of institutions. American Political Science Review 89, 582–600.

Dong, J., Zhang, H., Klapare, S., Wang, J., Luo, Y., 2019. Socioeconomic benefit of contaminated site remediation. In: Hou, D. (Ed.), Sustainable Remediation of Contaminated Soil and Groundwater: Materials, Processes, and Assessment. Elsevier.

George, C., 1999. Testing for sustainable development through environmental assessment. Environmental Impact Assessment Review 19, 175–200.

Gerhard, J., Grant, G., Torero, J.S.T.A.R., 2019. A uniquely sustainable in situ and ex situ remediation process. In: Hou, D. (Ed.), Sustainable Remediation of Contaminated Soil and Groundwater: Materials, Processes, and Assessment. Elsevier.

Gibson, R.B., 2006. Sustainability assessment: basic components of a practical approach. Impact Assessment and Project Appraisal 24, 170–182.

Griffiths, D.R., 2019. Controlling secondary pollution impacts during enhanced in-situ anaerobic bioremediation. In: Hou, D. (Ed.), Sustainable Remediation of Contaminated Soil and Groundwater: Materials, Processes, and Assessment. Elsevier.

Hacking, T., Guthrie, P., 2008. A framework for clarifying the meaning of triple bottom-line, integrated, and sustainability assessment. Environmental Impact Assessment Review 28, 73–89.

Harbottle, M.J., Al-Tabbaa, A., Evans, C.W., 2008. Sustainability of land remediation: Part I: overall analysis. Proceedings of the Institution of Civil Engineers-Geotechnical Engineering 161, 75–92.

Higgins, M.R., Olson, T.M., 2009. Life-cycle case study comparison of permeable reactive barrier versus pump-and-treat remediation. Environmental Science and Technology 43, 9432–9438.

Holling, C.S., 2001. Understanding the complexity of economic, ecological, and social systems. Ecosystems 4, 390–405.

Hou, D., Al-Tabbaa, A., 2014. Sustainability: a new imperative in contaminated land remediation. Environmental Science and Policy 39, 25–34.

Hou, D., Al-Tabbaa, A., Chen, H., Mamic, I., 2014a. Factor analysis and structural equation modeling of sustainable behaviour in contaminated land remediation. Journal of Cleaner Production 84, 439–449.

Hou, D., Al-Tabbaa, A., Guthrie, P., 2014b. The adoption of sustainable remediation behavior in the US and UK: a cross country comparison and determinant analysis. The Science of the Total Environment 490, 905–913.

Hou, D., Al-Tabbaa, A., Guthrie, P., Hellings, J., 2014c. Using a hybrid LCA method to evaluate the sustainability of sediment remediation at the London Olympic park. Journal of Cleaner Production 83, 87–95.

Hou, D., Al-Tabbaa, A., Luo, J., 2014d. Assessing effects of site characteristics on remediation secondary life cycle impact with a generalized framework. Journal of Environmental Planning and Management 57, 1083–1100.

Hou, D., Ding, Z., Li, G., Wu, L., Hu, P., Guo, G., et al., 2018a. A sustainability assessment framework for agricultural land remediation in China. Land Degradation and Development 29, 1005–1018.

Hou, D., Gu, Q., Ma, F., O'Connell, S., 2016a. Life cycle assessment comparison of thermal desorption and stabilization/solidification of mercury contaminated soil on agricultural land. Journal of Cleaner Production 139, 949–956.

Hou, D., Guthrie, P., Rigby, M., 2016b. Assessing the trend in sustainable remediation: a questionnaire survey of remediation professionals in various countries. Journal of Environmental Management 184, 18–26.

Hou, D., O'Connor, D., 2019. Green and sustainable remediation: past, present, and future developments. In: Hou, D. (Ed.), Sustainable Remediation of Contaminated Soil and Groundwater: Materials, Processes, and Assessment. Elsevier.

Hou, D., Song, Y., Zhang, J., Hou, M., O'Connor, D., Harclerode, M., 2018b. Climate change mitigation potential of contaminated land redevelopment: a city-level assessment method. Journal of Cleaner Production 171, 1396–1406.

Jin, F., 2019. Long-term effectiveness of in situ solidification/stabilization. In: Hou, D. (Ed.), Sustainable Remediation of Contaminated Soil and Groundwater: Materials, Processes, and Assessment. Elsevier.

Kalaiselvi, A., Roopan, S.M., Madhumitha, G., Ramalingam, C., Elango, G., 2015. Synthesis and characterization of palladium nanoparticles using Catharanthus roseus leaf extract and its application in the photo-catalytic degradation. Spectrochimica Acta Part A: Molecular and Biomolecular Spectroscopy 135, 116–119.

Kao, C.-M., Sheu, Y.-T., Ou, J.-H., Lin, W.-H., 2019. Application of slow release materials for in situ and passive remediation of contaminated groundwater. In: Hou, D. (Ed.), Sustainable Remediation of Contaminated Soil and Groundwater: Materials, Processes, and Assessment. Elsevier.

Kozma, G., Rónavári, A., Kónya, Z., Kukovecz, Á., 2016. Environmentally benign synthesis methods of zero-valent iron nanoparticles. ACS Sustainable Chemistry and Engineering 4, 291–297.

Lemming, G., Hauschild, M.Z., Chambon, J., Binning, P.J., Bulle, C., Margni, M., et al., 2010. Environmental impacts of remediation of a trichloroethene-contaminated site: life cycle assessment of remediation alternatives. Environmental Science and Technology 44, 9163–9169.

Leu, J., Hou, D., 2019. Remedial process optimization and sustainability benefits. In: Hou, D. (Ed.), Sustainable Remediation of Contaminated Soil and Groundwater: Materials, Processes, and Assessment. Elsevier.

Liang, D., Wang, S., 2017. Development and characterization of an anaerobic microcosm for reductive dechlorination of PCBs. Frontiers of Environmental Science & Engineering 11 (6).

Luo, F., Chen, Z., Megharaj, M., Naidu, R., 2014. Biomolecules in grape leaf extract involved in one-step synthesis of iron-based nanoparticles. RSC Advances 4, 53467–53474.

Ness, B., Urbel-Piirsalu, E., Anderberg, S., Olsson, L., 2007. Categorising tools for sustainability assessment. Ecological Economics 60, 498–508.

O'Connor, D., Hou, D., 2018. Targeting cleanups towards a more sustainable future. Environ Sci Process Impacts 20 (2), 266–269.

O'Connor, D., Hou, D., 2019. Sustainability assessment for remediation decision making. In: Hou, D. (Ed.), Sustainable Remediation of Contaminated Soil and Groundwater: Materials, Processes, and Assessment. Elsevier.

O'Connor, D., Peng, T., Li, G., Wang, S., Duan, L., Mulder, J., et al., 2018a. Sulfur-modified rice husk biochar: a green method for the remediation of mercury contaminated soil. The Science of the Total Environment 621, 819–826.

O'Connor, D., Peng, T., Zhang, J., Tsang, D.C., Alessi, D.S., Shen, Z., et al., 2018b. Biochar application for the remediation of heavy metal polluted land: a review of in situ field trials. The Science of the Total Environment 619, 815–826.

Pope, J., Annandale, D., Morrison-Saunders, A., 2004. Conceptualising sustainability assessment. Environmental Impact Assessment Review 24, 595–616.

Shen, Z., Hou, D., Zhao, B., Xu, W., Ok, Y.S., Bolan, N.S., et al., 2018. Stability of heavy metals in soil washing residue with and without biochar addition under accelerated ageing. The Science of the Total Environment 619, 185–193.

Shen, Z., Zhang, J., Hou, D., Tsang, D.C.W., Ok, Y.S., Alessi, D.S., 2019a. Synthesis of MgO-coated corncob biochar and its application in lead stabilization in a soil washing residue. Environment International 122, 357–362.

Shen, Z., Zhang, Y., McMillan, O., O'Connor, D., Hou, D., 2019b. The use of biochar for sustainable treatment of contaminated soils. In: Hou, D. (Ed.), Sustainable Remediation of Contaminated Soil and Groundwater: Materials, Processes, and Assessment. Elsevier.

Simon, J.A., 2019. Best management practices for sustainable remediation. In: Hou, D. (Ed.), Sustainable Remediation of Contaminated Soil and Groundwater: Materials, Processes, and Assessment. Elsevier.

Song, Y., Hou, D., Zhang, J., O'Connor, D., Li, G., Gu, Q., et al., 2018. Environmental and socio-economic sustainability appraisal of contaminated land remediation strategies: a case study at a mega-site in China. The Science of the Total Environment 610, 391–401.

Song, Y., Kirkwood, N., Maksimović, Č., Zhen, X., O'Connor, D., Jin, Y., Hou, D., 2019. Nature based solutions for contaminated land remediation and brownfield redevelopment in cities: A review. Science of the Total Environment 663, 568–579.

Tversky, A., Kahneman, D., Choice, R., 1981. The framing of decisions. Science 211, 453–458.

Upcraft, T., Guo, M., 2019. Phytoremediation value chains and modelling. In: Hou, D. (Ed.), Sustainable Remediation of Contaminated Soil and Groundwater: Materials, Processes, and Assessment. Elsevier.

USEPA, 2008. Green Remediation: Incorporating Sustainable Environmental Practices into Remediation of Contaminated Sites. EPA 542-R-08-002. USEPA. United States Environmental Protection Agency, Washington, DC.

Wang, L., Chen, L., Tsang, D.C.W., 2019a. Green remediation by using low-carbon cement-based stabilization/solidification approaches. In: Hou, D. (Ed.), Sustainable Remediation of Contaminated Soil and Groundwater: Materials, Processes, and Assessment. Elsevier.

Wang, Y., Li, Q., Zhang, P., O'Connor, D., Varma, R.S., Yu, M., Hou, D., 2019b. One-pot green synthesis of bimetallic hollow palladium-platinum nanotubes for enhanced catalytic reduction of p-nitrophenol. Journal of Colloid and Interface Science 539, 161–167.

Wang, Y., O'Connor, D., Shen, Z., Lo, I.M.C., Tsang, D.C.W., Pehkonen, S., et al., 2019c. Green synthesis of nanoparticles for the remediation of contaminated waters and soils: constituents, synthesizing methods, and influencing factors. Journal of Cleaner Production 226, 540–549.

Yang, H., Liu, X., Sun, S., Nie, Y., Wu, H., Yang, T., et al., 2016. Green and facile synthesis of graphene nanosheets/K3PW12O40 nanocomposites with enhanced photocatalytic activities. Materials Research Bulletin 78, 112–118.

Zhang, P., Hou, D., O'Connor, D., Li, X., Pehkonen, S.O., Varma, R.S., et al., 2018. Green and size-specific synthesis of stable Fe-Cu oxides as earth-abundant adsorbents for malachite green removal. ACS Sustainable Chemistry & Engineering 9229–9236.

Zhang, P., Lo, I., O'Connor, D., Pehkonen, S., Cheng, H., Hou, D., 2017. High efficiency removal of methylene blue using SDS surface-modified $ZnFe_2O_4$ nanoparticles. Journal of Colloid and Interface Science 508, 39–48.

Zhao, B., O'Connor, D., Zhang, J., Peng, T., Shen, Z., Tsang, D.C.W., et al., 2018. Effect of pyrolysis temperature, heating rate, and residence time on rapeseed stem derived biochar. Journal of Cleaner Production 174, 977–987.

Zheng, X., Kirkwood, N.G., 2019. Landscape architecture and sustainable remediation. In: Hou, D. (Ed.), Sustainable Remediation of Contaminated Soil and Groundwater: Materials, Processes, and Assessment. Elsevier.

Chapter 2

Green and sustainable remediation: past, present, and future developments

Deyi Hou, David O'Connor
School of Environment, Tsinghua University, Beijing, China

1. Introduction

Land contamination is a global issue, representing a serious threat to public health and the wider environment. In the United States, the United States Environmental Protection Agency (USEPA) has estimated the existence of hundreds of thousands of contaminated sites. At the current pace of cleanup, it will take another 100–200 years to remediate all of these sites (USEPA, 2004). In China, 16% of soils assessed during a recent official national soil quality survey were found to exceed environmental quality standards (MEP, 2014), triggering the Chinese national government to put forward a highly ambitious action plan. Around the world, millions of contaminated sites are in need of cleanup, requiring investment on the scale of trillions of US dollars (Hou and Li, 2017). There are tremendous pressures being placed on national governments to accelerate their remediation efforts so that contaminated sites can be redeveloped quickly. In the United States, the USEPA views it as one of its five most important strategic goals. In the United Kingdom, a task force commissioned by the former Department of the Environment, Transport, and Regions (DETR) proposed that all their polluted sites be restored by the year 2030 (DETR, 1999). This demand for acceleration comes at a stage when we have come to realize the potential environmental, social, and economic side effects of remediation.

The remediation industry has evolved through three broad stages (see Fig. 2.1). Initially, remediation practitioners were expected to "remove all" contaminants due to the unrealistic demands of regulators and public pressure brought by some high-profile cases. However, by the mid-1990s, practitioners had come to recognize that there were significant biogeophysical constraints that prevented the possibility of remediating many sites to pristine conditions.

Sustainable Remediation of Contaminated Soil and Groundwater. https://doi.org/10.1016/B978-0-12-817982-6.00002-1

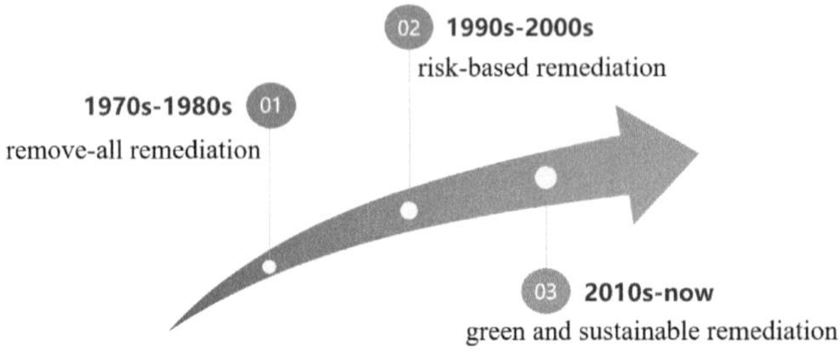

FIGURE 2.1 Three stages of development in the remediation field, 1970s—present.

Moreover, many nations recognized that the cost to remove all contaminants would considerably outweigh the perceived benefits to society. Instead a compromise solution was embraced, favoring remediation that makes land suitable for an intended use, but no further (O'Connor and Hou, 2018). This brought about the risk-based approach that is used today, allowing environmental cleanup standards to be set to "acceptable" levels and thus deciding what level of contamination could safely remain in situ after cleanup. In the latest era, a new imperative in remediation has arisen, demanding that the industry conduct remediation in a way that addresses the potential side effects associated with cleanup activities. This has led to movements known as "green remediation" in the United States and "sustainable remediation" in Europe (Hou and Al-Tabbaa, 2014). The term "green and sustainable remediation (GSR)" has been used to unite this concept.

Generally speaking, GSR can be divided into the three pillars of sustainability: environmental, economic, and social. Environmental sustainability of remediation stresses the importance of managing risks to human health and the wider environment, while minimizing the life-cycle environmental impacts associated with remediation operations, e.g., any emissions caused by the manufacture of treatment reagents or energy usage during remedial operations. The economic sustainability aspect includes life-cycle project costs and benefits and the effects that remediation can have on local economies, e.g., workers' wages, local employment, etc. The social aspect takes into account the social impacts of remediation projects on affected people's lives, including remediation workers, the local community, and vulnerable groups. This includes, for example, the assessment of the impacts on remediation: worker's health and safety, neighborhood impacts, stakeholder satisfaction, social inclusion, etc.

Demand for GSR has been prompted by multiple mechanisms. Firstly, based on scientific studies in the late 1990s and early 2000s (Blanc et al., 2004; Diamond et al., 1999; Volkwein et al., 1999), it was first brought to attention that remediation operations were themselves associated with adverse environmental impacts (e.g., secondary environmental emissions). These secondary impacts involve, for example, greenhouse gas (GHG) emissions, fossil fuel consumption, air pollution, eutrophication, etc. The second reason was that key remediation stakeholders, as well as society as a whole, were increasingly demanding "sustainability" in the progress of modern society. For instance, many of the large international corporations who are prominent contaminated site owners will now require their contractors to conduct sustainability assessments and track the performance of GSR indicators in remediation projects. Some consultants in the United States and the United Kingdom are purposefully developing GSR decision support tools, as well as conducting research and development into innovative technologies that meet GSR principles. Moreover, federal and state regulators in the United States are encouraging green remediation within their jurisdictions (DTSC, 2009; ITRC, 2011b; USEPA, 2008). Thirdly, there are new social norms and institutional pressures growing within the remediation industry, such as protocols issued by professional bodies, ethical principles, regulatory requirements, and so on. A number of professional organizations have been formed within the remediation industry who have the intention of promoting GSR. With this aim, they have published numerous white papers and technical guidelines for GSR implementation (Ellis and Hadley, 2009; Surf-UK, 2010; Surf-UK, 2014). In this chapter, we explore the development of GSR in the context of the past, present, and future, starting with the historical development of sustainable remediation in Europe and green remediation in the United States and ending with a discussion of future research needs, obstacles, and a suggested path forward.

2. The past of sustainable remediation (1990s–2010)

Although GSR has only recently been adopted by the remediation industry, there has been a growing interest in "sustainable infrastructure" since the early 1990s (Choguill, 1996; Dasgupta and Tam, 2005). This came about because civil systems such as sanitation, transportation, and utilities have wide-ranging implications for society beyond the normal realm of construction design and management. Remediation is also viewed as an infrastructure issue, often encompassing large-scale developments, and, therefore, the emergence of sustainable remediation may be viewed as a natural extension of the sustainable infrastructure movement. Naturally, the remediation field adopted lessons learned from the sustainable infrastructure movement. However, there were some specific challenges to applying sustainability in the remediation field. For instance, unlike the typically limited useful life of infrastructure, remediation can bring about a permanent change to the condition of a site. This poses a challenge in defining temporal boundaries in sustainability assessments.

The challenge of applying sustainability to remediation was first attempted by several professional bodies sponsored by the remediation industry in the late 1990s. These organizations included the Network for Industrially Contaminated Land in Europe (NICOLE, founded in 1995 in Europe), Contaminated Land:

Applications in Real Environments (CLAIRE, founded in 1999 in the United Kingdom), and Sustainable Remediation Forum (SuRF, founded in 2006 in the United States). Since 2002, NICOLE has published a series of publications on sustainable contaminated land management within the risk-based land management context (NICOLE, 2002; NICOLE, 2005). As a more recent initiative, SuRF has played an active role in promoting sustainable remediation since its foundation, with a series of white papers and guidance documents published over the last decade. In addition to industry-led bodies, governmental departments and agencies have also taken up initiatives on sustainable remediation. Among European policy makers and regulators, the COMMON FORUM on Contaminated Land in the European Union (a network of contaminated land policy maker and advisors from national ministries and environmental agencies) has embraced the sustainable remediation concept.

2.1 Sustainable management of contaminated land in Europe

The remediation industry in Europe began in the 1970s and early 1980s, driven by a number of high-profile contaminated site cases, such as Lekkerkerk in the Netherlands. This echoes the situation in the United States, where the Love Canal superfund site and other high-profile cases sparked the US remediation industry. The European remediation field grew fast in the 1980 and 1990s due to the introduction of various laws and regulations. An initial misperception that contaminated land was severe but only limited in scale meant that the determination of whether or not soil was contaminated was based on environmental quality criteria, such as the Dutch ABC and UK ICRCL intervention values. This often necessitated the complete removal of contaminated media, commonly by dig and haul (D&H) for contaminated soil, and pump and treat (P&T) for contaminated groundwater. In the mid-1990s, European remediation practitioners had realized that the contaminated land issue was so wide-spread that it was unrealistic to deal with all the sites in this way. In the Netherlands, the number of contaminated sites needing cleanup grew from 350 sites in 1981 to 300,000 sites in 1995 (Ferguson, 1999). In the United Kingdom, remediation experts complained about "overengineering" of remediation projects. The newly emerged "risk-based" approach started to take over, which was instigated with the introduction of Part 2A of the Environmental Protect Act (1990).

In the early 2000s, regulators and professional organizations in Europe called for a new paradigm in managing contaminated sites. The so-called "site specific risk-based contaminated land management" strategy, incorporating "sustainable land management" was developed and promoted (CLARINET, 2002; NICOLE, 2002). The risk-based method used by Europeans was akin to the widely used source−pathway−receptor model developed in the United States (e.g., ASTM standard for risk-based corrective action [RBCA]) (ASTM, 2010). In the Risk-Based Land Management (RBLM) framework developed by CLARINET, sustainability became a key objective, including the assessment and maximization of environmental, economic, and social benefits (NICOLE, 2005).

While the sustainability concept commenced among remediation regulators and professionals, researchers were developing one of the important aspects of sustainable remediation: life-cycle assessment (LCA). In the late 1990s, LCA was first used by researchers to quantify the environmental footprint of remediation operations (Diamond et al., 1999; Page et al., 1999; Volkwein et al., 1999). Following these efforts, remediation experts in various countries started to recognize the significance of the adverse effects associated with remediation activities. Researchers at the University of Cambridge, UK, stated that sustainable remediation must meet several basic principles (Al-Tabbaa et al., 2007a, 2007b): (1) future benefits outweigh the cost of remediation; (2) the environmental impacts of the remediation process are less than the impact of leaving the land untreated; (3) the environmental impact of the remediation process is minimal and measurable; (4) the timescale over which the environmental consequences occur, including intergenerational risk, is part of the decision-making process; and (5) the decision-making process includes an appropriate level of engagement with all stakeholders.

The emergence of the sustainable remediation movement in Europe led to significant changes in the remediation field. Some of the sustainable strategies that were adopted are shown in Fig. 2.2. Among others, sustainable land management in Europe encouraged technological innovation. For example, in the Netherlands, designers employed heat–cold storage facilities at contaminated sites and coupled groundwater energy with groundwater remediation, which not only reduced cost but also led to more effective cleanups (Slenders et al., 2010). UK designers used innovative power systems to recover organic

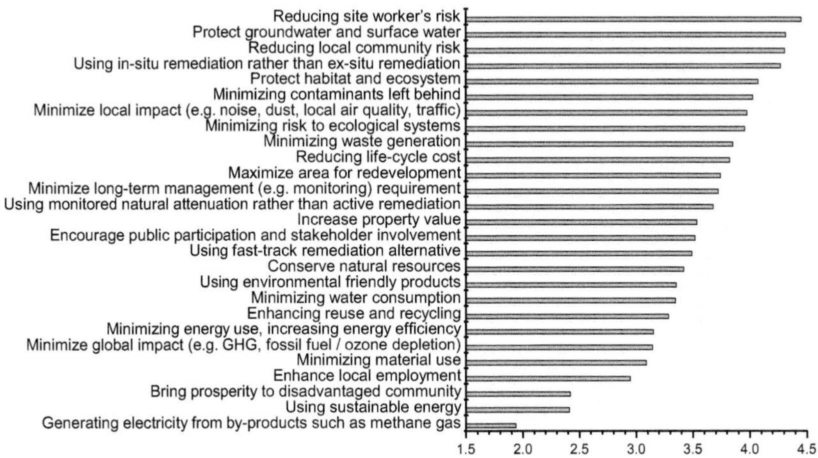

FIGURE 2.2 Adoption of 27 sustainability considerations by UK remediation practitioners.

contaminants from sites, which could be used as an energy source. They found that the system could reduce project costs by 20–30% and led to substantial GHG reductions (McLaren et al., 2009). Researchers also proposed to use brownfield sites to grow biomass as a biofuel feedstock and used LCA to show that it was more environmental-friendly than traditional remediation strategies (SNOWMAN, 2009; Witters et al., 2012a; Witters et al., 2012b).

The sustainable remediation movement in Europe had broader social impacts. In the United Kingdom, discussions regarding sustainable remediation occurred among regulators and practitioners. This was largely related to sustainable urban revival and project cost saving. In 1999, the Urban Task Force of the former Department of the Environment, Transport, and Region (DETR) proposed that all brownfield sites be cleaned up by 2030 (Rogers, 1999). Due to a public policy which required that 60% of new housing be built on brownfield sites, England built 79% of its new residential units on brownfield sites in 2008 (DCLG, 2009). Because remediation work was often conducted as a part of land development, time and cost became key considerations. An example is the preparation for the London 2012 Olympic Games (Hou et al., 2015). The London Olympic Park was built on a large and complex brownfield site. Such a large remediation project would typically take 5–15 years. However, the remediation team completed all soil remediation work and installed a groundwater remediation system in less than 3 years. A series of sustainable means were also employed at the Olympic site, including: (1) the project used different "suitable-for-use" cleanup goals for soft landscaping, hard landscaping, residential units, and stadia; (2) soil washing and complex sorting allowed a high percentage of reuse/recycling of demolition waste and grading materials; (3) social and economic sustainability considerations were taken into account during the entire process by engaging communities, enhancing local employment and businesses, and by protecting worker's health and well-being (Hou et al., 2015).

While there was a certain consensus regarding sustainable remediation in Europe, there were also considerable variations due to different socioeconomic and cultural backgrounds, as well as wide-ranging regulatory systems. In the early period of sustainable remediation (i.e., 1990s–2010), the United Kingdom had widely recognized sustainability and was applying it in practice, while some other European countries were less recognizant of sustainability (Maurer, 2009). The success in the United Kingdom can partly be attributed to two professional bodies—CLAIRE and SuRF-UK—who were active in promoting sustainable remediation. In 2010, they published a large set of sustainability indicators and developed a sustainability assessment framework for sustainable remediation, under the sponsorship of the Home and Communities Agency of the UK government (Surf-UK, 2009; Surf-UK, 2010). The SuRF-UK framework stressed "development" and "cost saving". It had a strong focus on "demonstrating the need not to implement unnecessary or unsustainable remediation measures," which is consistent with a regulation that encourages a "suitable for use" approach in remediation (DEFRA, 2006) and regulatory guidance that promotes sustainable remediation (EA, 2004).

The development of sustainable remediation in Europe led to the establishment of an ISO international standard: ISO/TC 190 ISO/NP Soil quality—Guidance on sustainable remediation. The guideline was designed to be informative rather than prescriptive, focusing on sustainability appraisal for remediation alternatives. It did not prescribe which methods or indicators should be used, instead, it provided standard terminology and informative advice, allowing users to address site-specific conditions with flexibility. The ISO standard was designed to promote the maximization of "overall environmental, social and economic benefits of the remediation work".

2.2 Green remediation in the United States

Green remediation is a more recent concept adopted by governments and remediation practitioners in the United States. The first Sustainable Remediation Forum was founded in the United States (SuRF, also called Surf-US outside the United States), promoting the importance of all three pillars of sustainability, i.e., social, economic, and environmental impacts (Ellis and Hadley, 2009). However, during the development of sustainable remediation concepts and frameworks, stakeholders from the USEPA recognized that the US congress only allowed it authority in the environmental domain (ITRC, 2011b). In other words, the USEPA did not have jurisdiction over the broad social and economic issues. Consequently, the environmental aspect was singled out, and the term "green remediation" was used.

The development of green remediation also occurred at a time when two important presidential executive orders were issued. The presidential Executive Orders (EO) 13423 and 13514, issued in January 2007 and October 2009, respectively, which mandated the promotion of sustainability in all US federal agency operations. As the USEPA is the US federal agency responsible for remediation regulatory enforcement and providing technological guidance under the CERCRL and RCRA statutes, it was able to publish a technology primer on green remediation that incorporated sustainable practice in 2008 (USEPA, 2008). This document stressed the importance of the secondary environmental impacts of remedial action and presented a series of best management practices (BMPs) to enhance the sustainability of remediation projects. The USEPA Office of Solid Waste and Emergency Response (OSWER), who oversee the Superfund and the Brownfields programs, went a step further in 2009 by setting policies that encourage sustainable remediation (USEPA, 2009).

The green remediation concept in the United States has primarily focused on the environmental footprint of remediation operations, particularly from a life cycle perspective. The USEPA's definition of green remediation solely refers to environmental effects and benefits, without reference to the social and

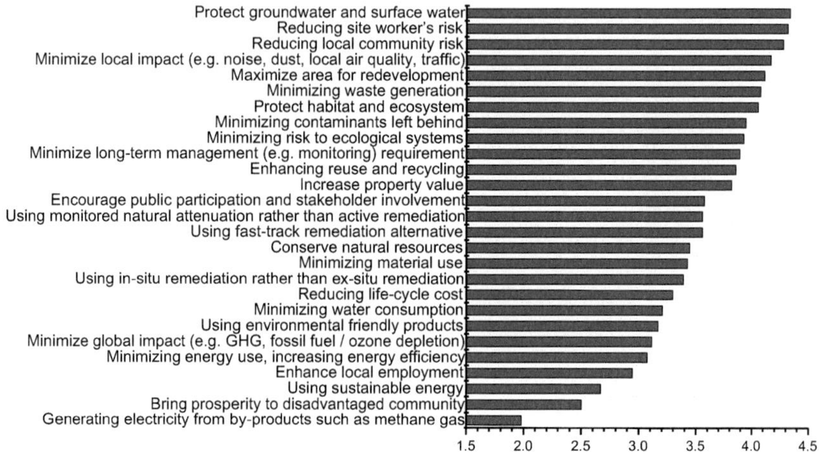

FIGURE 2.3 Adoption of 27 sustainability considerations by US remediation practitioners.

economic aspects of GSR. The USEPA's green remediation concept was based on the following core elements: (1) air, (2) water, (3) energy, (4) land and ecosystems, (5) materials, and (6) waste. Among other efforts, green remediation has focused on energy-efficient equipment and sustainable energy systems, such as solar, wind, and geothermal systems; low-energy remediation systems, such as enhanced in situ bioremediation, phytoremediation, and engineered wetlands; and sustainable construction techniques, such as recycling of demolition waste, capture and reuse of rainwater, and low-maintenance landscapes.

Fig. 2.3 shows the adoption of 27 sustainability considerations by US remediation practitioners. As can be seen, the most adopted practices align well with the protection of water receptors, i.e., groundwater and surface water, followed by reducing site worker's risk and reducing community risk. These three behaviors correspond with a focus on groundwater cleanup in contaminated site management, due to stringent groundwater remediation goals, the strong health and safety culture and regulatory requirement pertaining to hazardous waste operations and emergency response (HAZWOPER), and public participation requirements. The most frequently advocated and show-cased green remediation strategies, such as minimizing GHG emissions and using sustainable energy, were not so widely adopted.

The development of green remediation in the United States led to the establishment of an ASTM international standard: ASTM E2893—Standard Guide for Greener Cleanups. This standard "describes a process for evaluating and implementing activities to reduce the environmental footprint of a cleanup project in the United States while working within the applicable regulatory framework and satisfying all applicable legal requirements." Unlike the ISO standard, which covered all three sustainability pillars, the ASTM standard focused on the environmental aspect. Moreover, the ASTM standard primarily relied upon BMPs, rather than sustainability assessment, in guiding remediation practices. Please refer to Chapter 4, authored by John Simon, the lead author of the ASTM standard, for a detailed overview of green remediation BMPs.

3. The present state of green and sustainable remediation (2010–20)

The GSR movement blossomed in the 2010s. In the United States, following the publication of a green primer by the USEPA, several regional branches of the USEPA published guidelines and policies that promoted GSR (USEPA Region 2, 2012; USEPA Region 9, 2012; USEPA Region 10, 2012). Other federal agencies who hold large numbers of contaminated sites followed suit. For instance, the US Air Force developed an excel-based toolkit (the Sustainable Remediation Tool [SRT]) to calculate a range of GSR metrics for several commonly used remediation technologies (AFCEE, 2010); the US Army Corps of Engineer published guidance that described principles and considerations for incorporating GSR (USACE, 2010); the US Navy developed another excel-based and building-block-type program that calculates GSR metrics and incorporates footprint reduction analysis (NAVFAC, 2013).

In additional to national governments, local governments and the remediation industry also took action. In 2012, the State of Oregon published their green remediation policy (DEQ, 2012); the State of Illinois published guidelines on how to maximize environmental benefits of environmental remediation (EPA, 2012); and the State of Minnesota developed a GSR toolkit (PCA, 2012). SuRF and ITRC published several white papers and frameworks to guide the implementation of GSR (Favara et al., 2011; Holland et al., 2011; ITRC, 2011a), which have helped GSR to flourish (CLU-IN, 2012; Petruzzi, 2011). Some site owners have also started applying financial incentives to encourage their contractors to employ GSR practices.

3.1 Quantitative assessment and minimization of life-cycle environmental impacts

One of the greatest advancements made in the present time period (2010–20) has been in the quantitative assessment of GSR, most particularly LCA-based environmental impact evaluation. Prior to 2010, there were only a handful of studies that had examined life-cycle impact of environmental remediation. Subsequently, researchers have developed a wide variety of sustainability assessment methods for site-specific evaluations. A hybrid method was developed that combined traditional process-based LCA with input–output (IO) LCA (Hou et al., 2014a). The hybrid method renders several advantages: (1) expanding the system boundary and reducing truncation errors; (2)

incorporating consequential impacts of remediation (i.e., tertiary impacts); and (3) covering social and economic impacts in addition to environmental impacts. Another generalized model was also developed to compare the sustainability of remedial options, which was applied in a study involving P&T, enhanced in situ bioremediation (EIB), and permeable reactive barriers (PRBs) in a wide range of hydrogeological and biochemical conditions (Hou et al., 2014b). It was found that, in general, source zone treatment technologies (i.e., EIB and ISCR) tended to have less life-cycle impacts than containment technologies (i.e., P&T and PRB) on a life-cycle basis.

Remediation operation can result in significant amounts of GHG emissions. A single remediation project in New Jersey, USA, was calculated to have the potential to emit 2.7 million tons of CO_2 if D&H was employed (Garon, 2008). This would have been equivalent of ~2% of the total annual CO_2 emissions for the entire state. Based on a limited number of remediation LCA studies, it has been found that the cleanup of 1 kg of contaminants in soil can result in up to 5 tons of CO_2 emission, with a geometric mean of 0.015 tons of CO_2; moreover, the cleanup of 1 kg of contaminants in groundwater may result in 130 tons of CO_2 emissions, with a geometric mean of 1.3 tons of CO_2. High GHG emissions are mostly associated with the widely used traditional remediation technologies such as D&H and P&T. The D&H method is also known to generate large quantities of hazardous waste, while the P&T method disposes large quantities of valuable water resources. Other remediation technologies are also associated with secondary adverse effects, such as acidification, eutrophication, ozone depletion, ecological toxicity, etc.

A recent study has also assessed life-cycle environmental impacts of remediation on a larger scale. Hou et al. (2018b) examined the primary impacts associated with the physical state of San Francisco's brownfield and greenfield sites, the secondary impacts associated with remediation of the brownfield sites, and the tertiary impacts associated with postremediation usage and the avoided use of greenfield land (Hou et al., 2018b). It was found that brownfield land remediation and redevelopment in the city could lead to a net GHG reduction of 0.74 Mt CO_2 yr^{-1}, the equivalent of 14% of San Francisco's 5.3 Mt CO_2 eq. GHG emissions in 2010 (Hou et al., 2017).

Based on the philosophy of GSR that maximizes environmental benefit, quantitative assessment of life-cycle environmental impact can also provide a basis to search for optimal cleanup standards. For this, researchers have combined human health risk assessment with LCA, in an effort to identify "green" cleanup standards (Hou et al., 2017). For this, a case study was conducted at a lead-contaminated site. It was found that in comparison with the regulatory guidance value of 255 mg/kg applicable at the time of project implementation, an optimum cleanup level of 800 mg/kg could increase the net environmental benefit by ~3%, while reducing economic costs by more than a third (Hou et al., 2017).

In addition to academic research, much progress has been made by the industry. Since 2010, several technical guidelines for remediation sustainability assessment and LCA have been published by organizations such as SuRF-UK and SuRF-US (Favara et al., 2011; Surf-UK, 2010). Sustainability assessment tools based on multicriteria analysis have also been developed by various companies and professional organizations, e.g., GoldSET by Golder Associates, VHGFM by the Swedish Geotechnical Association (Brinkhoff, 2011; Golder, 2012), and SiteWise by Battelle (NAVFAC, 2013).

3.2 Social and economic impact of remediation and brownfield regeneration

Social and economic impacts form two of the three pillars of sustainability. However, in the early stages of GSR development, the consideration of social impacts was often lacking, and most evaluations of economic impacts were limited to project costs and lacked socioeconomic considerations, e.g., impacts to the local economy. This may be attributed to an institutional barrier: people working in the remediation industry usually have an environmentally focused technical background and lack awareness and perception of the social impacts caused by remediation. In the present stage, i.e., 2010–20, both professionals and researchers are now paying more attention to the social and economic impacts.

For many social and economic aspects, only a qualitative assessment can be performed. However, there are some aspects that can quite easily be assessed quantitatively. For instance, at the London 2012 Olympic Park site, seven social and economic indicators were identified that could be quantified: life-cycle cost, employment impact, increase in property value, compensation of employees, work-related fatality, work-related injuries, and impacts to local traffic. These indicators were selected based on three criteria: (1) they were strongly applicable to the present remediation project; (2) they could be quantified with existing data; and (3) they covered a broad range of social and economic concerns. The life-cycle cost was directly taken from project documentation. Employment impacts were estimated using an IO model with employment data extracted from national accounts. Worker wages were estimated using the IO model using worker compensation data. For work-related fatalities and injuries, data from the UK's Health and Safety Executive (HSE) were used to derive IO multipliers. To estimate the impact of waterway improvements to adjacent property values, data regarding the number and types of properties and their base prices were used. As Fig. 2.4 shows, the quantitative assessment results rendered an easy comparison of different remedial options.

Non-traditional channels have also been used in socioeconomic sustainability appraisal. In order to evaluate public acceptance of a remediation project at a mega site in China, Song et al. (2018) explored comments posted

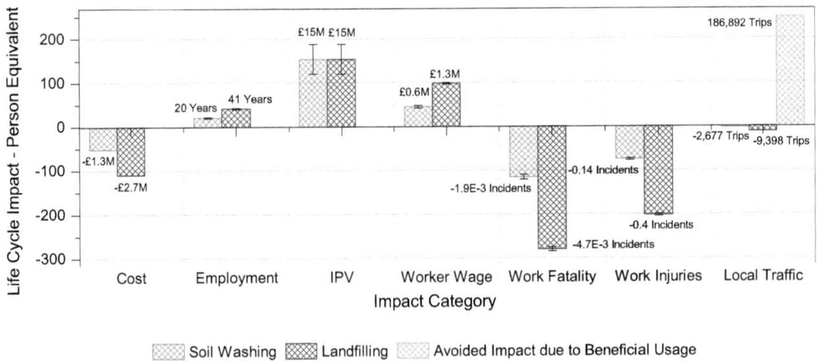

FIGURE 2.4 Typical quantitative assessment of social and economic impact of remediation activities.

in a social media forum established by local residents. They found that residents had concerns relating to a perceived health risk, even after validation of the site remediation had been granted by the local environmental agency (Song et al., 2018). For example, one commenter stated that because of the land contamination issue, they would be happier to buy a property on this land as an investment rather than as their own dwelling. Such socioeconomic factors could have implications for decision-making by site owners and land developers.

In Germany, market-oriented economic appraisal has been combined with sustainability evaluation to assess brownfield revitalization options (Schädler et al., 2010). Sustainable practices in the building industry are being used in the remediation industry, including recycling of demolition waste, capture and reuse of rainwater, low-maintenance landscape, and treatment building design. These practices follow similar principles as those defined in the United States Green Building Council (USGBC)—Leadership in Energy and Environmental Design (LEED) program. The LEED program becomes more relevant because it brings extra financial benefits when certification is granted, i.e., such sustainable behavior aligns with economic sustainability.

3.3 Barriers and promoting forces

There are barriers and promoting forces for the adoption of GSR. Such forces can be broadly grouped as either stakeholder pressures or institutional pressures. When it comes to stakeholder pressure, regulators tend to be the most influential, especially those who oversee remediation activities. Regulatory agencies range from national, to state/province, to local level. In the United States, there are national-level pressures originating from presidential executive orders, and in the United Kingdom, there are pressures from regulations that aim toward sustainable remediation. These pressures dissipate to the

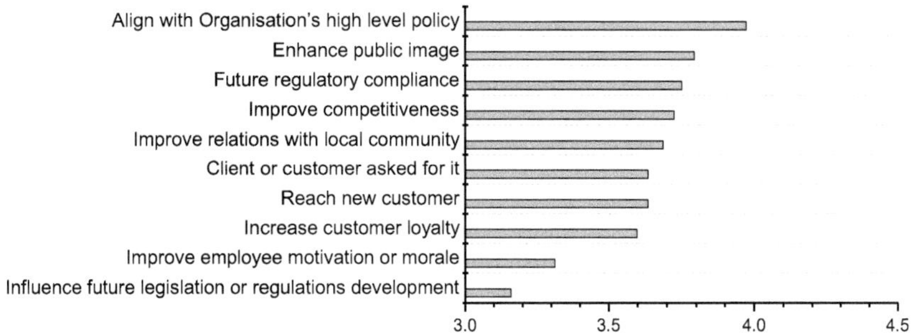

FIGURE 2.5 Ranking of factors motivating sustainable practice in remediation in the United States (Scale 1–5: 1 = not important, 5 = very important).

regulators who oversee individual remediation sites. In practice, the regulators can act as a coercive force by communicating the GSR message to contaminated site managers and workers, as well as providing technical guidance on GSR. The regulators may also relieve withholding pressures (i.e., eliminating impeding forces) in exchange for GSR practices. As Fig. 2.5 shows, "future regulatory compliance" was ranked the third among all potential promoting forces. On the other hand, when regulators are not supportive of GSR, or when regulators perceive that polluters intend to avoid effectively remediating contaminated sites by using a "green-washing" strategy, stakeholder pressure may become a serious barrier to GSR. As shown on Fig. 2.6, the lack of regulatory demand was ranked number two among all barriers.

Fig. 2.6 also shows that a lack of demand from clients, e.g., the site owner, was the number one barrier to the adoption of GSR. Site owners are key actors in the remediation industry, as they pay for remediation operations and usually make the final decisions. They naturally desire minimized costs while eliminating their remediation liabilities. Therefore, there are pressures to reduce remediation costs, or, in other words, to enhance economic sustainability in

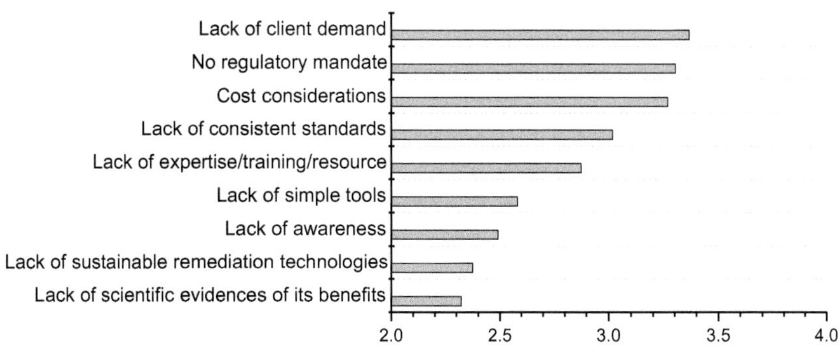

FIGURE 2.6 Ranking of barriers impeding sustainable practice in remediation in the United States (Scale 1–5: 1 = not at all, 5 = very significant).

their favor. On the other hand, land is a precious resource, and site owners often desire to clean up vacant contaminated land quickly so that it can be put to productive use. There is a trade-off to be found, i.e., between low cost/passive remediation (e.g., monitored natural attenuation (MNA) or containment) and high cost/active remediation (e.g., excavation and off-site disposal), with the solution often depending on stakeholder pressure.

Institutional pressures are more intangible than stakeholder pressures. For GSR, institutional forces may arise due to: (1) society as a whole increasingly viewing sustainability as socially desirable and watching for such sustainability labels and certifications; (2) a growing number of principles and guidance promulgated by government agencies as well as professional organizations; and (3) expanded reporting practices associated with corporate social responsibility and carbon accounting. In modern times, companies are often under corporate social responsibility pressures emphasizing the triple bottom line of sustainability. When it comes to remediation, companies tend to put a high priority on avoiding negative public perceptions, e.g., lack of action or inappropriate action. There are also incentives to present a good public image by showing sustainability initiatives (e.g., GSR). For instance, during preparations for the London 2012 Olympic Games, an independent assurance body known as the Commission for a Sustainable London 2012 (CSL) was established. The CSL influenced decision-making in favor of soil washing remediation, which saved approximately £68 million in comparison to landfilling cost (T. and D., 2013).

Institutional pressures may also be transmitted by stakeholders. Many large international oil and gas and chemical companies have contaminated sites around the world. As these companies adopt more sustainable behaviors under institutional pressures in developed countries, their corporate remediation branches will often handle contaminated sites in developing countries in a sustainable manner, thereby transmitting techniques across borders to subsidiaries and to other organizations. It has been observed that some contaminated land sites in China are being remediated by subsidiary companies following guidance from their multinational parent companies, mimicking remediation practice in the West.

3.4 Green and sustainable remediation in developing countries: China as a case study

The GSR movement is now seen to be dissipating to developing countries, although awareness and adoption of GSR in developed countries still remain much higher, particularly in the United States and the United Kingdom (Hou et al., 2016). Awareness of sustainable remediation at the national level was measured in a questionnaire survey by asking "use a scale of 1-5 to describe to what extent remediation practitioners in this country are aware of 'sustainable remediation'" on a Likert scale of 1–5. The United States and the United

Kingdom had ratings of 3.88 and 4.00, respectively. In comparison, developing countries had much lower ratings (2.50–2.60). The overall adoption of sustainable remediation was measured by the question "use a scale of 1-5 to describe how effective remediation practitioners in this country are adopting 'sustainable remediation'" The United States and the United Kingdom had ratings of 2.78 and 2.88, respectively, while developing countries only had a rating of 1.63.

For the development of GSR in developing countries, a notable country is China, which has severe soil and groundwater pollution issues and a fast-growing remediation market (Hou et al., 2018a; O'Connor et al., 2018b; Peng et al., 2019). China has become the world's second largest economy, with rapid economic growth over the past three decades at the cost of serious environmental pollution. A national soil survey published in 2014 indicated that 16.1% of all soils in the 6·3 million km^2 of land surveyed contained contaminants exceeding recommended soil quality standards (MEP, 2014). In response to social and environmental pressure, the Chinese government unleashed a plan on 31 May 2016, in order to curb soil pollution and clean up contaminated land. The plan, entitled Soil Pollution Prevention and Control Action Plan ("Action Plan"), creates a demanding schedule for national and local governments: finishing detailed soil investigation of agricultural land by 2018 and industrial land by 2020, cleaning up approximately 700,000 ha of seriously contaminated agricultural land by 2020 and utilizing 95% of the nation's contaminated land in a safe manner by 2030. It was estimated that the Action Plan would generate RMB 450bn (~$65bn) of revenue for the environmental industry by 2020 and would stimulate RMB 2.7 trillion (~$392bn) of gross domestic product growth (People's Daily, 2016).

So far, the emerging remediation market in China has been largely driven by land redevelopment rather than environmental regulatory compliance. This is similar to the UK approach, where large numbers of sites have been remediated through private sector redevelopment, but only a small proportion of cleanups have been enforced by environmental protection legislation. While the UK regulatory regime should not be considered insignificant, where its remit has indirectly yet significantly enhanced the perception of contaminated land responsibilities/liability among development stakeholders, however, in a China-specific context such development managerialism may lead to issues that range from underprotection of the environment to underuse of innovative in situ remediation technologies. Moreover, the development-driven remediation market in China has concentrated on the wealthiest areas of the country, such as Beijing and south-eastern provinces, and has primarily focused on soil remediation with a general ignorance of contaminated groundwater, etc.

Contaminated land remediation is an emerging market in China, but is now growing exponentially. Key drivers promoting this market are: (1) growing awareness of the risk of land contamination among the public, and the recognition of its seriousness by the national government; (2) a booming housing market leading to a spike in demand for redevelopment of former industrial land in city centers; and (3) the national government has passed a soil pollution prevention law and several key regulations, which demand rigorous soil investigation, risk management, and remediation if necessary. Interviews conducted with industry experts in China indicate that the remediation market may have now reached the equivalent of billions of US dollars in size, with the majority of land remediation projects driven by redevelopment, as mentioned previously. Because the new soil protection law pays special attention to groundwater pollution prevention and remediation, and several regulatory standards are being drafted regarding groundwater remediation, this situation may soon change.

Researchers based in China have conducted a series of studies aimed at the remediation of contaminated soils. Some of these studies have developed techniques that fit well with the GSR concept (O'Connor et al., 2018a; Shen et al., 2019a; Shen et al., 2018; Shen et al., 2019b; Wang et al., 2019; Zhang et al., 2018). For instance, researchers have identified pesticide degrading microorganisms and incorporated them into treatment reagents. A variety of microorganism species capable of degrading petroleum hydrocarbons have been identified. A microorganism species (*Paracoccus aminovorans*) has been isolated, which can be used to degrade PAH contamination. Researchers have also identified and enriched microorganism species that can degrade semi-volatile petroleum hydrocarbons and heavy-end petroleum hydrocarbons with high efficiency. In the phytoremediation field, researchers have also made significant progress. Plant species capable of accumulating zinc and cadmium (*Sedum plumbizincicola*) and cadmium (*Sedum jinianum*) have been identified and used to remediate heavy metal contamination. In groundwater remediation, new in situ bioremediation technologies have been developed, and researchers have combined physical separation, chemical oxidation, and biological degradation to increase removal efficiency and reduce treatment time for PAHs and large-molecule petroleum hydrocarbon, which remediation practitioners have historically found difficult to degrade.

In China, the adoption of GSR among industrial practitioners has been very limited. Based on literatures and questionnaire survey findings, we consider it imperative to strengthen the adoption of GSR in China. With proper design and cautious implementation, GSR could help to optimize the use of limited resources, improve China's environmental quality, enhance public health, curb social inequality, act as a springboard for remediation technology innovation, and help establish a mature remediation market. A sustainable remediation forum for China (SuRF-China) was founded in August, 2017; however, this organization has not been as active as other SuRF organizations. In 2018, the Chinese Ministry of Science and Technology (MOST) launched a major research funding theme in soil pollution prevention and control. As this 25 billion RMB (~US$3.7 billion) theme includes funding allocated to research and development of green remediation materials, sustainable remediation technologies, and GSR technical standards, it will stimulate a large number of research studies on GSR. Most recently, the first GSR technical standard in China has been distributed by the China Association of Environmental Protection Industry (CAEPI) for public comment. It is expected to be officially adopted in 2019. Overall, the development of GSR in developing countries, such as China, is lagging but holds much promise.

4. The future of sustainable remediation (2020–40)

4.1 Challenge in research

State-of-the-art practice in the remediation field still heavily relies upon risk assessment and standardized guideline values. Such risk assessment methods do not take into account secondary or tertiary impacts associated with remediation activities. More holistic and open-decision methodologies should be developed and adopted in order to align with sustainable development of society. Evidence-based scientific research should be the cornerstone of GSR. In general, the development of GSR in the past 20 years has largely been driven by policy demand. For future development and mainstreaming, it is critical to provide solid scientific evidence for all aspects of sustainability decision-making, as well as to provide technologies and tools that can meet the principles of GSR. However, a variety of challenges exist in the research field.

The adoption of different remediation technologies has many implications to GSR development on several fronts: (1) often a variety of remediation technologies are viable for any specific site condition, and these remedial alternatives can have very different levels of social (e.g., impacts to local traffic), economic (e.g., local employment and tax revenue), and environmental (e.g., fugitive emission and global warming effect) life-cycle impacts in a global boundary; (2) even when all remedial alternatives meet the same remedial objective, they may still leave different levels of residual contamination in situ and, therefore, have implications for future development choices, which in turn affects the sustainable development of local communities; (3) the adoption of remediation technologies at various cost levels eventually determines the total amount of contaminated sites being remediated, which in turn affects human health, the environment, as well as local socioeconomic conditions. Most existing studies have focused on sustainability assessment on a site-scale, which renders high practicality on the one hand, but renders limited transferability on the other hand. More studies are needed to provide more generic knowledge that can be used by practitioners managing various site conditions.

It is well accepted among GSR practitioners that one of the biggest challenges is the inclusion of social impacts in sustainability assessment. For social LCA, it is difficult to maintain consistency among standards to allow for a fair comparison. In this regard, more objective and quantitative indicators are important. Some researchers and practitioners intend to use the stakeholder theories and stakeholder engagement to address this issue. However, stakeholders tend to be optimistic about the issues that align the best with their organizational goals. This may reflect an overoptimism that can be detrimental in promoting GSR practices. For instance, policy makers who emphasize the need for recycling will often disregard any adverse effects from a life cycle and global perspective (Hou et al., 2015). Studies on environmental behavior have indicated that perceived control can be negatively associated with environmentally appropriate behavior (Grob, 1995). This is because overly optimistic people tend to take less action, or, on the other hand, people perceiving lower control are in general more concerned about environmental processes thus adapting their behavior accordingly. More empirical research is needed to assess the significance of this effect, and more theoretical research can be conducted to derive its implication to the effectiveness of policy instruments.

Putting the focus on stakeholders instead of remediation strategies may seem to be a counterintuitive approach to achieving sustainability. After all, sustainability lies in the technical decisions made in remediation. However, knowing how stakeholders may interact is critical knowledge for both policy makers and contaminated site managers in order to act strategically. For instance, remediation normally consists of three stages: investigation and planning, remediation system construction, and remediation system operation with long-term monitoring. In the first phase, consultants and regulators may be viewed as the primary stakeholders, with a strong influence on the focal firm's decision-making. In the second phase, the primary stakeholder may be the contractor, but they also receive influence from consultants, technology vendors, and regulators. In the last phase, the focal actors are consultants and regulators again. The variation of focal actors and stakeholders poses a challenge for the adoption of GSR practices. From this perspective, it is important to better understand the various stakeholder's views and to promote GSR practices with either technological development or socioeconomic and policy instruments.

In developing countries, there are also particular challenges. The continuation of a primarily redevelopment-driven remediation market in China could lead to several issues arising from the Chinese context: (1) poorer provinces and cities may be unable to clean up their contaminated sites, as remediation is focused on high-value land. This could lead to a social inequality issue, as sites in many densely populated yet poor mining cities remain unaddressed; (2) the most-high risk sites may not be of high priority in the cleanup agenda because high remediation costs make them economically unviable, and because many market externalities are not captured by land value, social well-being will be under-optimized; and (3) the remediation industry may bring little benefit to the most vulnerable rural communities, for example, those who live adjacent to polluting factories, or those who are reliant on seriously contaminated shallow groundwater as drinking water supplies. The GSR concept provides much promise in addressing these specific challenges; however, much scientific research is needed to support policy making as well as industrial practice.

4.2 Obstacles and promoting forces in practice

The past, present, and future developments of GSR also represent an evolving history of social learning. According to institutional theories, social and cultural pressures affect the behavior of organizational entities. Organizational decision-making is affected by three mechanisms: coercive force, mimetic force, and normative force. Coercive force is due to political influence and legitimacy consideration; mimetic force comes from a desire to reduce uncertainty; and normative force is based on professionalization. Institutional pressures may limit the remediation industries' ability to change. However, when it comes to critical points of intervention, old norms (e.g., risk-based remediation) can be complemented or replaced by new norms (e.g., GSR).

There are also interdisciplinary forces that can help drive changes. For example, the LEED program is a green design standard for new developments in the United States, which recognizes brownfield remediation as a credit toward sustainable development (USGBC, 2011). Based on the authors' own experience, it is not uncommon for land developers and remediation practitioners to work together in the reclamation of contaminated sites and to minimize waste disposal and maximize the reuse/recycling of demolished building waste in the process. This practice has multiple benefits encompassing various environmental, social, and economic aspects. This type of organizational behavior shows that new social norms in pertaining fields can be transmitted. The ongoing sustainable development movement in the larger society, coupled with the rise of GSR, may represent a critical intervention point when the remediation field can be reshaped.

On the other hand, despite the growing interest and large numbers of studies on GSR, there is a general gap between the desire to include "sustainability" and the "practicability" in industrial practice. A limited number of academic research studies have been conducted relating to this field and have mostly focused on the use of LCA and multicriterion analysis in sustainability evaluation for remedy selection. It is imperative that evidence-based scientific research be conducted on the subject of practicability, to support the uptake of GSR in the field.

4.3 Path forward

Pressures for an emphasis on sustainability in remediation derive from three general sources (Hou and Al-Tabbaa, 2014): (1) increasingly recognized secondary environmental impact from remediation operations (e.g., life-cycle greenhouse gas emission, air pollution, energy consumption, waste production), (2) stakeholder demand for economically sustainable brownfield restoration and "green" practices, and (3) institutional pressures (e.g., social norm and public policy) that promote GSR practices (e.g., sustainable energy, green building, recycling). The emergence of sustainable remediation represents a critical intervention point when the remediation field can be reshaped with new norms and standards being established for practitioners for years to come. GSR is becoming a new imperative in the environmental remediation field, with important implications for regulators, liability owners, consultants, contractors, and technology vendors. Sustainability studies provide implications to project management and/or policy making. Each industry or disciplinary field may be viewed as a self-organizing, complex, and adaptive system. In the ongoing sustainability movement, it is evident that these systems are attempting to make a transition toward a more sustainable future. To accelerate this transition, it is necessary to develop new sustainability assessment tools and better understand sustainable behavior, in order to catalyze interactions among researchers and actors on various levels in the industry.

The conception and adoption of GSR present a great opportunity, as well as challenge, for both the research community and practicing remediation community. The new GSR norm should not only assist in accelerating the efficient cleanup of existing contaminated land but also minimize the potential of future land contamination, and where land contamination is inevitable in the long run (e.g., in a statistical sense), it should assist in systematically planning for efficient future land remediation. GSR is neither simply a technological solution nor simply an evaluation criterion. Instead, it represents a new way of thinking. It should go beyond the simply inclusion of "sustainability" as an additional balancing factor in decision-making. Practitioners should incorporate sustainability considerations such as "cost" or "technical feasibility" considerations, in all phases of projects. Moreover, the successful adoption of GSR practices depends on adaptive management and institutional learning. The focus is to sustain the skills and learning processes necessary for the professionals to understand why and how to achieve sustainability, which on the other hand requires an expanded scientific knowledge base. In a sense, concerted action by the academia, government, and industry is needed to take full advantage of the ongoing GSR movement.

References

AFCEE, 2010. Sustainable Remediation Tool User Guide. AFCEE. Air Force Center for Engineering and the Environment, Lackland, Texas.

Al-Tabbaa, A., Harbottle, M., Evans, C., 2007a. Robust sustainable technical solutions. In: Dixon, T., Raco, M., Catney, P., Lerner, D.N. (Eds.), Sustainable Brownfield Regeneration. Blackwell Publishing.

Al-Tabbaa, A., Smith, S., De Munck, C., Dixon, T., Doak, J., Garvin, S., et al., 2007b. Climate change, pollutant linkage and brownfield regeneration. Sustainable Brownfield Regeneration: Liveable Places from Problem Spaces 263–314.

ASTM, 2010. E2081-00: Standard Guide for Risk-Based Corrective Action. Originally Approved in 1998. Re approved in 2010. American Society for Testing and Materials.

Blanc, A., Métivier-Pignon, H., Gourdon, R., Rousseaux, P., 2004. Life cycle assessment as a tool for controlling the development of technical activities: application to the remediation of a site contaminated by sulfur. Advances in Environmental Research 8, 613–627.

Brinkhoff, P., 2011. Multi-Criteria Analysis for Assessing Sustainability of Remedial Actions. Chalmers University of Technology, Göteborg, Sweden.

Choguill, C.L., 1996. Ten steps to sustainable infrastructure. Habitat International 20, 389–404.

CLARINET, 2002. Sustainable Management of Contaminated Land: An Overview. CLARINET. Contaminated Land Rehabilitation Network for Environmental Technologies (succeeded by Common Forum on Contaminated Land in the European Union).

CLU-IN, 2012. Green Remediation Focus: NASA Jet Propulsion Laboratory. JPL, Pasadena, CA.

Dasgupta, S., Tam, E.K.L., 2005. Indicators and framework for assessing sustainable infrastructure. Canadian Journal of Civil Engineering 32, 30–44.

DCLG, 2009. Land use Change Statistics (England) 2008 - Provisional Estimates (July 2009). DCLG. Department for Communities and Local Government, London, UK.

DEFRA, 2006. Defra Circular 01/2006, Environmental Protection Act 1990: Part 2A, Contaminated Land. Defra. Department for Environment, Food and Rural Affairs, London, UK.

DEQ O, 2012. Green Remediation Policy.

DETR, 1999. Towards an urban renaissance: report by the urban task force. In: Rogers, L. (Ed.), DETR. Department of the Environment, Transport and the Regions, London, UK.

Diamond, M.L., Page, C.A., Campbell, M., McKenna, S., Lall, R., 1999. Life-cycle framework for assessment of site remediation options: method and generic survey. Environmental Toxicology and Chemistry 18, 788–800.

DTSC, 2009. Interim Advisory for Green Remediation. DTSC. Department of Toxic Substances Control, Sacramento, CA.

EA, 2004. Model Procedures for the Management of Land Contamination, Contaminated Land Report 11. EA. Environment Agency, London, UK.

Ellis, D.E., Hadley, P.W., 2009. Sustainable remediation white paper—integrating sustainable principles, practices, and metrics into remediation projects. Remediation Journal 19, 5–114.

EPA I, 2012. Greener Cleanups: How to Maximize the Environmental Benefits of Site Remediation.

Favara, P.J., Krieger, T.M., Boughton, B., Fisher, A.S., Bhargava, M., 2011. Guidance for performing footprint analyses and life-cycle assessments for the remediation industry. Remediation Journal 21, 39–79.

Ferguson, C.C., 1999. Assessing risks from contaminated sites: policy and practice in 16 European countries. Land Contamination and Reclamation 7.

Garon, K.P., 2008. Sustainability Analysis for Improving Remedial Action Decisions, 2008 State Superfund Managers Symposium. Association of State and Territorial Solid Waste Management Offices, Scottsdale, AZ.

Golder, 2012. GoldSET: Fast and Reliable Sustainability Decision Support Tool.

Grob, A., 1995. A structural model of environmental attitudes and behaviour. Journal of environmental psychology 15 (3), 209–220.

Holland, K.S., Lewis, R.E., Tipton, K., Karnis, S., Dona, C., Petrovskis, E., et al., 2011. Framework for integrating sustainability into remediation projects. Remediation Journal 21, 7–38.

Hou, D., Al-Tabbaa, A., 2014. Sustainability: a new imperative in contaminated land remediation. Environmental Science & Policy 39, 25–34.

Hou, D., Al-Tabbaa, A., Guthrie, P., Hellings, J., Gu, Q., 2014a. Using a hybrid LCA method to evaluate the sustainability of sediment remediation at the London Olympic Park. Journal of Cleaner Production 83, 87–95.

Hou, D., Al-Tabbaa, A., Hellings, J., 2015. Sustainable site clean-up from megaprojects: Lessons from London 2012. Proceedings of the Institution of Civil Engineers: Engineering Sustainability 168 (2), 61–70.

Hou, D., Al-Tabbaa, A., Luo, J., 2014b. Assessing effects of site characteristics on remediation secondary life cycle impact with a generalized framework. Journal of Environmental Planning and Management 57, 1083–1100.

Hou, D., Guthrie, P., Rigby, M., 2016. Assessing the trend in sustainable remediation: a questionnaire survey of remediation professionals in various countries. Journal of Environmental Management 184, 18–26.

Hou, D., Li, F., 2017. Complexities surrounding China's soil action plan. Land Degradation and Development 28, 2315–2320.

Hou, D., Li, G., Nathanail, P., 2018a. An emerging market for groundwater remediation in China: policies, statistics, and future outlook. Frontiers of Environmental Science and Engineering 12.

Hou, D., Qi, S., Zhao, B., Rigby, M., O'Connor, D., 2017. Incorporating life cycle assessment with health risk assessment to select the 'greenest'cleanup level for Pb contaminated soil. Journal of Cleaner Production 162, 1157–1168.

Hou, D., Song, Y., Zhang, J., Hou, M., O'Connor, D., Harclerode, M., 2018b. Climate change mitigation potential of contaminated land redevelopment: a city-level assessment method. Journal of Cleaner Production 171, 1396–1406.

ITRC, 2011a. Green and Sustainable Remediation: A Practical Framework. Interstate Technology & Regulatory Council, Washington, DC.

ITRC, 2011b. Green and Sustainable Remediation: State of the Science and Practice. ITRC. Interstate Technology & Regulatory Council, Washington, DC.

Maurer, O., 2009. NICOLE's Shared Vision on Sustainable Remediation. Green Remediation Conference, Copenhagen, Denmark.

McLaren, S., Worboys, M., Speake, O., Mantell, P., 2009. Ex-situ Thermally Enhanced Coal Tar Recovery - A Low Carbon Option. Green Remediation Conference, Copenhagen, Denmark.

MEP, 2014. National Soil Contamination Survey Report. Ministry of Environmental Protection, Beijing, China.

NAVFAC, 2013. SiteWise Version 3 User Guide.

NICOLE, 2002. Need for Sustainable Land Management: Role of a Risk Assessment Based Approach. NICOLE. Network for Industrially Contaminated Land in Europe, Utrecht, The Netherlands.

NICOLE, 2005. The Impact of EU Directives on the Management of Contaminated Land. NICOLE. Network for Industrially Contaminated Land in Europe, Utrecht, The Netherlands.

O'Connor, D., Hou, D., 2018. Targeting cleanups towards a more sustainable future. Environmental Science: Processes & Impacts.

O'Connor, D., Hou, D., Ok, Y.S., Song, Y., Sarmah, A., Li, X., et al., 2018a. Sustainable in situ remediation of recalcitrant organic pollutants in groundwater with controlled release materials: a review. Journal of Controlled Release 283, 200–213.

O'Connor, D., Hou, D., Ye, J., Zhang, Y., Ok, Y.S., Song, Y., et al., 2018b. Lead-based paint remains a major public health concern: a critical review of global production, trade, use, exposure, health risk, and implications. Environment International 121, 85–101.

Page, C.A., Diamond, M.L., Campbell, M., McKenna, S., 1999. Life-cycle framework for assessment of site remediation options: case study. Environmental Toxicology and Chemistry 18, 801–810.

PCA, M., 2012. Toolkit for Greener Practices: Learn More about the Initiative-Toolkit Rationale and Concept.

Peng, T., O'Connor, D., Zhao, B., Jin, Y., Zhang, Y., Tian, L., et al., 2019. Spatial distribution of lead contamination in soil and equipment dust at children's playgrounds in Beijing, China. Environmental Pollution 245, 363–370.

People's Daily, 2016. "Soil Ten" Propose to Curb Deteriorating Soil Pollution by 2020 (2016) (in Chinese).

Petruzzi, N.M., 2011. A Case Study on the Evaluation and Implementation of Green and Sustainable Remediation Principles and Practices during a RCRA Corrective Action Cleanup. Ground Water Monitoring & Remediation.

Rogers, R.G., 1999. Towards an urban renaissance. Routledge, UK.

Schädler, S., Morio, M., Bartke, S., Rohr-Zänker, R., Finkel, M., 2010. Designing sustainable and economically attractive brownfield revitalization options using an integrated assessment model. Journal of Environmental Management 827–837.

Shen, Z., Hou, D., Jin, F., Shi, J., Fan, X., Tsang, D.C.W., et al., 2019a. Effect of production temperature on lead removal mechanisms by rice straw biochars. The Science of the Total Environment 655, 751–758.

Shen, Z., Zhang, J., Hou, D., Tsang, D.C.W., Ok, Y.S., Alessi, D.S., 2019b. Synthesis of MgO-coated corncob biochar and its application in lead stabilization in a soil washing residue. Environ Int 122, 357–362.

Shen, Z., Hou, D., Xu, W., Zhang, J., Jin, F., Zhao, B., et al., 2018. Assessing long-term stability of cadmium and lead in a soil washing residue amended with MgO-based binders using quantitative accelerated ageing. The Science of the Total Environment 643, 1571–1578.

Slenders, H.L.A., Dols, P., Verburg, R., de Vries, A.J., 2010. Sustainable remediation panel: sustainable synergies for the subsurface: combining groundwater energy with remediation. Remediation Journal 20, 143–153.

SNOWMAN, 2009. The Rejuvenate Decision-Making Approach-A Worked Example-Crop Based Systems for Sustainable Risk Based Land Management for Economically Marginal Degraded Land. Sustainable Management of Soil and Groundwater. SNOWMAN.

Song, Y., Hou, D., Zhang, J., O'Connor, D., Li, G., Gu, Q., et al., 2018. Environmental and socio-economic sustainability appraisal of contaminated land remediation strategies: a case study at a mega-site in China. The Science of the Total Environment 610, 391–401.

Surf-UK, 2009. A Review of Published Sustainability Indicator Sets: How Applicable are They to Contaminated Land Remediation Indicator-Set Development? Contaminated Land: Applications in Real Environments. CL:AIRE, London, UK.

Surf-UK, 2010. A Framework for Assessing the Sustainability of Soil and Groundwater Remediation. CLAIRE. Contaminated Land: Applications in Real Environments, London, UK.

Surf-UK, 2014. Sustainable Management Practices for Management of Land Contamination.

Maiden, T., Gray, D., 2013. Commission for a Sustainable London 2012 Independent Evaluation.

USACE, 2010. Decision Framework for Incorporation of Green and Sustainable Practices into Environmental Remediation Projects. USACE. United States Army Corps of Engineers, Washington, DC.

USEPA, 2004. Cleaning Up the Nation's Waste Sites: Markets and Technology Trends, 2004 Edition. United States Environmental Protection Agency, Washington, DC.

USEPA, 2008. Green Remediation: Incorporating Sustainable Environmental Practices into Remediation of Contaminated Sites. Office of Solid Waste and Emergency Response, Washington, DC.

USEPA Principles for Greener Cleanups, 2009. United States Environmental Protection Agency, Office of Solid Waste and Emergency Response.

USEPA Region 2, 2012. Region 2 Green Remediation.

USEPA Region 9, 2012. Clean Energy & Climate Change Green Site Cleanups.

USEPA Region 10, 2012. Green Cleanups.

USGBC, 2011. LEED 2009 for New Construction and Major Renovations (updated November 2011). USGBC. United States Green Building Council, Washington, DC.

Volkwein, S., Hurtig, H.W., Klöpffer, W., 1999. Life cycle assessment of contaminated sites remediation. The International Journal of Life Cycle Assessment 4, 263–274.

Wang, Y., Li, Q., Zhang, P., O'Connor, D., Varma, R.S., Yu, M., Hou, D., 2019. One-pot green synthesis of bimetallic hollow palladium-platinum nanotubes for enhanced catalytic reduction of p-nitrophenol. J Colloid Interface Sci 539, 161–167.

Witters, N., Mendelsohn, R., Van Passel, S., Van Slycken, S., Weyens, N., Schreurs, E., et al., 2012a. Phytoremediation, a sustainable remediation technology? II: economic assessment of CO_2 abatement through the use of phytoremediation crops for renewable energy production. Biomass and Bioenergy 39, 470–477.

Witters, N., Mendelsohn, R.O., Van Slycken, S., Weyens, N., Schreurs, E., Meers, E., et al., 2012b. Phytoremediation, a sustainable remediation technology? Conclusions from a case study. I: energy production and carbon dioxide abatement. Biomass and Bioenergy 39, 454–469.

Zhang, P., Hou, D., O'Connor, D., Li, X., Pehkonen, S.O., Varma, R.S., et al., 2018. Green and size-specific synthesis of stable Fe-Cu oxides as Earth-abundant adsorbents for malachite green removal. ACS Sustainable Chemistry & Engineering 9229–9236.

Chapter 3

Sustainability assessment for remediation decision-making

David O'Connor, Deyi Hou
School of Environment, Tsinghua University, Beijing, China

1. Introduction

Remediation based on health risk assessment is an established means to protect site users from exposure to land contamination. It provides a rationale for intervention—breaking source–pathway–receptor pollutant linkages—and provides a process to direct resources to the most hazardous sites. Therefore, some may assume that any remediation strategy following this approach is an inherently "sustainable" action. However, it should be acknowledged that environmental, social, and economic secondary impacts manifest as a result of remedial actions and that the adoption of different strategies at different sites gives rise to different impacts over their life cycle (O'Connor and Hou, 2018). The subject of this chapter is how to quantitatively assess these impacts, so that demonstrably greener and more sustainable remediation approaches can be selected.

Various sustainability assessment (SA) tools can be used to assess the impacts of remediation. Life-cycle assessment (LCA), as standardized by the ISO 14040 series, is generally considered to be the most comprehensive SA tool and is increasingly used by environmental consultants, public authorities, and researchers to support remediation decision-making. It is beneficial to perform SA using a quantitative process, such as LCA, so that nominal impact magnitudes can be estimated and compared objectively. However, quantitative assessments are not always applicable or appropriate for assessing every site, and this type of SA can be data and time intensive. Multicriteria assessment (MCA) can reduce the cost and complexity of decision-making (Bardos et al., 2016). Moreover, SA for certain sites will require qualitative approaches, such as MCA, due to large uncertainty in the input data, which cannot be adequately expressed by a quantitative approach (Suèr et al., 2004). Some assessors will follow a tiered approach, which begins with simplistic qualitative methods, such as MCA, and only proceed to full quantitative methods, such as LCA, if a robust decision cannot be made.

Sustainable Remediation of Contaminated Soil and Groundwater. https://doi.org/10.1016/B978-0-12-817982-6.00003-3

It should be noted that although LCA tools have recently become more holistic by incorporating social and economic metrics, environmental aspects still remain their predominant focus. In this chapter, we assume that all three sustainability pillars will be included in LCA; however, in practice, environmental LCA is often combined with socioeconomic MCA, for easier assessment.

Early SAs of remediation activities from a life-cycle perspective date back to the late 1990s. An example is the work of (Diamond et al., 1999), who developed a life-cycle management framework with the objective of considering environmental impacts beyond the site boundary on a prolonged time frame. SA studies conducted since have assessed a wide range of soil and groundwater remediation technologies, including: bioleaching (Blanc et al., 2004); enhanced in situ bioremediation (EIB) (Lemming et al., 2010); ex situ bioremediation (Toffoletto et al., 2005); in situ stabilization/solidification (S/S) (Hou et al., 2016); landfilling (Song et al., 2018); incineration (Mauko Pranjić et al., 2018); in situ chemical oxidation (ISCO) (Lemming et al., 2012); in situ chemical reduction (ISCR) (Hou et al., 2014); pump and treat (P&T) (Bayer and Finkel, 2006); permeable reactive barriers (PRBs) (Mak and Lo, 2011); soil washing (Hou et al., 2014); and thermal desorption (Hou et al., 2016). The findings and limitations of selected LCA studies on P&T, PRB, ISB, and in situ chemical treatment (ISCT) are presented in Table 3.1.

The development and improvement of LCA and other quantitative SA approaches have brought new findings that are unraveling the implications of remediation and are facilitating a more holistic view of remediation decision-making. For example, Hou et al. (2014) built an LCA model to compare the sustainability of four remedial options (P&T, EIB, PRB, and ISCR) under various site conditions. It was found that, in general, source zone treatment technologies (i.e., EIB and ISCR) will have lower life-cycle impacts over their life cycle. SA has also cast light on how remediation operations may cause significant emissions of greenhouse gases (GHGs), especially widely used traditional remediation technologies such as dig and haul (D&H) and P&T. This was exemplified by an LCA conducted for a proposed remediation project in New Jersey (NJ), which revealed the potential to emit 2.7 million tons of CO_2 if D&H was implemented (Garon, 2008). This is equivalent to ~2% of the entire NJ state's annual CO_2 emissions. It is becoming ever clearer that evaluating remediation impacts over the life cycle with SA tools, such as LCA, is of high value to the remediation field. Table 3.1 lists some other examples of LCA studies that have been conducted on remediation, revealing their relevant findings and study limitations.

2. Important concepts for sustainability assessment

2.1 Functional units

Impacts in SA are standardized by using a functional unit, which is associated with some quantity of remediation activity. For this, the system function needs to be described in a precise manner, then a quantifiable performance metric for

TABLE 3.1 Examples of LCA studies on P&T, PRB, ISB, and ISCT (Hou et al., 2014).

Reference	Technology	Findings	Limitations
Vignes (2001)	P&T	The study compared three different treatment options for wastewater generated in P&T: direct discharge to rivers, vacuum steam stripping operations, and GAC treatment. The study suggested changing the existing stripping system to GAC to reduce life-cycle environmental impact.	Vacuum steam stripping is rarely used for P&T; therefore, the results may only be applicable to uncommon site conditions.
Bayer and Finkel (2006)	P&T, PRB	Steel was found to be a main contributor of secondary impacts in a funnel and gate style PRB. Activated carbon was an additional determinant of the relative advantages of P&T and PRB. The relative advantages of the technologies depended on the extraction rate, activated carbon consumption, and construction method.	The study focused on identifying major contributors of secondary impacts rather than a direct comparison of P&T and PRB. The authors also noted that the use of iron in PRB and continuous wall construction may render different comparison results. Moreover, the uncertainty analysis was not linked to site characteristics, which would have determined extraction rates, GAC consumption, etc.
Cadotte et al. (2007)	P&T, ISCT	ISCT-based option was found to have the highest life-cycle impact; P&T has less impact than ISCT, but still much more impact than the biosparging-based option.	The study seems to have disqualified ISCT due to a number of considerations that are not consistent with general practice in the field: (1) P&T was assumed to run 300 years but

Continued

TABLE 3.1 Examples of LCA studies on P&T, PRB, ISB, and ISCT (Hou et al., 2014).—cont'd

Reference	Technology	Findings	Limitations
			system replacement was not included; (2) ISCT used over 100 permanent steel wells and a massive amount of oxidants, which does not seem realistic based on our project experiences.
Higgins and Olson (2009)	P&T, PRB	It found that the life-cycle impact of PRB is mainly from ZVI media demand and energy usage in construction, while the impact of P&T is mainly from operational energy demand. Overall, PRB is superior to P&T in all impact categories when ZVI longevity reaches 10 years.	The study identified media longevity as a major factor in determining the competitiveness of PRB over P&T; however, it does not identify other site characteristics that may have affected their relative competitiveness.
Lemming et al. (2010)	ISB	Using ISB would reduce the life-cycle impact of remediation remarkably compared to in situ thermal desorption or excavation with off-site disposal. However, there are time constraints, technological uncertainty, and local toxic by-products due to ISB.	Authors have noted that the LCA is very site-specific and result cannot be transferred to other sites.
Mak and Lo (2011)	PRB	The trench-based construction method was found to reduce life-cycle impact compared to the caisson-based method; ZVI and iron-oxide-coated sand can reduce environmental impact compared to ZVI and quartz sand mixture.	The PRB method was not compared with any other remedial methods; therefore, it is unclear whether the different construction methods and reactive media can

TABLE 3.1 Examples of LCA studies on P&T, PRB, ISB, and ISCT (Hou et al., 2014).—cont'd

Reference	Technology	Findings	Limitations
			change the relative desirability of PRB versus other technologies, such as the most commonly used P&T technique.

Note: *ISB,* in situ bioremediation; *ISCT,* in situ chemical treatment; *P&T,* pump and treat; *PRB,* permeable reactive barrier.

the system set. LCA studies will often report the impacts associated with the cleanup of 1 kg of contaminants in soil or groundwater, or the functional unit is set to equal the entire treatment of the site. In MCA, an aggregate score is calculated, which compares each option in terms of environmental, social, and economic impact. Results are presented as a visual diagram that illustrates the strengths and weaknesses of each option.

In all cases, the functional unit describes the relative performance of the system and provides a unified benchmark for data input and output to be compared. In other words, the functional unit identifies the quantitative aspect of the remediation and generally answers the "what," "how much," "how well," and "for how long" questions (Favara et al., 2011).

The "how much" question is answered if a quantity of media to be treated is specified. When the functional unit is defined in terms of a remediation target level, the equivalence is in the contaminant reduction to the required level, which answers the "what" and "how well" questions. When the time frame for a groundwater remedy is defined, the "how long" question is addressed (Favara et al., 2011). Defining the functional unit where different alternatives are compared can sometimes be challenging because the conditions of the remediation alternatives vary, such as treatment times, residual contaminant levels, value of the site, future use of the site.

In an LCA conducted by Song et al. (2018), the functional unit was set to 520,000 m^3 of soil treatment to the site cleanup goal, over a timeframe of 100 years. The functional unit for the hybrid LCA undertaken in (Hou et al., 2014) was based on managing ~2.5 km of waterways adjacent to the London Olympic Park for 100 years and was chosen as the removal and disposal of 30,000 m^3 of sediment when evaluating different treatment methods. In (Hou et al., 2017) the functional unit was the removal and treatment of Pb contaminated soil to the cleanup level, the time frame assumed for the LCA was 100 years, consistent with the time frame for health risk impact quantification. As can be seen, these examples describe the functional unit in terms of "what," "how much," "how well," and "for how long," though the level of detail for each element is variable, showing the variability of assigning functional units for remediation projects.

2.2 Project metrics

Project metrics in SA, whether qualitative or quantitative, form the sustainability indicator set. The data collected for these categories will be used to quantify the impacts of remediation. Therefore, the selection of appropriate sustainability indicators provides the basis for any SA to be conducted in a transparent and objective manner (Niemeijer and de Groot, 2008).

Various reference sustainability indicators for remediation SA have been published in a number of SA guidelines, e.g (SuRF-UK, 2011) (ITRC, 2011), and (ASTM, 2013), with different indicators being applicable depending on scope and specificity (Rizzo et al., 2016; Huysegoms and Cappuyns, 2017). When choosing sustainability indicators, a robust identification process should be followed so that the assessment is appropriate and incorporates a holistic view of the three environmental, social, and economic pillars of sustainability. Identification of relevant indicators requires an approach that regards systematic, quantitative, and comprehensive analysis of impacts (Page et al., 1999; Lesage et al., 2007; Suer and Andersson-Sköld, 2011; Busset et al., 2012). A range of elements can be found in the existing literature and guidelines. The indicators listed in the SuRF-UK, ITRC, and ASTM guidelines are summarized in Table 3.2. Metrics can also be selected by life-cycle tools, such as SiteWise. This simplifies metric selection, although the number of metrics consistent with remediation LCA goals may be limited (Favara et al., 2011). Some life-cycle tools allow the inclusion of additional metrics, for example, social and economic metric, that are sometimes lacking.

Some indicators included in guidelines are ambiguously defined, allowing flexibility and professional judgment in the approach of sustainability assessors. For example, the "soil quality" indicator included in the SuRF-UK guideline is so broad that it is hard to define specific parameters. In some guidelines, environmental, social, and economic sustainability indicators have been categorized into core elements and optional elements. These are selected according to specific conditions that fit with the context of the site and stakeholder preferences (Harclerode et al., 2015).

Social metrics involve worker safety, public acceptance, and impact to local communities, etc. Worker safety is a fundamental social sustainability indicator that considers the health and safety of remediation workers and the public. In most countries, worker safety is mandated by law, for example, the appointment of safety coordinators responsible for safe management of remediation activities is a legal requirement in Belgium (Cappuyns, 2016). Public acceptance, as

TABLE 3.2 Sustainability indicators used in different sustainable remediation guidelines.

Category	Sustainability indicators	SuRF-UK	ITRC	ASTM	Rationale
Environmental	Particulates	●	○	○	Common core element to describe impacts on air.
	Climate change/GHG emission	●	●	●	
	Photochemical oxidants	○	○	○	Commonly used to evaluate environmental impacts (e.g., footprint analysis). (USEPA, 2012)
	Ozone depletion	●	×	○	
	Water ecotoxicity	○	○	○	LCA-based indicator used to describe the impact on water.
	Water eutrophication	○	○	○	
	Flooding risk	●	×	×	Qualitative indicator refers to the LCA-based indicators in ecosystem category.
	Land transformation	×	×	○	
	Land occupation	×	×	○	
	Terrestrial acidification	○	×	○	
	Marine ecotoxicity	●	×	×	
	Biodiversity	○	●	×	Qualitative indicators refer to the LCA-based indicators in an ecosystem category.
	Habitat disturbance	●	●	●	
	Soil quality	●	×	○	
	Fossil depletion	○	○	○	Common core elements used to describe the impact on resources consumption.
	Metal depletion	○	○	○	
	Water depletion	●	●	●	

Continued

TABLE 3.2 Sustainability indicators used in different sustainable remediation guidelines.—cont'd

Category	Sustainability indicators	SuRF-UK	ITRC	ASTM	Rationale
	Waste generation	●	●	●	Common core elements incorporated into LCA method. (PRé, 2016)
	Energy use	●	●	●	
	Human toxicity	×	×	×	Essential LCA-based indicator, refers to the effects of toxic substances on the human environment.
	Ionizing radiation	×	×	×	In most cases, the impact of ionizing radiation will not be relevant.
Social	Work exposure/health risk	○	○	○	Common core elements in the social aspect.
	Work injuries/fatality	○	●	×	
	Community involvement	●	●	●	
	Community satisfaction	×	×	○	
	Equality	●	●	×	Equality is an element relating to affordability, disability, gender, ethnic, or cultural background, etc. (Paul Bardos, 2009)
	Local culture and vitality	●	●	●	
	Transparency	●	×	●	
	Regulatory approval	●	×	×	Regulatory approval is usually a prerequisite for the implementation of site remediation.
	Dust, noise, and odor	●	●	●	Common core elements used to describe the impact on local communities. Traffic relates to the off-site impacts such as air emission, noise pollution, etc.
	Traffic	●	○	●	
	Lighting	●	●	●	
	Visual impact	×	○	○	

	Site use	●	●	×	
	Built environment	●	○	○	
	Robustness	●	×	×	
Economic	Life-cycle cost of remediation	●	●	●	Measures the financial burden of all processes related to the remediation operation.
	Duration of remediation	○	×	○	Site owners and contractors are often inclined to choose remedial strategies with shorter time frames, to allow for quicker returns on investment.
	Project risk	●	×	×	The possibility of project failure leading to additional costs in the long-term
	Land value increase	●	×	○	Common core elements in economic aspect.
	Space creation	×	○	○	Different remediation strategies may lead to different amounts of land that is safe for use.
	Local property value increase	●	●	×	These are common elements in economic aspects. However, they are subjective indicators that may render large uncertainty in a sustainability assessment.
	Local economic vitality	●	○	●	
	Employment	●	●	●	
	Education and training	●	×	×	
	Reputation and brand value	×	×	×	
	Flexibility/resilience	●	×	×	This can be used as one of the criteria to evaluate the project risk.

●: the indicator is included; ○: a similar indicator is included (e.g., air pollutant is a similar indicator to particulates, photochemical oxidants, and ozone depletion); ×: the indicator is not included.

measured by community involvement and satisfaction, reflects stakeholder engagement with site remediation. The social aspect of remediation is improved when stakeholders participate at an early stage, before project decisions have been made. This could be achieved by communication with representative individuals or groups in a way that community satisfaction can be determined. In developing countries, community participation is not normally performed unless an environmental impact assessment (EIA) is required.

The social indicator "equality" is often included as an optional element in sustainability indicator sets, which relates to affordability, disability, gender, ethnic, or cultural background (SuRF-UK, 2009). It is recognized that equality will often be hard to measure directly; however, in certain circumstances, it will be possible to quantify it by participation and feedback from project stakeholders (Xie, 2011). The impact to local communities relates to the impact to neighborhoods or regions caused by the nuisance of dust, noise, light, and odor during remediation operations and congestion resulting from increased traffic (SuRF-UK, 2009). In addition, increased traffic volume also relates to off-site impacts such as harmful air emission, noise pollution, and road wear and tear (Nick, 2011). It should be noted that for most SA tools, impacts to the local environment are measured and specified in different manners (Mulligan; Döberl et al., 2013; Rosén et al., 2015).

Economic sustainability impacts can be quantified in SA by considering the measured benefits against costs. Direct economic benefits will include land value increases following remediation. Incorporation of indirect benefits such as employment and local business vitality is encouraged. In cost terms, the duration and risk of failure should be considered. Site owners and contractors will often want to choose remedial strategies with shorter time frames, to allow for rapid return on investment. For example, only 12.6% of remediation projects in China last longer than 500 days, with two-thirds completed in <200 days (IFEN, 2016). In the context of China, which has relatively little experience in remediation, the higher risk of technical failure should be considered when evaluating the economic sustainability of alternative remediation options.

A reference set of indicators for environmental and socioeconomic sustainability appraisal of contaminated land remediation strategies at mega-sites in China was presented in Song et al. (2018) and is illustrated in Fig. 3.1. This is useful because although technical guidelines on remediation practice have been developed to support soil and groundwater remediation, there is a lack of published guidance on how to incorporate SA.

2.3 Boundaries

According to the ISO 14040 series on LCA (ISO, 2006), an SA system boundary is determined by following an iterative process. This involves setting an initial system boundary, which is refined by adding unit processes that are later found to be significant (e.g., by sensitivity analysis).

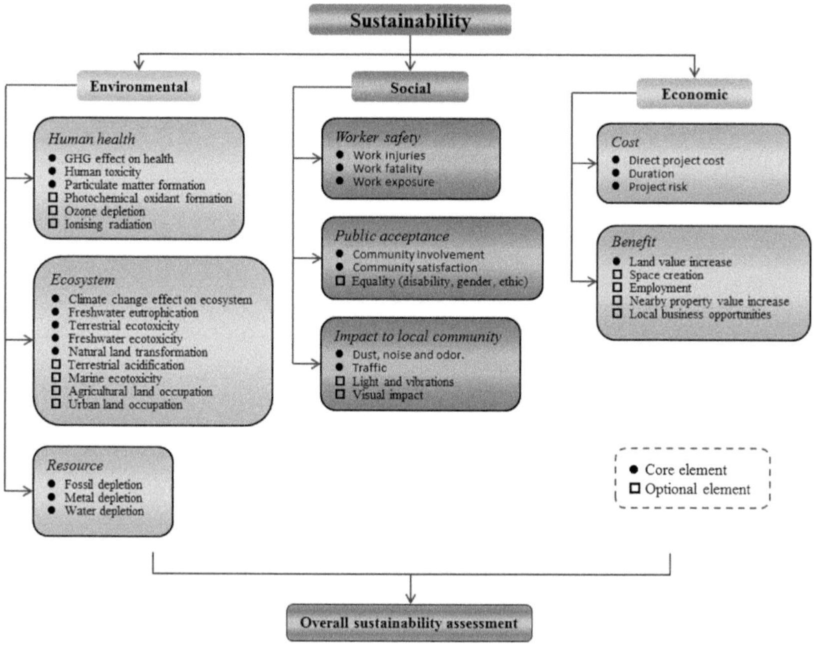

FIGURE 3.1 Indicators for environmental and socioeconomic sustainability appraisal of contaminated land remediation strategies at mega-sites in China (Song et al., 2018).

The general principle for setting an LCA initial system boundary is described in ISO 14040 Section 5.1.2.2 (corresponding to ISO 14041 5.3.3), which states that the system should be modeled so that inputs and outputs at its boundaries are elementary flows. Where an elementary flow is (i) material or energy entering the system being studied, which has been drawn from the environment without previous human transformation, or (ii) material or energy leaving the system being studied, which is discarded into the environment without subsequent human transformation. This requirement can only be satisfied in a process-based LCA if there are closed sets of processes that receive products and services only from the set of processes that they belong to. This makes compliance with ISO standards impossible without models containing loops (Suh et al., 2004); however, this problem is addressed by clause 5.3.3 in ISO 14041, which states that decisions shall be made regarding which unit processes shall be modeled by the study and the level of detail to which these unit processes shall be studied.

This means that efforts do not need to be expended quantifying inputs and outputs that will not significantly change the overall conclusions of a study. Though it should be noted that any decisions to omit life-cycle stages, processes, or inputs/outputs, should be clearly stated and justified. While traditional process-based LCA has truncation errors and tends to underestimate the

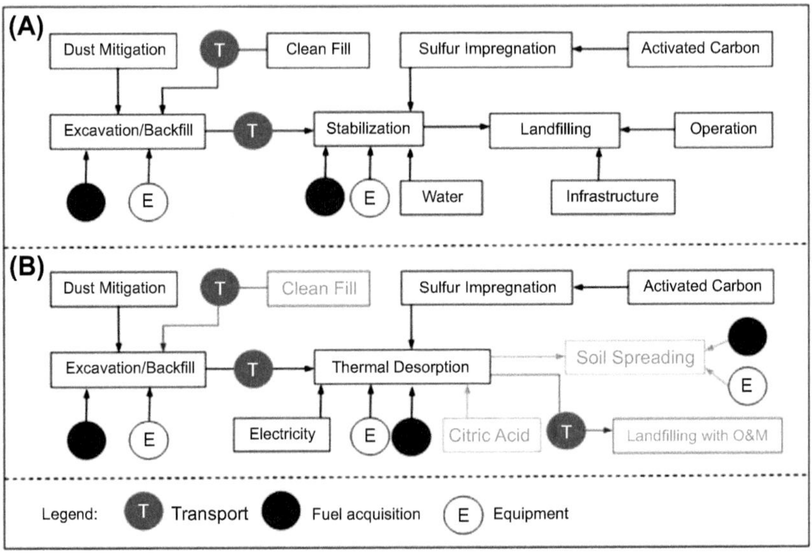

FIGURE 3.2 Remediation system boundaries: (A) S/S alternatives; (B) thermal desorption alternatives (processes in purple are specific to high-temperature thermal treatment alternative; and processes in green are specific to acid-facilitated thermal treatment alternative) (Hou et al., 2016).

overall impact of products and services, hybrid LCA minimizes such errors and is considered state of the art in LCA.

The process boundary of a comprehensive LCA will extend from "cradle" to "grave," including raw material and energy acquisition, equipment acquisition and usage, transportation, construction, and waste disposal. Although, as discussed earlier, this could be truncated if there are no significant differences between remediation alternatives. It is important to identify all pertinent aspects of the remediation system boundary and where the inputs and outputs lie. The boundary of the LCA conducted by (Hou et al., 2016) was from cradle to grave, including raw material and energy acquisition, equipment acquisition and usage, transportation, construction, and waste disposal (Fig. 3.2).

Moreover, there are three main types of boundaries to consider in SA: (i) process boundaries: the unity process involved in the assessment; (ii) geographical boundaries: the area covered in the assessment, and (iii) time boundaries: the assessed timescale. These will each depend on the assumptions made about each of the remediation alternatives selected, for example, Table 3.3 lists the assumptions for life-cycle time horizon used in (Hou et al., 2014)

3. Sustainability assessment tools

3.1 Multicriteria analysis tools

In qualitative assessments, sustainability impacts are given a relative classification score such as low, medium, or high, rather than applying measured data. A

TABLE 3.3 Assumptions used for the life-cycle time horizon in (Hou et al., 2014).

Remediation technology	Assumed time horizon
P&T	30 years of operation
EIB	Three injections in 2 years; continuous monitoring for 30 years
PRB	Assume that substrate is replenished when depleted (default at 15 years); continuous monitoring for 30 years
ISM	Continuous monitoring for 30 years

number of SA tools based on MCA have been developed by various companies or professional organizations, e.g., GoldSET by Golder Associates, VHGFM by the Swedish Geotechnical Association, and SiteWise by Battelle. These tools attempt to integrate economic, human, and environmental impacts across the entire supply chain to create a comprehensive SA (Mihelcic et al., 2003).

For example, the GoldSET evaluation tool includes five key steps. The user is asked to "set the scene" and identify the context in which the project and decision-making take place. Site conditions, project objectives, and key stakeholders and their issues are identified. The next step is to identify and elaborate on various remediation options that are suitable for the site specificities and project restrictions. Those options are assessed from an economic, social and environmental standpoint based on an indicator set (see Section 2.2) representative of a project's performance, impacts, and cost, in the context of its objectives and stakeholder concerns. MCA tools will often provide modules of preselected indicators, and indicators may be added or removed, depending on needs and context of each project. Each remediation option is subjectively ranked for each indicator according to a scoring scheme determined by the user. An aggregate score is calculated to compare options in terms of environmental, social, and economic impacts. Results are presented as visual diagrams (e.g., spider diagrams), which illustrate the strengths and weaknesses of each option. The evaluation process can be iterative, and further refining can be accomplished if additional information becomes available. Sensitivity analysis can also be performed on the results to improve the reliability of the findings (Golder Associates, 2019).

3.2 Life-cycle assessment tools

Remediation practitioners are increasingly using commercial LCA tools, and comprehensive databases (e.g., Ecoinvent), to conduct assessments. These approaches provide professional standards, tools, and databases to address the limitation of inconsistency between remediation LCA studies. Commercial LCA software packages include SimaPro, GaBi, and Economical Input–Output Life-Cycle Assessment (EIO-LCA). However, many remediation practitioners consider these approaches too costly to implement. Various other tools have been developed by governments and professional organizations, which are available in the public domain (for instance, the USEPA's "Methodology for Understanding and Reducing a Project's Environmental Footprint" available at https://clu-in.org/greenremediation/methodology/#SEFA). Though it should be noted that these tools are not as powerful as commercial software packages.

SimaPro is a commercial LCA software package that was first introduced in 1990 by the Dutch product ecology advisory group and has since been updated to the current seventh version. SimaPro 7 can be applied for various applications such as carbon footprint calculation, products and ecology design, environmental products statement, environmental impacts assessment of products or services, environment report and key performance indicator development. SimaPro can use large environmental load databases, such as Environment v.2, US LCI, ELCD, US InPut–OutPut, EU and Danish InPut–OutPut, Swiss InPut–OutPut, LCA Food, Industry data v.2; and provides a variety of evaluation methods, including ReCiPe, 99, Eeo-indicator 99, USEtox, IPCC 2007, EPD, ImPact 2002+, CML-IA, Traci 2, BEES, Ecological FootPrint, EDIP 2003, Ecological scarcity 2006, EPS 2000, and Greenhouse Gas protocol. In addition, the software has placeholders within the program to input social and economic impacts for integration into the sustainability assessment.

Another LCA software package, GaBi, is a specialist tool, jointly developed by the LBP institute of the University of Stuttgart and PE. It can be used to conduct LCA, GHG emission calculation, carbon footprint calculation, corporate carbon accounting, material flow analysis, and energy efficiency analysis. GaBi 5 has 4700 LCI (life-cycle list) data and renews the assessment method of LCA, such as improved water footprint assessment and waste modeling, improved carbon footprint and carbon balance modeling.

EIO-LCA is an assessment tool developed by the Green Design Institution of Carnegie Mellon University. EIO-LCA can calculate the environmental and economic impacts according to the material and energy consumption and GHG production by the whole remediation project. EIO-LCA can also achieve a rapid calculation of the LCA of a product or services as well as its supply chain. The core indicator is material consumption and GHG emission of these products, services, and industries in the supply chain. EIO-LCA is one of the most wildly used LCA tools, and the application of EIO-LCA in remediation field is increasing in recent years.

Life-cycle impact assessment (LCIA) can be evaluated using models, such as ReCiPe (Goedkoop et al., 2013), which utilize USES-LCA (Van Zelm et al., 2009) to simulate multimedia fate, exposure, and effects. The ReCiPe model can derive both midpoint and endpoint indicators. The endpoint method integrates results for various impact categories to render a single final score, thus allowing straightforward interpretation in management and decision-making-oriented settings. However, the endpoint method may result in more uncertainty in comparison with the middle-point method (Hauschild, 2005). Another recognized LCIA model is the UNEP-SETAC USEtox method (Rosenbaum et al., 2008). It should be noted that LCIA models rely on a number of assumptions that need to be confirmed as relevant for each site assessed, for example, that terrestrial species are exposed to contaminants only within the top 20 cm of soil (Goedkoop et al., 2013).

4. Life-cycle assessment approach

LCA is an established method to assess sustainability involving certain procedures that must be followed by assessors (Hou et al., 2014). Importantly, the International Organization for Standardization (ISO) introduced the ISO 14040 series on LCA (ISO, 2006), which sets the international standard to which LCA studies should conform—although it should be noted that the ISO standards only cover the assessment of environmental impacts, and not social or economic impacts.

LCA is conducted in accordance with the ISO standards by following a four-step process: (i) setting the goal and scope definition, (ii) inventory analysis, (iii) impact assessment, and (iv) interpretation. This provides a systematic approach for conducting an LCA and its interpretation. These are discussed in the following sections.

4.1 Goal and scope

The initial step in an LCA is to define the goal and scope. This involves firstly determining the SA purpose, detailing the expected and potential applications, and identifying the people involved. At this point, it is also a good idea to determine whether LCA or another approach is the most suitable way to assess sustainability for the site.

Setting a clear LCA goal allows remediation practitioners to set the purpose of the study, define the questions needed to be answered, the decisions that will be supported by the assessment results, the level of detail necessary, and the details for performing the study. In most cases, the goal of an LCA involves appraising different remediation options, in order to choose the most sustainable option. The scope setting aspect of this step helps determine the study methodology, quality, reporting, and review in accordance with the study goals, the intended applications, and the results. The scope ought to include all significant on-site and off-site processes for each remediation alternative to be assessed, including site preparation, construction, on-site operations, and waste disposal activities. SuRF-UK recommends extending the use of SA to site planning and project design (SuRF-UK, 2010), and ASTM recommends combining sustainability assessment with best management practices (BMPs) (ASTM, 2013).

It is necessary to determine whether primary, secondary, or tertiary impacts will apply to the study. Primary impacts are the health risk associated with the contaminants on-site during cleanup and residual contaminants remaining after cleanup. Secondary impacts derive from cleanup activities, for example, due to energy and material usage. Tertiary impacts relate to the use of the land after remediation, which is important if different remediation alternatives lead to different future land uses (Lesage et al., 2006). Primary impacts are not usually included in LCA if it is assumed that the site poses no exposure risk after remediation, although this may not always be the case. Tertiary impacts are almost always avoided in LCAs due to assessment complexity and data requirements.

Importantly, the LCA scope should depend the goal chosen, and vice versa. Defining the LCA scope in different ways will lead to different LCA outcomes, and, therefore, should be chosen judiciously. SuRF and SuRF-UK have discussed the importance of scope setting and its acceptance by relevant stakeholders (Bardos et al., 2011). Initial well-defined goals and scopes are also important for selecting the project metrics (Section 2.4). In (Hou et al., 2016), the LCA goal was to evaluate and compare the life-cycle impacts of thermal desorption and S/S in the remediation of mercury contaminated soil. The effects of acid facilitation in thermal desorption and using biochar in S/S were also evaluated. In (Song et al., 2018), the goal was to assess the environmental sustainability of site remediation options by using an LCA-based assessment method.

4.2 Inventory

The project inventory stage of an LCA involves collecting input and output inventory data associated with the sustainability indicators within the boundary of the LCA. The inventory is based on elements that make up the project life-cycle phases and their input and output flows. These may include the following (based on (Favara et al., 2011)):

- Chemical reagents—chemical oxidants (e.g., peroxide), pH buffers, electron donors, stabilization reagents, microbial cultures, etc.
- Construction materials—sand, cement, aggregates, bentonite, etc.
- Equipment and fixtures—drill rigs, excavators, generators, pumps, piping, tanks, etc.
- Energy—grid electricity, solar power, etc.
- Fuels—diesel, natural gas, petroleum, etc.
- Transportation—workers, equipment, samples, waste arisings, etc.
- Land use—conversion to residential/industrial use, etc.
- Contamination end-of-life—landfill, incineration, recycled, etc.

Data quality is of paramount importance for a robust project inventory, which determines the accuracy of the SA. Therefore, it is crucial to make appropriate assumptions regarding the process life cycle in accordance with site conditions, industry standards, and professional judgment. For example, process life-cycle inventory assumptions applied by (Hou et al., 2014) for comparing P&T, EIB, PRB, ISM remediation options at the London 2012 Olympic site are listed in Table 3.4.

For upstream and downstream material flows and emissions, data can often be collected firsthand from bench-scale experiments or pilot trials or from existing sources of information such as published literature, pertaining industrial consultants, contractors, equipment vendors, etc. Examples of specific data sources are listed below (based on (Favara et al., 2011) and (US EPA,

TABLE 3.4 Process life-cycle inventory assumptions for the London 2012 Olympic site (Hou et al., 2014).

Relevant alternatives	Assumptions
P&T, EIB, PRB, ISM	Drilling and installation of monitoring wells involve the use of a hollow stem auger drill rig with a production rate of 20 m/day. Well diameters are assumed to be 5.1 cm (2 inches); and borehole diameters are assumed to be 20.3 cm (8 inches).
P&T, EIB, PRB, ISM	Eight groundwater samples/day are collected by a two-member crew.
P&T, EIB, PRB, ISM	Drilling operations involve a four-person field crew.
P&T	80% in-well drawdown and 25% water table drawdown are assumed.
P&T	Above ground construction involves 4-inch Schedule 80 PVC conveyance piping from each well. The treatment system is proportional to the system described in (Higgins and Olson 2009) based on extraction flow rates, but limited to 20%–500% of the example system.
P&T	O&M is conducted by one person. Three O&M samples are collected each month from the influent, system, and effluent.
EIB	Injections involve a three-person field crew. Water/oil mixture is injected at 2 gpm per well in five wells simultaneously using a 3-hp pump.
PRB	The backfilling rate is 200 m^2/hours, with an additional 2 hours needed for setting up.

2006)); proprietary databases, such as Ecoinvent, are also useful for compiling project inventories.

- Equipment information—meter readings, process specifications, daily logs, etc.
- Engineering designs—drafts, estimates, invoices, etc.
- Vendor environmental product declarations or material safety data sheets
- Industry data reports, databases, and consulting knowledge
- Laboratory test reports
- Government and nongovernmental organization documents, reports, and databases
- Journals, papers, reference books, and patents
- Trade associations
- Previous LCA studies
- Local energy providers
- Professional judgment

As an example, the consumption and emission inventory sources and assumptions used in (Hou et al., 2014) are listed in Table 3.5. The process inventory data presented were collected according to ISO 14040. The system boundary encompassed acquisition of raw material and energy, transportation to and from the site, on-site operation, and waste disposal (i.e., cradle-to-grave). An average daily commute distance of 24 km is used for the London metropolitan area, and it is assumed that 85% of the construction workforce use public transport according to ODA. Physical flows in and out of the site were identified by reviewing project documents from the Olympic Delivery Authority (ODA).

For an economic input—output (IO) LCA (see Section 5.2), the data gathered may incorporate processes that would otherwise not be captured by a process-based LCA. Such processes include renting machinery, technical consulting, project management, mobilization and demobilization equipment, and many minor materials and processes such as personal protection equipment and courier services. For example, in (Hou et al., 2014), an IO-LCA model was developed that consisted of 123 different economic IO tables for the UK, which were mostly collected from literature published by the Office for National Statistics (ONS), including sector-level energy consumption, environmental emissions, and other socioeconomic impact vectors such as employment and compensation for employees. Worker fatalities and injury data were collected from literature published by the UK's health safety executive (HSE).

4.3 Impact assessment

This step involves calculating the life-cycle impacts associated with the project inventory. For this, the life-cycle inventory quantities are characterized in impact categories using characterization factors, so that a broad range of

TABLE 3.5 Consumption and emission inventory sources and assumptions applied for the London 2012 Olympic site by (Hou et al., 2014), which were mainly based on the ecoinvent database.

Process	Emission inventory source and assumptions
Electricity	Electricity mix is US-specific, based on 2004 data
Steel	High alloyed steel, stainless steel, chromium steel 18/8, at plant, add transportation to a regional storage site: transport, lorry >16 t, fleet average: 0.1 tkm Transport, freight, rail: 0.2 tkm
Cement	Cement, unspecified, at plant, add transportation to a regional storage site: transport, lorry >16 t, fleet average: 0.1 tkm Transport, freight, rail: 0.2 tkm
PVC	Mix of the two important PVC types: Suspension and emulsion PVC; contains an average transport from the production site to a regional storage site. transport, lorry >16 t, fleet average: 0.1 tkm Transport, freight, rail: 0.2 tkm
Sand/gravel	Gravel, unspecified, at mine, mix of round and crushed gravel: 79% round gravel and 21% crushed gravel, add transportation to a regional storage site: transport, lorry >16 t, fleet average: 0.1 tkm Transport, freight, rail: 0.2 tkm
Aluminum	Aluminum, primary, at plant; includes cast aluminum ingot production, transports of materials to the plant and then to a regional storage site, and the disposal of the wastes. Transport, lorry >16 t, fleet average: 0.1 tkm Transport, freight, rail: 0.2 tkm
Vegetable oil	Vegetable oil is modeled as soybean oil, at oil mill, data from an industrial oil mill in the United States, add transportation to a regional storage site: transport, lorry >16 t, fleet average: 0.1 tkm Transport, freight, rail: 0.2 tkm
Iron	ZVI is modeled as gray cast iron (Higgins and Olson 2009). 35% scrap and 65% pig iron assumed as iron input. Transport, lorry >16 t, fleet average: 0.1 tkm Transport, freight, rail: 0.2 tkm
Granulated activated carbon (GAC)	GAC is modeled as charcoal, at plant, production of charcoal from forest wood including emissions, add transportation to a regional storage site: transport, lorry >16 t, fleet average: 0.1 tkm Transport, freight, rail: 0.2 tkm

Continued

TABLE 3.5 Consumption and emission inventory sources and assumptions applied for the London 2012 Olympic site by (Hou et al., 2014), which were mainly based on the ecoinvent database.—cont'd

Process	Emission inventory source and assumptions
Chemical analysis	Energy and material use and emission from chemical analysis are based on USEPA standard method 8260.
Transportation by truck	Transport, lorry >16 t, fleet average, including operation of vehicle; production, maintenance, and disposal of vehicles; construction and maintenance and disposal of road.
Transportation by rail	Transport, freight, rail, diesel, average US diesel train freight transport, extrapolated from European data, include operation of vehicle, production, maintenance, and disposal of vehicles, construction and maintenance and disposal of railway tracks.
Site worker, pick-up truck	Light-medium duty pick-up trucks are modeled as van <3.5 ton in the Ecoinvent database, includes operation of vehicle; production, maintenance, and disposal of vehicles; construction and maintenance and disposal of road.
Off-road emission	Excavation, hydraulic digger, includes infrastructure, lubricating oil and fuel consumption, and some measured air emission, impact is based on unit volume of earth removed. Since well drilling rig is much less efficient in removing earth than an excavator, a factor of 146 is applied to estimate emission from a drilling rig based on *SiteWise User Guide* (NAVFAC, 2011). Similarly, a factor of 0.42 is used to estimate emission from dozer operations, and a factor of 6 is used to estimate emission from in situ soil mixing.

inventory data is presented in common units or indictors, in order that inventory information is presented concisely (Favara et al., 2011). It should be noted that for an environmental footprint analysis, only the climate change potential (CCP) is evaluated.

Environmental impact categories usually include three main impact categories: ecosystem, human health, and resources. Ecosystem includes factors such as agricultural land occupation, climate change, ecosystems, freshwater, ecotoxicity, freshwater eutrophication, marine ecotoxicity, natural land transformation, terrestrial acidification, terrestrial ecotoxicity, and urban land

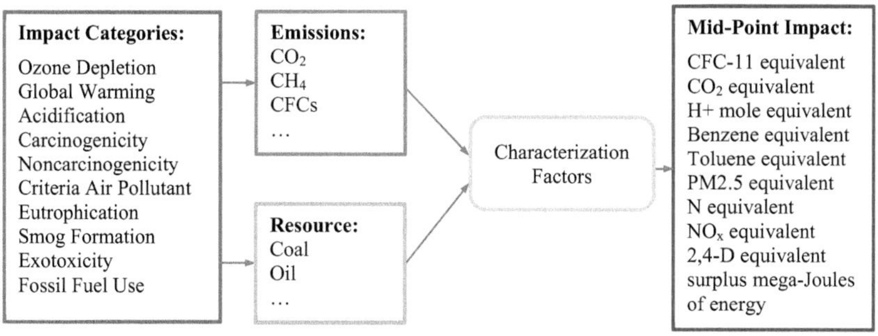

FIGURE 3.3 Schematic of the USEPA tool for reduction and assessment of chemical and other environmental impacts (TRACI) model.

occupation. Human health includes climate change, human toxicity, ionizing radiation, ozone depletion, particulate matter formation, and photochemical oxidant formation. Resources include fossil depletion and metal depletion.

The procedure for impact assessment includes three steps:

- choosing the impact type, parameter type, and characterization model;
- classifying the result of the life cycle; and
- calculating the parameter result.

The choice of LCIA method should pay regard to the study goal, for instance, midpoint methods will derive less uncertainty, whereas endpoint methods allow for easier managerial decision-making. The methodology used by (Hou et al., 2014) was adapted from the USEPA Tool for Reduction and Assessment of Chemical and Other Environmental Impacts (TRACI) system (Fig. 3.3).

Normalizing impact results helps with their interpretation. For example (Hou et al., 2014), normalized impact results as follows: The life-cycle costs were normalized by using the per capita gross domestic product. The increase in property value was normalized by the median housing price in the Newham Borough of London. The compensation of employees (i.e., worker wage) was normalized by the average yearly household income. Work fatalities and work injuries were normalized by the total fatality and injuries per year divided by total population. The impact to local traffic was normalized to the total yearly traffic in and out of the Lower Lea Valley Area divided by total population in that area. All these normalization factors are listed in Table 3.6.

4.4 Interpretation

By interpreting LCA results, the major factors on sustainability can be found. Life-cycle interpretation is the final stage of an LCA and provides the output for remediation decision-making. The interpretation of SA results should reflect the fact that they are relative and only indicate potential impacts, and

TABLE 3.6 Normalization factors.

Impact category	Normalization value	Units	Description and reference
Environmental impact			
Ecosystems	1.81E-04	species.yr/p/yr	ReCiPe normalization values for Europe (Goedkoop et al., 2013)
Human health	2.02E-02	DALY/p/yr	ReCiPe normalization values for Europe (Goedkoop et al., 2013)
Resources	3.08E+02	$/p/yr	ReCiPe normalization values for Europe (Goedkoop et al., 2013)
Social impact			
Worker fatality	1.7E-06	Incidents	Number of occupational fatality per capita (HSE, 2013)
Worker injuries	2.0E-03	Incidents	Number of occupational injuries per capita (HSE, 2013)
Local traffic	753	Vehicle trips	Yearly traffic in and out of the Lower Lea Valley area divided by total population in that area (ODA, 2007b)
Economic impact			
Life-cycle cost	£22,942	Pounds sterling (£)	GDP per capita in 2007 (ONS, 2010b)
Increase in property value	£94,467	Pounds sterling (£)	Median housing price in the Newham Borough of London in 2007 divided by average household size in London (DCLG, 2013)
Worker wages	£14,255	Pounds sterling (£)	Average per capita income in 2007 (ONS, 2010b)

that they do not predict actual impacts on category endpoints, the exceeding of thresholds, or factors of safety or risk (ISO, 2006). This means that SA is most suitable as a means to compare different options, i.e., to find the most sustainable approach to remediate a site among a number of potentially viable options, as opposed to a tool to calculate absolute impacts—although LCA can be used for this purpose if a sufficiently robust and comprehensive LCA is completed.

The interpretation of LCA results can be seen an opportunity to identify the most significant sustainability issues, evaluate the results, draw conclusions, and make recommendations (Favara et al., 2011). Significant issues reflected in the elements that contribute most impacts and may include one or more of the following (Favara et al., 2011):

- Inventory parameters—processes, products, wastes, etc.
- Impact category indicators—climate change potential, land use, acidification, ozone depletion, etc.
- Key contributors to the results among remediation project life-cycle processes—activated carbon renewal, waste transport, chemical oxidant production, etc.

To understand the effectiveness and limitations of assessment result, the analyses of sensitivity and uncertainty are required. This is because LCA results depend significantly on the assessor's choices and assumptions made regarding the methods and data used. An uncertainty analysis can be conducted following the procedures described in (Busset et al., 2012) where inventory data sets are qualitatively evaluated and assigned uncertainty factors for reliability (U1), exhaustivity (U2), temporal correlation (U3), geographical correlation (U4), technological correlation (U5), sample size (U6), and basic uncertainty (U7). Qualitative scores and corresponding uncertainty factors can be assigned following guidance described in published literature such as (Frischknecht et al., 2007; Weidema and Wesnæs, 1996). The standard deviation corresponding to each data set is calculated using the following equations:

$$V_{95\%} = \exp\sqrt{\sum_{i=1}^{7} \ln^2 \mu_i}$$

$$SD_{95\%} = \mu\sqrt{V_{95\%} - 1}$$

where V represents variance, SD represents standard deviation, and μ represents mean value. Monte Carlo simulation can be used to derive the probability distribution of the overall impacts. This involves random sampling of probability distribution functions for each data set. Usually, a total of 1000 scenarios are generated for each remedial alternative.

Sensitivity analysis is conducted to further evaluate the importance of several key parameters, to render implications for different sites conditions. For example, this may include changing the distance to the nearest landfill site, changing the contamination level or the soil moisture content, etc.

In (Hou et al., 2014), several factors were considered key parameters and were used to determine the range of uncertainty analysis. The minimum and maximum impact values for each impact category were then calculated based on the range of these tested parameters and used to draw error bars on result figures. Because the effect of each parameter being tested on any impact category is mathematically monotonic, the minimum and maximum impact values were calculated by using the "worst" and "best" combinations of minimum and maximum parameter values. A −30%/+50% factor was applied to distance to landfills based on reported values in other studies (Harbottle et al., 2007; Suer and Andersson-Sköld, 2011). The uncertainty associated with distance to a sand quarry (22 km) was tested by applying a +50% factor given that sand quarry farther away may have to be used to meet the quantity requirement at the Olympic site. The study assumed that 78% of the soil in soil washing was reused based on the grain distribution of sediment samples; however, in practice, this percentage could vary. A range of 70%−85% was tested in the uncertainty analysis. The allocated capital equipment cost was applied at a factor of −50/+50%, because under different market conditions the capital equipment's utilization and capital recovery period may differ dramatically (McNally, 2012). For social and economic impacts, all IO-based impact categories were tested by adjusting the economic input structure: by applying −30/+30% factor for capital equipment allocation, −20/+20% for technical consulting and project management, and −15/+15% to construction work. For property value, there are many factors that can affect property prices. It is difficult to isolate the price premium attributable to a single factor such as improved waterways. Therefore, uncertainty analysis is important for the impact on property value. The Olympic site study tested the uncertainty by adjusting the impact factor by ±30% and adjusting property unit price by ±20%.

5. Advanced sustainability assessment methods

5.1 Estimating primary, secondary, and tertiary impacts: using greenhouse gas as an example

GHG emissions can be assessed by LCA or footprint calculations. It should be noted that recent studies have revealed that tertiary impacts can be significant; therefore, a comprehensive life-cycle sustainability assessment ought to include these impacts (Hou et al., 2014, 2018). However, in practice, sustainability assessors will often omit tertiary impacts due to the practical difficulties of data collection and quantification for tertiary impact assessment. This section discusses an approach to calculate the primary, secondary, and tertiary GHG emission associated with remediation.

The primary GHG impact of redeveloping a number of brownfield sites can be estimated as follows (Hou et al., 2018):

$$\Delta \mathrm{GHG}_p = \sum_{i=1}^{\mathrm{n}} \left(\alpha_i M_i + \beta_i A_i \mathrm{T}_p - \chi A_i^r T_p^r \right)$$

where n is the total number of contaminated sites; α_i is the conversion factor for contaminant degradation at the i_{th} site contributing to GHG emissions, in kg/kg; M_i is the total amount of organic contaminants at the i_{th} site, in kg; β_i is the emission factor difference at the i_{th} brownfield site contributing to net GHG emissions due to land use change, in kg/ha-year; A_i is the total area of the i_{th} contaminated site, in hectares (ha); T_p is the duration of primary impact associated with landscape change, in years; A_i^r is total area of the i_{th} greenfield site, in hectares (ha), replacing the i_{th} contaminated site; χ_i is the emission factor difference at the i_{th} greenfield site contributing to net GHG emissions due to land use change; T_p^{r} is the duration of the primary impact associated with the landscape change at the greenfield site.

The impact associated with an individual site can be simplified by reducing n to 1; and if land use is not accounted for, the equation can be simplified by removing the terms associated with greenfield sites. It should be note that methods used to estimate primary impacts for other environmental categories will be similar to this, but the effects of site contaminants are more complicated.

Process-based LCA is the most commonly used method to estimate secondary impacts; however, if secondary impacts are to be estimated for a number of sites, a hybrid LCA method combining IO and process-based methods is recommended (Hou et al., 2014, 2018). Secondary impacts mainly derive from the use of energy (e.g., fuel and electricity) and materials (e.g., the manufacture of iron-ZVI for PRBs and substrates for EIB); but postremediation long-term monitoring can also contribute significantly to secondary impacts, especially for less certain remediation technologies that require long-term monitoring (Hou et al., 2014).

Remediation tertiary impacts have only been accounted for in a limited number of LCA studies to date, but they have been found to be potentially very large, and, therefore, an important component for deciding whether or not to conduct remediation and redevelopment. The tertiary impact in GHG emission for a number of sites can be estimated using the following equation (Hou et al., 2014):

$$\Delta GHG_t = \sum_{i=1}^{n}\Big[G_u\left(U_i^b - U_i^g\right) + G_R\left(R_i^b - R_i^g\right) + G_C\left(C_i^b - C_i^g\right)T_C + G_E\left(E_i^b - E_i^g\right)T_E\Big]$$

where G_u represents the emission factor for construction of new utilities during redevelopment (ton CO_2 eq per \$); U_i^b and U_i^g represent the construction costs of new utilities (\$) in brownfield redevelopment and greenfield development, respectively; G_R represents the emission factor for construction of new roads during redevelopment (ton CO_2 eq per \$); R_i^b and R_i^g represent the construction costs of new roads (\$) in brownfield redevelopment and greenfield development, respectively; G_C represents the emission factor for the average daily residential commute to/from the redeveloped site to/from work (ton CO_2 eq per km); C_i^b and C_i^g represent the average residential daily commute distance (km/day) to/from work for the brownfield redevelopment and greenfield development, respectively; T_c represents total number of commuting days within the study period; G_E represents the emission factor for energy use by residents living at the site (ton CO_2 eq per unit energy); E_i^b and E_i^g represent average annual energy consumption by residents of brownfield re-developments and greenfield developments, respectively; T_E represents the total number of years of residential energy consumption within the study period. Impacts for an individual site could be simplified by reducing n to 1, and assessment of environmental impacts other than GHG is similar.

5.2 Economic input–output–based LCA

The economic IO-based LCA model was developed by Joshi at the end of 1990s as a means to analyze product systems (i.e., comparing steel vs. plastic fuel tank systems) (Joshi, 1999). It has since been adapted to be used widely for estimating sustainability impacts for civil engineering projects (Sharrard et al., 2008). The main advantage of IO-based LCA is that it can be used to integrate economic, human, and environmental impacts across the entire supply chain for a comprehensive SA (Mihelcic et al., 2003). For social and economic impacts, seven indicators were identified for economic IO LCA by (Hou et al., 2014a,b): life-cycle cost, employment impact, increase in property value, compensation of employees, work-related fatality, work-related injuries, and impact to local traffic. These indicators were selected based on the following criteria: (1) they are strongly applicable to the remediation project; (2) they can be quantified with existing data; and (3) they cover a broad range of social and economic concerns.

Economic IO also has the benefit of being able to incorporate processes that would otherwise not be captured by process-based LCA. These processes include renting of machinery, technical consulting and project management, mobilization and demobilization, and many minor materials and processes (e.g., personal protection equipment and courier service). The IO model avoids double counting, but it does not include end-of-life impacts, as the IO tables only include preconsumer-stage industrial activities; therefore, additional end-of-life impacts are necessary for product LCA analysis if using this approach (Suh et al., 2004). For remediation LCA, consumptions are usually made by intermediate activities, and therefore, the end-of-life is accounted by including life-cycle costs in IO analysis.

It should be noted that IO-based hybrid models can be associated with a number of uncertainties. First of all, economic data is retrospective, and there may be several years before published data is updated. As some industries undergo rapid technology evolution, IO coefficients may change rapidly. However, the remediation industry is considered a relatively stable industry, especially as the largest costs are often associated with either earth moving (e.g., excavation and transportation) or landfilling. Economic data and environmental effect data also have inherent uncertainty due to published data not perfectly representing the actual situation. Limited data availability is another concern. While atmospheric emissions are often well reported, pollutant emissions to water and soil are often lacking in national statistics. Consequently, the IO-based LCA may be biased to impacts in categories that originate from atmospheric emissions. IO-LCA has also been reported to overestimate impacts if methane production is based on industrial average data, this is because typical municipal waste has high organic content and can cause significant methane emission. Therefore, (Hou et al., 2014) corrected the methane production data based on the organic content of sediment materials.

5.3 Hybrid LCA

Traditional process-based LCA is associated with truncation errors and tends to underestimate overall impacts. The hybrid LCA offers a flexible tool for remediation practitioners to perform life-cycle-based sustainability appraisal with different levels of project process data, which minimizes such errors (Finnveden et al., 2009; Wiedmann et al., 2011). A hybrid method was developed (Hou et al., 2014) to combine the traditional process-based LCA and IO-based LCA. The method rendered several advantages: (1) expanding the system boundary and reducing truncation errors; (2) incorporating consequential benefits (i.e., tertiary impact) of remediation; and (3) covering social and economic impact in addition to environmental impacts. By using the total project cost as an encompassing system boundary, the hybrid LCA approach rendered a consistent and complete system boundary, thus mitigating truncation errors. The hybrid method can also incorporate tertiary benefits and social/economic impacts into the overall SA, in an attempt to extend SA in three dimensions: spatial, temporal, and impact categories. The functional unit of the hybrid LCA is to achieve a predefined remedial objective at any specific contaminated site.

6. Suggested path forward for sustainability assessment

All too often, remediation engineers are not involved in site development planning (SuRF-UK, 2010); this means that they only have an opportunity to improve sustainability at the remediation options appraisal stage. Therefore, the opportunity to holistically optimize sustainability is lost. Ideally, SA of site investigation and remediation phases will be conducted in alignment with development planning, so that greater sustainability is achieved and relevant sustainability indicators fully optimized throughout the full process of site development.

Very few studies have assessed life-cycle tertiary impacts for multiple remediation sites on large scales, e.g., at a city or regional level, with the implications regarding climate change mitigation only just starting to be unraveled. For example, Hou et al. (2018) found that favoring brownfield land remediation and development over greenfield sites could lead to a net GHG reduction of 51.9 Mt CO_2 eq. over a 70-year period, or 0.74 Mt CO_2 yr^{-1}, the equivalent of 14% of San Francisco's 5.3 Mt CO_2 eq. GHG emission in 2010. This was mainly attributed to tertiary impacts associated with different post-remediation use of land, which are often overlooked in LCA (Hou et al., 2017).

Based on the philosophy of sustainable remediation to maximize the net environmental benefit (NEB), quantitative assessment could potentially be used as a basis to set optimal cleanup standards in combination with health risk assessment. For example, (Hou et al., 2017) conducted case study for the remediation of lead-contaminated soil where it was discovered by LCA that an optimal cleanup level of 800 mg/kg (compared with the regulatory guidance value of 255 mg/kg) could increase the NEB by 3%, while reducing economic costs by 36% (Hou et al., 2017).

References

ASTM, 2013. Standard Guide for Integrating Sustainable Objectives into Cleanup. E2876-13. ASTM International, West Conshohocken, PA.

Bardos, P., Bone, B., Boyle, R., Ellis, D., Evans, F., Harries, N.D., Smith, J.W.N., 2011. Applying sustainable development principles to contaminated land management using the SuRF-UK Framework. Remediation Journal 21 (2), 77–100.

Bardos, R.P., Bone, B.D., Boyle, R., Evans, F., Harries, N.D., Howard, T., Smith, J.W., 2016. The rationale for simple approaches for sustainability assessment and management in contaminated land practice. The Science of the Total Environment 563–564, 755–768.

Bayer, P., Finkel, M., 2006. Life cycle assessment of active and passive groundwater remediation technologies. Journal of Contaminant Hydrology 83 (3–4), 171–199.

Blanc, A., Métivier-Pignon, H., Gourdon, R., Rousseaux, P., 2004. Life cycle assessment as a tool for controlling the development of technical activities: application to the remediation of a site contaminated by sulfur. Advances in Environmental Research 8 (3–4), 613–627.

Busset, G., Sangely, M., Montrejaud-Vignoles, M., Thannberger, L., Sablayrolles, C., 2012. Life cycle assessment of polychlorinated biphenyl contaminated soil remediation processes. International Journal of Life Cycle Assessment 17 (3), 325–336.

Cadotte, M., Deschênes, L., Samson, R., 2007. Selection of a remediation scenario for a diesel-contaminated site using LCA. The International Journal of Life Cycle Assessment 12 (4), 239.

Cappuyns, V., 2016. Inclusion of social indicators in decision support tools for the selection of sustainable site remediation options. Journal of Environmental Management 184 (Part 1), 45–56.

DCLG, 2013. Housing Market: Median House Prices Based on Land Registry Data, by District, from 1996. Department for Communities and Local Government.

Diamond, M.L., Page, C.A., Campbell, M., McKenna, S., Lall, R., 1999. Life-cycle framework for assessment of site remediation options: method and generic survey. Environmental Toxicology & Chemistry 18 (4), 788–800.

Döberl, G., Ortmann, M., Frühwirth, W., 2013. Introducing a goal-oriented sustainability assessment method to support decision making in contaminated site management. Environmental Science & Policy 25 (1), 207–217.

Favara, P.J., Krieger, T.M., Boughton, B., Fisher, A.S., Bhargava, M., 2011. Guidance for performing footprint analyses and life-cycle assessments for the remediation industry. Remediation Journal 21 (3), 39–79.

Finnveden, G., Hauschild, M.Z., Ekvall, T., Guinee, J., Heijungs, R., Hellweg, S., Koehler, A., Pennington, D., Suh, S., 2009. Recent developments in life cycle assessment. J. Environ. Manage. 91, 1–21.

Frischknecht, R., Jungbluth, N., Althaus, H.J., Bauer, C., Doka, G., Dones, R., Hischier, R., Hellweg, S., Humbert, S., Kollner, T., et al., 2007. Implementation of Life Cycle Impact Assessment Methods:Data v2.0. Ecoinvent Report No. 3. Swiss Centre for Life Cycle Inventories.

Garon, K.P., 2008. Sustainability Analysis for Improving Remedial Action Decisions. 2008 State Superfund Managers Symposium. Association of State and Territorial Solid Waste Management Offices, Scottsdale, AZ.

Goedkoop, M., Heijungs, R., Huijbregts, M., De Schryver, A., Struijs, J., van Zelm, R., 2013. ReCiPe 2008, a Life Cycle Impact Assessment Method Which Comprises Harmonised Category Indicators at the Midpoint and the Endpoint Level, first ed. Report I: Characterisation.

Golder Associates, 2019. Methodology. https://golder.goldset.com/portal/methodology.aspx.

Harbottle, M.J., Al-Tabbaa, A., Evans, C.W., 2007. A comparison of the technical sustainability of in situ stabilisation/solidification with disposal to landfill. J. Hazard. Mater 141, 430e440.

Harclerode, M., Ridsdale, D.R., Darmendrail, D., Bardos, P., Alexandrescu, F., Nathanail, P., Pizzol, L., Rizzo, E., 2015. Integrating the social dimension in remediation decision-making: state of the practice and way forward. Remediation Journal 26 (1), 11–42.

Hauschild, M.Z., 2005. Assessing environmental impacts in a life-cycle perspective. Environ. Sci. Technol. 39, 81A–88A.

Higgins, M.R., Olson, T.M., 2009. Life-cycle case study comparison of permeable reactive barrier versus pump-and-treat remediation. Environmental science & technology 43 (24), 9432–9438.

Hou, D., Al-Tabbaa, A., Guthrie, P., Hellings, J., Gu, Q., 2014a. Using a hybrid LCA method to evaluate the sustainability of sediment remediation at the London Olympic Park. Journal of Cleaner Production 83, 87–95.

Hou, D., Al-Tabbaa, A., Luo, J., 2014b. Assessing effects of site characteristics on remediation secondary life cycle impact with a generalised framework. Journal of Environmental Planning and Management 57 (7), 1083–1100.

Hou, D., Gu, Q., Ma, F., O'Connell, S., 2016. Life cycle assessment comparison of thermal desorption and stabilization/solidification of mercury contaminated soil on agricultural land. Journal of Cleaner Production 139, 949–956.

Hou, D., Qi, S., Zhao, B., Rigby, M., O'Connor, D., 2017. Incorporating life cycle assessment with health risk assessment to select the 'greenest'cleanup level for Pb contaminated soil. Journal of Cleaner Production 162 (9), 1157–1168.

Hou, D., Song, Y., Zhang, J., Hou, M., O'Connor, D., Harclerode, M., 2018. Climate change mitigation potential of contaminated land redevelopment: a city-level assessment method. Journal of Cleaner Production 171, 1396–1406.

HSE, 2013. Health and Safety Executive Annual Statistics Report. Health and Safety Executive.

Huysegoms, L., Cappuyns, V., 2017. Critical review of decision support tools for sustainability assessment of site remediation options. Journal of Environmental Management 196, 278–296.

IFEN, 2016. Survey of Land Contamination at Changzhou. Retrieved from http://news.ifeng.com/a/20160428/48624891_0.shtml.

ISO, 2006. Environmental Management - Life Cycle Assessment - Principles and Framework 14040–2006. International Organization for Standardization.

ITRC, 2011. Green and Sustainable Remediation: A Practical Framework. Interstate Technology & Regulatory Council, Washington, DC. ITRC.

Joshi, S., 1999. Product environmental life-cycle assessment using input-output techniques. J. Ind. Ecol 3, 95–120.

Lemming, G., Chambon, J.C., Binning, P.J., Bjerg, P.L., 2012. Is there an environmental benefit from remediation of a contaminated site? Combined assessments of the risk reduction and life cycle impact of remediation. Journal of Environmental Management 112, 392–403.

Lemming, G., Hauschild, M.Z., Chambon, J., Binning, P.J., Bulle, C., Margni, M., Bjerg, P.L., 2010. Environmental impacts of remediation of a trichloroethene-contaminated site: life cycle assessment of remediation alternatives. Environmental Science and Technology 44 (23), 9163–9169.

Lesage, P., Ekvall, T., Deschênes, L., Samson, R., 2006. Environmental assessment of brownfield rehabilitation using two different life cycle inventory models. The International Journal of Life Cycle Assessment 12 (7), 497.

Lesage, P., Ekvall, T., Deschênes, L., Samson, R., 2007. Environmental assessment of brownfield rehabilitation using two different life cycle inventory models Part 2: case study. International Journal of Life Cycle Assessment 12 (7), 497–513.

Mak, M.S.H., Lo, I.M.C., 2011. Environmental life cycle assessment of permeable reactive barriers: effects of construction methods, reactive materials and groundwater constituents. Environmental Science and Technology 45 (23), 10148–10154.

Mauko Pranjić, A., Oprčkal, P., Mladenovič, A., Zapušek, P., Urleb, M., Turk, J., 2018. Comparative life cycle assessment of possible methods for the treatment of contaminated soil at an environmentally degraded site. Journal of Environmental Management 218, 497–508.

McNally, J., 2012. Drivers of Cost and Price for Dredging [WWW Document]. URL. https://www.westerndredging.org/index.php/2012-pacific-weda-presentations?download¼535:drivers-of-cost-a-price-for-dredging.

Mihelcic, J.R., Crittenden, J.C., Small, M.J., Shonnard, D.R., Hokanson, D.R., Zhang, Q., Chen, H., Sorby, S.A., James, V.U., Sutherland, J.W., Schnoor, J.L., 2003. Sustainability science and engineering: the emergence of a new metadiscipline. Environ. Sci. Technol. 37, 5314e5324. https://doi.org/10.1021/es034605h.

Mulligan, C.N. Case studies for a tool for evaluating the sustainability of site remediation treatment systems.

NAVFAC, 2011. SiteWise version 3 user guide, Naval Facilities Engineering Command. Engineering Service Center Port.

Nick, M.P., 2011. A case study on the evaluation and implementation of green and sustainable remediation principles and practices during a RCRA corrective action cleanup. Ground Water Monitoring & Remediation 31 (2), 63–71.

Niemeijer, D., de Groot, R.S., 2008. A conceptual framework for selecting environmental indicator sets. Ecological Indicators 8 (1), 14–25.

O'Connor, D., Hou, D., 2018. Targeting cleanups towards a more sustainable future. Environ Sci Process Impacts 20 (2), 266–269.

ONS, 2010b. Annual Abstract of Statistics. Office for National Statistics.

Page, C.A., Diamond, M.L., Campbell, M., McKenna, S., 1999. Life-cycle framework for assessment of site remediation options: case study. Environmental Toxicology & Chemistry 18 (4), 801–810.

Paul Bardos, A.L., Willenbrock, N., 2009. A Review of Published Sustainability Indicator Sets: How Applicable are they to Contaminated Land Remediation Indicator-Set Development?.

PRé, E.A., 2016. SimaPro Database Manual Methods Library.

Rizzo, E., Bardos, P., Pizzol, L., Critto, A., Giubilato, E., Marcomini, A., Albano, C., Darmendrail, D., Döberl, G., Harclerode, M., 2016. Comparison of international approaches to sustainable remediation. Journal of Environmental Management 184 (Pt 1), 4.

Rosén, L., Back, P.E., Söderqvist, T., Norrman, J., Brinkhoff, P., Norberg, T., Volchko, Y., Norin, M., Bergknut, M., Döberl, G., 2015. SCORE: a novel multi-criteria decision analysis approach to assessing the sustainability of contaminated land remediation. The Science of the Total Environment 511C, 621–638.

Rosenbaum, R.K., Bachmann, T.M., Gold, L.S., Huijbregts, M.A.J., Jolliet, O., Juraske, R., Koehler, A., Larsen, H.F., MacLeod, M., Margni, M., others, 2008. USEtox e the UNEP e SETAC toxicity model: recommended characterisation factors for human toxicity and freshwater ecotoxicity in life cycle impact assessment. Int. J. Life Cycle Assess 13, 532e546.

Sharrard, A.L., Matthews, H.S., Ries, R.J., 2008. Estimating construction project environmental effects using an input-output-based hybrid life-cycle assessment model. Journal of Infrastructure Systems 14 (4), 327–336.

Song, Y., Hou, D., Zhang, J., O'Connor, D., Li, G., Gu, Q., Li, S., Liu, P., 2018. Environmental and socio-economic sustainability appraisal of contaminated land remediation strategies: a case study at a mega-site in China. The Science of the Total Environment 610, 391–401.

Suer, P., Andersson-Sköld, Y., 2011. Biofuel or excavation? - life cycle assessment (LCA) of soil remediation options. Biomass and Bioenergy 35 (2), 969–981.

Suèr, P., Nilsson-Pågledal, S., Norrman, J., 2004. LCA for site remediation: a literature review. Soil and Sediment Contamination 13 (4), 415–425.

Suh, S., Lenzen, M., Treloar, G.J., Hondo, H., Horvath, A., Huppes, G., Jolliet, O., Klann, U., Krewitt, W., Moriguchi, Y., Munksgaard, J., Norris, G., 2004. System boundary selection in life-cycle inventories using hybrid approaches. Environmental Science and Technology 38 (3), 657–664.

SuRF-UK, 2009. A Review of Published Sustainability Indicator Sets: How Applicable are they to Contaminated Land Remediation Indicator-Set Development? AIRE, CL.

SuRF-UK, 2010. A Framework for Assessing the Sustainability of Soil and Groundwater Remediation. AIRE, CL.

SuRF-UK, 2011. Annex 1: The SuRF-UK Indicator Set for Sustainable Remediation Assessment. AIRE, CL.

Toffoletto, L., Deschênes, L., Samson, R., 2005. LCA of ex-situ bioremediation of diesel-contaminated soil. International Journal of Life Cycle Assessment 10 (6), 406–416.

US EPA, 2006. Life Cycle Assessment: Principles and Practice. US Environmental Protection Agency, Washington, DC.

USEPA, 2012. Green Remediation Best Management Practices: overview of EPA's Methodology to Address the Environmental Footprint of Site Cleanup.

Van Zelm, R., Huijbregts, M.A.J., van de Meent, D., 2009. USES-LCA 2.0da global nested multimedia fate, exposure, and effects model. Int. J. Life Cycle Assess 14, 282–284.

Vignes, R.P., 2001. Use limited life-cycle analysis for environmental decision-making. Chemical engineering progress 97 (2), 40–54.

Weidema, B., Wesnæs, M., 1996. Production of Pulses, Grains and Oilseeds: Guidelines for LCA practitioners. J. Clean. Prod 4, 167–174.

Wiedmann, T.O., Suh, S., Feng, K., Lenzen, M., Acquaye, A., Scott, K., Barrett, J.R., 2011. Application of Hybrid Life Cycle Approaches to Emerging Energy Technologies – The Case of Wind Power in the UK. Environmental Science & Technology 45 (13), 5900–5907.

Xie, L., 2011. Environmental Justice in China's Urban Decision-making. Taiwan in Comparative.

Chapter 4

Best management practices for sustainable remediation

John A. Simon
Nathan Associates, Inc., Remediation Journal, John Wiley & Sons, Arlington, VA, United States

1. Introduction

The advent of sustainable remediation resulted in the need for frameworks providing practitioners direction on implementing this relatively new practice area within the field of remediation. The frameworks developed over the past decade generally have a common theme—the use of best management practices or "BMPs." Best management practice is a common civil engineering term that is frequently referred to in the design and implementation of stormwater management systems (for example, USEPA, 1992). In the context of stormwater management, BMPs are identified and selected to minimize the volume of stormwater runoff as well as pollutants in the stormwater runoff. The founders and, thereafter, leaders of sustainable remediation adopted the general concept of BMPs with the objective of using BMPs to help implement concepts that integrate sustainability into remediation projects.

The various definitions of sustainable remediation BMPs are explained further next, followed by a discussion on selecting and implementing BMPs, green product resources, and a case study illustrating the application of sustainable remediation BMPs in the field. For the remainder of this chapter, the term BMP specifically refers to sustainable remediation BMPs.

2. What is a sustainable remediation best management practice?

The sustainable remediation guidance documents with the most definitive definitions and guidance on selecting and implementing BMPs were developed by ASTM International, Inc., (ASTM) workgroups. The ASTM guidance development process involves a group of volunteers who are dedicated to a subject, in this case, sustainable remediation, who work in a conciliatory manner to prepare proposed standards that are then vetted through a consensus

Sustainable Remediation of Contaminated Soil and Groundwater. https://doi.org/10.1016/B978-0-12-817982-6.00004-5

process before being accepted for publication. Two ASTM work groups developed sustainable-remediation-related standards, the Greener Cleanup Work Group, which only focused on the environmental aspects of sustainable remediation, and the Integrating Sustainable Objectives into Cleanup Work Group, which addressed the social and economic aspects of sustainable remediation in addition to the environmental aspect. Two ASTM standards were ultimately published as a result of these work groups' efforts: *Standard Guide for Greener Cleanup (E2893-16;* ASTM, 2016) and *Standard Guide for Integrating Sustainable Objectives into Cleanup* (ISOC; E2876-13; ASTM, 2013).

Each of these ASTM guides includes a definition of BMPs, as follows:

- ASTM Standard Guide for Greener Cleanup—activities that, if applicable to the situation, typically will reduce the environmental footprint of a cleanup activity.
- ASTM Standard Guide for Integrating Sustainable Objectives into Cleanup—an activity that, under most situations, improves one or more sustainable aspects (environmental, social, economic) of a cleanup at a specific site.

These definitions are relatively similar with the major difference of the Greener Cleanup Standard Guide having a narrower definition limited to the environmental aspects, whereas the ISOC Standard Guide includes the social and economic elements in the BMP definition.

3. Selecting sustainable remediation best management practices

The ASTM ISOC Standard Guide provides a framework that allows and encourages remediation professionals to "… integrate sustainable objectives into cleanup while working within applicable regulatory criteria." (ASTM, 2013, P. 1). The guide's framework establishes a stepwise process for identifying, selecting, and implementing BMPs. Effectively integrating sustainable BMPs into remediation projects is not as simple as reviewing a list or database of BMPs and selecting those that potentially may be beneficial to a cleanup project. Instead, sophisticated, sustainable remediation professionals follow a process that is rational, cost-effective, and documented. The ASTM Greener Cleanup Standard Guide provides a similar framework and process; however, as stated earlier, it is limited in its application to the environmental aspects of a sustainable remediation project.

A key to the various guidance documents for sustainable remediation is that sustainable remediation projects must meet applicable regulations and be protective of human health and the environment. Sustainable remediation cannot be an excuse for conducting a less thorough or less protective remedy. The ASTM ISOC Standard Guide explicitly states: "*This guide is not intended to justify the avoidance of regulatory requirements or any applicable cleanup standards.*" (Section 4.12.4).

In developing the ASTM guides, the task groups determined that selecting BMPs is a stepwise structured process that involves logic and analysis. One way to approach this process is by considering the BMP selection process within the realm of multicriteria decision analysis or "MCDA" (also commonly referred to as multicriteria decision-making). There is a large body of knowledge in the literature about MCDA and the "decision-making paradox." When examined carefully, the decision-making process for a remediation project is distilled down to first defining the project needs or goals and then how to best employ the appropriate technologies to reach those goals given the site conditions. The BMP identification and selection process in the ASTM guides closely follows MCDA theory.

First and foremost, when considering BMPs, the question to ask is whether the management practice is potentially applicable to the project's needs and goals. Second, the decision-making process involves evaluating which BMPs can cost-effectively reduce the environmental footprint of the remedy or provide social or economic benefits. Deciding which BMPs have the greatest impact on the project's sustainability typically can be conducted through a basic comparison of potentially applicable BMPs; however, for some remedies, the process will involve calculating metrics. In certain situations, it may be beneficial to forecast the generation of energy savings (and the associated greenhouse gas emission reductions) when comparing two environmental BMPs and then to evaluate the net present value cost over time to implement each BMP. For example, when optimizing a pump-and-treat system, a common BMP to consider is installing variable-frequency drives to reduce the energy consumption of pumps. The ultimate decision can be assisted by calculating the energy savings in light of both cost and the desire to reduce energy consumption. If the present value capital and operating cost of the BMP over time is less than the present value of the cost savings within the desired return-on-investment period, then the decision can be made based on economics. If the cost is higher, then the decision must be made based on the ultimate responsible party's (the entity paying for the cleanup) social responsibility goals. In other words, is the responsible party willing to pay more for a sustainable cleanup?

As previously mentioned, the ISOC Standard Guide and the Greener Cleanup Standard Guide provide the most authoritative and detailed guidance on selecting BMPs. These guides, which have nearly identical BMP selection processes, provide remediation professionals a clear, stepwise approach to implementing sustainable remediation projects and are the most direct guidance available. Industry experience shows that sustainable remediation with the full complement of sustainability factors (i.e., environmental, social, and economic factors) is optimal for larger projects that involve redevelopment or repurposing of the site, whereas more straightforward projects may only be suitable for green remediation. The ISOC guide should be followed when performing a sustainable remediation project involving all three sustainability aspects—the environmental, social, and economic aspects. Some remediation projects are not particularly well suited for sustainable remediation; however, the environmental aspect of sustainable remediation applies to almost all cleanup projects. The Greener Cleanup Standard Guide should be followed when conducting a project that only involves the environmental aspect of sustainable remediation.

An illustrative example of when not to utilize all three aspects of sustainability is an underground storage tank cleanup project, which may not have many opportunities for social or economic BMPs and, therefore, is better suited for only considering the environmental BMPs. On the other hand, robust brownfield projects that bring jobs and improve the social fabric of a community are ideal sustainable remediation project candidates with all three BMP categories incorporated into the project. However, it is important to recognize that if only environmental BMPs are utilized that the project should not be designated as a sustainable remediation project but can be designated as a green remediation project.

The ASTM task groups endeavored to design guidance applicable to any site, from straightforward excavation projects to complicated multimedia, multi-contaminant sites. Either guide can be applied to any stage of the cleanup process from site characterization through remedy implementation and operation, maintenance, and monitoring. The guides are especially adept for users considering remedy optimization. However, the guides may not be appropriate for certain projects, such as some emergency response actions, because there may not be sufficient time for integrating sustainability concepts into the project.

Both ASTM guides incorporate concepts already established by the U.S. Environmental Protection Agency (USEPA), such as tracking environmental footprint reductions across the following five core elements: (1) minimize total energy use and maximize use of renewable energy; (2) minimize air pollutants and *greenhouse gas* emissions; (3) minimize water use and impacts to water resources; (4) reduce, reuse, and recycle material and waste; and (5) protect land and ecosystems. By utilizing these preestablished core elements, the user is better able to rely on USEPA guidance for green remediation, such as the 12 BMP guides published by the Agency and available on its Clu-in Green Remediation website at http://www.clu-in.org/greenremediation/. Obviously, the ISOC Standard Guide includes additional information on social and economic BMPs.

3.1 Stepwise approach for selecting and implementing sustainable BMPs

The ISOC Standard Guide provides a six-step decision-making process that generally follows the MCDA theory. The underlying concept of this theory is that, when making decisions, it is advantageous to identify potential solutions and then prioritize those that have the most significant benefit for the costs incurred. Then, selecting the alternatives that are most cost-effective and beneficial becomes simpler because there is a process to follow. The Guide's six-step process follows this theory and provides the following instructional path for selecting sustainable BMPs:

Step 1: **Identify Potentially Applicable BMPs**—identify BMPs that are potentially applicable to a project.
Step 2: Evaluate BMPs—identify and implement a BMP evaluation method, such as a ranking or rating system or some other type of MCDA model.
Step 3: Select a Practicable Set of BMPs—select BMPs based on the evaluation performed in Step 2; the BMPs should substantially benefit each of the sustainable aspects without causing a significant negative impact to any of the other sustainable aspects (i.e., ensure that each BMP does not have an unintended consequence of negatively impacting a sustainability factor in the process of providing a benefit to another factor).
Step 4: **Implement BMPs**—implement the selected BMPs and document BMPs not implemented in the situation where a BMP is found to be ineffective, impracticable to implement, or cost prohibitive.
Step 5: Quantify BMP Results (optional)—if desired and deemed beneficial, the user may choose to quantify the impacts of the BMPs implemented; however, some BMPs, particularly those focused on the social aspect, may not be quantifiable.
Step 6: Documentation—the activities and evaluations performed must be documented.

The following discussion provides additional information on how each of these steps is implemented in practice.

3.1.1 Step 1: Identify potentially applicable BMPs

The first step in a robust, sustainable remediation project is to compile a list of BMPs potentially applicable to the specific project. This step involves identifying the selected remedy or potentially applicable remedial technologies as well as the social and economic setting for the project. This list should identify the attributes of each BMP from across the realm of sustainable remediation.

One widely debated subject is whether the sustainable remediation process should be used to help select remedies. Generally speaking, the most cost-effective solution that can achieve regulatory standards is going to be selected and implemented. However, this does not mean that the sustainable remediation process cannot be implemented because a "green technology" or "sustainable remedy is not being employed." Sustainable remediation leaders conclude that the opposite is true and that sustainable remediation can be overlaid on top of almost any remedy by incorporating sustainable remediation

BMPs into remedies. Stated another way, sustainable remediation BMPs can be identified and implemented for almost any remedy and, therefore, sustainable remediation can be integrated into almost any remediation project.

When applying the BMP selection process, the remediation practitioner should first consult lists of potentially applicable remedies. Resources for these lists include, but are not limited to:

- ASTM ISOC Standard Guide
- ASTM Greener Cleanup Standard Guide, which has a sophisticated spreadsheet of more than 180 BMPs organized into 10 remediation categories
- U.S. Sustainable Remediation Forum (SURF) BMP database[1]
- USEPA's Green Remediation Best Management Practices webpage that provides a series of BMP fact sheet within the following categories:
    - Site investigation
    - Excavation and surface restoration
    - Soil vapor extraction and air sparging technologies
    - Pump and treat technologies
    - Bioremediation
    - In situ thermal technologies
    - Landfill cover systems and associated energy production
    - Leaking underground storage tank systems
    - Mining sites
    - Renewable energy applications
    - Clean fuel and emission technologies
    - Materials and waste management
    - USEPA's methodology for environmental footprint assessment

The ASTM Greener Cleanup BMP Table indicates which technologies each specific BMP may apply to as well as providing the applicable core elements for each BMP. ASTM offers the Greener Cleanup BMP Table spreadsheet in Excel format as an adjunct purchase that enables remediation practitioners to sort the table by key elements, such as applicable technologies. The sorting process substantially expedites the use of the standard by allowing the user to sort among any of the major categories, including technologies. For example, if in situ thermal treatment is the selected remedy, a sort can be conducted that will list all the BMPs that are potentially applicable to that technology. Then, the sort can also be expanded to the USEPA's five core elements.

A resource for social and economic BMPs is the ASTM ISOC Standard Guide. Table X1.1 of the guide provides 61 social BMPs and 16 economic BMPs.

1. Only members of the U.S. Sustainable Remediation Forum can access this database.

These resources should be considered a starting point. The remediation practitioner should also seek additional project-specific, bespoke BMPs to consider.

An important concept at this stage in the remediation process is that all potentially applicable BMPs should be included without regard for implementability, cost, or other considerations related to whether a BMP should be considered for the specific project.

3.1.2 Step 2: Evaluate BMPs

The second step in the six-step sustainable remediation process is identifying and implementing a BMP evaluation method, such as a ranking or rating system or MCDA model. The practitioner should determine how detailed and in depth of an evaluation process is warranted based on the number of BMPs, cost, and project complexity. For most projects, a basic low, medium, and high or 1 through 5 ranking system should suffice to help the practitioner prioritize the BMPs identified in Step 1.

For more complicated and expensive projects, it may be worthwhile to invest in developing a more sophisticated ranking system with perhaps different weights for more important metrics. Roughly, this is somewhat similar to the USEPA's feasibility process, which puts more weight on some of the nine criteria than others; however, it is not a fully quantitative process. Similarly, MCDA could be employed to help make remedial decisions. Havranek (2019) provides guidance on using MCDA for making remedy decisions, and the concepts employed in his work are readily applicable to sustainable remediation projects.

Finally, the holy grail for the BMP evaluation process is to involve the public in the community's preferred outcome. This evaluation can be accomplished through surveys (Harclerode et al., 2015), charrettes (USEPA, 2010), or public meetings (USEPA, 2017). Public participation is the ultimate social BMP because it increases the potential for an outcome that the community values—a critical tenant of the social aspect of sustainable remediation.

3.1.3 Step 3: Select a practicable set of BMPs

Step 3 is the selection of BMPs based on the evaluation performed in Step 2. As stated previously, the BMPs should substantially benefit each of the sustainable aspects without causing a significant negative impact to any of the other sustainable aspects (i.e., ensure that each BMP does not have the unintended consequence of negatively impacting a sustainability factor in the process of providing a benefit to another factor).

In this step, the practitioner develops a list of BMPs that will benefit the project from a sustainability perspective taking into account the critical decision-making factors, which generally are reducing the environmental footprint relative to other pertinent factors, such as implementability, effectiveness, reliability, short-term risks, community concerns, cost, and potential for environmental trade-offs. The practitioner considers the unwanted transfer of contaminants from one environmental media to another or negative effects on one core element from implementing a BMP with a positive impact on another core element. An important issue is that the use of either guide does not mandate an expense beyond the costs incurred for consulting fees or internal labor costs related to the BMP selection process and documentation. Costs are considered a reasonable factor for eliminating a BMP. Many practitioners find that using the guides *saved* money. If any of these factors is not ideal, then it is appropriate not to retain a BMP for implementation.

3.1.4 Step 4: Implement BMPs

Step 4, implement the selected BMPs, seems relatively straightforward. However, this is the stage where it is essential to properly identify sources of BMPs, whether they involve equipment and materials or social-responsible-sourced services. The environmental aspect of sustainable remediation requires incorporating green products and concepts into the project design. Typically, the social and economic aspects of sustainable remediation require integrating an approach, which may require a particular type of specialty, but not green products to implement the project.

Identifying green products is covered in more detail in the section, Green Products Resources, but, in short, it is important to avoid products and services tainted by "green washing." Furthermore, there are various green product certifications that should be considered to provide the designer confidence that the environmental aspects of the projects are truly sustainable. In addition to products, many environmental BMPs are conceptual in nature (i.e., not a product you can purchase). A good example of a conceptual BMP is using recycled materials from the site such as chipping vegetation removed from the site and using it as mulch instead of disposing of it off-site.

Implementing social and environmental aspects can range from relatively simple applications of a concept to requiring special services that are outside the skill set of most remediation practitioners. For example, the project team incorporating sustainable concepts into a project may consider hiring a public relations or consulting firm with expertise in public charrette-style meetings to engage stakeholders.

It is important to document the BMPs implemented during the project implementation for purposes of later documentation. For example, if local labor is used, each laborer's city of residence should be documented, as they may not be able to be contacted in the future or could move out of the area. This recordation will vary between BMPs; however, it is important that the project field notes include the BMP implementation.

It may not be possible to implement every BMP identified. In this case, it is critical for the project team to be confident in the decision and record the reason a BMP cannot be implemented. Both ASTM guides identify this potential situation and allow excluding BMPs at the implementation stage provided the rationale for not including a previously identified BMP for a project is properly documented with the reason that the BMP could not be implemented.

3.1.5 Step 5: Quantify BMP results (optional)

The first five steps described earlier are primarily qualitative in nature in that the practitioner does not need to make calculations regarding how the BMPs evaluated and selected could reduce the environmental footprint, social impacts, or economic implications. However, as an option, practitioners may elect to enhance the sustainable remediation project by quantifying the BMP results or, if the remedy has not been implemented, the predicted BMP results. This optional quantification is offered by both ASTM standard guides.

Quantifying BMPs from an environmental perspective can be accomplished using web-based software developed by the Air Force (Sustainable Remediation Tool or "SRT") or Battelle (Sitewise, which was funded by the US Army and Navy). More sophisticated software, such as SimaPro, is available for users skilled in life-cycle assessment (LCA) principles. The quantitative approach involves considerably more effort and, thus, additional cost but offers a numerical basis to assist with the decision-making process. In general, the quantitative evaluation is applied at Step 3 of the aforementioned process but can also be applied during the remedy implementation in Step 4. Quantifying social and economic impacts is BMP-specific and may require the input of a sociologist or economist.

The quantitative evaluation process offers three potential uses:

1) provide information on the most significant contributions or activity's environmental footprint relative to the core elements;
2) provide information on potential benefits achieved by specific BMPs to facilitate BMP selection (in other words, compare the environmental footprint reductions of competing BMPs); and
3) provide a quantitative evaluation of BMPs implemented as a method of measuring various metrics as the sustainable remediation project is implemented.

Again, the quantitative evaluation is an optional activity, and the practitioner can opt to apply it for any of the aforementioned three purposes (or none at all).

Quantitative evaluations tend to be best suited for three phases of cleanups: remedy selection, remedy design/implementation, and remedy optimization. However, the project designer may apply a quantitative evaluation process during any stage of cleanup. Generally speaking, quantitative evaluations provide the most significant potential benefit for more complex sites where the time and expense of conducting the evaluation could lead to better decisions related to the BMP selection and also, depending on the responsible party's needs, provide numerical results that many organizations utilize to quantify and benchmark their sustainable remediation programs.

3.1.6 Step 6: Documentation

An overriding concern of sustainable remediation practitioners is the potential for practitioners to overstate the sustainable nature of a project. For green remediation, this is referred to as "green washing." Consistent documentation and reporting increase transparency by requiring the results to be reported to the public; thereby, minimizing the overstatement of sustainability results. The ASTM guides offer a series of required elements to be included in the documentation and reporting. The Greener Cleanup Standard Guide includes an example reporting template in an Appendix (appendices of ASTM documents are, by definition, optional). Also, there are many avenues for public reporting, including, but not limited to, reports submitted to regulatory agencies; public repositories, such as libraries; and, publicly accessible websites.

4. Green product resources

From a products perspective, BMPs fall into two simple categories: (1) BMPs that are concepts, such as using local labor or hosting community charrettes; and (2) BMPs that are product-related (e.g., a product manufactured using recycled materials). When working with BMPs that fall into Category 2, the remediation practitioner must ensure that the products selected are considered "green." While there are numerous definitions for "green remediation," there is not a common industry definition for "green remediation products." As a result, remediation practitioners should vet that the green products identified meet a threshold of credibility. This credibility threshold is typically referred to as "greenwashing." While there is no established industry-wide definition of greenwashing, AZoCleantech, an online publisher, provides the following practical definition of greenwashing:

> *The misleading act of companies, industries, governments, organizations, and individuals trying to promote unjustified environmentally friendly practices, products, and services through branding, mislabeling, packaging, or public relations.*

For more information on greenwashing, see the SourceWatch greenwashing webpage at: https://www.sourcewatch.org/index.php/Greenwashing.

Two available tools to consider when evaluating whether a product is green are an environmental product declaration (EPD) and ecolabeling. Each of these is briefly discussed next.

4.1 Environmental product declaration

USEPA's definition of an EPD is:

> *"[A] declaration of a product's performance with regard to different environmental parameters during the product's life cycle. An EPD requires the gathering of quantified environmental data for a product with preset categories of parameters (raw material, energy use, etc.)."*

The International Organization for Standardization (ISO), ASTM, and NSF Sustainability (a division of NSF International) each publish standards for gathering data and information for EPDs. EPDs provide data and information for comparing the environmental footprint of various products. The types of information generated include: material content, recycled content, service life, global warming potential, water consumption, emissions to air and water, waste generation, ozone depletion potential, and respiratory effects (this list is from ASTM). EPD reports generally summarize the information listed above; however, the information is often also provided on a product label or in a product's brochure. There are product-specific standards for preparing EPDs (typically called "product category rules"). These rules specify an LCA process to develop an EPD for each product category. For example, there is an ASTM product category rule specific to precast concrete that must be followed to prepare an ASTM-compliant EPD for this product type.

EPD International maintains an international database of EPDs and product category rules that can be accessed at: http://environdec.com. The caution is that the existence of an EPD does not necessarily mean a product is certified to be green, only that the information developed about the product is certified. Industry green certification programs are discussed next.

4.2 Green products accreditation and labeling

A common approach to investigating whether a green product is reputable is to determine if it meets a standard established by an independent standards-setting organization. When doing so, it is suggested to only consider product claims validated following industry-accepted standards that are established by a standards-setting organization, such as American National Standards Institute (ANSI) or ISO. For example, ISO has three ecolabeling standards (ISO, 2012).

UL LLC (formerly Underwriters Laboratories) hosts a database of sustainable products called "SPOT." The SPOT website has over 100,000 sustainable products and can be accessed at: https://spot.ul.com. This website provides a wealth of product information, and UL is considered a reputable organization.

5. SURF green products and sustainable remediation services technical initiative

Recognizing the need to provide a tool to assist in the design of green and/or sustainable remediation projects, the U.S. Sustainable Remediation Forum (SURF) established a technical initiative to develop a database that includes green remediation products and sustainable remediation services. The SURF database follows the basic outline of the BMPs described in the ASTM standard guides. The database lists green products and services, provides links to the associated company websites, and also indicates the applicable technologies, BMP categories, and USEPA greener cleanup core elements, which relate to the products in the database.

The database, which is a downloadable Excel file, is designed to be a flexible tool that can be sorted by remediation technology, BMP categories, or USEPA core elements. For example, if implementing a groundwater pump and treat project, the user could search the database for products that may apply to groundwater extraction systems to develop a list of potentially applicable resources that includes piping, energy sources, concrete, building components, etc.

A subset of green remediation products is green technologies that may not necessarily be manufactured using a greener approach but can reduce the environmental footprint of a cleanup nonetheless. The best example of this concept is equipment powered by alternative energy (typically wind or solar) or other attributes that lower the environmental footprint of the technology application. These types of products are included in the database.

In compiling the database, the technical initiative team identified the following general attributes of green remediation projects:

- Recycled product content—products manufactured with at least a portion of, and many 100%, recycled materials; includes not only wood-based products but also piping, plastics, concrete, lubricants, engine oil, drums, vehicle filters, and asphalt
- Alternative-energy-powered equipment—solar and/or wind-powered pumps, skimmers, soil vapor extraction systems, and remote samplers
- Biodegradable products not intended for long-term use—sampling equipment (protective gloves, bailers, and soil sampling tube liners, packaging) and degradable erosion control products (biomats and silt fences)
- Products manufactured with renewable resources—such as the U.S. Department of Agriculture's BioPreferred products (generally an alternative to synthetic materials)
- Products manufactured using green chemistry—cleaners of all types
- Products manufactured using reduced environmental footprints—various products (an example is Plast-O-Matic Valves, see http://plastomatic.com/pomgreen.html)

The SURF green remediation products database can be accessed from the organization's website at: http://www.sustainableremediation.org; however, only current members can access the database.

6. Case study for sustainable remediation best management practices

The following case study illustrates how the ASTM standard guides can be applied to a site undergoing cleanup. The case study site is the Pharmacia and Upjohn Company LLC site in New Haven, Connecticut. The Upjohn Company produced specialty and industrial chemicals, including pharmaceutical, dye, pigment, and photographic intermediates, agricultural treatment chemicals, ultraviolet curing initiators, coating and adhesive additives, and flavor and fragrance components. Pfizer assumed responsibility for the site in April 2003 as a result of its acquisition of the Pharmacia Corporation.

When Pfizer acquired the site, the manufacturing operations had been discontinued, the buildings had been demolished, and the site was an idle, vacant property, except for the groundwater treatment building.

Contamination at the site consisted of a dense nonaqueous-phase liquid (DNAPL) area, with volatile organic compounds (VOCs; predominately benzene, chlorobenzene, toluene, and tetrachloroethene [PCE]); semivolatile organic compounds (SVOCs; predominately, dichlorobenzenes, azobenzene, 2-chloroaniline, and 3,3′-dichlorobenzidine); and polychlorinated biphenyls (PCBs). There were large piles with wastewater treatment residuals that contained VOCs, SVOCs, PCBs, and metals, and groundwater contaminated with VOCs, SVOCs, PCBs, and metals (lead). Furthermore, sediment in the Quinnipiac River and its creeks contained chemicals with bioaccumulation potential that included 3,3- dichlorobenzidine, benzidine, and PCBs.

Pfizer was obligated to remediate the site under a corrective action administrative order issued by USEPA under the Resource Conservation and Recovery Act (RCRA).

The site cleanup involved the following corrective measures:

- Remediation of Quinnipiac River sediment in two areas of tidal mudflats and in a small stretch of an adjacent creek
- In situ thermal remediation system to treat DNAPL
- Construction of caps or barrier covers in the vicinity of former manufacturing operations to allow safe commercial/light industrial redevelopment
- Enhancement of the east side ecological habitat, including the creation of higher value uplands and wetlands habitat on the southern portion of the site

- Constructing walking trails for interpretative environmental education for guided viewing of the enhanced habitats, which will be maintained as an ecological area
- The long-term operation, monitoring, and maintenance of installed remediation systems, including long-term groundwater monitoring

Now that the site remediation is complete, the east side of the site is a nature conservatory, and the west side of the site is available for commercial or industrial development.

As part of the remedy selection process, Pfizer involved the community by engaging stakeholders. In addition to Pfizer, USEPA, and the Connecticut Department of Energy and Environmental Protection, the site has had active stakeholder participation by the following organizations: North Haven Citizens' Advisory Panel, Quinnipiac River Watershed Association, North Haven Land Trust, Regional Growth Partnership, North Haven Trail Association, and North Haven Board of Selectmen.

The project goals included instructing Pfizer's consultants to promote sustainable remediation concepts during the implementation of the project and Pfizer's retention of a specialty consultant to provide expertise in community relations pertaining to environmental issues and identifying ecological enhancements to incorporate into the project. The remedial planning process for the site included selecting and implementing social and economic BMPs. The BMP evaluation process followed the ASTM process, including the evaluation of social and economic BMPs.

The project included integrating 87 green BMPs, 9 social BMPs, and 5 economic BMPs into the site remedy. Some of the more notable environmental, social, and economic BMPs are summarized next.

6.1 Environmental BMPs

The environmental BMPs included the following:

- Remedy consolidated majority of contaminated materials under engineered caps, which avoided a substantial amount of truck traffic (which was a community concern as well)
- Creating on-site wetlands and related habitat
- Utilizing recycled blast furnace slag for hydraulic barrier wall construction
- Operating the groundwater extraction and treatment system in pulsed mode and adjusting pumping to match the river's tidal elevation changes
- Selecting revegetation plantings that required minimal mowing
- Managing drilling wastes on-site under caps instead of off-site landfill disposal
- There were 15 environmental BMPs for thermal treatment alone; examples include:
    - Insulating piping

- ○ Recovering electrodes
- ○ Using local labor and labs, when possible
- ○ Employing variable-frequency drives
- ○ Segregating drilling waste and placing under the cap (if possible)

6.2 Social BMPs

As described earlier, Pfizer worked to integrate the community into the decision-making process. The social BMPs implemented included:

- Implemented strategies to develop a more collaborative relationship with stakeholders beyond existing regulatory requirements to the extent possible, for example, by engaging the stakeholders and increasing the transparency of operations at the site. This included conducting community workshops and interviews to solicit and incorporate input into the remedial decision process; with community input and support, selected and implemented a remedy incorporating in situ thermal treatment for source control and utilizing on-site containment cells to limit the transport of waste throughout the community.
- Involved town Selectmen in the remedial decision process to ensure the local government supported the proposed remedy.
- Restored site surroundings so that they are visually attractive. Constructed ecological preserve on the east side of the property consisting of nature trails and wetlands; allowed community access on a guided basis for interpretative environmental education and guided viewing of the enhanced habitats.
- Provided the public with adequate and timely information concerning forthcoming actions or decisions by developing the North Haven Informational Website that ensured that the community was fully informed about corrective action plans and progress.
- Communicated site activities to stakeholders in a nontechnical fashion so that stakeholders understood the issues at hand and the process of evaluating public health risk.
- Implemented a local education program about site impacts and remediation impacts.
- Ensured that the site was secure. Hired a full-time security guard and installed physical barriers, such as a fence to prevent unauthorized entry to the site and protect remediation equipment from vandalism or theft. A full-time security guard and/or physical barriers protected site trespassers from risks associated with site activities, from tripping hazards to exposure to hazardous materials.
- Conducted on-site citizen training sessions for the local community that directly related to site assessment and cleanup efforts.

- Retained a third-party specialist to assist with the management of communications with the community.
- Prepared community relations plan to ensure that the aforementioned measures were implemented in a coordinated manner.

6.3 Economic BMPs

The economic BMPs that Pfizer incorporated into the remedy implementation included the following activities:

- Conducted market analysis to evaluate optimum future land use.
- Integrated redevelopment plans into the cleanup, which enabled appropriate infrastructure design to be incorporated into the remedial design.
- Local buying commitment, including a web-based form on the website that local businesses can use to provide information on local services and products to Pfizer.
- Created local jobs both during the remediation and as a result of the future land use integrated into the remedy.
- Outreach to the real estate market and community stakeholders to ensure that reuse potential was integrated with the remedy.

The ASTM Technical Summary Form for the site is accessible from the USEPA's clu-in.org website at: https://clu-in.org/greenremediation/docs/Pharmacia-Upjohn_North_Haven_CT_Pfizer.pdf.

7. Conclusion

Integrating BMPs for sustainable remediation projects is best accomplished following methodologies established by ASTM. The ASTM guides, *Standard Guide for Greener Cleanups* and *Standard Guide for Integrating Sustainability into Cleanups*, provide a clear path for identifying, evaluating, selecting, implementing, and documenting sustainable remediation BMPs. The process is repeatable, defensible, and prevents greenwashing. There are several resources for identifying BMPs, including ecolabeled products and the U.S. SURF database of green remediation products. Incorporating sustainable principles into cleanup projects provides a mechanism for remediation specialists to participate in the important concept of making the environmental remediation industry more sustainable.

John A. Simon is the editor-in-chief of *The Remediation Journal*, a John Wiley and Sons publication. He is also a Senior Vice President at Nathan Associates (Arlington, Virginia), a consulting firm specializing in providing advisory services for environmental, economic, and international matters. Mr. Simon frequently consults private industry on the assessment and remediation of hazardous sites, as well as environmental liability and risk-transfer issues.

He is the President Emeritus of the U.S. Sustainable Remediation Forum and led the ASTM Greener Cleanups Task Group. Mr. Simon received his BE in civil and environmental engineering from Vanderbilt University and his MS in civil engineering, environmental engineering, and science from Stanford University.

References

ASTM International Inc, 2013. Standard Guide for Integrating Sustainable Objectives into Cleanup, E2876 – 13. ASTM International Inc., Conshohocken, PA.

ASTM International Inc, 2016. Standard Guide for Greener Cleanup, E2893 −16. ASTM International Inc., Conshohocken, PA.

Harclerode, M., Ridsdale, Darmendrail, D., Bardos, P., Alexandrescu, F., Nathanail, P., Pizzol, L., Rizzo, E., 2015. Integrating the social dimension in remediation decision-making: state of the practice and way forward. Remediation 26 (1), 11−42.

Havranek, T.J., 2019. Multi-criteria decision analysis for environmental remediation: benefits, challenges, and recommended practices. Remediation 29 (2), 93−108.

International Organization for Standardisation (ISO), 2012. Environmental Labels and Declarations: How ISO Standards Help. ISO, Geneva, Switzerland.

U.S. Environmental Protection Agency (USEPA), 1992. October. Storm Water Management for Construction Activities, Developing Pollution Prevention Plans and Best Management Practices, Summary Guidance. USEPA, Washington, DC. EPA 833-R-92-001.

U.S. Environmental Protection Agency (USEPA), 2010. The Charrette: Redevelopment by Design, an Introduction to Reuse Planning, Workshops for Superfund Sites. USEPA, Washington, DC. Retrieved from. https://clu-in.org/conf/tio/charrette/Charrette-Online-Tutorial.pdf.

U.S. Environmental Protection Agency (USEPA), 2017. Public Participation Guide. USEPA, Washington, DC. Retrieved from. https://www.epa.gov/international-cooperation/public-participation-guide.

Chapter 5

Green remediation by using low-carbon cement-based stabilization/solidification approaches

Dr Lei Wang, Mr Liang Chen, Dr Daniel C.W. Tsang
Department of Civil and Environmental Engineering, Hong Kong Polytechnic University, Kowloon, Hong Kong, China

List of Abbreviations

AAC Alkali-activated cement
AFm Alumina, ferric oxide, monosulfate
BFS Blast furnace slag
BYF Belite-Ye'elimite-Ferrite
CCR Calcium carbide residue
CCSC Carbonatable Calcium Silicate clinkers
CH Calcium hydroxide
CO_2 Carbon dioxide
CSA Sulfoaluminate cement
C–S–H Calcium silicate hydrates
GGBS Granulated blast furnace slag
ISSA Incinerated sewage sludge ash
LC^3 Limestone calcined clay cement
MC MgO cement
MK Metakaolin
MOC Magnesium oxychloride (Sorel) cement
MOS Magnesium oxysulfate cement
MOMS Magnesium Oxides from Magnesium Silicates
MKPC Magnesium potassium phosphate cement
MPC Magnesium phosphate cement
OPC Ordinary Portland cement
PFA Pulverized fly ash
RBPC Reactive Belite-rich Portland cement
SCMs Supplementary cementitious materials
SF Silica fume
S/S Stabilization/solidification

1. Introduction

Most soil contamination is associated with industrial or agriculture activities. The soil may contain different types of toxic elements, such as metals,

Sustainable Remediation of Contaminated Soil and Groundwater. https://doi.org/10.1016/B978-0-12-817982-6.00005-7

chlorides, and sulfates. The concentration of toxic elements in the soil varies greatly and may exceed the acceptable limits of the environment (Bolan et al., 2014; Akcil et al., 2015). Therefore, soil remediation is required to mitigate potential environmental impacts. Some in situ and ex situ remediation (e.g., chemical-enhanced washing, electrokinetic extraction, and bioremediation) shows a high effectiveness in the treatment of contaminated sediment (Akcil et al., 2015). Nevertheless, in view of its full-scale application, considerable cost and time implication should be taken into consideration. Cement-based stabilization/solidification (S/S) has been used as one of the most effective methods with the advantages of relatively low cost, time efficiency, and high compatibility to treat various contaminated soils (Hou and A1-Tabbaa, 2014; Wang et al., 2015).

In the S/S approach, stabilization, involving the chemical interaction between soil and binder, can convert toxic elements into more stable forms, chemically altering contaminated soil to produce less toxic or less fluid forms (Wang et al., 2016a, 2019a). Moreover, with the treatment of solidification, soil, semisolid sludge, or powder solidification can be converted into a monolithic or granular material, making landfill processing and transportation relatively easy (Chen et al., 2019; Ucaroglu and Talinli, 2012). Ordinary Portland cement (OPC) is one of the most conventional materials for S/S of toxic elements (Bolan et al., 2014; Wang et al., 2016b). OPC is a cheap and robust binder for soil treatment, which can transform contaminated soil into useful S/S materials with high compressive strength, high durability, and low permeability.

The production of OPC is usually associated with abundant carbon dioxide (CO_2) emission. Cement industry contributes a significant source of the total CO_2 release. In 2016, the estimated value was 1.45 Gt CO_2. That is, about 8% of the global CO_2 emission (Andrew, 2017). As the awareness of climate change increases, the cement industry is facing a challenge of reducing its CO_2 emissions. However, the calcination of limestone at 1400−1450°C occupied approximately 50%−60% of cement-production-related CO_2 emissions, which cannot be achieved by improving the energy efficiency (Unluer and Al-Tabbaa, 2013; Wang et al., 2017a).

Therefore, developing alternative low-carbon binders is considered as an important way to reduce CO_2 release (Flatt et al., 2012). Low-carbon supplementary cementitious materials (SCMs) are considered as the latent substitutes of OPC. The application of SCMs such as ground granulated blast furnace slag (GGBS), pulverized fly ash (PFA), silica fumes (SF) illustrates an alternative solution to partially substitute OPC (Wang et al., 2018a, 2018b). By using such materials, CO_2 emissions of cementitious materials significantly reduced because of no additional clinkering process and an approach to recycle by-products from industrial manufacturing processes (Suraneni and Weiss, 2017). The addition of SCMs could promote the generation of additional calcium silicate hydrates (C−S−H) gel by spontaneous hydration reaction or pozzolanic reaction, which benefits mechanical strength enhancement, long-term durability, and contaminants immobilization (Jin et al., 2016).

CO_2 curing is regarded as an eco-friendly means, to facilitate carbon sequestration with the purpose of reducing carbon footprint and restraining the process of global warming by storing carbon in the cementitious products (Xuan et al., 2016; Wang et al., 2017b) and to further contribute to accelerate cement reaction in cement setting/hardening and early-stage formation of calcium (Wang et al., 2017c). The method of CO_2 curing could not only immobilize the heavy metals but also lighten the environmental burden (Li et al., 2016).

Alkali-activated cements (AACs) are one kind of promising binder for civil and environmental applications (Provis and Bernal, 2014). They can be fully or partially used as an alternative or supplemental binder, depending on the availability of local raw materials (Provis, 2017). Compared with OPC-based binders, AACs show some superior properties, such as lower permeability, better acid and sulfate resistance (Aiken et al., 2017), lower shrinkage and creep, higher compatibility with heavy metals. These advantages can be achieved by custom binder systems—materials that are well designed and have a low water/binder ratio may achieve most of the aforementioned characteristics (Provis and Bernal, 2014).

Reactive MgO cement (MC) is another alternative of OPC, which has some advantages over OPC due to its low-CO_2 emission (lower production temperatures) and high compatibility with contaminants (Al-Tabbaa, 2013). The S/S mechanisms of MC mainly include physical encapsulation and chemical precipitation by the hydration product brucite ($Mg(OH)_2$) (Wu et al., 2018). MC also has good compatibility with toxic elements (Jin et al., 2014) and organic matter (Wang et al., 2016b, 2017d) due to abundance of $Mg(OH)_2$ formed from cement hydration. The practical application demonstrated that MgO-based S/S products reduce the risk of contaminants leaching and degradation in the field and underground environment (Nied et al., 2016).

In S/S processes, some special cementitious materials have been used to deal with hazardous wastes. Sulfoaluminate cement (CSA) was produced by adding gypsum to CSA clinkers (Shi et al., 2011). As a low-cost and low-carbon footprint cement, it shows the characteristics of fast gain in strength and good compatibility with toxic elements. The hydration product of CSA provides a moderate internal pore solution, where the pH is relatively lower than OPC. Ettringite, which is the main hydration product of CSA, can adsorb a lot of ions into its crystal structure, making it an ideal substitute for the immobilization of pollutants (Niu et al., 2018). Limestone calcined clay cement (LC^3) has the advantages of comparable mechanical strength development, better chloride resistance, lower gas permeability and capillary water absorption, higher durability. LC^3, which has lower net global CO_2 emissions, is an alternative to OPC clinker (Gartner and Hirao, 2015). Although there are few applications and researches about treating S/S contaminated soil by using LC^3, LC^3 has the potential to be good binder to immobilize contaminants since the clay has superior compatibility with soils.

In view of the aforementioned superior properties, this study summarizes the low-carbon S/S of contaminated soil by using different types of low-carbon cement-based materials. This chapter explores the immobilization mechanisms of toxic elements by different cement-based materials; evaluates the different factors affecting the effectiveness of contaminated soil remediation; and introduces the benefits and challenges of these cement-based materials in practical use.

2. Ordinary portland cement-based S/S

2.1 Mechanisms

OPC is a mature and reliable material for the S/S of toxic elements in contaminated soils. Possible immobilization mechanisms for toxic elements are: (1) adsorption, (2) chemical bonding (surface complexation, coprecipitation, precipitation, divalent), and (3) micro- or macro-encapsulation (Guo et al., 2017a). The hydration product of OPC is C−S−H gel, a mixture of crystalline minerals in disordered phase. The product C−S−H gel shows a layered structure, which looks like an imperfect 1.4 nm tobermorite layered structure (Richardson, 2008). C−S−H gel could immobilize toxic elements by ion exchange or adsorption. Vespa et al. (2014) indicated that three possible positions in the C−S−H structure probably occupied by toxic elements with reference to the structure of tobermorite. Fig. 5.1 shows the potential positions occupied by toxic elements.

However, soil and cement are very complex systems. Therefore, the S/S efficiency varies, and it is associated with the variability of the soil, the equilibrium of the solution, and the kinetics of the cement phase of the surface and near surface (Guo et al., 2017a).

The manufacturing process of OPC is one of the most energy-cost procedures, accounting for global man-made carbon dioxide emissions. The carbon dioxide emission of OPC during its life cycle (Shen et al., 2016) is showed in Fig. 5.2. The life-cycle CO_2 emission of OPC includes six parts: raw material, fuel-derived, electricity-derived, transportation, concrete life cycle, and carbon sinking in concrete (Fig. 5.3). In China, the CO_2 emission of OPC industry accounts for approximately 12% of all industrial CO_2 release (Chen et al., 2014). Upon emission, the life cycle of CO_2 includes its release from raw materials through processing, manufacturing, distribution, use, repair, and maintenance of raw materials, as well as treatment or recycling (Shen et al., 2016). Thus, more attention should be paid to greener soil remediation for the development of low-carbon S/S technology.

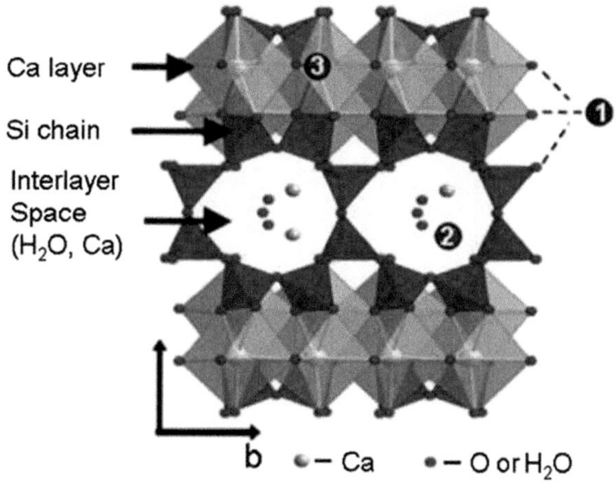

FIGURE 5.1 The approximate structure of C–S–H and the potential positions occupied by toxic elements (Vespa et al., 2014).

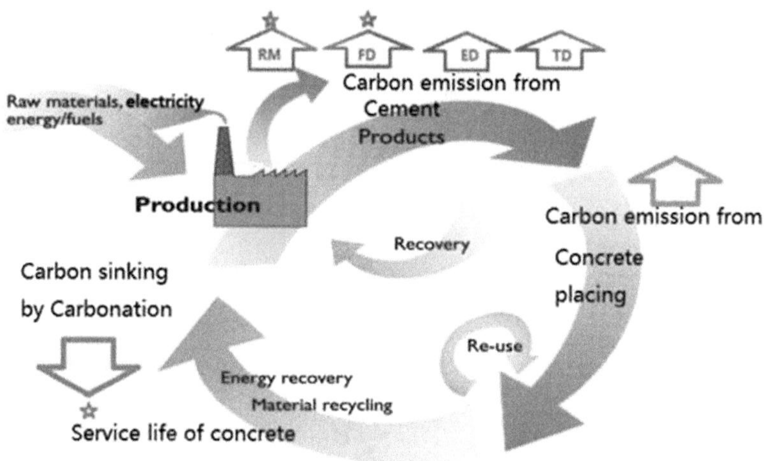

FIGURE 5.2 The life-cycle process of cement (Shen et al., 2016).

2.2 Supplementary cementitious materials

Approximately one ton of CO_2 is generated with the manufacture of one ton of OPC. CO_2 has been recognized as the main source of greenhouse gases, which intensify global warming. In order to decrease the carbon footprint of the cement-based materials, the OPC can be partially replaced by low-carbon SCMs, such as GGBS, PFA, SF, etc. According to the calcium-to-silica ratio, SCMs can be classified into two groups: hydraulic materials (containing calcium and spontaneously reacting with H_2O) and pozzolanic materials

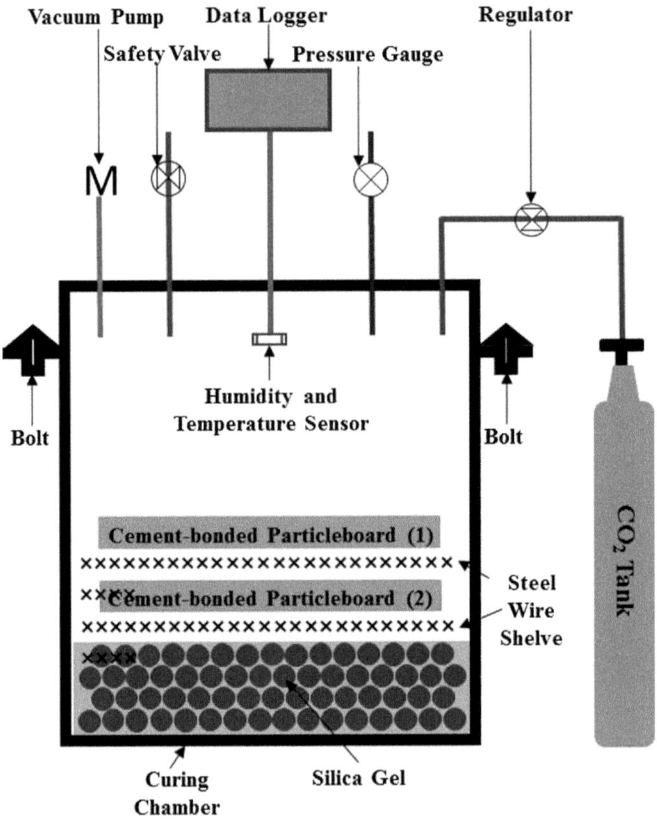

FIGURE 5.3 Experimental facility of static CO_2 curing (Wang and Tsang, 2018).

(having slight or no cementation value, but chemically reacting with $Ca(OH)_2$). Most SCMs, including SF, PFA, and GGBS, have hydraulic or pozzolanic properties (Tashima et al., 2016; Suraneni and Weiss, 2017).

2.2.1 Pulverized fly ash

PFA is a by-product from power plants and is a solid material extracted from the flue gas of bituminous coal powder furnace by electrostatic and mechanical methods (Xiao et al., 2015). It is likely that polycyclic aromatic hydrocarbon compounds and a small amount of concentrations of toxic elements exist in PFA. Therefore, PFA is considered as a kind of waste, and the inappropriate disposal of PFA will not only occupy the land but also put a great burden on the environment (Guo et al., 2017b). Over the last few decades, PFA has been used as an effective stabilizer that offers less energy consumption and more economical heavy metal removal process in S/S treatment (Wang et al., 2013, 2019b). The spherical PFA particles can improve flowability and strength by the effects of pozzolanic reaction, micro-aggregate filling, and friction rolling. The effect of using PFA to repair contaminated sites is significant that the adsorption of metals by exchangeable metal components in soil and the diffusion flux of free metals into solution is substantially reduced (Tsang et al., 2014). Based on the Ca content, the PFAs can be ranked to Class-C PFA (Ca-rich) and Class-F PFA (low-Ca). The Class-C PFA can provide extra alkaline for precipitation of metals, whereas Class-F PFA can produce secondary C−S−H gel for metals immobilization via pozzolanic reaction (Wang et al., 2018b). Thus, PFA was widely used as an alternative of OPC for the S/S of contaminated soil.

2.2.2 Silica fumes

SF is a silica-rich material, which contains as high as 99% reactive SiO_2. OPC hydration produces many hydration products, including C−S−H gel and CH. When added to the fresh cement, SF chemically reacts with the CH to form additional C−S−H gel via pozzolanic reaction. The additional C−S−H gel generated from SF shows higher resistance to aggressive chemicals than the less strong calcium hydroxide, increasing the durability of S/S products. Moreover, some metal may dissolve in the overhigh pH environment of OPC system (Li et al., 2014). The addition of SF reduces the overhigh pH and shows a favorable immobilization efficiency since most of the metals show the lowest solubility at the pH range around 11−13. The third role of SF in the S/S process is a physical filling effect. The particle size of SF is 100−150 times smaller than OPC size, making it possible to fill the pores created by free water. This process, known as particle packing, refines the microstructure of product, making a dense pore structure and decreasing the leachability of toxic elements (Goodarzi et al., 2016). Thus, SF-incorporated binder system can significantly reduce the leachability of toxic elements and enhance the compressive strength for engineering application (Chen et al., 2019).

2.2.3 Ground granulated blast-furnace slag

GGBS is one of the by-products of iron and steel-making. It is economical in large quantities, so it is suitable for premixed concrete, bulk on-site concrete production, and precast product manufacturing (Samad et al., 2017). GGBS produces lower heat during the hydration process, enhances resistance from chloride and sulfate attack compared with OPC. In addition, it benefits to environmental protection as it reduces usage of OPC (Samad et al., 2017). As a green SCM, GGBS can react with CH and form additional C−S−H gel during hydration process. The process significantly enhances heavy metal immobilization and mechanical strength of the S/S products (Li et al., 2016; Wang et al., 2018c). Besides, GGBS of small particle size can also function as fillers for densifying pore structure, facilitating the low permeability as well as physical encapsulation (Cheah et al., 2016). Besides, GGBS is an Al-rich material (10%–20% Al_2O_3), as Al_2O_3 is common in soil. The Al_2O_3-rich material usually presents high compatibility with soil, while the nature of amphoteric oxide could bond with toxic cation and anion in contaminated soil (Li et al., 2016).

2.2.4 *Incinerated sewage sludge ash*

ISSA is a by-product from sludge incineration facilities, with certain pozzolanic property for its semiactive alumina and silica content (Li et al., 2017). ISSA is an alternative to replace OPC in the cement-based S/S of heavy metals contaminated soil. In the ISSA-incorporated mixture, an interlocking structure immobilizes heavy metals in the contaminated soil during hydration process (Wang et al., 2018d). The heavy metal leaching is lower due to the small surface area of the immobilized soil exposed to the environment. Previous results have indicated that ISSA has a large surface area and porosity, which can remove heavy metals (Smith et al., 2009). Therefore, the incorporation of ISSA can enhance the treatment characteristics and reduce the impact on the environment through chemical and physical adsorption, chemical incorporation (surface complexation, ion exchange, precipitation), and microcapsule immobilization of heavy metals (Li et al., 2017).

2.2.5 *Calcium carbide residue*

Calcium carbide residue (CCR) is one of the by-products of acetylene gas and iron production and is considered as a hydraulic material because it contains a relatively high amount of calcium (Namarak et al., 2017). The generation of calcium hydroxide (CH) in spontaneous hydration process can alleviate the interference of pollutants in hydration process (Jin et al., 2016). More importantly, the combining use of CCR and other silicate-rich materials has been a promising choice. The CCR can provide sufficient alkalinity to activate other SCMs (e.g., PFA, SF, ISSA) to generate additional C–S–H gel for S/S. However, the rate of pozzolanic reaction is relatively low compared to the rate of OPC. Thus, the early-term properties of SCMs-modified products is a big issue. Besides, mixing proportion of CCR and other SCMs, the curing conditions should be further optimized before practical application (Jin et al., 2016).

2.3 CO_2 curing

The method of utilization and absorption of CO_2 in cement-based materials is usually conducted in the curing phase (Wang and Tsang, 2018). In the accelerated CO_2 curing process, pressurized CO_2 gas reacts with CH to generate stable carbonation product in a short period of time (2–24 h) (Wang and Tsang, 2018). Accelerated CO_2 curing approach includes static curing and flow-through curing. For the static CO_2 curing method (Fig. 5.3), specimens

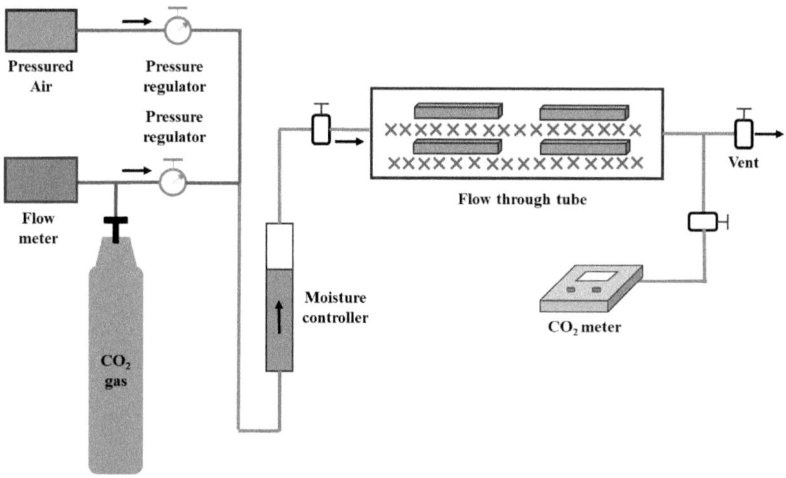

FIGURE 5.4 Experimental facility of flow through CO_2 curing (Wang and Tsang, 2018).

are usually placed in a chamber. CO_2 concentration, temperature, air pressure, humidity can be well controlled in the chamber. Therefore, static CO_2 curing method displays the advantage of simplicity and controllability that can be fully used for carbonation. The flow-through curing method (Fig. 5.4) can also monitor the CO_2 concentration, temperature, humidity, except for the air pressure. As exhaust gas can be the sources of CO_2, from power plants, cement kilns, and steel mills, etc. CO_2 curing for cement-based products is a promising technology.

This rapid reaction can compensate for the negative effects of calcium-organic complexation and pollutants. Therefore, the utilization of CO_2 curing also enables CO_2 sequestration of cement materials to decrease the carbon footprint and mitigate global warming (Wang et al., 2016b, 2017c). Furthermore, a large amount of experimental evidence indicates the specific impact of CO_2 curing on the performance of cement-based products, i.e., the change of chemical bonding structure and the densification of microstructure, enhancement of pollutants immobilization (Wang et al., 2018a).

2.4 Benefits and challenges

Cement-based S/S soil is susceptible to chemical and physical degradation in the long term. The vulnerability is mainly determined by factors such as permeability, chemical and mineralogical composition, and micro-structure. Cement-based products are attacked by CO_2, chloride, sulfate, water, and organic degradation materials when exposed to air. Further chemical processes such as hydration, carbonization, perovskite formation can result in changes in porosity, volume, and solidified soil properties. They can cause swelling, cracking, and loss of strength. Aggressive solution erosion can lead to secondary mineralization and ultimately decomposition of treated soil (Chen et al., 2009). Furthermore, the decalcification of C–S–H gel easily occurs in long-term service life. Although the incorporation of SCMs could provide additional C–S–H gel, the risk of C–S–H gel decalcification still exists in SCMs-blended products.

CO_2 erosion may be the most common degradation mechanism of cement solidified soil. The main component of cement, hydrated calcium silicate (C–S–H), is thought to decalcify in the long-term exposure to atmospheric carbon dioxide, eventually becoming silica gel and losing its binding capacity in the process (Chen et al., 2009). The toxic elements lead to the reduced content of CH in cement paste, which also increases its vulnerability to carbonization and acid corrosion (Chen et al., 2009). Therefore, the durability of cement-based S/S products should be further improved.

3. Alkali-activated cement-based S/S

3.1 Mechanisms

OPC-based binders have been widely used for S/S due to their availability and cost efficiency. In the consideration of strict environmental requirements and global carbon emission control, a priority target of the OPC manufacturing and construction industry is to find novel cement binders and increase the environmental sustainability (energy saving and low greenhouse gas emissions) (Duxson et al., 2007). Geopolymer is such a kind of material, which is a synthetic alkali aluminosilicate formed by combining solid aluminosilicate with a high-concentration alkali hydroxide or silicate aqueous solution. Geopolymers typically use aluminosilicate-based (i.e., pozzolanic) wastes that provide high mechanical properties, low drying shrinkage, thermal conductivity, high fire resistance and acid resistance. In addition, the production cost of geopolymer is lower than that of OPC (Duxson et al., 2007).

Furthermore, OPC is slightly ineffective in immobilizing some metals such as Hg, As, Cu, Pb, Cr, etc. One of the substitute materials to OPC is AAC, which shows the advantages in environmental and cost aspects over the traditional OPC (Rasaki et al., 2019). It has been studied that toxic elements interfere much less with the hydration of AAC than OPC. The results published by Shi et al. (2011) confirm that AAC can be effectively used for S/S of various polluting materials due to the abundant alumina, ferric oxide, monosulfate (AFm) phases for element fixation. During the OPC hydration, C–S–H gel exhibits a satisfactory effect for heavy metal immobilization. AAC has a lower capillary porosity and higher gel pore volume than those of OPC matrix. Therefore, AAC is predicted to be a suitable curing medium of metals (Rasaki et al., 2019).

Two main compositions are required in alkaline activation, a solid precursor rich in alumina and silica (with the highest possible degree of amorphization) and an activator solution based on an alkali metal (i.e., sodium or potassium) or alkaline earth metal (i.e., calcium). Compared with OPC caused by mixing with water, alkali activation requires adding alkaline components to water. Thus, it is necessary to introduce the properties of alkaline solutions. The solutions discussed are hydroxides and concentrated alkali metal silicates (Provis and Bernal, 2014). The hydroxide solution most commonly used as an alkali activator is sodium hydroxide; potassium hydroxide sees some use in specialized applications, whereas lithium, rubidium, and cesium hydroxides are of limited large-scale application. Silicate-activated counterparts tend to show lower permeability than hydroxide-activated binders, whether based on fly ash or on blast furnace slag (BFS). Nevertheless, hydroxide-activated mixes can show more favorable workability than silicate-activated binders because the viscosity of concentrated alkali hydroxide solutions is much lower than that of alkali silicate solutions of comparable concentration (Provis et al., 2014). The main precursors used to produce AAC are PFA, BFS, and metakaolin (MK). When dissolved in alkaline solutions, these precursors condense and polymerize to form AAC or geopolymers depending on the chemical composition of the starting material used (rich or poor in CaO or Al_2O_3) and the final component of the main reaction product. These materials are regarded as a new type of inorganic polymers with properties of cement and ideal materials for refractory panels and the encapsulation of toxic elements (Shi et al., 2011).

3.2 State-of-the-art of AAC S/S

3.2.1 Blast furnace slag-based AAC

Blast furnace slag is a calcium-silicate-based product removed from the top of molten iron during its extraction from ore in a blast furnace. Usually, it is rapidly cooled to a glassy state and ground for use in construction materials (Provis et al., 2015). By the step of water quenching and out of rapid decrease in temperature, most of the compounds have not reached a stable phase. The slag retaining its former structure as a vitreous body shows good activity. From statistics, 0.3−1 tonne BFS is produced during the production of 1 tonne of pig iron. The discharge and accumulation of BFS not only require much labor force, raw materials, and financial investment but also pollute the environment. Thus, preparing geopolymers by using BFS for solid waste recycling, lower energy cost, and CO_2 emissions is of great significance (Wang et al., 2011). In addition, the existence of BFS shows the property of reducing the permeability of the penetrant, thus enhancing the durability of geopolymers and favoring the long-term capture of heavy metals in the solidified matrix. Reactive CaO, SiO_2, and Al_2O_3 in BFS were beneficial to the generation of geopolymers. The BFS in the AAC reacts with a soluble sodium silicate activator to form poorly crystallized C−S−H gel. The reaction of BFS with water is very slow, forming a hardening binder (Huang et al., 2016). The role of the alkaline activator is to accelerate the reaction, allowing the material to harden and strengthen within hours to days. This process produces the indispensable early strength in S/S products (Wang et al., 2011).

Recent studies have shown that AAC could immobilize toxic elements very well. Toxic elements, such as Zn^{2+}, Pb^{2+}, Cd^{2+}, and Cr^{6+}, could be effectively immobilized in NaOH, Na_2CO_3, and sodium-silicate-activated slag cements. The AAC shows superior compatibility with toxic elements (Sargent et al., 2014). Huang et al. (2016) studied the stabilization of Cr^{6+} by using BFS-based geopolymer. They conclude that a series of reactions such as adsorption, physical immobilization, ion exchange, neutralization, precipitation, reduction, and complex formation may be involved during the S/S process. During these reactions, the decrease of S^{2-} in BFS makes a great difference in Cr^{6+} stabilization. The combined effects significantly immobilize Cr^{6+}.

3.2.2 Pulverized fuel ash-based AAC

Researchers have reported using the alkali-activated PFA cements as effective binders for S/S of the toxic elements in contaminated soil. This binder sometimes also named as one of geopolymers since it has a polymeric structure (Shi et al., 2011). Theoretically, geopolymer is a product of any aluminosilicate materials with alkali activation. The geopolymer method offers a new green approach to the recycling of PFA, avoiding its negative burden on the environment. The alumina and silica in PFA can be activated by alkali to generate geopolymer. In addition, the toxic elements in contaminated soil can be immobilized in the geopolymer framework (Zhang et al., 2016). It has a three-dimensional (3D) aluminosilicate network framework. The toxic elements can be tightly immobilized in the 3D structure of PFA-based geopolymer. And physical encapsulation and chemical stabilization are the major mechanisms of metal immobilization in PFA-based geopolymer (Izquierdo et al., 2009).

The existing metals slow down the conventional cement hydration. However, Shi et al. (2011) revealed that PFA-based geopolymer could result in the mobilization of Pb. There are two different explanations for the S/S mechanisms of metals, e.g., Pb. One is that Pb may be combined with the amorphous geopolymer gel, implying that the Pb is amorphous. The other explanation is that Pb is stabilized in the structure of Pb_3SiO_5, which is an insoluble crystalline phase (Zhang et al., 2016). Zhang et al. (2008) observed the Pb-rich regions in the SEM micrographs of the geopolymer with 0.5% Pb as $Pb(NO_3)_2$ and found that the composition of the Pb-rich regions and the Pb_3SiO_5 phase was highly consistent with each other, thus indicating that the immobilization effect of heavy metal in geopolymer is determined by the form of the pollutants. The difference between the two explanations is the existing Pb phase bonded to the geopolymer. In addition, Guo et al. (2017b) found that the behavior of Pb ions in geopolymer is determined by the chemical characteristic of the Pb compounds. Pb is chemically absorbed into the amorphous 3D framework of the geopolymer. After added with PbO, Pb is dispersed throughout the geopolymer to form lead-containing geopolymer.

In OPC-based systems, chromium can be immobilized in the generation of Ca_2CrO_4, meanwhile improving the mechanical strength of the product. However, in geopolymer system, since $Na_2CrO_4 \cdot 4H_2O$ is very soluble that all Cr^{6+} in PFA-based geopolymer could be leached out (Shi et al., 2011). The generation of $Na_2CrO_4 \cdot 4H_2O$ has a negative impact on the reactivity of PFA (Shi et al., 2011) as well. Studies show that extremely low leaching concentrations from mixed metakaolin/PFA AAC containing 1000 mg/kg doses of Cu^{2+}, Cd^{2+}, Cr^{3+}, and Pb^{2+}. Cu^{2+} was somewhat less immobilized than Pb^{2+} in PFA-based geopolymers resulting from Cu's smaller radius. This in turn reduces the tendency to be immobilized in the nanopore structure (Provis et al., 2014).

3.2.3 Metakaolin-based AAC

Various aluminosilicate raw materials have been conducted as precursors for the synthesis of geopolymer. The most widely used material is MK. MK is a thermal treatment (500–800°C) product from a natural clay named kaolin, which is from a tetrahedral silicon layer and one octahedral aluminum layer (Shi et al., 2011). Compared with fly ash that is not derived from a well-defined starting material, MK has a more reproducible chemical compound. However, MK-based geopolymer demands more water than fly-ash-based geopolymer (Vance and Perera, 2009). Results show that the mixing of sodium hydroxide and sodium silicate presents good effect on activating MK and its geopolymer. However, NaOH alone is not as effective as Na-silicate (Zhang et al., 2016). Using the AAC-based S/S technology to reduce the mobility of heavy metals shows much more economical efficiency for metal-contaminated soils (El-Eswed et al., 2015). Researchers concluded that (Pb^{2+}, Cu^{2+}, Cd^{2+}, and Cr^{3+}) metals could be effectively immobilized in MK-based geopolymer. According to the leachability, conductivity, and XRF analysis, heavy metals were believed to balance the negative charge of tetrahedral Al in unreacted MK-based geopolymer structure. Studies found that during the preparation of the geopolymer, the amount of alkali metal ions (Na^+ and K^+) leached increased with the addition of heavy metals (El-Eswed et al., 2015). MK-based geopolymer is a promising material for the S/S of contaminated soils.

3.3 Benefits and challenges

3.3.1 Stabilization/solidification of radioactive materials in contaminated soils

The sustainable and safe immobilization approach of radioactive materials contaminated soils has been a big challenge for researchers. Currently, the most commonly used immobilization matrixes are based on OPC. For the more difficult to stabilize nuclides, such as cesium (Cs^+), researchers have been studying alternative immobilization matrix, of which AAC is a very promising option. Cs is one of the most difficult radionuclides to be immobilized in radioactive materials contaminated soil. The stabilization of cesium is particularly challenging because of its high solubility in both alkaline and acidic solution (Bar-Nes et al., 2008). Mineral additives such as PFA, slag, and SF are often added to OPC-based matrix for the conditioning of low-level radioactive soil. These additives can be alkaline activated in OPC-based materials, resulting in products that are chemically similar to those found in alkali-activated cements and have been shown to improve the overall performance of waste, especially Cs, and more specifically Cs immobilization (Tian and Sasaki, 2019). In addition, alkaline-activated aluminosilicate materials have been reported to be more durable than OPC in corrosive environments. AACs have shown a promising effect in the immobilization of radionuclides such as cesium and strontium. Therefore, AACs produced by alkaline activation of these and similar pozzolanic materials are also expected to effectively immobilize different types of waste (Vandevenne et al., 2018). As an example, the superior Cs^+ retention of AAC was shown by Shi et al., (2011), who accessed the leaching of Cs^+ and Sr^{2+} from AAC-containing zeolites and/or metakaolin (MK) as additives. Results illustrate that contaminant shows much less interference with the hydration of AAC than that of OPC.

3.3.2 Anticorrosion

Corrosion is one of the major issues influencing the durability of cement-based materials. It shortens the designated service life of buildings and infrastructure and causes huge economic losses every year. The main reason of corrosion in concrete structures, especially in the saliferous environment, is the intrusion of chloride ions (Ghafoori et al., 2013). Ganesan et al. (2015) investigated the chloride diffusion of 100% fly-ash-based AAC and concluded that chloride diffusion of AAC and that of OPC are similar. AAC showed much less corroded depths than the conventional OPC. The apparent chloride diffusion coefficient of blended slag and that of fly ash geopolymer are lower than that of OPC (Tennakoon et al., 2017). The AAC showed favorable durability of S/S products against corrosion, especially applications in the saliferous environment (Provis et al., 2017).

4. Magnesium-rich cement-based S/S

As the increasing awareness to the CO_2 emissions, a new research role has emerged on the development of an alternative cement with lower CO_2 emission. MgO-based cement is such kind of cement. Mechanisms and applications of S/S by using reactive magnesium oxide cement (MC), magnesium phosphate cement (MPC), magnesium oxychloride (Sorel) cement (MOC), and magnesium oxysulfate cement (MOS) are introduced in this section.

4.1 Reactive magnesium oxide cement

MgO cement is regarded as an eco-friendly cementitious substitute for contaminated soil remediation. The S/S process immobilizes the contaminants in the soils and decreases their bioavailability and leachability. MgO-based binders have the good compatibility with contaminants, since most of the toxic elements have the lowest solubility at the pH range of 9−12 (Mo et al., 2014; Wang et al., 2019c). The S/S mechanism of MgO-based cement works by encapsulating contaminants into its hydration products (e.g., brucite) and precipitating the hydroxides. The benefits of using MgO-based cement for S/S include: (1) The equilibrium pH of magnesium oxide after hydration is 10.5, which has a better precipitation than OPC/lime because many toxic elements have the lowest solubility in this pH range; (2) Brucite can hold a variety of heavy metal and organic pollutants through hydrogen bonding or ion exchange (Wang et al., 2016b).

Prof Al-Tabbaa's group (2016b) investigated the potential application of the novel MgO binders in the immobilization of contaminants. Results indicated that MgO binders have a good immobilization effect on heavy metals and organics, but the mechanical strength is weak, together with its poor water resistance accounts for its limited practical application in engineering. $Mg(OH)_2$ gel presents weak durability and dissolves in water. In recent years, a method of converting $Mg(OH)_2$ gel into insoluble magnesium silicate hydrate (M−S−H) using activated silica has been studied. Wang et al. (2019b) studied the impact of Si-rich materials with different properties in the generation of M-S-H gel and evaluated the effect of the S/S treatment by using the M−S−H gel. The Si-rich materials reacted with the $Mg(OH)_2$ to generate M-S-H gel via a "pozzolanic reaction," and the M-S-H gel formed and polymerized with curing time. The batch-scale experiments illustrated that Si-modified MgO cement can transform contaminated waste into construction materials with satisfactory mechanical strength and toxic element stabilization capabilities, especially the PFA-modified samples with an Mg/Si molar ratio of 20:1. Studies showed that M−S−H has the potential in immobilization of nuclear-contaminated soil, with particular attention paid to the less-alkaline pH properties of M−S−H compared with those of the traditional OPC. This is regarded as a negative impact in reinforced concrete, where the mild pH of M−S−H cannot provide effective mild rebar passivation in the same way as the high pH in the OPC pore solution. However, the same moderate pH allows M−S−H to be applied in the immobilization of nuclear waste (Walling and Provis, 2016).

Besides, to improve the mechanical properties and elements immobilization, using the reactive MgO cement with GGBS for soil S/S has been studied to provide a variety of mechanical and chemical benefits compared to OPC or lime–slag blends. This is because the hydration products of MgO–GGBS mixtures are C–S–H/M–S–H and hydrotalcite-like (Ht) phases. The relatively high MgO/CaO ratio in this mixture increases the homogeneity of C–S–H gel, leading to a reduced Ca/Si ratio of C–S–H gel and the additional amount of voluminous Ht phases, which could effectively fill the pores in the mixture (Gu et al., 2015). MgO–GGBS blends at a ratio of 1:4 showed the most effective mixes in stabilizing organics. It should be noted that high water content in situ has a negative effect on the efficiency of mass immobilization process (Gu et al., 2015). In summary, M–S–H can be a potential cementitious material for remediation of contaminated soil.

4.1.1 CO_2 curing in MC-based system

Previous studies have shown that MC hydration generates a significant amount of $Mg(OH)_2$. And $Mg(OH)_2$ can form magnesium carbonates and/or hydrated magnesium carbonates under accelerated CO_2 curing, including magnesite, magnesium calcite, nesquehonite, dypingite, and hydromagnesite (Unluer et al., 2014). Accelerated CO_2 curing boosted strength development, densified the pore structure, improved mechanical properties as well as formed thermodynamically stable carbonate products for long-term durability (Al-Tabbaa, 2013). In addition, accelerated carbonation can also mitigate interference of organic contaminants and stabilize heavy-metal-contaminated soil by facilitating the cement hydration and carbonation in the porous matrix (Pan et al., 2015). Thus, CO_2 curing in MC system presents an eco-friendly approach for contaminated soil remediation and CO_2 absorption to decrease carbon footprint and mitigate climate change.

4.2 Magnesium phosphate cement

MPC is generated by an acid–base reaction between MgO and soluble phosphoric acid (usually ammonium or potassium phosphate) to form a magnesium phosphate salt with cementitious characteristics (Wang et al., 2017c). This kind of materials is known broadly as MPC (Walling and Provis, 2016). Compared with conventional OPC, MPC has the advantages of fast setting and hardening, rapid strength gain, good corrosion resistance, low dry shrinkage, low permeability, good compatibility with impurities and contaminants (Wang et al., 2018f).

A typical MPC, such as the magnesium potassium phosphate cement (MKPC), is more effective for Pb stabilization compared to OPC as Pb can form highly insoluble pyromorphite and lead phosphate in a phosphate-rich environment. Pyrophosphate and lead phosphate are much less soluble than $Pb(OH)_2$, but highly soluble in high alkali binders. The excellent acid–base corrosion resistance of MPC has also been widely recognized (Debela et al.,

2013), making it an ideal binder for the immobilization of Pb-contaminated soil. Wang et al. (2018f) found that the major mechanism for OPC-based S/S of soluble Pb was physical fixation, whereas the main mechanisms for MKPC-based S/S are chemical stabilization through phosphate precipitation and physical encapsulation. Thus, MKPC could be a more effective and chemically stable inorganic binder for the S/S of Pb-contaminated soil.

MPC can be hardened in a short time and applied in cold environments that is possibly the constraints of other cements. Different retarders have been used to extend the setting times, and borate is currently the most popular retarder. In the recent 15–20 years, people prefer potassium-struvite cement more than MPC, mainly because of the reduction in thermal output and the elimination of ammonia liberation in hydration. MPC has its uniqueness as a promising nuclear soil stabilization material, particularly for the S/S of reactive metals that are not suitable for conventional OPC. However, due to the relatively low internal pH of MPC and the pretty high cost, large-scale application will continue to be limited (Walling and Provis, 2016).

4.3 Magnesium oxychloride cement and magnesium oxysulfate cement

Magnesium oxychloride cement (MOC), also named Sorel's cement, is based on the aqueous reaction between $MgCl_2$ and MgO. The major hydration products Phase5 ($5\ Mg(OH)_2 \cdot MgCl_2 \cdot 8H_2O$) and Phase3 ($3\ Mg(OH)_2 \cdot MgCl_2 \cdot 8H_2O$) are the main sources of mechanical strength (Walling and Provis, 2016). This cement provides the advantage of high early strength, good abrasion resistance, and good fire resistance. On the other hand, magnesium oxysulfate cement (MOS) belongs to the similar system of MOC cement, except that $MgSO_4$ is used instead of $MgCl_2$. Major hydration products are $3\ Mg(OH) \cdot MgSO_4 \cdot 8H_2O$ and $5\ Mg(OH) \cdot MgSO_4 \cdot 2H_2O$ (Walling and Provis, 2016). Attention was paid to MOS cements because they have better water resistance compared with MOC. It is more difficult to generate MOS cement due to the limited solubility of $MgSO_4 \cdot 7H_2O$ at room temperature. The main reason limiting the applicability of MOS and MOC is due to their lower water resistance and their high price compared with those of the existing water-resistant cement or gypsum plasters. In moist conditions, the main hydration product would dissolve in a solution, leading to the loss of the strength (He et al., 2017). Although there are few applications and literature about using MOC and MOS cement in S/S contaminated soil, they may be useful binders that can mitigate the hindrance of pollutants to cement hydration because of rapid hydration capability.

4.4 Benefits and challenges

Reactive MgO cement has some advantages over OPC. However, there are certain concerns in the manufacture and application of MgO-based cements.

These include the unfamiliarity of MgO cement, few reliable records, and insufficient literature of MgO, and low availability of raw materials, while the traditional OPC receives high validation and market confidence. In addition, most of the MgO-based cements are costlier to produce than OPC because the reactive sources of MgO are relatively expansive, despite some such cement can reduce the CO_2 emissions. MOC and MOS appear to be easily dissolved in water and therefore not be used in structural cements. Even so, these cements will still have a place in special applications, where OPC is not suitable for the physical or chemical requirements (Walling and Provis, 2016).

5. Special cement-based S/S

5.1 Calcium sulfoaluminate cements

Using BFS or fly ash to replace OPC in cement systems can stabilize/solidify low- and intermediate-level radioactive-contaminated soil. However, the high pH of these cement systems will corrode heavy metals in soils releasing hydrogen and generating expansive corrosion products. CSA cement systems can offer different hydration chemistry that makes heavy metals to be immobilized with lower activity. It has several advantages including cost-effective, low-carbon with rapid setting (up to 40 MPa within 24 h under the conditions of high humidity), and compatibility with contaminants (Al-Horr et al., 2017). CSA cement is produced from calcined natural materials such as limestone, bauxite, and clay with gypsum or an anhydride component at 1250–1350°C calcination of calcium sulfoaluminate or ye'elimite ($3CaO \cdot 3Al_2O_3 \cdot CaSO_4$). Based on the raw materials, there may be dicalcium silicate (Belize) or an aluminate or ferrite phase (Al-Horr et al., 2017). In addition, ettringite as the main hydrate from CSA hydration may be favorable for S/S process. This hydration product could incorporate toxic elements in its framework. This advantage plays an important role in the S/S process even if it is difficult to obtain a stable, low-consumption solid. The major obstacle is how to keep the heavy metals in the solidified matrix. This difficulty is directly related to the solubility presented by many metals of amphoteric behavior (Al-Horr et al., 2017). Besides, it has high retention capacity for zinc ions and is dimensionally stable. When the waste contains high concentrations of borate and lithium ions, it is feasible to use 70% CSA and 30% OPC mixture for immobilization (Coumes et al., 2017).

5.2 Calcined clay limestone cement

Calcined clay limestone cement (LC^3), which is a promising new type of cement with lower CO_2 emission, is a ternary binder system made of Portland cement clinker, limestone, and calcined clay. In the previous section, we introduced several widely known and well-studied SCMs, but their availability is not within the scope of large-scale cement manufacture. However, limestone and clays are the most abundant natural materials in the world. The wide use of calcined clays can meet the growing demand for cement alternative materials. The production of calcined clays involves similar equipment to OPC and approximate cost of investment. Lower energy is consumed in calcination at lower temperatures of clay (750−850°C) than that of clinker (1450°C) (Scrivener et al., 2018). In addition, compared with OPC production in CO_2 emissions, the LC^3 usually emits around 30% less CO_2 (Scrivener et al., 2018). Because of the existing limestone, the reactive calcined clay generates an additional amount of AFm phases (i.e., monocarboaluminate hemi-car-boa-luminate and monocarboaluminate), resulting in a complementary reaction between calcined clay and limestone. As a highly reactive pozzolan, calcined clay can improve its microstructure and pore structure at an early stage (Tironi et al., 2013). Dhandapani et al. (2018) studied the mechanical strength and durability performance of concretes with LC^3. Results showed that when compared with OPC, LC^3 binder showed comparable mechanical strength development, better chloride resistance, lower gas permeability and capillary water absorption, early enhancement of durability parameters. Although there are few applications and research about S/S of contaminated soil by using the LC^3, LC^3 is likely to be a good binder to immobilize contaminants because its lower permeability would allow less contaminants leaching.

5.3 Alternative cement clinkers

This section introduces four kinds of reactive binder components while giving low CO_2 emission: Reactive Belite-rich Portland cement (RBPC) clinkers, Belite-Ye'elimite-Ferrite (BYF) clinkers, Carbonatable Calcium Silicate clinkers (CCSC), and Magnesium Oxides are produced from Magnesium Silicates (MOMS). RBPC and BYF are hydraulic materials that harden by hydraulic reaction. CCSC is a clinker that can be hardened by absorbing CO_2, whereas MOMS falls into both categories.

According to the clinker mineralogy, RBPC belongs to the same system with OPC as they are in the $C_2S-C_3S-C_3A$-C_4AF system. They are also named high Belite cement. Clinker composition in Belit/Alite is the major difference between RBPC and OPC. In RBPC, the bainite content exceeds 40% and the bainite content is usually less than 35%, This shows that bainite is the most abundant form in RBPC, as opposed to Alite in OPC. Since Alite consumes more energy to form and cause more CO_2 emission than Belite, the production of RBPC results in lower energy cost and carbon footprint. However, the potential reduction of CO_2 is limited by RBPC approach and usually does not exceed 10% that of OPC. Compared with OPC, RBPC cements generally have, lower water demand, similar setting times, good compatibility with most water-reducing agents (mainly due to lower C_3A content), lower drying shrinkage, thermal evolution, and early strength but higher late strength. RPBC is also well tolerated by sulfates and chlorides, mainly because of the small content of portlandite in the hydration products (Gartner and Sui, 2018). Thus, the alternative cement clinkers have potential to be better and low-CO_2 binders for the S/S of contaminated soil.

The BYF approach can be regarded as an intermediate between CSA cement and conventional OPC. The major difference between the BYF method and the existing CSA cement is that CSA cement is currently targeted at the special cement application and is required to meet high standards regarding the special properties of cement such as rapid setting and compensation of shrinkage. The intention to develop the BYF method is to reduce the manufacturing cost based on CSA burnt blocks, to allow the use of less costly aluminum-rich raw materials, to make the method more suitable than existing CSA cement, and concrete production with a significantly lower CO_2 emission (Barcelo et al., 2014). Therefore, BYF could be an efficient and green binder for the S/S of contaminated soil like CSA cement.

CCSC is based on the well-studied approach that calcium silicates could be enhanced by CO_2 curing. The advantages for CCSC approach include: 30% lower CO_2 emissions than OPC in the production process because of the low calcium content of the clinker; absorption of a large content of CO_2 during the curing phase; higher early-term strength; consumption of very little water overall. CCSC could mitigate the hindrance of pollutants to cement hydration in the early 24 h, which makes it a promising binder in the S/S technology (Atakan et al., 2014).

Compared with the aforementioned three technologies, where limestones are mainly used as the major supplementary raw materials, the MOMS method uses magnesium silicate as the major raw material, making it no chemically bound CO_2. This procedure also provides the potential absorption of a large content of CO_2, like the CCSC technology. However, MOMS is still in early research stage, for their application and durability are still unclear (Gartner et al., 2017). Although MgO has been proved to be an alternative binder in the S/S, the emergence of MOMS approach makes it better being a green binder.

5.4 Benefits and challenges

Compared with OPC, the advantage of CSA cement is its higher early strength. This advantage is obvious since many contaminants are serious retarders of OPC hydration (Niu et al., 2018). LC^3 binder shows some advantages including comparable mechanical strength development, better chloride resistance, denser microstructure, early enhancement of durability. Alternative clinker system showed satisfactory strength, lower global CO_2 emissions in use or during the manufacturing process, better resistance to chloride ion. However, for the alternative clinker system, making an estimate of either application or manufacture is too immature since there is no well-defined production process to study (Gartner et al., 2017).

6. Conclusion

This chapter illustrates an overview of green S/S approach of contaminated soil by using low-carbon cement-based materials including SCMs-incorporated cement, AAC, MgO-based cement, and special cements. The S/S mechanism, state-of-the-art, benefits, and challenges of these cements have been systematically discussed. Green remediation by using low-carbon cement-based S/S is a novel and meaningful approach, which lessens environmental burdens and generates valuable products. Using industrial by-products or green cements for soil S/S solves the problems of contaminated soil while mitigating global warming and landfill disposal. Even if these green materials cannot completely replace OPC currently, they can be applied in specified areas where OPC is not suitable due to some special physical or chemical requirements. For example, SCMs could be made to further improve mechanical strength, durability, contaminants immobilization. AAC shows better acid and sulfate resistance, higher compatibility with metals and radioactive waste. MgO-based binders can be made to achieve satisfactory environmental-friendly properties through CO_2 curing, excellent compatibility with contaminants and heavy metals. MPC can be applied in environments requiring near-neutral-pH cements and for rapid repair, especially in cold environments. Furthermore, MPC contains potassium and phosphorus, which can be reused as nutrients to promote plants growth. CSA cement is a novel product with low-energy consumption and low CO_2 emission and shows high hydration activity and compatibility with contaminants. LC^3 has the advantages of comparable setting/hardening, better chloride resistance, lower gas permeability and capillary water absorption, higher durability. Alternative clinker systems show properties to OPC but with less CO_2 emission, although they are restrained by some technical and economic issues in commercial and practical applications. Therefore, green and sustainable remediation of contaminated soil by effective, economical, environmentally friendly S/S technologies has their niche in the market.

References

Aiken, T.A., Sha, W., Kwasny, J., Soutsos, M.N., 2017. Resistance of geopolymer and Portland cement based systems to silage effluent attack. Cement and Concrete Research 92, 56–65.

Akcil, A., Erust, C., Ozdemiroglu, S., Fonti, V., Beolchini, F., 2015. A review of approaches and techniques used in aquatic contaminated sediments: metal removal and stabilisation by chemical and biotechnological processes. Journal of Cleaner Production 86, 24–36.

Al-Horr, Y., Elhoweris, A., Elsarrag, E., 2017. The development of a novel process for the production of calcium sulfoaluminate. International Journal of Sustainable Built Environment 6, 734–741.

Al-Tabbaa, A., 2013. Reactive magnesia cement. In: Eco-Efficient Concrete, pp. 523–543.

Andrew, R.M., 2017. Global CO_2 emissions from cement production. Earth System Science Data Discussions. https://doi.org/10.5194/essd-2017-77. In review.

Atakan, V., Sahu, S., Quinn, S., Hu, X., De Cristofaro, N., 2014. Why CO2 matters - advances in a new class of cement. ZKG International 67, 60–63.

Bar-Nes, G., Katz, A., Peled, Y., Zeiri, Y., 2008. The mechanism of cesium immobilization in densified silica-fume blended cement pastes. Cement and Concrete Research 38, 667–674.

Barcelo, L., Kline, J., Walenta, G., Gartner, E., 2014. Cement and carbon emissions. Materials and Structures 47, 1055–1065.

Bolan, N., Kunhikrishnan, A., Thangarajan, R., Kumpiene, J., Park, J., Makino, T., Kirkham, M.B., Scheckel, K., 2014. Remediation of heavy metal(loid)s contaminated soils - to mobilize or to immobilize. Journal of Hazardous Materials 266, 141–166.

Cheah, C.B., Chung, K.Y., Ramli, M., Lim, G.K., 2016. The engineering properties and microstructure development of cement mortar containing high volume of inter-grinded GGBS and PFA cured at ambient temperature. Construction and Building Materials 122, 683–693.

Chen, Q.Y., Tyrer, M., Hills, C.D., Yang, X.M., Carey, P., 2009. Immobilisation of heavy metal in cement-based solidification/stabilisation: a review. Waste Management 29, 390–403.

Chen, W., Hong, J., Xu, C., 2014. Pollutants generated by cement production in China, their impacts, and the potential for environmental improvement. Journal of Cleaner Production 103, 61–69.

Chen, L., Wang, L., Cho, D.W., Tsang, D.C.W., Tong, L.Z., Zhou, Y.Y., Yang, J., Hu, Q., Poon, C.S., 2019. Sustainable stabilisation/solidification of municipal solid waste incinerator fly ash by incorporation of green materials. Journal of Cleaner Production (in press).

Coumes, C., Dhoury, M., Champenois, J.B., Mercier, C., Damidot, D., 2017. Combined effects of lithium and borate ions on the hydration of calcium sulfoaluminate cement. Cement and Concrete Research 97, 50–60.

Debela, F., Arocena, J.M., Thring, R.W., Whitcombe, T., 2013. Organic acids inhibit the formation of pyromorphite and Zn-phosphate in phosphorous amended Pb- and Zn-contaminated soil. Journal of Environmental Management 116, 156–162.

Dhandapani, Y., Sakthivel, T., Santhanam, M., Gettu, R., Pillai, R.G., 2018. Mechanical properties and durability performance of concretes with limestone calcined clay cement (LC^3). Cement and Concrete Research 107, 136–151.

Duxson, P., Fernández-Jiménez, A., Provis, J.L., Lukey, G.C., Palomo, A., Van Deventer, J.S.J., 2007. Geopolymer technology: the current state of the art. Journal of Materials Science 42, 2917–2933.

El-Eswed, B.I., Horr, R.I., Alshaaer, M., Hamadneh, I., Al-Gharabli, S.I., Khalili, F., 2015. Stabilisation/solidification of heavy metals in kaolin/zeolite based geopolymers. International Journal of Mineral Processing 137, 34–42.

Flatt, R.J., Roussel, N., Cheeseman, C.R., 2012. Concrete: an eco material that needs to be improved. Journal of the European Ceramic Society 32, 2787–2798.

Ganesan, N., Abraham, R., Deepa Raj, S., 2015. Durability characteristics of steel fibre reinforced geopolymer concrete. Construction and Building Materials 93, 471–476.

Gartner, E., Hirao, H., 2015. A review of alternative approaches to the reduction of CO_2 emissions associated with the manufacture of the binder phase in concrete. Cement and Concrete Research 78, 126–142.

Gartner, E., Sui, T., 2018. Alternative cement clinkers. Cement and concrete research 114, 27–39.

Ghafoori, N., Najimi, M., Sobhani, J., Aqel, M.A., 2013. Predicting rapid chloride permeability of self-consolidating concrete: a comparative study on statistical and neural network models. Construction and Building Materials 44, 381–390.

Goodarzi, A.R., Akbari, H.R., Salimi, M., 2016. Enhanced stabilisation of highly expansive clays by mixing cement and silica fume. Applied Clay Science 132–133, 675–684.

Gu, K., Jin, F., Al-Tabbaa, A., Shi, B., Liu, C., Gao, L., 2015. Incorporation of reactive magnesia and quicklime in sustainable binders for soil stabilisation. Engineering Geology 195, 53–62.

Guo, B., Liu, B., Yang, J., Zhang, S., 2017. The mechanisms of heavy metal immobilization by cementitious material treatments and thermal treatments: a review. Journal of Environmental Management 193, 410–422.

Guo, B., Pan, D., Liu, B., Volinsky, A.A., Fincan, M., Du, J., Zhang, S., 2017. Immob Unluer utilization mechanism of Pb in fly ash-based geopolymer. Construction and Building Materials 134, 123–130.

He, P., Poon, C.S., Tsang, D.C.W., 2017. Effect of pulverized fuel ash and CO_2 curing on the water resistance of magnesium oxychloride cement (MOC). Cement and Concrete Research 97, 115–122.

Hou, D., Al-Tabbaa, A., 2014. Sustainability: a new imperative in contaminated land remediation. Environmental Science & Policy 39, 25–34.

Huang, X., Huang, T., Li, S., Muhammad, F., Xu, G., Zhao, Z., Yu, L., Yan, Y., Li, D., Jiao, B., 2016. Immobilization of chromite ore processing residue with alkali-activated blast furnace slag-based geopolymer. Ceramics International 42, 9538–9549.

Izquierdo, M., Querol, X., Davidovits, J., Antenucci, D., Nugteren, H., Fernández-Pereira, C., 2009. Coal fly ash-slag-based geopolymers: microstructure and metal leaching. Journal of Hazardous Materials 166, 561–566.

Jin, F., Al-Tabbaa, A., 2014. Evaluation of novel reactive MgO activated slag binder for the immobilisation of lead and zinc. Chemosphere 117, 285–294.

Jin, F., Wang, F., Al-Tabbaa, A., 2016. Three-year performance of in-situ solidified/stabilised soil using novel MgO-bearing binders. Chemosphere 144, 681–688.

Li, J., Poon, C.S., 2017. Innovative solidification/stabilisation of lead contaminated soil using incineration sewage sludge ash. Chemosphere 173, 143–152.

Li, X., Chen, Q., Zhou, Y., Tyrer, M., Yu, Y., 2014. Stabilisation of heavy metals in MSWI fly ash using silica fume. Waste Management 34, 2494–2504.

Li, P., Pan, S.Y., Pei, S., Lin, Y.J., Chiang, P.C., 2016. Challenges and perspectives on carbon fixation and utilization technologies: an overview. Aerosol and Air Quality Research 16, 1327–1344.

Mo, L., Deng, M., Tang, M., Al-Tabbaa, A., 2014. MgO expansive cement and concrete in China: past, present and future. Cement and Concrete Research 57, 1–12.

Namarak, C., Satching, P., Tangchirapat, W., Jaturapitakkul, C., 2017. Improving the com- pressive strength of mortar from a binder of fly ash-calcium carbide residue. Construction and Building Materials 147, 713–719.

Nied, D., Enemark-Rasmussen, K., L'Hopital, E., Skibsted, J., Lothenbach, B., 2016. Properties of magnesium silicate hydrates (M-S-H). Cem. Concrete Research 79, 323–332.

Niu, M., Li, G., Wang, Y., Li, Q., Han, L., Song, Z., 2018. Comparative study of immobilization and mechanical properties of sulfoaluminate cement and ordinary Portland cement with different heavy metals. Construction and Building Materials 193, 332–343.

Pan, S.Y., Chen, Y.H., Chen, C.D., Shen, A.L., Lin, M., Chiang, P.C., 2015. High-gravity carbonation process for enhancing CO_2 fixation and utilization exemplified by the steel-making industry. Environmental Science and Technology 49, 12380–12387.

Provis, J.L., 2017. Alkali-activated materials. Cement and Concrete Research 114, 40–48.

Provis, J.L., Bernal, S.A., 2014. Geopolymers and related alkali-activated materials. Annual Review of Materials Research 44, 299–327.

Provis, J.L., Palomo, A., Shi, C., 2015. Advances in understanding alkali-activated materials. Cement and Concrete Research 78, 110–125.

Rasaki, S.A., Bingxue, Z., Guarecuco, R., Thomas, T., Minghui, Y., 2019. Geopolymer for use in heavy metals adsorption, and advanced oxidative processes: a critical review. Journal of Cleaner Production 213, 42–58.

Richardson, I.G., 2008. The calcium silicate hydrates. Cement and Concrete Research 38, 137–158.

Samad, S., Shah, A., 2017. Role of binary cement including Supplementary cementitious material (SCM), in production of environmentally sustainable concrete: a critical review. International Journal of Sustainable Built Environment 6, 663–674.

Sargent, P., 2014. The development of alkali-activated mixtures for soil stabilisation. In: Handbook of Alkali-Activated Cements, Mortars and Concretes, pp. 555–604.

Scrivener, K., Martirena, F., Bishnoi, S., Maity, S., 2018. Cement and concrete research calcined clay limestone cements (LC^3). Cement and Concrete Research 114, 49–56.

Shen, W., Cao, L., Li, Q., Wen, Z., Wang, J., Liu, Y., Dong, R., Tan, Y., Chen, R., 2016. Is magnesia cement low carbon-Life cycle carbon footprint comparing with Portland cement. Journal of Cleaner Production 131, 20–27.

Shi, C., Fernández Jiménez, A., Palomo, A., 2011. New cements for the 21st century: the pursuit of an alternative to Portland cement. Cement and Concrete Research 41, 750–763.

Smith, K.M., Fowler, G.D., Pullket, S., Graham, N.J.D., 2009. Sewage sludge-based adsorbents: a review of their production, properties and use in water treatment applications. Water Research 43, 2569–2594.

Suraneni, P., Weiss, J., 2017. Examining the pozzolanicity of supplementary cementitious materials using isothermal calorimetry and thermogravimetric analysis. Cement and Concrete Composites 83, 273–278.

Tashima, M.M., Soriano, L., Payá, J., Monzó, J., Borrachero, M.V., 2016. Assessment of pozzolanic/hydraulic reactivity of vitreous calcium aluminosilicate (VCAS). Materials and Design 96, 424–430.

Tennakoon, C., Shayan, A., Sanjayan, J.G., Xu, A., 2017. Chloride ingress and steel corrosion in geopolymer concrete based on long term tests. Materials and Design 116, 287–299.

Tian, Q., Sasaki, K., 2019. A novel composite of layered double hydroxide/geopolymer for co-immobilization of Cs^+ and $SeO_{42}-$ from aqueous solution. Science of The Total Environment 695, 133799.

Tironi, A., Trezza, M.A., Scian, A.N., Irassar, E.F., 2013. Assessment of pozzolanic activity of different calcined clays. Cement and Concrete Composites 37, 319–327.

Tsang, D.C.W., Yip, A.C.K., Olds, W.E., Weber, P.A., 2014. Arsenic and copper stabilisation in a contaminated soil by coal fly ash and green waste compost. Environmental Science & Pollution Research 21, 10194–10204.

Ucaroglu, S., Talinli, I., 2012. Recovery and safer disposal of phosphate coating sludge by solidification/stabilisation. Journal of Environmental Management 105, 131–137.

Unluer, C., Al-Tabbaa, A., 2013. Impact of hydrated magnesium carbonate additives on the carbonation of reactive MgO cements. Cement and Concrete Research 54, 87–97.

Unluer, C., Al-Tabbaa, A., 2014. Enhancing the carbonation of MgO cement porous blocks through improved curing conditions. Cement and Concrete Research 59, 55–65.

Vance, E.R., Perera, D.S., 2009. Geopolymers for nuclear waste immobilisation. In: Geopolymers: Structures, Processing, Properties and Industrial Applications, pp. 401–420.

Vandevenne, N., Iacobescu, R.I., Carleer, R., Samyn, P., D'Haen, J., Pontikes, Y., Schreurs, S., Schroeyers, W., 2018. Alkali-activated materials for radionuclide immobilisation and the effect of precursor composition on Cs/Sr retention. Journal of Nuclear Materials 510, 575–584.

Vespa, M., Dähn, R., Wieland, E., 2014. Competition behaviour of metal uptake in cementitious systems: an XRD and EXAFS investigation of Nd- and Zn-loaded 11Å tobermorite. Physics and Chemistry of the Earth 70–71, 32–38.

Wang, Y., Tsang, D.C.W., 2013. Effects of solution chemistry on arsenic(V) removal by low-cost adsorbents. Journal of Environmental Sciences 25, 2291–2298.

Wang, Q., Ding, Z.Y., Zhang, J., Qiu, L.G., Sui, Z.T., 2011. Study on slag-based geopolymer hydration process. Key Engineering Materials 477, 67–71.

Wang, L., Tsang, D.C.W., Poon, C.S., 2015. Green remediation and recycling of contaminated sediment by waste-incorporated stabilisation/solidification. Chemosphere 122, 257–264.

Wang, L., Chen, S.S., Tsang, D.C.W., Poon, C.S., Shih, K.M., 2016a. Value-added recycling of construction waste wood into noise and thermal insulating cement-bonded particleboards. Construction and Building Materials 125, 316–325.

Wang, L., Chen, S.S., Tsang, D.C.W., Poon, C.S., Shih, K., 2016b. Recycling contaminated wood into eco-friendly particleboard using green cement and carbon dioxide curing. Journal of Cleaner Production 137, 861–870.

Wang, L., Chen, S.S., Tsang, D.C.W., Poon, C.S., Dai, J.G., 2017a. CO_2 curing and fibre reinforcement for green recycling of contaminated wood into high-performance cement-bonded particleboards. Journal of CO2 Utilization 18, 107–116.

Wang, L., Yu, I.K.M., Tsang, D.C.W., Li, S., Poon, C.S., 2017b. Mixture design and reaction sequence for recycling construction wood waste into rapid-shaping magnesia-phosphate cement particleboard. Industrial & Engineering Chemistry Research 56, 6645–6654.

Wang, L., Yeung, T.L.K., Lau, A.Y.T., Tsang, D.C.W., Poon, C.S., 2017c. Recycling contaminated sediment into eco-friendly paving blocks by a combination of binary cement and carbon dioxide curing. Journal of Cleaner Production 164, 1279–1288.

Wang, L., Yu, I.K.M., Tsang, D.C.W., Li, S., Li, J.S., Poon, C.S., Wang, Y.S., Dai, J.G., 2017d. Transforming wood waste into water-resistant magnesia-phosphate cement particleboard modified by alumina and red mud. Journal of Cleaner Production 168, 452–462.

Wang, L., Chen, L., Tsang, D.C.W., Yeung, T.Y.K., Li, J.S., Ding, S.M., Poon, C.S., 2018a. Green remediation of contaminated sediment by stabilisation/solidification with industrial by-products and CO_2 utilization. Science of the Total Environment 631–632, 1321–1327.

Wang, L., Zou, F.L., Fang, X.L., Tsang, D.C.W., Poon, C.S., Leng, Z., Baek, K., 2018b. A novel type of controlled low strength material derived from alum sludge and green materials. Construction and Building Materials 165, 792–800.

Wang, L., Yu, K.Q., Li, J., Tsang, D.C.W., Poon, C.S., Yoo, J.C., Baek, K., Ding, S.M., Hou, D.Y., Dai, J.G., 2018c. Low-carbon and low-alkalinity stabilisation/solidification of high-Pb contaminated soil. Chemical Engineering Journal 351, 418–427.

Wang, L., Chen, L., Tsang, D.C.W., Li, J., Poon, C.S., Baek, K., Hou, D.Y., Ding, S.M., 2018d. Recycling dredged sediment into fill materials, partition blocks, and paving blocks: technical and economic assessment. Journal of Cleaner Production 199, 69–76.

Wang, L., Tsang, D.C.W., 2018. Carbon dioxide sequestration on composites based on waste wood. In: Carbon Dioxide Sequestration in Cementitious Construction Materials, pp. 431–450.

Wang, L., Yu, I.K.M., Tsang, D.C.W., Yu, K.Q., Li, S., Poon, C.S., Dai, J.G., 2018f. Upcycling wood waste into fibre-reinforced magnesium phosphate cement particleboards. Construction and Building Materials 159, 54–63.

Wang, L., Cho, D.W., Tsang, D.C.W., Cao, X.D., Hou, D.Y., Shen, Z.T., Alessi, D.S., OK, Y.S., Poon, C.S., 2019a. Green remediation of as and Pb contaminated soil using cement-free clay-based stabilisation/solidification. Environment International 126, 336–345.

Wang, L., Chen, S.S., Sun, Y.Q., Tsang, D.C.W., Yip, A.C.K., Ding, S.M., Hou, D.Y., Baek, K., Ok, Y.S., 2019b. Efficacy and limitations of low-cost adsorbents for in-situ stabilisation of contaminated marine sediment. Journal of Cleaner Production 212, 420–427.

Wang, L., Chen, L., Cho, D.W., Tsang, D.C.W., Yang, J., Hou, D.Y., Baek, K., Kua, H.W., Poon, C.S., 2019c. Novel synergy of Si-rich minerals and reactive MgO for stabilisation/solidification of contaminated sediment. Journal of Hazardous Materials 365, 695–706.

Walling, S.A., Provis, J.L., 2016. Magnesia-based cements: a journey of 150 Years, and cements for the future. Chemical Reviews 116, 4170–4204.

Wu, H.L., Jin, F., Bo, Y.L., Du, Y.J., Zheng, J.X., 2018. Leaching and microstructural properties of lead contaminated kaolin stabilized by GGBS-MgO in semi-dynamic leaching tests. Construction and Building Materials 172, 626–634.

Xiao, Z., Yuan, X., Li, H., Jiang, L., Leng, L., Chen, X., Zeng, G., Li, F., Cao, L., 2015. Chemical speciation, mobility and phyto-accessibility of heavy metals in fly ash and slag from combustion of pelletized municipal sewage sludge. The Science of the Total Environment 536, 774–783.

Xuan, D., Zhan, B., Poon, C.S., Zheng, W., 2016. Carbon dioxide sequestration of concrete slurry waste and its valorisation in construction products. Construction and building materials 113, 664–672.

Zhang, J., Provis, J.L., Feng, D., Deventer, J.S.J. Van, 2008. Geopolymers for immobilization of Cr^{6+}, Cd^{2+}, and Pb^{2+}. Journal of Hazardous Materials 157, 587–598.

Zhang, Z.H., Zhu, H.J., Zhou, C.H., Wang, H., 2016. Geopolymer from kaolin in China: an overview. Applied Clay Science 119, 31–41.

Chapter 6

The use of biochar for sustainable treatment of contaminated soils

Zhengtao Shen[1], Yunhui Zhang[2], Oliver McMillan[2], David O'Connor[3], Deyi Hou[3]

[1]*Department of Earth and Atmospheric Sciences, University of Alberta, Edmonton, AB, Canada;* [2]*Department of Engineering, University of Cambridge, Cambridge, United Kingdom;* [3]*School of Environment, Tsinghua University, Beijing, China*

1. Introduction

The increasing atmospheric CO_2 levels, resulting from fossil fuel burning, deforestation, and other anthropogenic and natural sources, is regarded as a major cause of climate change (Edenhofer, 2015). Climate change is expected to bring environmental problems such as warming temperatures, increasing frequency of extreme weather events, rising sea levels, and expansion of deserts (Lu et al., 2007). In order to mitigate climate change, efforts have been made globally for carbon capture and storage (CCS). Since 2006 (Lehmann et al., 2006), biochar technology has been regarded as a promising technology to store carbon with many additional benefits in waste management, energy production, and soil improvement.

Biochar is the solid, carbon-rich product of heating biomass in the absence of air in a process called pyrolysis (Sohi 2012). Biochar production converts the labile biomass to a recalcitrant form and therefore can store carbon for hundreds to thousands of years (Lehmann et al., 2008). After production, biochar is typically added to soil to improve the agricultural productivity of the soil. Some research has recently shown that biochar may also be useful for immobilizing contaminants in contaminated soils (Beesley and Marmiroli, 2011b). The contaminants in the soil generally include organic and inorganic pollutants generated through human activities such as oil drilling, mining, chemical processing, and waste disposal (Zhang et al., 2013b). These contaminants in the soil can be transported via pathways (e.g., groundwater and rivers) and present risks to human health (Berman and Timoshenko, 2017).

Sustainable Remediation of Contaminated Soil and Groundwater. https://doi.org/10.1016/B978-0-12-817982-6.00006-9

The application of biochar in soil remediation has the following environmental benefits: (1) biochar can be produced using agricultural, household, and industrial wastes (e.g., crop residues, food scraps, manure, and sludge) as the feedstocks (Ronsse et al., 2013), and the reuse of wastes leads to biochar's low carbon footprint; (2) the coproducts (syngas and bio-oil) produced during pyrolysis can be used as green energy and therefore offset energy use associated with producing biochar (Lehmann and Environment, 2007); (3) biochar can immobilize contaminants in soils and therefore reduce their environmental risks to humans (Cao et al., 2011); (4) biochar can help retain the nutrients and improve the water holding capacity and resilience of soils, thus aiding the greening or revegetation of contaminated land (Beesley et al., 2011); and (5) the recalcitrant form of carbon in biochar can remain in soil for hundreds to thousands of years, effectively removing it from the carbon cycle (Lehmann and Environment, 2007). Therefore, the application of biochar in soil remediation is regarded as a green and sustainable approach. This chapter introduces the origins and concept of sustainable biochar technology, the production and properties of biochar, the interactions between biochar and contaminants, the applications of biochar in soil remediation, and the conclusions and implications for sustainable remediation.

2. Sustainable biochar technology

Human activities since the Industrial Revolution, such as fossil fuel burning and deforestation, have increased the concentrations of greenhouse gases (CO_2, methane, and tropospheric ozone) in the atmosphere. The abrupt increase in the atmospheric CO_2 since 1950 is regarded as the main cause of climate change, which brought significant climatic problems to the planet (Lu et al., 2007). In order to prevent the onset of climate change, by reducing greenhouse gas concentrations in the atmosphere, an international treaty called "the Kyoto Protocol" was adopted in 1997 and entered into force in 2005, which now covers 192 parties. Therefore, mitigating climate change calls for efforts on CCS. In the late 1990s and early 2000s, a range of research activities on the local soil in Amazonia inspired people to store carbon in soils through biochar technology.

Amazonian Dark Earth (ADE) (locally known as *terra preta*) is a dark soil with high nutrient and organic matter contents widely distributed throughout Amazonia, South America (Arroyo-Kalin, 2017; Neves et al., 2003). Extensive studies on ADE found that the organic matter content in ADE mainly consists of charcoal and highly aromatic humic substances, which were likely formed due to the incomplete combustion of organic materials (crop residues and food waste) by the local residents 500–2500 years ago (Arroyo-Kalin, 2017; Neves et al., 2003). ADE contained much more charcoal (as much as 70 times) than the surrounding soils, and the charcoal has been stable in ADE since it was produced, due to its relatively high degree of aromatization (Glaser et al.,

2003). According to Glaser et al. (2003), approximately 147–506 Mg C/ha m is stored in ADE around Manaus and in the Belterra area compared with the adjacent soils, which only contain 72–149 Mg C/ha m.

In tropical areas (e.g., Amazonia), the nutrients in the topsoils can be easily leached by the heavy rainfall. The slow oxidation on the edge of the aromatic structure of the charcoal in ADE resulted in active functional groups (e.g., carboxylic groups), which can significantly retain the nutrients in the soil against rainfall wash and consequently enhance crop production (Neves et al., 2003). Therefore, studies of ADE demonstrate that charcoal within the soil can persist for thousands of years, aiding carbon sequestration, and provide benefits to the soil in terms of nutrient control and crop production (Glaser et al., 2003).

These findings inspired researchers and agricultural workers to put charcoal into soil, as a new strategy to store carbon as well as to improve soil conditions. Based on this evidence, the term "biochar" was developed in 2006 with a purpose of explicitly referring to soil management and carbon storage using specifically designed charcoal (Lehmann et al., 2006), and the principle of carbon sequestration using biochar was illustrated. Briefly, plants absorb CO_2 from the atmosphere and transfer the carbon to biomass through photosynthesis. After death, plant biomass will finally enter into soil and become soil organic matter (SOM), which decomposes by soil microbes with a mean residence time of commonly less than 4 years (Tiessen et al., 1994), releasing the carbon back to the atmosphere as CO_2. Therefore, the net carbon withdrawal from atmosphere of these processes is approximately neutral. In contrast, through biochar technology, the plant biomass (regularly agricultural waste) is pyrolyzed and the resulting solid products (biochar) is put back to the soil, where it has a mean residence time of hundreds to thousands of years (Glaser et al., 2001). In this case, the carbon fixed by plants through photosynthesis can be stored in the soil as a carbon sink and is effectively removed from the carbon cycle. Therefore, biochar technology links four benefits including waste management, mitigation of climate change, energy production, and soil improvement. Recently, biochar was proposed to have a huge potential to be applied in soil remediation (Beesley et al., 2011; Bian et al., 2014; Tong et al., 2014; Zhang et al., 2013b). The external benefits such as reuse of waste material, carbon sequestration, and energy generation give biochar technology additional values beyond remediating contaminated sites.

3. Biochar production, properties, and its influencing factors

3.1 Biochar production

Pyrolysis is the most typical way to produce biochar (Sohi, 2012). Pyrolysis is the process of heating (usually under controlled temperature) biomass with the

FIGURE 6.1 Biochar image before and after pyrolysis. *Sohi, S.P., 2012. Agriculture. Carbon storage with benefits. Science 338(6110), 1034–1035.*

exclusion of oxygen (in an oxygen-limited environment, or an oxygen-free environment by adding a protective gas such as N_2 or CO_2), with the products being biochar, bio-oil, and syngas (Sohi, 2012) (Fig. 6.1). Pyrolysis consists of slow pyrolysis, fast pyrolysis, and flash pyrolysis. Slow pyrolysis is the most typically used pyrolysis method to produce biochar (Jahirul et al., 2012). It has a longer heating duration and higher biochar yield compared with fast and flash pyrolysis.

Modern slow pyrolysis units are designed to control production parameters and consequently design the properties of biochar, as well as to capture the volatile components for the production of bio-oil and syngas (Manya, 2012). Fig. 6.2 shows a typical design for a modern slow pyrolyzer (Laird et al., 2009). The highest heating temperature (HHT), heating rate, and residence time can be controlled by the combustor. The residence time of feedstock in combustor is long with a low heating rate, which will result in a relatively high biochar yield for slow pyrolysis. The HHT for slow pyrolysis is typically lower compared with fast and flash pyrolysis (Jahirul et al., 2012). The heat and syngas produced during slow pyrolysis can offset the energy consumed for biochar production. Increasing pressure and decreasing HHT help to enhance the yield of biochar for slow pyrolysis (Manya, 2012). Furnace was also actively used to produce biochar in scientific research (Shen et al., 2019a, 2019b).

Considering biochar yield and suitability, slow pyrolysis may be an optimal method for biochar production (Manya, 2012). It is actually the most commonly used traditional and modern method to produce biochar (Lehmann and Joseph, 2015). A range of modern slow pyrolyzers have been developed to produce biochar for research and engineering uses (Manya, 2012). The carefully controlled modern slow pyrolysis procedure will enhance the

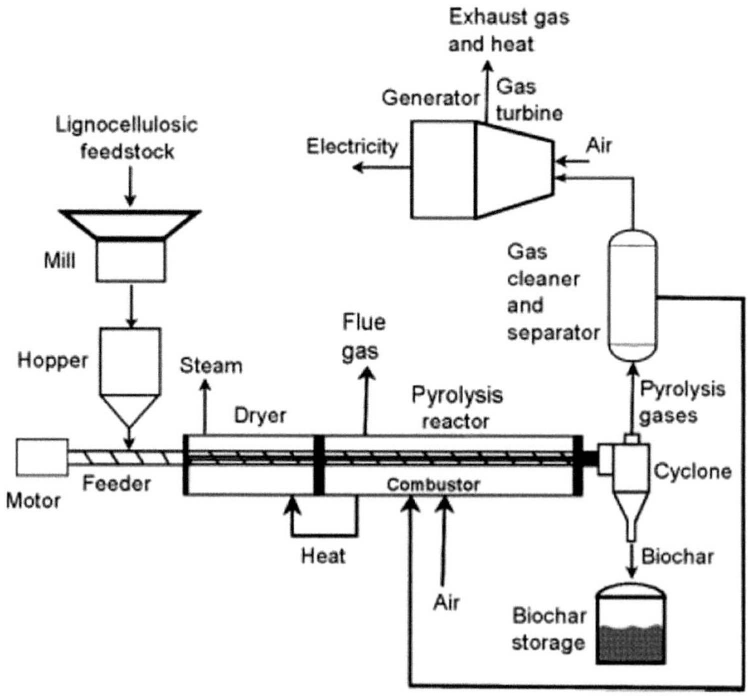

FIGURE 6.2 One possible design for a slow pyrolyzer. *Laird, D.A., Brown, R.C., Amonette, J.E., Lehmann, J., 2009. Review of the pyrolysis platform for coproducing bio-oil and biochar. Biofuels, Bioproducts and Biorefining 3(5), 547–562.*

reproducibility of biochar, which is very important for quality control and assurance of scientific research.

3.2 Biochar properties and its influencing factors

3.2.1 Biochar properties

The physicochemical properties of biochar related to its performance in soil remediation are surface area, pH, cation-exchange capacity (CEC), elemental composition, particle size, ash content, aromaticity, and volatile matter (VM) content, in which the first three are the most important properties determining its environmental benefits (Tag et al., 2016). Surface area is the total area that the surface of a solid object occupies (Mukherjee et al., 2011). It is determined by the pore structure of the solid object (Mukherjee et al., 2011). In general, a more porous structure will result in a higher surface area for solids, with micropores generally contributing the greatest proportion (Kwon and Pignatello, 2005). The surface area of biochar is closely related to its physical adsorption capacity, with higher surface area generally resulting in higher physical adsorption capacity for heavy metals and organic contaminants. pH represents the alkalinity or acidity of biochar. A higher pH of biochar can significantly aid its adsorption of heavy metals by a range of mechanisms. CEC is the ability for biochar to exchange cations from the environment with its own cations. A higher CEC of biochar will aid the retention of nutrients (e.g., N, P, and K) as well as the adsorption of contaminants when applied to soil. The value of CEC for a biochar is closely dependent on its exchangeable cations such as H^+, K^+, Na^+, and Ca^{2+}.

In addition to the three properties mentioned earlier, other physicochemical properties were found to affect the environmental performance of biochar. Rees et al. (2014) observed that the short-term effects of biochar (produced from 80% coniferous and 20% hardwood chips at 450°C) on soil heavy metal mobility are controlled by intraparticle diffusion, and biochar with fine particles (0–0.2 mm) adsorbed significantly more Cu^{2+}, Cd^{2+}, and Ni^{2+} than that with coarse particles (0.2–2 mm). A higher aromaticity of biochar will aid its adsorption of heavy metals through enhanced cation–π interactions. A higher VM content typically indicates more acid surface functional groups of biochar (Mukherjee et al., 2011), which may aid its adsorption of heavy metals through enhanced CEC and complexation capacity.

3.2.2 *Influence of feedstock and production temperature on biochar properties*

Biochar can be produced from a wide range of feedstocks, which typically come from agricultural and industrial wastes (Khan et al., 2013; Lehmann and Environment, 2007). In practice, agricultural wastes such as woods, grasses, trees, and crop residues are most commonly used in producing biochar. Biomass contains cellulose, hemicellulose, and lignin; however, the content of each varies depending on the feedstock used (Jahirul et al., 2012). Wood typically contains more lignin than straw, and leaves may have more hemicellulose than wood and straw.

Biochar properties and its environmental benefits depend on the thermal decomposition of these components during production and their contents. The thermal decompositions of lignin, hemicellulose, and cellulose are different (Jahirul et al., 2012). Hemicellulose is the easiest to be thermally decomposed, which occurs between ~240 and 400°C (Jahirul et al., 2012). The thermal decomposition of cellulose occurs between ~320 and 410°C (Jahirul et al., 2012). Compared with hemicellulose and cellulose, lignin is resistant to thermal decomposition, which hardly decomposes upon heating up to 500°C (Jahirul et al., 2012). Yang et al. (2006) observed that bio-oil was mainly derived from the cellulose component of biomass (at ~ 500°C), whereas biochar mainly came from the lignin. Therefore, biochar typically has an elemental composition close to that of lignin (Axelsson et al., 2012).

Keiluweit et al. (2010) investigated the dynamic molecular structure of plant biomass-derived biochars (a wood and a grass) produced at different temperatures at 100–700°C and used Fourier Transformed Infrared Spectroscopy (FT-IR) to indicate the change of functional groups of biochar. As shown in Fig. 6.3, no significant changes occurred on FT-IR spectra before 300°C. After 300°C, stronger C=O (1700 cm^{-1}) and C=C (1600 cm^{-1}) stretching vibrations were observed, indicating the formation of ketones, anhydrides, esters, carboxylic and aromatic C for grass biochar due to the decomposition of biomass. At 400°C, the intensity of hydrogen-bonded O–H stretching (3500-3200 cm^{-1}) decreased, indicating a dehydration process. Meanwhile, peaks associated with aromatic C (1600, 885, 750 cm^{-1}) became more significant, which suggests a highly developed aromatic structure for both wood and grass biochars. Further heating (400–600°C) resulting in the decomposition of functional groups (associated with water, oxygenated substituents, aliphatic C, and aromatic C=C stretch) as the relevant peaks decreased. The peaks at 885–750 cm^{-1} slightly increased due to that the small aromatic units condense into larger sheet. At 700°C, no significant peaks for wood and grass were observed and both FT-IR spectra were similar to graphite. The difference of the dynamic molecular structure between wood and grass is due to the different thermal decomposition processes of biomass, which contains cellulose, hemicellulose, lignin, and earth elements. In the research of Keiluweit et al. (2010), the thermal decomposition of cellulose and hemicellulose occurs rapidly at relatively low temperatures (230–400°C). In contrast, lignin is more resistant to heating as its decomposition temperature varies from 160 to 900°C (Keiluweit et al., 2010).

A typical thermal decomposition process of wheat straw was shown by Dodson (2011). During heating (400–600°C), the carboxylate on the edge of the aromatic rings decomposed and the small aromatic units condense and form a larger aromatic sheet, during which the aromaticity of biochar increased. Meanwhile, the earth alkaline elements (e.g., K) decomposed from the carboxylate and formed alkaline minerals (e.g., K_2O), which increased the pH of biochar. Therefore, the aromaticity and pH generally increased with increasing HHT. The influence of HHT on CEC is complicated. The increase of HHT at low temperatures (<400°C) will aid the formation of acidic functional groups and therefore increase exchangeable H^+ of biochar. At high temperatures (≥400°C), further increase of HHT will cause the decomposition of acidic functional groups (e.g., carboxylic and phenolic groups) and therefore decrease exchangeable H^+ of biochar. However, the increase of HHT at high temperatures will promote the formation of minerals, which can offer more exchangeable K^+, Na^+, and Ca^{2+}. The influence of HHT on biochar CEC is a combination of these two opposing effects. During heating, the organics in the biomass decompose, resulting in a porous structure of biochar. Higher HHT resulted in a higher degree of decomposition and consequently a more porous structure and higher surface area (Suliman et al., 2016).

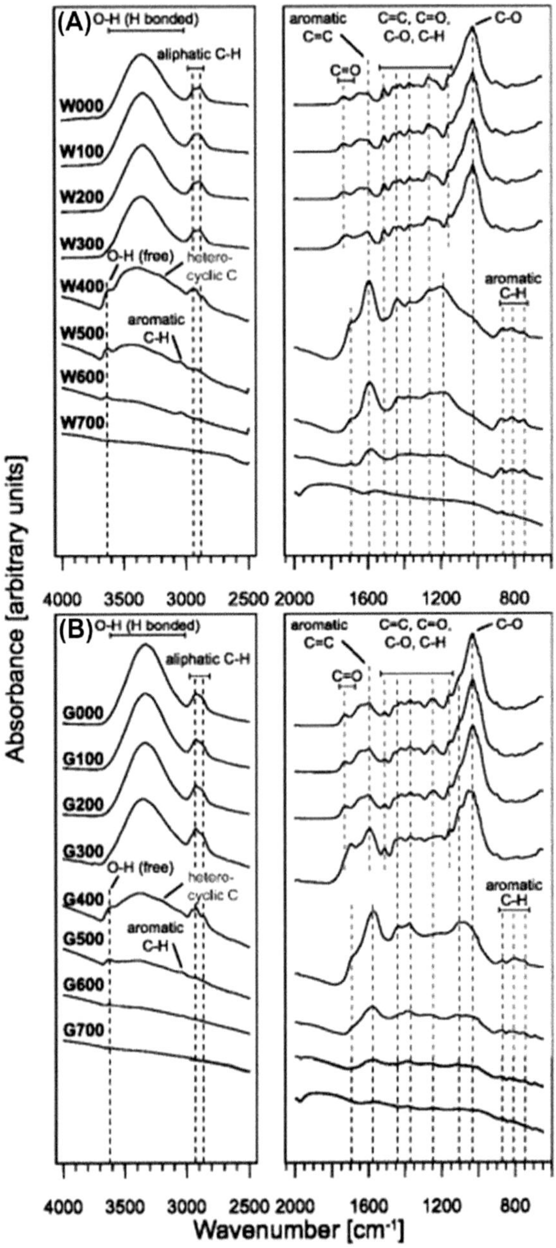

FIGURE 6.3 FT-IR spectra of (A) wood and (B) grass biochar during heating. *Keiluweit, M., Nico, P.S., Johnson, M.G., Kleber, M., 2010. Dynamic molecular structure of plant biomass-derived black carbon (biochar). Environmental Science & Technology 44(4), 1247–1253.*

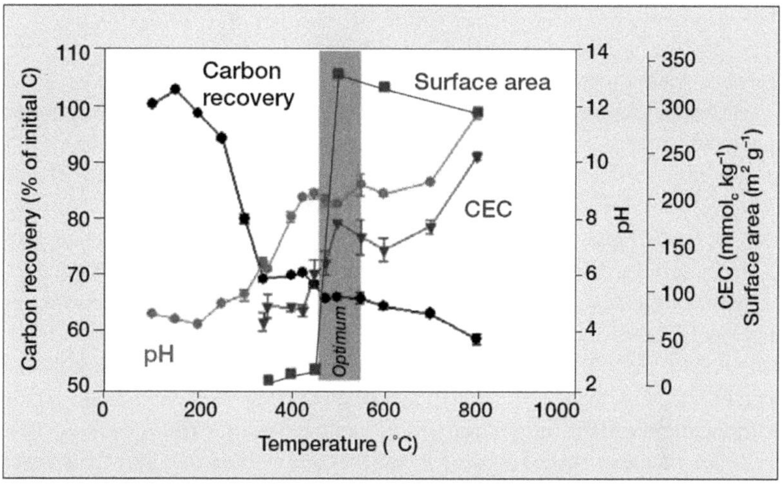

FIGURE 6.4 Influence of highest heating temperature on a wood biochar property. *Lehmann, J., 2007. Bio-energy in the black. Frontiers in Ecology and the Environment 5(7), 381–387.*

Lehmann (2007a) investigated the influence of HHT on biochar properties for one type of wood (Fig. 6.4). The pH of this biochar increased with HHT within 100–800°C. The CEC reached a local maximum at around 500°C in the range of 300–600°C and increased in 600–800°C, which is very likely the outcome of the two different effects mentioned earlier. The surface area of the biochar reached an optimum at around 500°C. The slight decrease of surface area with increasing temperature from 500 to 800°C was likely due to the macropores being destroyed and the micropores being formed during heating. Therefore, although the total surface area of the biochar may have increased, the surface area of macropores decreased. Lehmann (2007b) used the Brunauer–Emmett–Teller (BET) N_2 method, which can only detect the surface area of macropores, to determine the surface area of the biochar and therefore observed a slight decrease of surface area of the biochar at high temperatures. In general, the BET surface area of biochar has a positively linear relationship with its micropore volume (Lehmann and Joseph, 2015). Lehmann (2007a) suggested an optimal temperature of around 500°C after considering the surface area, CEC, pH, carbon recovery, and energy cost during production.

In contrast, Tag et al. (2016) investigated the influence of HHT on the properties of four biochars derived from poultry litter, vine pruning, orange pomace, and algal biomass and observed that increasing HHT increased the pH while decreased the CEC of the biochars, with surface area (N_2 method) hardly changing at 250–600°C. Ronsse et al. (2013) found that increasing HHT increased the surface area (N_2) and pH of a woody biochar at 300–750°C.

Shen et al. (2019a) also observed that the surface area (N_2) and pH of rice straw biochar increased with HHT at 300–700°C.

There seems to be a consensus that increasing HHT will increase biochar pH. However, the influence of HHT on biochar CEC and surface area varies among different studies and different feedstocks (Shen et al., 2019a). Higher surface area, pH, and CEC will generally aid the immobilization of heavy metals for biochar. The selection of feedstock and HHT may aim to obtain biochars with higher surface area, pH, and CEC in order for a better performance in soil remediation. However, some other factors are also needed to be considered. Considering the costs in transportation, biomass from local traditional plants may be the best choice for biochar production, such as hardwood in the UK (Thurkettle, 1997), wheat straw and rice straw in China (Bian et al., 2014; Qin et al., 2013), and palm in Malaysia (Som et al., 2013) In addition, higher HHT may result in biochar with high pH; however, higher HHT will increase the energy cost and decrease the solid yield for biochar production.

4. Interactions between biochar and contaminants

The interactions between biochar and contaminants lay the foundation of the application of sustainable biochar technology in soil remediation. This section introduces the interactions between biochar and heavy metals and organic contaminants. The interactions between biochar and heavy metals include physical adsorption, cation exchange, cation–π interactions, surface precipitation, and surface complexation. The interactions between biochar and organic contaminants include partitioning, pore filling, π–π interactions, hydrophobic effects, and hydrogen bonding.

4.1 Interactions between biochar and heavy metals

4.1.1 Physical adsorption

Heavy metals can be adsorbed to biochar's surface by diffusional movement into biochar pores. This is defined as physical adsorption here, which results from the physical interactions between heavy metals and biochar, such as van der Waals' forces. No chemical bonds are formed during this process. Surface area is an important factor controlling physical adsorption, so biochar favors physical adsorption due to its high surface area and pore volumes (Inyang et al., 2015). Physical adsorption represents a weak interaction between heavy metals and biochar and is reversible. It may not be effective in reducing heavy metal mobility and bioavailability in soil as the physically bonded heavy metals are soluble and bioavailable (Filgueiras et al., 2002). However, it may be effective in waste water treatment as the adsorbed heavy metals can be easily desorbed from biochar after the treatment, and therefore the biochar can be repetitively used. Animal bone biochars were suggested to adsorb Cd^{2+},

Cu^{2+}, and Zn^{2+} through physical adsorption (Choy and McKay, 2005); however, most existing literature found that biochar adsorbs heavy metals mainly through chemical processes (Inyang et al., 2015; Shen et al., 2017a, 2017b, 2018e). Physical adsorption usually takes longer than chemisorption (Inyang et al., 2015). The rest four mechanisms discussed next belong to chemisorption.

4.1.2 Cation exchange

Biochar can exchange its own cations (e.g., H^+, K^+, Na^+, and Ca^{2+}) with heavy metals and therefore form a chemical bond to retain the heavy metals on its surface. The exchangeable cations of biochar come from two fractions: H^+ from its acidic functional groups such as carboxyl and phenolic groups; and alkaline minerals (e.g., K_2CO_3/K_2O) in biochar, which is formed during production.

The heavy metals adsorbed on biochar through cation exchange are regarded as a readily bioavailable fraction, which can be easily uptaken by plants (Filgueiras et al., 2002). They may also desorb from biochar and pose environmental risks when pH significantly changes. Therefore, the adsorption of heavy metals on biochar through cation exchange may not be suitable for immobilizing heavy metals in contaminated soils in the long term. However, as the exchangeable heavy metals on biochar are relatively easily desorbed by changing pH or compulsive exchange (Gillman and Sumpter, 1986), this adsorption mechanism has the potential to be applied in water treatment due to reusability, as with physical adsorption.

4.1.3 Cation−π interactions

Heavy metals can be adsorbed on the aromatic structure of biochar electrostatically through cation−π interactions. In a benzene molecule, two δ dipole moments of a C atom are aligned "tail-to-tail" above and below the nucleus and therefore create uneven charge distributions between the δ π−electro clouds above and below the δ^+ ring face (Fig. 6.5A). The electron-rich π clouds can act as a π-donor due to the high density of electrons within this region (Fig. 6.5B), which can therefore electrostatically adsorb positively charged heavy metals. More condensed rings (i.e., a larger aromatic lattice) will result in stronger π-donors (Fig. 6.5C) and consequently a stronger adsorption of heavy metals through cation−π interactions. Some O-containing functional groups on the edge of aromatic ring will draw electrons away from π clouds, resulting in the π cloud becoming δ^+ due to the low electron density, which becomes a π-acceptor instead of a π-donor and can therefore no longer adsorb cations (Fig. 6.5C). The proposed cation−π interaction energies vary from several to approximately 30 kJ/mol (Keiluweit and Kleber, 2009). Increasing HTT during production generally results in a biochar with a more

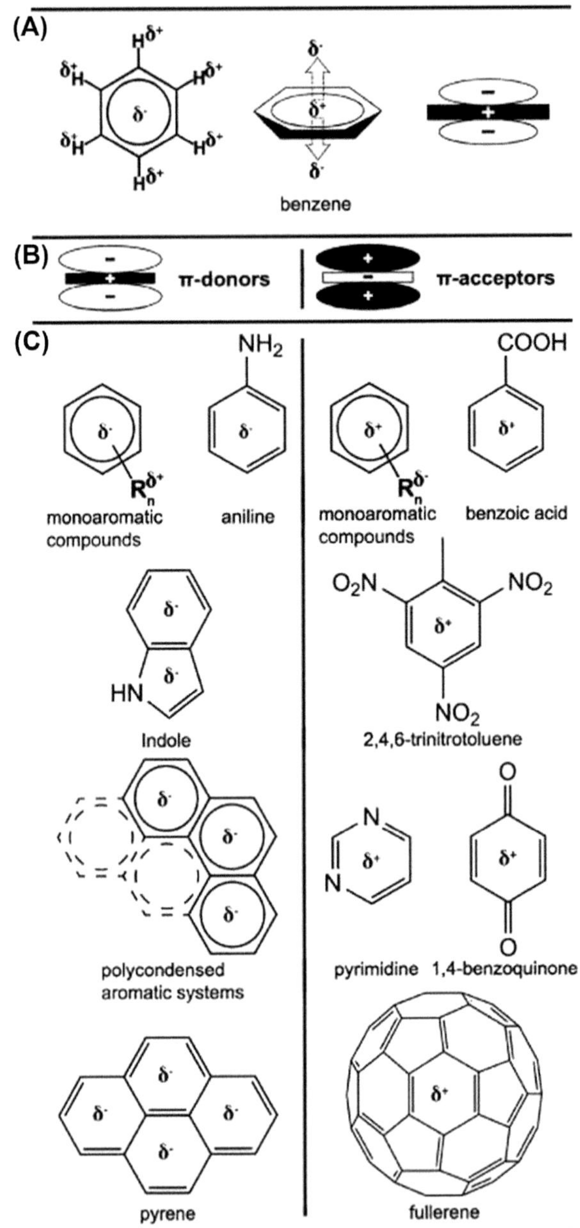

FIGURE 6.5 Aromatic π-donors and -acceptors. (A) Schematic representation of the quadrupole moment of benzene. (B) Quadrupoles of aromatic π-donors and -acceptors. (C) Effects of substituents on electron density in monoaromatic π-donor and -acceptor systems. Effects of heteroatoms on electron density in aromatic π-donor and -acceptor systems (center). Effects of additional fused rings on electron density of polyaromatic π-donor and -acceptor systems (below) (Keiluweit and Kleber, 2009).

condensed aromatic structure (Keiluweit et al., 2010) and therefore will aid its adsorption of heavy metals through cation−π interaction.

The heavy metals adsorbed on biochar through cation−π interaction is regarded as potentially bioavailable fraction (Filgueiras et al., 2002). They are generally stable however may be mobilized and pose environmental risks when surrounding pH significantly decreases. The point of zero charge (pH_{pzc}) can be used as an indicator of the state of this fraction. pH_{pzc} represents the pH value at which the sorbent's surface is neutrally charged (Violante et al., 2010). The sorbent's surface is negatively charged and can electrostatically adsorb heavy metals (e.g., cation−π) when the pH is higher than pH_{pzc}. In contrast, when the pH decreases to lower than pH_{pzc}, the adsorbed heavy metals will dissociate from biochar surface and pose hazards to the environment.

4.1.4 Surface precipitation

The pH of biochar is typically high due to the formation of alkaline minerals during production. These minerals can create an alkaline environment on the biochar's surface that will aid the precipitation or coprecipitation of heavy metals. In addition, the compounds such as CO_3^{2-} and PO_4^{3-} from biochar will also aid the precipitation or coprecipitation of heavy metals (Cao et al., 2011; Yang et al., 2014). For a certain pH, metals with intermediate ionization potentials (2.5−9.5) (e.g., Ni^{2+}, Zn^{2+}, Cu^{2+}, and Pb^{2+}) are more likely to form precipitates on biochar's surface compared with other elements (Inyang et al., 2015). For instance, Pb^{2+} can be precipitated as cerussite ($PbCO_3$) and hydrocerussite $Pb_3(CO_3)_2(OH)_2$ by alternanthera philoxeroides (alligator weed) biochar (Yang et al., 2014). Dairy manure biochar was reported to precipitate Pb^{2+} as β-$Pb_9(PO_4)_6$ in aqueous solutions (Cao et al., 2009) and hydroxypyromorphite $Pb_5(PO_4)_3(OH)$ in soil (Cao et al., 2011).

The mobility of the heavy metals adsorbed on biochar surface through precipitation is affected by the surrounding pH and the type of precipitates formed. For instance, cerussite ($PbCO_3$) and hydrocerussite $Pb_3(CO_3)_2(OH)_2$ can be dissolved in acetic acid and dilute nitric acid (Noel et al., 2014). Therefore, these fractions of Pb^{2+} are regarded as potentially bioavailable as they may mobilize and pose environmental risks when surrounding pH significantly decreases, which may occur in normal field soil conditions. In contrast, the solubility of $Pb_5(PO_4)_3(OH)$ (Fig. 6.6) is lower and normal pH

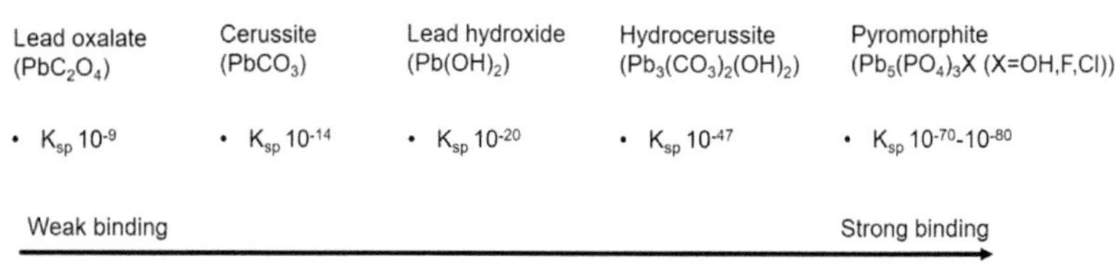

FIGURE 6.6 Solubility of different Pb precipitates.

changes by rainfall and groundwater in common soils cannot lead to the dissociation of Pb^{2+} from $Pb_5(PO_4)_3(OH)$ (Noel et al., 2014), which makes this fraction of Pb^{2+} nonbioavailable and of low environmental risk. Pyromorphite may be the least soluble precipitates for Pb in soil environment. Therefore, the formation of pyromorphite for Pb is always preferred for stabilization-based remediation (e.g., using apatite and manure) (Shen et al., 2018d). It is important that the biochar and the site are both suitably characterized so that the precipitates, and their likely risks, can be predicted and designed for.

4.1.5 Surface complexation

Heavy metals can form complexes with ligands on the biochar's surface. For example, O-containing functional groups such as carboxylic, phenolic, and lactonic groups favor surface complexation with heavy metals (Inyang et al., 2015). Biochars produced at relatively low temperatures contain more oxygen functional groups and therefore have a relatively strong ability to adsorb heavy metals through surface complexation. When applied to contaminated land, biochar can be slowly oxidized and more carboxyl groups can be created in the long term, which will aid the heavy metal complexation by biochar on site. On the other hand, the organic fraction of biochar can be slowly decomposed or destroyed by soil microbes (Balogh et al., 2011; Uchimiya et al., 2012b), which may decrease its ability to form complexes and immobilize heavy metals on site. Therefore, the O-containing functional groups on the aromatic fraction will likely have the greatest long-term influence on immobilization through complexation. The heavy metals adsorbed on biochar through surface complexation are generally regarded as a nonbioavailable fraction, the environmental risk of which is lower compared with the fractions bonded through physical adsorption, cation exchange, cation$-\pi$ interaction, and weak precipitation (precipitates that can be dissolved in HOAc/NaOAc (pH 5)).

4.2 Interactions between biochar and organic contaminants

Biochar has a strong adsorption capacity toward organic contaminants due to its unique pore structure and surface chemical properties. In general, the adsorption capacity and bond strength of biochar toward organic contaminants are significantly higher than nonpyrogenic material or soils (Zhang et al., 2013b). The sorption of organic contaminants to biochar is typically dependent on the surface area and pore size distribution (Chen and Chen, 2009), while for heavy metals, sorption is also dependent on CEC (Shen et al., 2015), though other properties such as biochar polarity and functionality, ash content and composition, and pH will also have an influence. Sorption to biochar may be one to three orders of magnitude higher than that of typical soils and the original feedstock material (Yaning et al., 2003; Zheng et al., 2013). Many papers have reported biochar as having a high sorption capacity for a range of polar and nonpolar organic compounds.

The sorption of organic contaminants to biochar is controlled by multiple processes occurring simultaneously, as demonstrated in Fig. 6.7. Partitioning describes the distribution of a contaminant molecule between the aqueous and solid phases and contributes to total sorption. Pore filling is a process through which adsorbate molecules are transported to sorption sites. Partitioning is not technically a form of adsorption—it is therefore necessary to use the term *sorption* to refer to the combined effects of adsorption and partitioning. The three main adsorption mechanisms are π−π electron donor−acceptor (π−π EDA) interactions, hydrophobic effects (or condensation), and hydrogen bonding. Together, these five processes have been consistently identified in the literature, although to varying degrees among different studies. Further, the relevance of these processes to actual remediation outcomes has received very little attention. This section will therefore determine the extent to which each

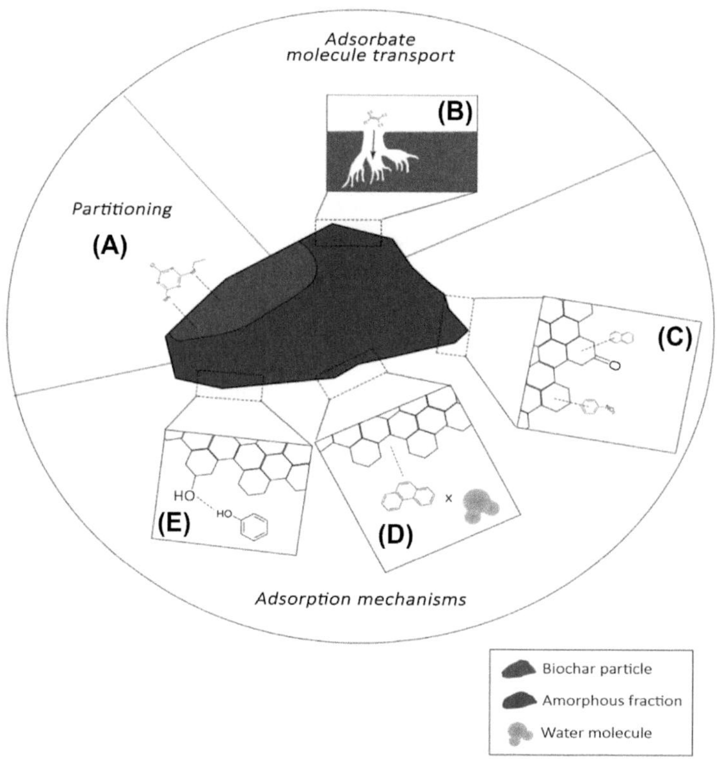

FIGURE 6.7 Schematic of the sorption processes of organic contaminants to biochar; (A) partitioning to the amorphous phase; (B) pore diffusion; (C) π−π EDA interactions; (D) hydrophobic effects; (E) hydrogen bonding.

of these mechanisms occurs in the sorption of organic contaminants to biochar and subsequently determine their role in remediation by considering their responses to biochar aging in the soil profile.

4.2.1 Partitioning

Partitioning refers to the sorption of organic compounds to the amorphous, organic phase of biochar (Chen and Chen, 2009; Chiou et al., 2015; Zhang et al., 2011; Zheng et al., 2013). It is therefore the dominant form of sorption to low-temperature biochars, where the amorphous phase exists in high proportions, and to noncarbonized materials (Xing et al., 1997). Partitioning typically occurs linearly with respect to aqueous concentration, in contrast to adsorption, which is highly nonlinear (Chen et al., 2008, 2012b; Xing et al., 1997; Zhang et al., 2011). A number of studies have reported increasing nonlinearity of isotherms as the HHT increases for a range of biochar feedstocks and contaminants (Chen and Chen, 2009; Zhang et al., 2011; Zheng et al., 2013). This increase in nonlinearity is correlated with the removal of the amorphous phase and an increasing prevalence of adsorption over partitioning at increased pyrolysis temperatures.

The extent of partitioning may be measured using a partitioning coefficient, K_P. K_P describes the linear relationship between the amount of an adsorbate partitioned to a solid phase and its aqueous concentration (Jung et al., 2013). The partitioning component can be mathematically separated from the isotherm using methods such as the dual-mode sorption model (Huang and Chen, 2010), as shown in Fig. 6.8 for the sorption of nitrobenzene to a straw ash charcoal. The dual-mode sorption model assumes that there is negligible increase in adsorption at high concentrations, and therefore the slope of the isotherm at high concentrations represents K_P. The dual-mode sorption model therefore allows for the determination of separate partitioning and adsorption isotherms.

In the context of soil remediation, the significance of partitioning will depend on its contribution to sorption at environmentally relevant concentrations, its reversibility, and the long-term stability of the amorphous phase. These three factors are examined in the following paragraphs and suggest that partitioning will have a relatively small contribution to the remediation of organic contaminants using biochar.

Partitioning is unlikely to be present in significant quantities at realistic environmental concentrations due to its high linearity. The high linearity means that it contributes relatively little to total sorption at low concentrations, though its contribution increases at higher concentrations. However, at realistic soil conditions, partitioning is likely to have very little contribution to the total sorption. For example, the soil guideline value (SGV) for naphthalene in the UK is 7−1400 mg/kg, depending on the land use. Even at the highest of these limits, partitioning is likely to be an insignificant contributor to the overall

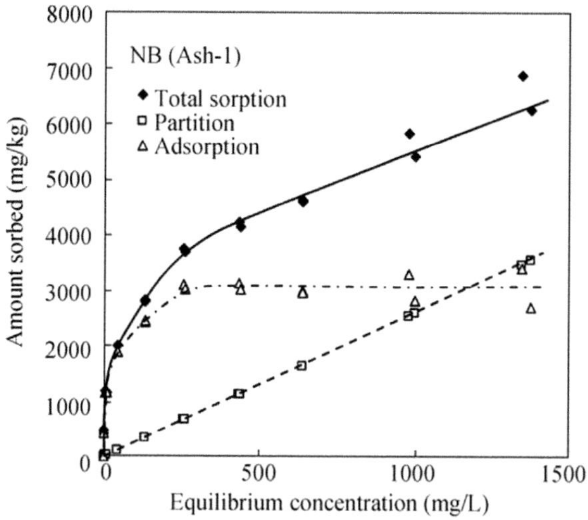

FIGURE 6.8 Separation of partitioning and adsorption for the sorption of nitrobenzene to a straw ash charcoal. *Replicated from Huang, W., Chen, B., 2010. Interaction mechanisms of organic contaminants with burned straw ash charcoal. Journal of Environmental Sciences 22(10), 1586–1594.*

sorption. Huang and Chen (Huang and Chen, 2010) showed that at a sorbed concentration of 1400 mg/kg of naphthalene, partitioning comprises <1% of the total sorption to a straw ash charcoal. Other papers have reported much higher adsorption capacities of naphthalene including 80,000 mg/kg (Chen and Chen, 2009; Singh et al., 2017), 800 mg/kg (Chen et al. 2008, 2012a) and 187,800 mg/kg (Nguyen et al., 2007) (based on Freundlich isotherm at Ce = Csaturation), to which partitioning is likely to be even less significant at low concentrations. This relationship is also true for other organic contaminants (Alvarez and Fuiman, 2005; Hayes et al., 2002; Knight et al., 2013). The high nonlinearity of other previous isotherms also suggests that partitioning is unlikely to play a significant role at low concentrations. Therefore, partitioning is expected to represent a small component of total sorption of organic contaminants in soil and will therefore play a relatively small role.

The influence of partitioning will also depend on the reversibility of contaminants immobilized through this mechanism. Partitioning describes the phase distribution of an organic molecule between the aqueous and solid phases and can therefore be reversed under conditions where the aqueous phase reduces (e.g., rainfall, groundwater flow). Sorption–desorption tests at a range of initial concentrations may provide further insights into this relationship for a given biochar, as different contributions of partitioning and adsorption will likely result in different amounts of desorption. Similarly, sorption–desorption tests with and without an adsorption inhibitor, such as p-nitrophenol (Chiou et al., 2015), may also provide insights into the reversibility when partitioning is the dominant form of sorption. However, these results provide evidence that the partitioning mechanism may be highly reversible in soil.

Finally, contaminants immobilized through partitioning are likely to be released when the amorphous component of biochar degrades in soil, resulting in little long-term immobilization through partitioning. The net result is that partitioning will not have a significant effect in the long term and may actually result in the release of some contaminants, although this has not been thoroughly studied.

The combined results suggest that partitioning is likely to play a negligible role in the remediation of contaminated soil using biochar. Partitioning may contribute significantly to the overall sorption while providing little actual environmental benefit and therefore should be separated from an isotherm using a tool such as the dual-mode sorption model (Huang and Chen, 2010; Xing et al., 1997; Zhang et al., 2011) for a more accurate representation of the capacity of a biochar for soil remediation.

4.2.2 Pore filling

Pore filling has been thoroughly investigated for the sorption of organic compounds to activated carbon (Ismadji and Bhatia, 2001; Pelekani and Snoeyink, 2000), and this knowledge has been used previously to infer insights into how it behaves in biochar. Its relative role in determining total adsorption is a product of the contaminant molecule's size and the physical structure of the biochar, while its long-term role will depend largely on pore accessibility in the presence of SOM. Pore filling is a significant process contributing to the sorption of many organic contaminants. It is often described as one of the most significant sorption processes for biochar (Lattao et al., 2014; Zhang et al., 2013a).

Pore filling typically results in highly nonlinear sorption profiles, suggesting that it will play a significant role in sorption at low concentrations (Chen et al., 2012c; Zhang et al., 2011) due to the combined result of two processes. First, pore filling occurs primarily into micropores (<20 nm) whose sizes are comparable to those of contaminant molecules, leading to interactions (such as hydrophobic effects, hydrogen bonding, or $\pi-\pi$ EDA interactions) with both walls of the pore (Nguyen et al., 2007). At low contaminant concentrations, the high-energy micropores are preferentially filled by contaminant molecules (Ji et al., 2011; Nguyen et al., 2007; Zheng et al., 2013). Second, at higher contaminant concentrations, these micropores become saturated with adsorbed compounds, after which no further adsorption can occur at these sites. Together, these two processes result in high adsorption at low concentrations but little further adsorption as the aqueous concentration increases, and therefore nonlinear adsorption.

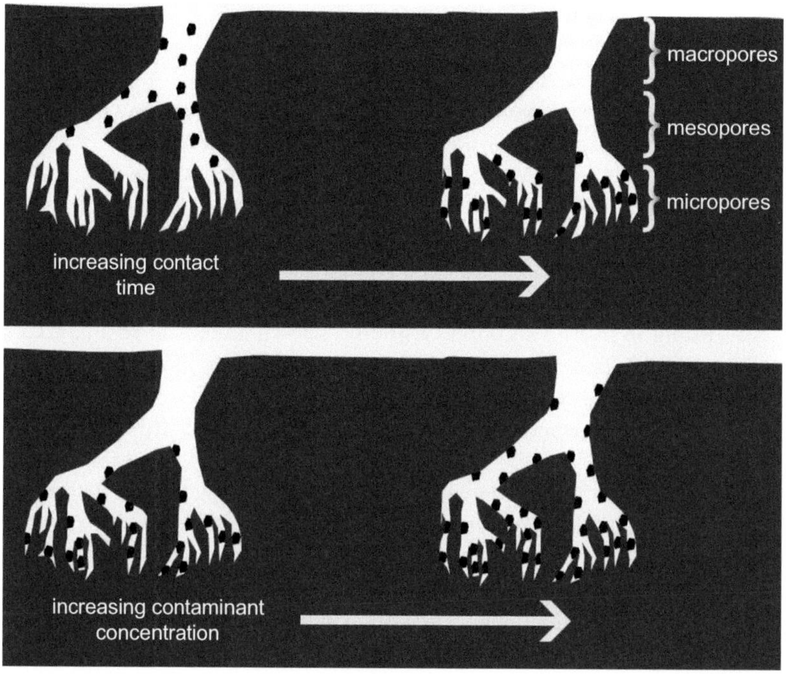

FIGURE 6.9 Contaminant molecule distributions within pores as (A) contact time and (B) concentration increase in the context of transport via pore diffusion.

The configuration of the pore network can also influence the effectiveness of pore filling (Fig. 6.9). Though micropores contribute the most to total sorption and to the total number of sorption sites (Manya, 2012), a number of studies have also demonstrated the significance of mesopores to the transport of adsorbate molecules. For example, Lattao et al. (2014) demonstrated that the best predictor of sorption for benzene, naphthalene, and 1,4-dinitrobenzene to a range of wood chars was a weighted sum of the volumes of micropores and mesopores. Ahmad et al. (2013) suggested that trichloroethylene is held within the mesopores of pine needle biochar, then adsorbed into micropores through capillary forces. Therefore, while adsorption generally correlates with micropore volume, an overall well-developed pore network is also significant in order to make these adsorption sites accessible.

The extent of pore filling may be reduced for large contaminant molecules due to the size-exclusion effect. The size-exclusion effect (also called steric hindrance) describes the process of bulky adsorbate molecules such as tetracycline (Ji et al., 2011) and humic acids (Kasozi et al., 2010) being unable to access a fraction of the micropores. The steric effect has been reported to play a significant role in controlling the sorption of 1,2-dichlorobenzene, 1,2,4-trichlorobenzene, 1,4-dichlorobenzene (Nguyen et al., 2007), and polycyclic aromatic hydrocarbons (PAHs) (Sun and Zhou, 2008). Similar effects have also been proposed for the sorption of heavy metals to biochar (Park et al., 2016). Together, these studies show that pore filling occurs more appreciably for smaller contaminant molecules than larger ones and therefore may only be significant for contaminants with small molecular diameters.

Pore blockages by SOM are likely to significantly reduce the extent of pore filling in the soil profile. Long-chain SOM molecules can adsorb to the surface of biochar and prevent access to microporous regions (Ren et al., 2016; Wang et al., 2010; Zhang et al., 2010). The significant decreases in sorption due to competition with SOM highlight the significance of surface area and the role of pore filling in overall sorption. While no known studies have analyzed the pore size distribution after incubation with SOM, it is likely that access to micro- and mesopores will be significantly reduced. These studies together suggest that the influence of pore filling may be significantly reduced in the soil profile.

Overall, the literature to date suggests that pore filling may be a significant process influencing the immobilization of contaminants in soil, due to its high binding strength and low reversibility. However, its effects in influencing remediation may be limited due to the significant reduction in the adsorption of molecules greater than approximately 1 nm, its potentially slow diffusion rate into micropores, reliance on the biochar having both micropores and mesopores, and these pores potentially being blocked by SOM. In particular, high-temperature, highly microporous biochars are likely to adsorb contaminants via pore filling, although theses biochars may result in some leaching in the short term due to the slow diffusion of contaminants into the micropores.

4.2.3 π−π interactions

π−π interactions refer to chemical interactions between the π electrons on the aromatic surface of the carbonized component of biochar and the π electrons of an aromatic adsorbate compound (Keiluweit and Kleber, 2009; Lattao et al., 2014). These are discussed here as EDA interactions, in which one of the aromatic rings acts as an electron donor while the other acts as an electron acceptor, as illustrated in Fig. 6.10. Biochar can act as a π-electron donor and a π-electron acceptor simultaneously (Zhu and Pignatello, 2005), while aromatic organic contaminants are typically either an acceptor or a donor. Few papers discuss in depth the role of π−π EDA interactions in remediation using biochar. Its overall contribution will depend on the chemical properties of the adsorbent and adsorbate, while its long-term stability will be influenced by the oxidation of biochar in soil during aging.

The capacity of an adsorbate to donate or accept π electrons depends on the compound's molecular structure. A compound is a π-acceptor if it has an electron-withdrawing group, such as−OH, -NH2, =O, or −Cl, which has a higher electronegativity than C and H and therefore attracts electrons away

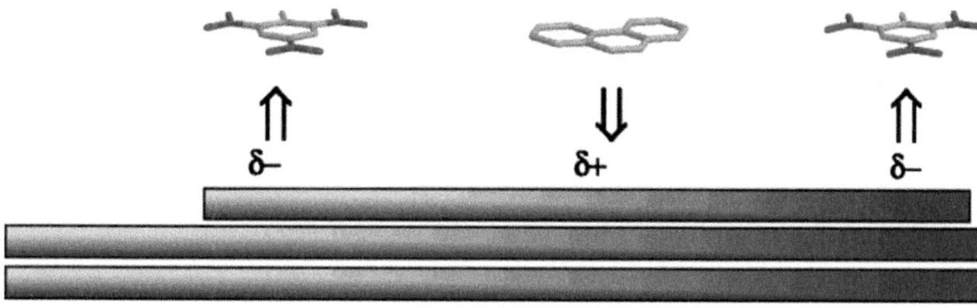

FIGURE 6.10 Schematic of π–π EDA interactions between a graphitic sheet of charcoal, acting as a π-donor with π-acceptor molecules (e.g., trinitrotoluene) and as a π-acceptor with π-donor molecules (e.g., phenanthrene) (Zhu and Pignatello, 2005).

from the aromatic ring (Zhu et al., 2005), as shown in Fig. 6.11. This attraction away from the aromatic ring results in the π-electron groups, which are perpendicular to the aromatic ring, having a slight electron deficiency and therefore a capacity to accept electrons from other sources. Conversely, a compound is a π-donor if no electron-withdrawing group is present, such as in PAHs, resulting in a π-electron-rich aromatic ring and therefore the capacity to donate electrons to nearby molecules (Zhu and Pignatello, 2005). The strength of a π-donor is proportional to the number of carbon atoms in the aromatic system, while the strength of a π-acceptor is proportional to the electronegativity of the electron-withdrawing group (Zhu et al., 2005). It has also been suggested that this effect is more prominent for adsorbate molecules, which are more planar 179. Therefore, π–π EDA effects are expected to be the strongest for large PAHs and planar aromatic compounds containing highly electronegative groups (McMillan, 2018).

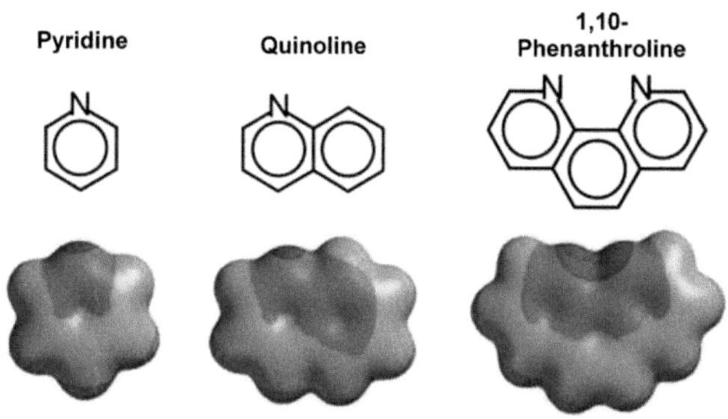

FIGURE 6.11 Electron density of three π-acceptor compounds. Red (dark gray in print version) represents electron-rich regions while green/blue (gray in print version) represents electron-poor regions. *Adapted from Xiao, F., Pignatello, J.J., 2015. π(+)-π interactions between (hetero)aromatic amine cations and the graphitic surfaces of pyrogenic carbonaceous materials. Environmental Science & Technology 49(2), 906*

For large PAHs, such as benzo[a]pyrene, some of the pores may be inaccessible due to the size-exclusion effect, suggesting that strong influences of pore filling and π-acceptance are unlikely to occur simultaneously. This suggests that adsorption through π−π EDA forces in the mesopore and macropore regions may dominate the overall adsorption. In contrast, small compounds with nitrogen or chlorine groups are expected to adsorb strongly to biochar due to the combined effects of pore filling and π−π EDA interactions.

The significance of π−π EDA interactions for the sorption of a given contaminant will also depend on the aromaticity and polarity of a biochar (Hale et al., 2016). Similarly, high-temperature biochars will promote the adsorbance of π-electron acceptors due to the increasing degree of aromatization and reduction in polarity, resulting in biochar acting as a π-donor (Chen and Chen, 2009; Chen et al., 2008; Zheng et al., 2013). A highly aromatic biochar will therefore promote π−π EDA interactions, as the planar surface promotes sorption of π-acceptor compounds while edge-functional groups promote the sorption of π-donor compounds. The significance of π−π forces in determining adsorption to biochar has been noted several times (Sun and Zhou, 2008; Xin et al., 2014). Zhu et al. (2005) demonstrated that increased adsorption of several aromatic compounds could only be attributed to π−π interactions and proposed an electron−donor acceptor mechanism. π−π EDA interactions are therefore likely to exist in significant quantities for the sorption of aromatic compounds to various biochars.

The long-term role of π−π EDA interactions will be influenced by surface changes of biochar over time. The surface of the biochar is expected to be oxidized over time due to biological and chemical processes in the soil, resulting in an increased presence of O-groups. These O-groups draw electrons away from the carbonized sheets, resulting in them being π-electron deficient and increasing the capacity of biochar to act as a π-acceptor, resulting in increased sorption of π-donor molecules and decreased sorption of π-acceptor molecules. This therefore suggests that the long-term role of π−π EDA interactions will be dependent on the properties of the adsorbate molecule; however, little data is available on the reversibility of π−π EDA interactions to biochar in realistic environmental conditions.

Recent research has also shown that π−π interactions can occur as $\pi^+-\pi$ interactions, which occur between biochar and cationic aromatic compounds such as sulfamethazine 191 or aromatic amine cations (Xiao et al., 2015), including many herbicides. In these cases, π−π EDA interactions occur simultaneously with a charge-assisted hydrogen bond (CAHB), which is particularly prevalent at low pH values where zwitterionic compounds exist largely as cations. The presence of $\pi^+-\pi$ EDA interactions suggests that π−π interactions for cationic or zwitterionic compounds are unlikely to desorb if the environmental pH levels decrease, although this should be confirmed experimentally in each case.

Overall, the literature to date suggests that π−π EDA interactions are likely to be significant for the sorption of aromatic compounds to biochar. In particular, highly carbonized and aromatic biochars will promote these interactions, while oxygen groups will increase the capacity of biochar to act as a π-acceptor. π−π EDA interactions will be particularly effective for polar contaminants to high-temperature biochars and for large PAHs to low-temperature biochars, though interactions with the pore filling mechanism should also be considered. However, few studies have considered the role of biochar π−π EDA interactions in the long-term soil profile, although the effects of aging suggest that the sorption of π-acceptor molecules may decrease while the sorption of π-donor molecules will increase.

4.2.4 Hydrophobic effects

Hydrophobic effects refer to the nonspecific adsorption of a hydrophobic organic compound (HOC) to the hydrophobic surface of an adsorbate (Inyang and Dickenson, 2015). It is primarily driven by the adsorbate molecule's aversion to dissolving in water (Moreno-Castilla, 2004), although is ultimately a function of the properties of both the adsorbent and adsorbate (Liao et al., 2012; Wang et al., 2010); both adsorbate molecules and biochar may exhibit hydrophobicity. Its relevance in soil remediation will depend on its total contribution to sorption to given biochar, while its long-term significance will depend on the effects of biochar oxidation in soil.

The hydrophobicity of biochar is commonly described as one of its properties most favorable to adsorption, and many studies have noted the significance of hydrophobic effects in immobilizing organic contaminants. Typically, high-temperature biochars have low polarities (i.e., low O:C ratios) and therefore demonstrate greater hydrophobic effects. Liao et al. (2012) investigated the sorption of two dyes to a series of bamboo charcoals before and after microwave modification and found that the sorption of each dye correlated with biochar hydrophobicity, suggesting that hydrophobic effects may be significant in adsorbing these compounds. Hydrophobic effects are most prevalent to biochars with high aromaticities (i.e., low H:C ratios) and low polarities.

The properties of the adsorbate molecule will also determine the extent of hydrophobic effects. In particular, hydrophobic effects are expected to be significant for HOCs such as PAHs (Sun and Zhou, 2008; Yang et al., 2016a). Hydrophobic effects are therefore most relevant for contaminants with low K_{ow} values, which typically applies to nonpolar organic molecules.

The long-term role of hydrophobic effects will also depend on changes to the properties of biochar over time; in particular, surface oxidation of biochar in soil may reduce their influence. Many studies (Cheng et al., 2006; Ren et al., 2016) have noted that biochar is likely to be oxidized in the soil over the long term, which will increase polarity. While this author is not aware of any studies that have investigated the presence of hydrophobic effects after biochar aging, many have commented on the role of polarity in determining hydrophobicity of a biochar, as previously noted. Therefore, it is expected that the increased oxidation will reduce the presence of hydrophobic effects.

4.2.5 Hydrogen bonding

Hydrogen bonding occurs between an O, N, or F atom on one molecule and an H atom, which is bonded to an O, N, or F atom on the other molecule (Fang et al., 2014; Qiu et al., 2009; Ren et al., 2016). Several studies have suggested that hydrogen bonding may occur between a contaminant molecule and phenol or carboxyl groups in the biochar, resulting in contaminants being tightly bound. The role of hydrogen bonding in soil remediation will depend on its overall contribution to sorption, the influence of water molecules, and the effects of biochar oxidation in soil.

Hydrogen bonding is likely to be a significant mechanism controlling the sorption of organic contaminants to biochar, particularly for polar organic compounds (Ahmed et al., 2017; Jung et al., 2013; Qiu et al., 2009). Hydrogen bonding is also likely to tightly bound contaminants, stronger than hydrophobic effects in particular (Lima et al., 2010). Hydrogen bonding is significant in a range of cases and therefore should be considered when designing biochar for environmental remediation.

The influence of competition with water may be significant in determining the influence of hydrogen bonding in practice. Water molecules can also form hydrogen bonds with the polar surface of biochar and therefore may compete for sorption sites with contaminant molecules (McMillan, 2018). While this effect is difficult to appropriately characterize, care should be taken when working with polar contaminants to ensure that this effect is captured if attempting to optimize highly polar adsorbents for their removal from the soil.

The influence of hydrogen bonding is likely to increase over time in the soil due to increased surface oxidation of the biochar. As previously noted, the biochar surface is likely to be oxidized in the soil, resulting in an increased number of O-groups and therefore a greater capacity for hydrogen bonding. However, no known studies have discussed the role of aging on the long-term significance of hydrogen bonding, though many have noted increased hydrogen bonding after artificially increasing the oxygen content. For example, Franz et al. (2000) noted increased sorption of a number of aromatic compounds to an activated carbon after increasing oxygen content, particularly in cyclohexane where the influence of water is removed. On the contrary, many papers have also noted decreased sorption after increasing the oxygen content, even for molecules capable of forming hydrogen bonds, due to interactions with water molecules. The long-term significance of hydrogen bonding will therefore be dependent on the relative sorption properties of the given contaminant and water molecules and may result in either increased or decreased adsorption. The role of hydrogen bonding in determining long-term remediation outcomes is therefore likely to be significant. Due to its high contribution to overall sorption and strong binding strength, it is likely to effectively immobilize polar organic contaminants. However, the increased abundance of oxygen-containing functional groups on the surface of biochar over time will promote hydrogen bonding with both contaminant and water molecules, and it is not trivial to determine the net effect of these for a given contaminant.

5. Applications of biochar in soil remediation

Currently, the main application of biochar in soil remediation is in situ amendment. Although the application of biochar in permeable reactive barrier (PRB) and phytoremediation has been limitedly investigated, it is suggested that biochar has huge potential in the application in these two areas. Therefore, this section introduces the concept and application of biochar in in situ amendment, PRB, and phytoremediation.

5.1 Application of biochar in in situ immobilization

5.1.1 In situ amendment concept

Soil remediation aims to control, modify, or destroy pollutant linkages that present unacceptable risks. Based on the source–pathway–receptor model (Lynch and Moffat, 2005), soil remediation can be achieved by (1) isolating or removing the contaminated source, (2) interrupting or managing the pathway, or (3) protecting the receptor or modifying its exposure. Most remediation options are focused on (1) and (2).

Soil amendment uses biochar, activated carbon, compost, clay minerals, and other adsorptive materials to immobilize heavy metals on site and improve soil conditions for greening (Beesley and Marmiroli, 2011b; Beesley et al., 2010; Karami et al., 2011; Shen et al., 2018c; Smith, 2009; Uchimiya et al., 2011). Compared with activated carbon, biochar has great advantages in the cost of production. Compared with compost, which can be gradually degraded by soil microbes in the field, biochar is recalcitrant and can stay stable for hundreds to thousands of years (Lehmann, 2007a). The adsorption capacity of biochar is typically stronger than that of clay minerals (Bradl, 2004; Inyang et al., 2015). Therefore, biochar has the potential to be applied in soil remediation using the amendment method.

Organic and/or inorganic pollutants typically exist in contaminated land. Unlike organic contaminants that can be completely removed from the contaminated source by thermal, chemical, or microbial methods, heavy metals cannot be degraded in soil and will persist for a long time (Zhang et al., 2013b). To immobilize the heavy metals by adding soil amendments is an effective way and typically is used to reduce their leachability and bioavailability, which is a pathway-based treatment (Wang et al., 2014; Zhang et al., 2013b). Biochar has a high capability to immobilize heavy metals due to its properties being favorable toward cation exchange, electrostatic adsorption, surface complexation, and precipitation. Therefore, biochar has the potential to be applied as a soil amendment to contaminated soils to immobilize heavy metals and reduce their environmental risks.

In addition to immobilizing heavy metals by biochar itself through adsorption, as biochar is generally alkaline, the released alkaline earth elements and increased soil pH (if biochar pH is higher than soil pH) after biochar treatment can aid the bonding of heavy metals to soil matrix (liming effect) (Beesley et al., 2011; Jiang et al., 2012). Moreover, biochar can enhance the soil fertility and resilience and therefore aid the greening and revegetation on site, which is important for the long-term stability of the contaminated land. These benefits suggest that biochar may be a promising remedial material for contaminated land treatment. Therefore, a range of laboratory and field trials have been carried out in recent years applying different biochars in soil remediation (Table 6.1). It can be seen from Table 6.1 that biochar was effective in immobilizing heavy metals and reducing their environmental risks in different contaminated sites.

As shown in Table 6.1, soil texture and contaminants vary among different sites. The effectiveness of biochar in immobilizing heavy metals on site is also dependent on factors such as soil texture, site conditions, and local climate. In addition, the long-term effectiveness of the immobilization of heavy metals by biochar is always a safety concern for contaminated sites, which has not been well studied. In order to aid greening and revegetation on site, compost was applied with biochar to supply the nutrients for plant growth in previous studies. Therefore, it is important to understand the effects of both biochar and compost, as well as the mixture on heavy metal immobilization in contaminated land.

5.1.2 Performance and influencing factors of biochar in in situ amendment

5.1.2.1 Influence of soil texture

Soil comprises of gravel, sand, silt, and clay particles. In laboratory analysis, the soil is typically sieved to less than 2 mm particle size. As summarized in Table 6.1, the majority of studies to date have investigated the performance of biochar in sand-dominated soils and generally observed a positive performance of biochar in immobilizing heavy metals. In comparison, Puga et al. (2015a) applied a sugar cane straw biochar to a soil containing 42.7% clay and observed that the application reduced both diethylenetriaminepentaacetic acid (DTPA) extractable Cd^{2+}, Pb^{2+}, and Zn^{2+} in soil and the uptake of Cd^{2+}, Pb^{2+}, and Zn^{2+} by two plants, with higher biochar dosage resulting in a better performance at 0%−5% w/w. Puga et al. (2015b) used a sugar cane straw

TABLE 6.1 Application of biochar for heavy-metal-associated soil remediation and its impact on heavy metal mobility.

Biochar feedstock + compost if applied	Pyrolysis temperature (°C)	Soil		Heavy metals and contamination level (mg/kg)	Treatment duration (day (d), week (w), month (m), or year (y))	Impact on heavy metal mobility	Reference
		pH	Texture				
Wheat straw	700	9.14	10.2% silt, 89.7% clay	Pb^{2+} (3410)	14 d	Little impact	Shen et al. (2018b)
Broadleaf hardwood	600	4.78	65% silt, 35% clay	Pb^{2+} (501000)	28 d	No impact	Shen et al. (2016b)
Wheat straw	550	6.07	N.G.	Cd^{2+} (22.65), Pb^{2+} (621.31)	5 y	Continuous immobilization	Cui et al. (2016)
Corn straw, oak wood	300	4.43	26.2% clay	Cd^{2+} (2.2), Cu^{2+} (452)	3 y	Immobilization	Li et al. (2016)
Soybean stover, pine needle	300, 700	5.8	83.4% sand, 8.6% silt, 8.0% clay	Pb^{2+} (1945)	90 d	Immobilization	Ahmad et al. (2016)
Chicken manure	500	5.76	Sandy loam	Cu^{2+} (500)	6.5 m	Immobilization	Meier et al. (2015)
Oat hull	300						

Continued

TABLE 6.1 Application of biochar for heavy-metal-associated soil remediation and its impact on heavy metal mobility.—cont'd

Biochar feedstock + compost if applied	Pyrolysis temperature (°C)	Soil		Heavy metals and contamination level (mg/kg)	Treatment duration (day (d), week (w), month (m), or year (y))	Impact on heavy metal mobility	Reference
		pH	Texture				
Sugar cane straw	700	6.1	Technosol (33.6% sand, 23.7% silt, 42.7% clay)	Cd^{2+} (6.4), Pb^{2+} (3688), Zn^{2+} (2027)	150 d	Immobilization	Puga et al. (2015c)
Bamboo	750	5.7	Sandy loam (51% sand, 38.9% silt, 9.6% clay)	Cd^{2+} (1.4), Cu^{2+} (693), Pb^{2+} (527), Zn^{2+} (1471)	1 y	Immobilization	Yang et al. (2016b)
Rice straw	500						
Sugarcane straw	400, 500, 600, 700	5.8	Technosol (22.4% sand, 15.0% silt, 62.6% clay)	Cd^{2+} (1.23), Pb^{2+} (134.0), Zn^{2+} (55.8)	120 d	Immobilization	Puga et al. (2015c)
Corn cob + rice straw compost	450	7.62	30.27% sand, 45.54% silt, 24.19% clay	Cu^{2+}, Zn^{2+}, Cd^{2+}, Pb^{2+}	60 d	Immobilization except Cu^{2+} with compost-only amendment	Zeng et al. (2015)

Wine lees	600	6.25	Paddy soil	$Cr^{3+/6+}$ (114), Ni^{2+} (26), Cu^{2+} (39), Zn^{2+} (1579), Cd^{2+} (13), Pb^{2+} (87)	4 w	Immobilization	Zhu et al. (2015)
Wheat straw	550	5.36	Paddy soil (42% −59% sand, 10% −20% clay)	Cd^{2+} (5), Pb^{2+} (100)	3 y	Continuous immobilization	Bian et al. (2014)
80% coniferous and 20% hardwood	450	5.9 −8.1	Redoxic cambisols (23.5%−33.4% sand, 46.9% −60.1% silt, 16.4% −19.7% clay)	Cd^{2+} (17.6 −18.6), Pb^{2+} (1080-1120), Cu^{2+} (45.1 −90.8), Ni^{2+} (16.0−22.5), Zn^{2+} (1380-3170)	Not given	Immobilization	Rees et al. (2014)
Orchard prune + olive mill waste compost	500	3.6	N.G.	Cd^{2+} (74), Pb^{2+} (4170), Cu^{2+} (2940), Zn^{2+} (13,200)	4 w	Immobilization	Beesley et al. (2014)
Dairy manure	350	7.2	90.2% sand + silt, 9.8% clay	Cd^{2+} (60), Pb^{2+} (60), Zn^{2+} (165)	56 d	Immobilization	Liang et al. (2014)
Wood	450	6.8	Sandy clay loam	As (11.6), Cd^{2+} (0.7), Pb^{2+} (24.9), Cu^{2+} (15.1), Ni^{2+}	3 y	No significant influence	Lucchini et al. (2014a)

Continued

TABLE 6.1 Application of biochar for heavy-metal-associated soil remediation and its impact on heavy metal mobility.—cont'd

Biochar feedstock + compost if applied	Pyrolysis temperature (°C)	Soil		Heavy metals and contamination level (mg/kg)	Treatment duration (day (d), week (w), month (m), or year (y))	Impact on heavy metal mobility	Reference
		pH	Texture				
				(12.1), Zn^{2+} (88.1)			
Wood	550	6.02	Sandy clay loam, sandy loam	Cd^{2+} (11.49), Pb^{2+} (16.71), Cu^{2+} (11.49), Ni^{2+} (9.18), Zn^{2+} (37.56)	45 d	Immobilization	Lucchini et al. (2014b)
Wheat straw	350, 550	5.01 −6.31	Redoxic cambisols (23.5%–33.4% sand, 46.9% −60.1% silt, 16.4% −19.7% clay)	Cd^{2+} (0.16 −21.84)	2-3 y	Immobilization	Bian et al. (2013)
Miscanthus straw	600	6.57	64% sand, 24% silt, 12% clay	Cd^{2+} (24.0), Pb^{2+} (3110), Zn^{2+} (2980)	56 d	Immobilization	Houben et al. (2013)
Cottonseed hulls	800	6.27	Sandy soil	Pb^{2+} (14,847), Cu^{2+} (1419), Zn^{2+} (4.3), Sb^{2+} (33)	Not given	Immobilization	Uchimiya et al. (2012a)

Dairy manure	450	6.89 −7.09	87.7%–92.3% sand, 7.7%–12.3% silt + clay	Pb^{2+} (502070,400)	210 d	Immobilization	Cao et al. (2011)
Willow stems	400	5.7	N.G.	Cd^{2+} (0.41), Pb^{2+} (12.1), Cu^{2+} (1.22), Zn^{2+} (65.6)	Not given	No significant influence	Trakal et al. (2011)
Hardwood + green waste compost	400	6.2	83% sand, 16% silt, <1% clay	As (96), Cd^{2+} (119), Zn^{2+} (249)	8 w	Immobilization	Beesley and Marmiroli (2011b)
British oak + green waste compost	N.G.	5.4	N.G.	Pb^{2+} (20,703), Cu^{2+} (603)	3 m	Immobilization	Karami et al. (2011)
Activated wood	550	6.38	Sandy soil (87.7% sand, 4.8% silt, 7.5% clay)	Cd^{2+}, Pb^{2+}, Cu^{2+}, Zn^{2+}	10 w	Immobilization	Namgay et al. (2010)

N.G. = not given, sand: 2–0.05 mm, silt: 0.05–0.002 mm, clay: 0.002–0 mm.

biochar to amend a contaminated soil containing 62.6% clay and observed a reduction of the heavy metals both in soil solution and in plant. On the contrary, Shen et al. (2016b) applied a hardwood biochar to a Pb-contaminated kaolin (35% clay) and observed that the biochar had no impact on the immobilization of Pb during a 28-day laboratory incubation study. Shen et al. (2018b) used a wheat straw biochar to stabilize a soil washing residue (89.2% clay). They found that biochar only slightly reduced the mobility of Pb in the soil during a 14-day laboratory incubation study, and this effect diminished after accelerated aging.

The gravel, sand, and silt particles in soil are relatively chemically inert (Bergaya and Lagaly, 2013). In contrast, the layered structure of silicate and gibbsite clay minerals results in active chemical properties such as relatively high CEC values and surface areas for clay particles (Bergaya and Lagaly, 2013). Therefore, the clay minerals have the ability to retain heavy metals on their surfaces through cation exchange and reduce their leaching to the environment under field conditions (Alloway, 2012). However, the retained heavy metals on clay particles, which are bioavailable fractions, carry the risk of being uptaken by the plants grown on clay soil, bringing further hazards to human beings (Alloway, 2012; Bergaya and Lagaly, 2013).

Therefore, it is important to understand the performance of biochar on clay-rich soils contaminated with heavy metals and the mechanisms involved. The comparison of biochar performance in sand-rich and clay-rich soils is important for the potential field applications of biochar in heavy-metal-associated soil remediation. If soil texture plays a significant role in biochar immobilization of heavy metals, then large-scale biochar applications must take soil texture into account, and laboratory treatablity studies using an appropriate soil texture are crucial prior to field application.

5.1.2.2 Short-term and long-term performances

Biochar has shown to immobilize heavy metals in soils and reduce their bioavailability in a range of short-term (less than 3 years) studies (Table 6.1). One exception comes from Trakal et al. (2011), who used willow stem biochar to treat a Zn^{2+}-contaminated soil; however, no significant effects were found. The failure of biochar to immobilize Zn^{2+} was attributed to the insufficient dosage of biochar (1% and 2% w/w).

There are limited long-term (3 years or more) studies investigating the impact of biochar on heavy metal mobility in contaminated soils (Table 6.1). Bian et al. (2014) conducted a 3-year study using wheat straw biochar to treat agricultural land contaminated with Cd^{2+} and Pb^{2+}. During the 3 years after the treatment, Cd^{2+} and Pb^{2+} concentrations were continuously reduced in both $CaCl_2$ and DTPA extractions as well as in crop biomass. The chemical and SEM/EDX analysis indicated that the heavy metals were likely bound to the mineral phases (Al, Fe, and P) of biochar and through cation exchange.

This study was followed by Cui et al. (2016), who investigated the mobility and speciation of the Cd^{2+} and Pb^{2+} on the same site 5 years after amendment with biochar. The findings of Cui et al. (2016) is in line with Bian et al. (2014) that biochar continuously immobilized Cd^{2+} and Pb^{2+} in the soil throughout the 5 years after the treatment. Shen et al. (2016a) carried out a field trial using a hardwood biochar to amend a multicontaminated industrial land. They observed that the leachability and mobility of Ni^{2+} and Zn^{2+} significantly reduced 3 years after the treatment. In contrast, Lucchini et al. (2014c) conducted a 3-year study using wood biochar to treat lightly contaminated agricultural land and observed no significant decreases in total metal concentrations in the soil or plant matter, though biochar did not necessarily have a negative effect.

In the long term, the aging of biochar due to surface oxidation and microbial decomposition may affect the immobilization of heavy metals by biochar and therefore increase the uncertainty of its effectiveness (Hale et al., 2011). Especially in field conditions, factors such as rainfall washing, groundwater flow, wet–dry cycles, and disturbance by plant roots and earthworms may affect the long-term biochar performance. Considering the very limited long-term field trials and the conflicting findings between Cui et al. (2016) and Lucchini et al. (2014c), the long-term performance of biochar in immobilizing heavy metals in soils and the mechanisms involved remain unclear and need further investigation. Therefore, it is crucial to continue such trials to verify the feasibility of biochar application in practical remediation projects and give validation to further applications. Artificially accelerated aging in laboratory is suggested to be a useful method to predict the long-term effectiveness of biochar performance in heavy metal immobilization in soil (Shen et al. 2018a, 2018b); however, it remains a challenge in its correlation with real field aging.

5.1.2.3 Influence of biochar and compost

As shown in Table 6.1, compost can be applied with biochar in practical applications for soil remediation. Beesley et al. (2014) amended a soil contaminated by mining waste with orchard pruning biochar (dosage of 10% v/v), compost (a mixture of olive husk and cow manure) (dosage of 10% v/v), and biochar + compost (5% + 5% v/v). They observed a significant reduction of Cd^{2+}, Cu^{2+}, Pb^{2+}, and Zn^{2+} in pore water 1 and 4 weeks after all treatments compared with the control soil. Biochar treatment showed the highest reduction for all heavy metals, and the addition of compost inhibited the immobilization of heavy metals by biochar. Zeng et al. (2015) applied corn cob biochar (5% w/w), rice straw compost (5% w/w), and biochar + compost (2.5% + 2.5% w/w) to a field contaminated soil. After 60 days, all treatments reduced the mobility of Zn^{2+}, Cd^{2+}, and Pb^{2+} in soil, while compost did not significantly affect the immobilization of heavy metals by biochar. Meanwhile, they observed that the mobility of Cu^{2+} increased 60 days after the individual compost treatment and attributed this to the mobilization of Cu^{2+} by humic acid from compost. Compared with individual compost application, biochar and biochar + compost decreased the mobility of Cu^{2+}, suggesting that biochar can offset the mobilization of compost and immobilize Cu^{2+} in the soil. Beesley et al. (2010) observed an increase of Cu^{2+} in pore water 56 days after amending a multielement-contaminated site with hardwood biochar (50% v/v), green waste compost (50% v/v), and biochar + compost (100% + 100% v/v). In addition, although Cd^{2+} and Zn^{2+} concentrations were reduced, and compost did not significantly affect the performance of biochar. They attributed the increase of Cu^{2+} to the increased dissolved organic carbon and pH after the three treatments. Karami et al. (2011) applied biochar (derived from British oak, ash, sycamore, and birch) (20% v/v), green waste compost (30% v/v), and biochar + compost (20% + 30% v/v) to a former copper mine site. They observed a reduction of Cu^{2+} and Pb^{2+} in pore water 3 months after each treatment. No significant differences were found among the three treatments, and the influence of compost on biochar performance was insignificant.

Therefore, the performance of biochar in immobilizing heavy metals in contaminated land in the presence of compost is conflicting among existing literature and remains unclear. Compost may either inhibit or hardly affect the immobilization of heavy metals by biochar. The performance of biochar + compost or compost alone in immobilizing heavy metals in soils is also related to the metal type. Cu^{2+} seems different from the other heavy metals investigated earlier. Very limited studies have been carried out to investigate the specific mechanism involved. Rodríguez-Vila et al. (2015) conducted a greenhouse experiment applying oak wood biochar + compost (rabbit and horse manure, pruning residues, seaweeds, and fruit waste) and vegetation to a mine soil. The sequential extraction on the soil samples 90 days after the treatment showed that the treatment decreased the mobile fractions of Cu^{2+}, Ni^{2+}, and Pb^{2+}, but increased the Zn^{2+} concentration in the amended soils. This study did not isolate the immobilization of biochar from biochar + compost mixture, making it impossible to understand the mechanistic influence of compost on biochar in terms of heavy metal immobilization. The complex mechanisms of how biochar + compost will affect the mobility and speciation of heavy metals in contaminated soils and how compost will affect the performance of biochar need further investigation.

5.2 Application of biochar in permeable reactive barrier

5.2.1 Permeable reactive barrier concept

Groundwater contamination has become a serious matter of concern due to the increasing mining, smelting, and industrial activities. In addition, the groundwater contamination always occurs along with soil contamination. In the past few decades, a broad range of ex situ and in situ remediation technologies have been developed to treat the contaminated groundwater, such as pump-and-treat (P&T), chemical reduction/oxidation, and PRBs.

The concept of the PRB was first developed by the University of Waterloo in the early 1990s. It was defined as "an emplacement of reactive materials in the subsurface designed to intercept a contaminant plume, provide a flow path through the reactive media, and transform the contaminant(s) into environmentally acceptable forms to attain remediation concentration goals downgradient of the barrier" (Powell et al., 1998). PRB is an effective in situ technology for aquifer and groundwater remediation. Barriers filled with reactive materials are constructed across the flow path of a contaminant plume (Gibert et al., 2011). As the fluid moves across PRB, contaminants are degraded or trapped by reactive materials through physical, chemical, and/or biological processes. This technique has progressed rapidly over the past decade from laboratory bench-scale studies to full-scale implementation. There are two general types of PRB designs for field applications: funnel-and-gate and continuous gate.

PRBs have several advantages over conventional P&T methods for groundwater remediation: (1) PRBs can degrade or immobilize contaminants in situ without bringing them up to the surface. Therefore, it is economic with no need to install ground facilities for storage, treatment, transport, or disposal other than monitoring wells. Also, there are no potential cross-media contamination, no loss of groundwater, and no exposure of contaminants; (2) PRBs do not require continuous input of energy, because the natural gradient of groundwater flow will force contaminants to flow through the reactive zone. Only periodic replacement or rejuvenation of the reaction medium might be required after its reactive capacity has been exhausted or it has been clogged by precipitants and/or microorganisms. However, the drastically reduced operating costs offset the higher construction costs of PRBs, resulting in an overall reduction in the life-cycle costs; (3) the site can be in use while the treatment is occurring; (4) technical and regulatory problems related to ultimate discharge requirements of effluent from P&T systems are avoided with the PRB technology.

However, PRBs also have their limitations like other techniques. The primary potential failure mechanisms include mineral precipitation, biofouling, gas clogging, competition between contaminants or between contaminants and other dissolved species, desorption/remobilization, temperature effects, and geochemical conditions (Bone, 2012). In addition, the site characterization for PRBs is very complex and there is a need to consider a wide range of hydrogeological, biochemical, and geochemical variables of the site. In addition, PRBs require long-term monitoring by installing monitoring wells.

5.2.2 Application of biochar as a reactive medium

The reactive medium is the key component of PRBs, and its selection is dependent on the nature of the target contaminants and hydrogeological site conditions. Various materials, such as zerovalent iron (ZVI), activated carbon, lime, and zeolites, have been used individually or as a mixture as the reactive media to decontaminate groundwater by adsorption, reduction, precipitation, or degradation (Faisal et al., 2015; Henderson and Demond, 2007; Thiruvenkatachari et al., 2008; Zhang et al., 2018; Zhou et al., 2014a).

Biochar has been investigated widely and is regarded as a sustainable sorbent for the remediation of water and soil contaminated with heavy metals as well as organics (Shen et al., 2015). However, research on the use of biochar as the reactive material in PRBs is limited. Generally, biochar can be applied as a reactive medium by two approaches.

Firstly, biochar can be used to modify other materials, such as ZVI, and produce more sustainable reactive materials for remediation. ZVI is the earliest and most prevalent material to clean up both metals and organic contaminants, such as halogenated hydrocarbons and chromate, by chemical reduction and further precipitation or degradation (Chen et al., 2011; Henderson and Demond, 2007). However, the formation of surface coating or changes of site conditions may lead to decreased reactivity of ZVI, resulting in reduction of the longevity of ZVI, clogging, and reduced permeability of the PRBs. In addition, current ZVI production methods carry a significant climate impact. Therefore, the exploration of the sustainable production methods of ZVI is a necessity for remediation. It has been demonstrated that ZVI composite with biochar exhibits better remediation kinetics than conventional ZVI. The production of biochar–ZVI composite by the pyrolysis of biorenewable biomass and iron-containing sources is therefore potentially a greener and more sustainable pathway for environmental remediation. In one-step pyrolysis, biomass is pretreated with a Fe-rich source, such as $FeCl_3$ and $FeSO_4$, and then pyrolyzed. It should be noted that the composition of different biomaterials can affect the reduction of Fe and the quality of ZVI in biochar. Therefore, the selection of suitable feedstock is a key step when exploring the pyrolysis-based production of biochar–ZVI composite. In addition, the calcination temperature and the types of iron source also need be further studied. Various methods have been developed to synthesize biochar–ZVI composite. To name a few, Zhou et al. (2014b) used chitosan as an "organic glue" to attach fine ZVI particles onto biochar to make biochar-supported ZVI; Shang et al. (2017) obtained biochar first and then dissolved it in the $FeSO_4$ solution, followed by the dropwise addition of $NaBH_4$ solution. It is generally believed that the adsorption and reduction of pollutants significantly increased when biochar is mixed chemically with ZVI and composites are formed. However, from the microstructural observations, some pores of biochar may be blocked by the ZVI particles.

Secondly, biochar or modified biochar can be applied as reactive medium alone or mixed physically with other materials. The applications of biochar in field-scale PRBs are very limited to date. A field-scale PRB (6 $\times$1 $\times$3 m) filled with rice husk biochar was installed in 2013 in Korea, and reduced concentrations of ammonium-nitrogen ($NH4^+$-N) leaching from animal carcass disposal site was observed (Yoon et al., 2017). However, increasing studies exist on the laboratory-scale batch and column tests to simulate the PRBs using biochars in PRBs for the removal of various pollutants, such as U(VI), ammonium-nitrogen ($NH4^+$-N) (Kim et al., 2016), chlorobenzene (Zhang et al., 2017), and heavy metals (Beiyuan et al., 2017) in groundwater.

In spite of the limited studies, biochar holds huge potential in the application as the reactive material in PRBs. Biochar is a cost-effective adsorbent for metals, organic pollutants, and antibiotics due to its high cation-exchange capacity, oxygen-containing functional groups, porous and stable structure, and large surface area. With the introduction of biochar, it became possible to use inexpensive and environmentally friendly filling materials to enhance the effectiveness of PRBs. The application of biochar in PRBs requires a good understanding of the adsorption characteristics of contaminants onto biochars, the physical and chemical reactions involved in the complex wall system, the technical feasibility, etc. In order to popularize the application of biochar in PRBs, some suggestions are proposed as follows: Firstly, more sustainable and economical production and modification methods of biochar and engineered/designer biochar need to be further explored. It would be promising if biochar and engineered/designer biochar can be produced with high adsorption capacity targeting certain contaminant(s). Secondly, after the production and characterization of biochar, a series of batch and column tests are strongly suggested to be conducted to explore their detailed adsorption dynamics toward different contaminants and provide more accurate information for design. The literature studies about the batch adsorption tests on biochar are extensive, providing guidance for their application in wastewater treatment. However, the lack of further column tests, let alone the permeability and technical evaluation tests, limits the development of biochar in PRBs for groundwater remediation. Thirdly, the combination of biochar with other materials, such as ZVI, zeolites, clays, lime, and sand, is a trend and is suggested to be more efficient than single materials in many cases. More research on combined materials needs to be conducted to optimize the proportions of materials for a higher removal efficiency of targeting contaminants.

5.3 Application of biochar in phytoremediation

5.3.1 Phytoremediation concept

Phytoremediation is an in situ treatment technique by the use of vegetation to remove contaminants from the environment or transform them into less harmful forms for remediation of low-to-medium contamination (Mench et al., 2010). It includes phytoextraction, phytofiltration, phytostabilization, phytovolatilization, phytodegradation, rhizodegradation, and phytodesalination (Paz-Ferreiro et al., 2014). It is a relatively new technique and was defined by Raskin et al. (1997) as the use of green plants to remove pollutants from the environment or to render them harmless. This technology has been proven to be effective and inexpensive in a number of full-scale and pilot-scale studies, and heavy metals (Lu et al., 2015; Paz-Ferreiro et al., 2014), inorganic salts (Zeng et al., 2013), organic and nutrient contaminants (Mrozik and Piotrowska-Seget, 2010; Schnoor et al., 1995), etc., can be treated in the rhizosphere and by root uptake. It is shown to cause minimal environmental disturbance, and the introduction of plants can help to prevent erosion or contaminant leaching. Phytoremediation also has some limitations, such as the relatively slow decontamination, limited tree species suitable for remediation, potential risks of the contaminated plants, and the application depth (typically <5 m).

5.3.2 Combination of biochar application and phytoremediation

An innovative in situ remediation technique has been recently developed via the application of biochars combined with fast-growing tree species in contaminated sites (Fellet et al., 2014; Yoon et al., 2017). In an animal carcass burial site in Korea, biochar PRBs in combination with phytoremediation were used to decontaminate groundwater and surface water in 2013. Biochar PRBs targeted ammonium-N, and in situ phytoremediation achieved by the plantation of *Populus euramericana* reduced electrical conductivity and ammonium-N by contaminant uptake (Yoon et al., 2017). However, very few researchers studied the combination of biochar application and phytoremediation for contaminated soils.

As biochar has a strong adsorption capacity toward various pollutants as mentioned earlier, its amendment to contaminated soils can reduce the bioavailability and leachability of heavy metals and further reduce phytotoxicity of polluted soils. This is potentially valuable in the context of contaminated but nutrient-rich soils, such as where sewage sludge has been applied (Karami et al., 2011). In addition, biochar has also been demonstrated to be able to improve soil quality and plant growth due to increased water and nutrient retention, improved biological properties and CEC, alteration of soil microbial community, effects on nutrient cycling and turnover, and improvements in soil pH (Ahmad et al., 2012). These properties make biochar a suitable option to amend soil in order to enhance phytoremediation. Therefore, the combination of biochar with phytoremediation may show synergistic effects and can promote the revegetation and restoration of degraded or infertile lands. The application of biochar may rapidly immobilize the heavy metals in the soil and reduce the risks of the contaminated site in the mean time. In the longer term, biochar can aid the growth of accumulator and therefore aid the extraction of the heavy metals from the soil.

As increasing attention is paid to the bioavailability and mobility of contaminants and the consequent effects on the environment rather than just their total concentrations in soil (Beesley and Marmiroli, 2011a), the combination of biochar with phytoremediation for soil remediation is a promising technique and may attract more attention in the next decades. After phytoremediation, the potential environmental risks of the plants used in phytoremediation should also be considered to avoid secondary pollution. For example, the extraction plants can be made to biochar by pyrolysis for reutilization as an adsorbent (Gong et al., 2017, 2018), but the potential risks of heavy metals in the resulting biochar remain unknown and warrant further investigation. A recent study by Huang et al. (2018) produced biochar from the extraction plants and found that the increase of pyrolysis temperature decreased the soluble/exchangeable heavy metal fraction and alleviated the bioavailability of heavy metals. When the temperature was over 550°C, the adsorbed Zn, Pb, and Cd were turned into oxide forms and concentrated in biochar with more stable forms exhibiting lower potential ecological risk.

6. Conclusions and implications for sustainable remediation

Research into biochar for environmental applications began in approximately 2006 (Lehmann et al., 2006) and has become increasingly popular. A number of studies focusing on the application of biochar for sustainable remediation of contaminated soil have been conducted. However, a range of research gaps have not been filled.

Biochar has been applied to contaminated soils under both laboratory and field conditions. However, the majority of studies to date have investigated the performance of biochar in sand-rich soils. The influence of soil texture on biochar's performance in heavy metal immobilization in in situ amendment remains unclear. Although biochar has generally been successful in immobilizing heavy metals, the long-term effectiveness and the mechanisms of the immobilization remain unclear. A combination of biochar and compost is recommended in soil remediation for its benefits in revegetation and heavy metal immobilization. However, the contradictory performances among different studies using biochar + compost suggest that there are a number of influences that will affect the practical performance. The mechanisms involved in this have not been well investigated.

In order to popularize the application of biochar in PRBs, some suggestions are proposed as follows: Firstly, more sustainable and economical production and modification methods of biochar and biochar related materials need to be further explored. Secondly, after the production and characterization of biochar, a series of batch and column tests are suggested to be conducted to explore their detailed adsorption features and adsorption dynamics toward different typical contaminants and provide more accurate information for design. Thirdly, more research on combined materials needs to be conducted to optimize the proportions of materials with targeting contaminants.

The combination of biochar application and phytoremediation is suggested, and more research is needed in the future. The application of biochar is shown to be able to improve the soil fertility and quality and aid the extraction of metals. Therefore, the combination of biochar and phytoremediation may show synergistic effects and can promote the revegetation and restoration of degraded or infertile lands. This technique holds huge potential due to the fact that it will not only reduce the total concentrations of contaminants but also reduce their bioavailability and mobility. Further studies are suggested to: (1) explore the interactions between biochar and plants; (2) match different plant species with biochars to obtain the optimal growth conditions of the plants and removal efficiency toward contaminants; (3) find more suitable plant species, e.g., hyperaccumulator, and explore their accumulation mechanisms; (4) study the transport and fate of contaminants from environmental media to biochar and plant tissues; (5) explore the activities and transport of biochar around the roots; (6) conduct more well-designed and large-scale field tests; (7) evaluate the long-term performance; (8) figure out the treatment of polluted plants, such as the reutilization as biochar feedstocks, and evaluate their potential risks.

References

Ahmad, M., Lee, S.S., Rajapaksha, A.U., Vithanage, M., Zhang, M., Cho, J.S., Lee, S.-E., Ok, Y.S., 2013. Trichloroethylene adsorption by pine needle biochars produced at various pyrolysis temperatures. Bioresource Technology 143, 615–622.

Ahmad, M., Ok, Y.S., Rajapaksha, A.U., Lim, J.E., Kim, B.Y., Ahn, J.H., Lee, Y.H., Al-Wabel, M.I., Lee, S.E., Lee, S.S., 2016. Lead and copper immobilization in a shooting range soil using soybean stover- and pine needle-derived biochars: chemical, microbial and spectroscopic assessments. Journal of Hazardous Materials 301, 179–186.

Ahmad, M., Soo Lee, S., Yang, J.E., Ro, H.M., Han Lee, Y., Sik Ok, Y., 2012. Effects of soil dilution and amendments (mussel shell, cow bone, and biochar) on Pb availability and phytotoxicity in military shooting range soil. Ecotoxicology and Environmental Safety 79, 225–231.

Ahmed, M.B., Zhou, J.L., Ngo, H.H., Guo, W., Johir, M.A.H., Belhaj, D., 2017. Competitive sorption affinity of sulfonamides and chloramphenicol antibiotics toward functionalized biochar for water and wastewater treatment. Bioresource Technology 238, 306–312.

Alloway, B.J., 2012. Heavy Metals in Soils: Trace Metals and Metalloids in Soils and Their Bioavailability. Springer Science & Business Media.

Alvarez, M.D., Fuiman, L.A., 2005. Environmental levels of atrazine and its degradation products impair survival skills and growth of red drum larvae. Aquatic Toxicology 74 (3), 229–241.

Arroyo-Kalin, M., 2017. Amazonian Dark Earths. Archaeological Soil and Sediment Micromorphology, pp. 345–357.

Axelsson, L., Franzén, M., Ostwald, M., Berndes, G., Lakshmi, G., Ravindranath, N., 2012. Perspective: jatropha cultivation in southern India: assessing farmers' experiences. Biofuels Bioprod Biore-fining 6, 246–256.

Balogh, J., Pintér, K., Fóti, S., Cserhalmi, D., Papp, M., Nagy, Z., 2011. Dependence of soil respiration on soil moisture, clay content, soil organic matter, and CO2 uptake in dry grasslands. Soil Biology and Biochemistry 43 (5), 1006–1013.

Beesley, L., Inneh, O.S., Norton, G.J., Moreno-Jimenez, E., Pardo, T., Clemente, R., Dawson, J.J., 2014. Assessing the influence of compost and biochar amendments on the mobility and toxicity of metals and arsenic in a naturally contaminated mine soil. Environmental Pollution 186, 195–202.

Beesley, L., Marmiroli, M., 2011a. The immobilisation and retention of soluble arsenic, cadmium and zinc by biochar. Environmental pollution 159 (2), 474–480.

Beesley, L., Marmiroli, M., 2011b. The immobilisation and retention of soluble arsenic, cadmium and zinc by biochar. Environmental Pollution 159 (2), 474–480.

Beesley, L., Moreno-Jimenez, E., Gomez-Eyles, J.L., 2010. Effects of biochar and greenwaste compost amendments on mobility, bioavailability and toxicity of inorganic and organic contaminants in a multi-element polluted soil. Environmental Pollution 158 (6), 2282–2287.

Beesley, L., Moreno-Jimenez, E., Gomez-Eyles, J.L., Harris, E., Robinson, B., Sizmur, T., 2011. A review of biochars' potential role in the remediation, revegetation and restoration of contaminated soils. Environmental Pollution 159 (12), 3269–3282.

Beiyuan, J., Awad, Y.M., Beckers, F., Tsang, D.C., Ok, Y.S., Rinklebe, J., 2017. Mobility and phytoavailability of as and Pb in a contaminated soil using pine sawdust biochar under systematic change of redox conditions. Chemosphere 178, 110–118.

Bergaya, F., Lagaly, G., 2013. Handbook of Clay Science. Newnes.

Berman, M., Timoshenko, A., 2017. Greening International Institutions. Routledge, pp. 38–54.

Bian, R., Chen, D., Liu, X., Cui, L., Li, L., Pan, G., Xie, D., Zheng, J., Zhang, X., Zheng, J., Chang, A., 2013. Biochar soil amendment as a solution to prevent Cd-tainted rice from China: results from a cross-site field experiment. Ecological Engineering 58, 378–383.

Bian, R., Joseph, S., Cui, L., Pan, G., Li, L., Liu, X., Zhang, A., Rutlidge, H., Wong, S., Chia, C., Marjo, C., Gong, B., Munroe, P., Donne, S., 2014. A three-year experiment confirms continuous immobilization of cadmium and lead in contaminated paddy field with biochar amendment. Journal of Hazardous Materials 272, 121–128.

Bone, B., 2012. Review of UK Guidance on Permeable Reactive Barriers, pp. 611–768.

Bradl, H.B., 2004. Adsorption of heavy metal ions on soils and soils constituents. Journal of Colloid and Interface Science 277 (1), 1–18.

Cao, X., Ma, L., Gao, B., Harris, W., 2009. Dairy-manure derived biochar effectively sorbs lead and atrazine. Environmental Science and Technology 43 (9), 3285–3291.

Cao, X., Ma, L., Liang, Y., Gao, B., Harris, W., 2011. Simultaneous immobilization of lead and atrazine in contaminated soils using dairy-manure biochar. Environmental Science and Technology 45 (11), 4884–4889.

Chen, B., Chen, Z., 2009. Sorption of naphthalene and 1-naphthol by biochars of orange peels with different pyrolytic temperatures. Chemosphere 76 (1), 127–133.

Chen, B., Yuan, M., Qian, L., 2012a. Enhanced bioremediation of PAH-contaminated soil by immobilized bacteria with plant residue and biochar as carriers. Journal of Soils and Sediments 12 (9), 1350–1359.

Chen, B., Zhou, D., Zhu, L., 2008. Transitional adsorption and partition of nonpolar and polar aromatic contaminants by biochars of pine needles with different pyrolytic temperatures. Environmental Science and Technology 42 (14), 5137–5143.

Chen, L., Liu, F., Liu, Y.L., Dong, H.Z., Colberg, P.J.S.J., 2011. Benzene and toluene biodegradation down gradient of a zero-valent iron permeable reactive barrier. Journal of Hazardous Materials 188 (1), 110–115.

Chen, Z., Chen, B., Chiou, C.T., 2012b. Fast and slow rates of naphthalene sorption to biochars produced at different temperatures. Environmental Science and Technology 46 (20), 11104–11111.

Chen, Z., Chen, B., Zhou, D., Chen, W., 2012c. Bisolute sorption and thermodynamic behavior of organic pollutants to biomass-derived biochars at two pyrolytic temperatures. Environmental Science and Technology 46 (22), 12476–12483.

Cheng, C.H., Lehmann, J., Thies, J.E., Burton, S.D., Engelhard, M.H.J.O.G., 2006. Oxidation of black carbon by biotic and abiotic processes. Organic Geochemistry 37 (11), 1477–1488.

Chiou, C.T., Cheng, J., Hung, W.-N., Chen, B., Lin, T.-F., 2015. Resolution of adsorption and partition components of organic compounds on black carbons. Environmental Science and Technology 49 (15), 9116–9123.

Choy, K.K., McKay, G., 2005. Sorption of cadmium, copper, and zinc ions onto bone char using Crank diffusion model. Chemosphere 60 (8), 1141–1150.

Cui, X., Hao, H., Zhang, C., He, Z., Yang, X., 2016. Capacity and mechanisms of ammonium and cadmium sorption on different wetland-plant derived biochars. The Science of the Total Environment 539, 566–575.

Dodson, J., 2011. Wheat Straw Ash and its Use as a Silica Source. University of York.

Edenhofer, O., 2015. Climate Change 2014: Mitigation of Climate Change. Cambridge University Press.

Faisal, A.A.H., Hmood, Z.A.J.D., Treatment, W., 2015. Groundwater protection from cadmium contamination by zeolite permeable reactive barrier. Desalination and Water Treatment 53 (5), 1377–1386.

Fang, Q., Chen, B., Lin, Y., Guan, Y., 2014. Aromatic and hydrophobic surfaces of wood-derived biochar enhance perchlorate adsorption via hydrogen bonding to oxygen-containing organic groups. Environmental Science and Technology 48 (1), 279–288.

Fellet, G., Marmiroli, M., Marchiol, L.J., 2014. Elements uptake by metal accumulator species grown on mine tailings amended with three types of biochar. Science of The Total Environment 468–469, 598–608.

Filgueiras, A.V., Lavilla, I., Bendicho, C., 2002. Chemical sequential extraction for metal partitioning in environmental solid samples. Journal of Environmental Monitoring 4 (6), 823–857.

Franz, M., Arafat, H.A., Pinto, N.G., 2000. Effect of chemical surface heterogeneity on the adsorption mechanism of dissolved aromatics on activated carbon. Carbon 38 (13), 1807–1819.

Gibert, O., Rötting, T., Cortina, J.L., Pablo, J.D., Ayora, C., Carrera, J., Bolzicco, J.J., 2011. In-situ remediation of acid mine drainage using a permeable reactive barrier in Aznalcóllar (Sw Spain). Journal of Hazardous Materials 191 (1), 287–295.

Gillman, G., Sumpter, E., 1986. Modification to the compulsive exchange method for measuring exchange characteristics of soils. Soil Research 24 (1), 61–66.

Glaser, B., Guggenberger, G., Zech, W., Ruivo, M.D.L., 2003. Amazonian Dark Earths. Springer, pp. 141–158.

Glaser, B., Haumaier, L., Guggenberger, G., Zech, W., 2001. The'Terra Preta'phenomenon: a model for sustainable agriculture in the humid tropics. Naturwissenschaften 88 (1), 37–41.

Gong, X., Huang, D., Liu, Y., Zeng, G., Wang, R., Wei, J., Huang, C., Xu, P., Wan, J., Zhang, C., 2018. Pyrolysis and reutilization of plant residues after phytoremediation of heavy metals contaminated sediments: For heavy metals stabilization and dye adsorption. Bioresource Technology 253, 64–71.

Gong, Y.P., Ni, Z.Y., Xiong, Z.Z., Cheng, L.H., Xu, X.H., 2017. Phosphate and ammonium adsorption of the modified biochar based on Phragmites australis after phytoremediation. Environmental Science and Pollution Research 24 (9), 1–10.

Hale, S.E., Arp, H.P.H., Kupryianchyk, D., Cornelissen, G., 2016. A synthesis of parameters related to the binding of neutral organic compounds to charcoal. Chemosphere 144, 65–74.

Hale, S.E., Hanley, K., Lehmann, J., Zimmerman, A., Cornelissen, G., 2011. Effects of chemical, biological, and physical aging as well as soil addition on the sorption of pyrene to activated carbon and biochar. Environmental Science & Technology 45 (24), 10445–10453.

Hayes, T.B., Collins, A., Lee, M., Mendoza, M., Noriega, N., Stuart, A.A., Vonk, A., 2002. Hermaphroditic, demasculinized frogs after exposure to the herbicide atrazine at low ecologically relevant doses. Proceedings of the National Academy of Sciences of the United States of America 99 (8), 5476–5480.

Henderson, A.D., Demond, A.H.J.E.E.S., 2007. Long-term performance of zero-valent iron permeable reactive barriers. A Critical Review 24 (4), 401–423.

Houben, D., Evrard, L., Sonnet, P.J.B., Bioenergy, 2013. Beneficial effects of biochar application to contaminated soils on the bioavailability of Cd, Pb and Zn and the biomass production of rapeseed (Brassica napus L.). Biomass and Bioenergy 57 (11), 196–204.

Huang, H., Yao, W., Li, R., Ali, A., Du, J., Guo, D., Xiao, R., Guo, Z., Zhang, Z., Awasthi, M.K., 2018. Effect of pyrolysis temperature on chemical form, behavior and environmental risk of Zn, Pb and Cd in biochar produced from phytoremediation residue. Bioresource Technology 249, 487–493.

Huang, W., Chen, B., 2010. Interaction mechanisms of organic contaminants with burned straw ash charcoal. Journal of Environmental Sciences 22 (10), 1586–1594.

Inyang, M., Dickenson, E., 2015. The potential role of biochar in the removal of organic and microbial contaminants from potable and reuse water: a review. Chemosphere 134, 232–240.

Inyang, M.I., Gao, B., Yao, Y., Xue, Y., Zimmerman, A., Mosa, A., Pullammanappallil, P., Ok, Y.S., Cao, X., 2015. A review of biochar as a low-cost adsorbent for aqueous heavy metal removal. Critical Reviews in Environmental Science and Technology 46 (4), 406–433.

Ismadji, S., Bhatia, S.K., 2001. A modified pore-filling isotherm for liquid-phase adsorption in activated carbon. Langmuir 17 (5), 1488–1498.

Jahirul, M., Rasul, M., Chowdhury, A., Ashwath, N., 2012. Biofuels production through biomass pyrolysis—a technological review. Energies 5 (12), 4952–5001.

Ji, L., Wan, Y., Zheng, S., Zhu, D., 2011. Adsorption of tetracycline and sulfamethoxazole on crop residue-derived ashes: implication for the relative importance of black carbon to soil sorption. Environmental Science and Technology 45 (13), 5580–5586.

Jiang, J., Xu, R.K., Jiang, T.Y., Li, Z., 2012. Immobilization of Cu(II), Pb(II) and Cd(II) by the addition of rice straw derived biochar to a simulated polluted Ultisol. Journal of Hazardous Materials 229–230, 145–150.

Jung, C., Park, J., Lim, K.H., Park, S., Heo, J., Her, N., Oh, J., Yun, S., Yoon, Y., 2013. Adsorption of selected endocrine disrupting compounds and pharmaceuticals on activated biochars. Journal of Hazardous Materials 263, 702–710.

Karami, N., Clemente, R., Moreno-Jimenez, E., Lepp, N.W., Beesley, L., 2011. Efficiency of green waste compost and biochar soil amendments for reducing lead and copper mobility and uptake to ryegrass. Journal of Hazardous Materials 191 (1–3), 41–48.

Kasozi, G.N., Zimmerman, A.R., Nkedi-Kizza, P., Gao, B., 2010. Catechol and humic acid sorption onto a range of laboratory-produced black carbons (biochars). Environmental Science and Technology 44 (16), 6189–6195.

Keiluweit, M., Kleber, M., 2009. Molecular-level interactions in soils and sediments: the role of aromatic pi-systems. Environmental Science and Technology 43 (10), 3421–3429.

Keiluweit, M., Nico, P.S., Johnson, M.G., Kleber, M., 2010. Dynamic molecular structure of plant biomass-derived black carbon (biochar). Environmental Science and Technology 44 (4), 1247–1253.

Khan, S., Wang, N., Reid, B.J., Freddo, A., Cai, C., 2013. Reduced bioaccumulation of PAHs by Lactuca satuva L. grown in contaminated soil amended with sewage sludge and sewage sludge derived biochar. Environmental Pollution 175, 64–68.

Kim, G.-H., Chowdhury, S., Bolan, N., 2016. Assessing the Efficiency of Permeable Reactive Barrier to Reduce the Groundwater Contamination from Carcass Burial, pp. 9–14.

Knight, L.A., Christenson, M.K., Trease, A.J., Davis, P.H., Kolok, A.S., 2013. The spring runoff in Nebraska's (USA) Elkhorn River watershed and its impact on two sentinel organisms. Environmental Toxicology and Chemistry 32 (7), 1544–1551.

Kwon, S., Pignatello, J.J., 2005. Effect of natural organic substances on the surface and adsorptive properties of environmental black carbon (char): pseudo pore blockage by model lipid components and its implications for N2-probed surface properties of natural sorbents. Environmental Science and Technology 39 (20), 7932–7939.

Laird, D.A., Brown, R.C., Amonette, J.E., Lehmann, J., 2009. Review of the pyrolysis platform for coproducing bio-oil and biochar. Biofuels, Bioproducts and Biorefining 3 (5), 547–562.

Lattao, C., Cao, X., Mao, J., Schmidt-Rohr, K., Pignatello, J.J., 2014. Influence of molecular structure and adsorbent properties on sorption of organic compounds to a temperature series of wood chars. Environmental Science and Technology 48 (9), 4790–4798.

Lehmann, J., 2007a. Bio-energy in the black. Frontiers in Ecology and the Environment 5 (7), 381–387.

Lehmann, J., 2007b. A handful of carbon. Nature 447 (7141), 143.

Lehmann, J., Gaunt, J., Rondon, M., 2006. Bio-char sequestration in terrestrial ecosystems—a review. Mitigation and Adaptation Strategies for Global Change 11 (2), 403–427.

Lehmann, J., Joseph, S., 2015. Biochar for Environmental Management: Science, Technology and Implementation. Routledge.

Lehmann, J., Skjemstad, J., Sohi, S., Carter, J., Barson, M., Falloon, P., Coleman, K., Woodbury, P., Krull, E., 2008. Australian climate—carbon cycle feedback reduced by soil black carbon. Nature Geoscience 1 (12), 832–835.

Lehmann, J.J.F.i.E., Environment, t., 2007. Bio-energy in the black 5 (7), 381–387.

Li, H., Ye, X., Geng, Z., Zhou, H., Guo, X., Zhang, Y., Zhao, H., Wang, G., 2016. The influence of biochar type on long-term stabilization for Cd and Cu in contaminated paddy soils. Journal of Hazardous Materials 304, 40–48.

Liang, F., Li, G.-t., Lin, Q.-m., Zhao, X.-r., 2014. Crop yield and soil properties in the first 3 years after biochar application to a calcareous soil. Journal of Integrative Agriculture 13 (3), 525–532.

Liao, P., Ismael, Z.M., Zhang, W., Yuan, S., Tong, M., Wang, K., Bao, J., 2012. Adsorption of dyes from aqueous solutions by microwave modified bamboo charcoal. Chemical Engineering Journal 195, 339–346.

Lima, D.L.D., Schneider, R.J., Scherer, H.W., Duarte, A.C., Santos, E.B.H., Esteves, V.I., 2010. Sorption-desorption behavior of atrazine on soils subjected to different organic long-term amendments. Journal of Agricultural and Food Chemistry 58 (5), 3101–3106.

Lu, H., Li, Z., Fu, S., Méndez, A., Gascó, G., Paz-Ferreiro, J.J.C., 2015. Combining phytoextraction and biochar addition improves soil biochemical properties in a soil contaminated with Cd. Chemosphere 119, 209–216.

Lu, J., Vecchi, G.A., Reichler, T., 2007. Expansion of the Hadley cell under global warming. Geophysical Research Letters 34 (6).

Lucchini, P., Quilliam, R., DeLuca, T.H., Vamerali, T., Jones, D.L., 2014a. Does biochar application alter heavy metal dynamics in agricultural soil? Agriculture, Ecosystems & Environment 184, 149–157.

Lucchini, P., Quilliam, R., DeLuca, T.H., Vamerali, T., Jones, D.L., 2014b. Increased bioavailability of metals in two contrasting agricultural soils treated with waste wood-derived biochar and ash. Environmental Science and Pollution Research 21 (5), 3230–3240.

Lucchini, P., Quilliam, R.S., Deluca, T.H., Vamerali, T., Jones, D.L., 2014c. Increased bioavailability of metals in two contrasting agricultural soils treated with waste wood-derived biochar and ash. Environmental Science and Pollution Research International 21 (5), 3230–3240.

Lynch, J.M., Moffat, A.J., 2005. Bioremediation–prospects for the future application of innovative applied biological research. Annals of Applied Biology 146 (2), 217–221.

Manya, J.J., 2012. Pyrolysis for biochar purposes: a review to establish current knowledge gaps and research needs. Environmental Science and Technology 46 (15), 7939–7954.

McMillan, O., 2018. Characteristics and Mechanisms of Atrazine Sorption to Biochar for Land Remediation. University of Cambridge.

Meier, S., Curaqueo, G., Khan, N., Bolan, N., Cea, M., Eugenia, G.M., Cornejo, P., Ok, Y.S., Borie, F., 2015. Chicken-manure-derived biochar reduced bioavailability of copper in a contaminated soil. Journal of Soils and Sediments 17 (3), 741–750.

Mench, M., Lepp, N., Bert, V., Schwitzguébel, J.P., Gawronski, S.W., Schröder, P., Vangronsveld, J.J.J.o.S., Sediments, 2010. Successes and limitations of phytotechnologies at field scale: outcomes, assessment and outlook from. COST Action 859 (6), 1039–1070.

Moreno-Castilla, C., 2004. Adsorption of organic molecules from aqueous solutions on carbon materials ☆. Carbon 42 (1), 83–94.

Mrozik, A., Piotrowska-Seget, Z.J., 2010. Bioaugmentation as a strategy for cleaning up of soils contaminated with aromatic compounds. Microbiological Research 165 (5), 363–375.

Mukherjee, A., Zimmerman, A., Harris, W., 2011. Surface chemistry variations among a series of laboratory-produced biochars. Geoderma 163 (3–4), 247–255.

Namgay, T., Singh, B., Singh, B.P., 2010. Influence of biochar application to soil on the availability of As, Cd, Cu, Pb, and Zn to maize (Zea mays L.). Soil Research 48 (7), 638–647.

Neves, E.G., Petersen, J.B., Bartone, R.N., Da Silva, C.A., 2003. Amazonian Dark Earths. Springer, pp. 29–50.

Nguyen, T.H., Cho, H.-H., Poster, D.L., Ball, W.P., 2007. Evidence for a pore-filling mechanism in the adsorption of aromatic hydrocarbons to a natural wood char. Environmental Science and Technology 41 (4), 1212–1217.

Noel, J.D., Wang, Y., Giammar, D.E., 2014. Effect of water chemistry on the dissolution rate of the lead corrosion product hydrocerussite. Water Research 54, 237–246.

Park, J.-H., Ok, Y.S., Kim, S.-H., Cho, J.-S., Heo, J.-S., Delaune, R.D., Seo, D.-C., 2016. Competitive adsorption of heavy metals onto sesame straw biochar in aqueous solutions. Chemosphere 142, 77–83.

Paz-Ferreiro, J., Lu, H., Fu, S., Méndez, A., Gascó, G., 2014. Use of phytoremediation and biochar to remediate heavy metal polluted soils: a review. Solid Earth 5 (1), 65–75.

Pelekani, C., Snoeyink, V.L., 2000. Competitive adsorption between atrazine and methylene blue on activated carbon: The Importance of Pore Size Distribution. Carbon 38 (10), 1423–1436.

Powell, R.M., Puls, R.W., Blowes, D., Gillham, R., Schultz, D.J.N., 1998. Permeable Reactive Barrier Technologies for Contaminant Remediation. (19990008853).

Puga, A., Abreu, C., Melo, L., Beesley, L., 2015a. Biochar application to a contaminated soil reduces the availability and plant uptake of zinc, lead and cadmium. Journal of Environmental Management 159, 86–93.

Puga, A., Abreu, C., Melo, L., Paz-Ferreiro, J., Beesley, L., 2015b. Cadmium, lead, and zinc mobility and plant uptake in a mine soil amended with sugarcane straw biochar. Environmental Science and Pollution Research 22 (22), 17606–17614.

Puga, A.P., Abreu, C.A., Melo, L.C., Paz-Ferreiro, J., Beesley, L., 2015c. Cadmium, lead, and zinc mobility and plant uptake in a mine soil amended with sugarcane straw biochar. Environmental Science and Pollution Research International 22 (22), 17606–17614.

Qin, G., Gong, D., Fan, M.-Y., 2013. Bioremediation of petroleum-contaminated soil by biostimulation amended with biochar. International Biodeterioration and Biodegradation 85, 150–155.

Qiu, Y., Zheng, Z., Zhou, Z., Sheng, G.D., 2009. Effectiveness and mechanisms of dye adsorption on a straw-based biochar. Bioresource Technology 100 (21), 5348–5351.

Raskin, I., Smith, R.D., Salt, D.E., 1997. Phytoremediation of metals: Using Plants to Remove Pollutants from the Environment. Current Opinion in Biotechnology 8 (2), 221.

Rees, F., Simonnot, M.O., Morel, J.L., 2014. Short-term effects of biochar on soil heavy metal mobility are controlled by intra-particle diffusion and soil pH increase. European Journal of Soil Science 65 (1), 149–161.

Ren, X., Sun, H., Wang, F., Cao, F., 2016. The changes in biochar properties and sorption capacities after being cultured with wheat for 3 months. Carbon 144, 2257–2263.

Rodríguez-Vila, A., Asensio, V., Forján, R., Covelo, E.F., 2015. Chemical fractionation of Cu, Ni, Pb and Zn in a mine soil amended with compost and biochar and vegetated with Brassica juncea L. Journal of Geochemical Exploration 158, 74–81.

Ronsse, F., van Hecke, S., Dickinson, D., Prins, W., 2013. Production and characterization of slow pyrolysis biochar: influence of feedstock type and pyrolysis conditions. GCB Bioenergy 5 (2), 104–115.

Schnoor, J.L., Licht, L.A., Mccutcheon, S.C., Wolfe, N.L., Carreira, L.H., 1995. Phytoremediation of organic and nutrient contaminants. Journal of Environmental Science and Technology 29 (7), 318A.

Shang, J., Zong, M., Yu, Y., Kong, X., Du, Q., Liao, Q.J., 2017. Removal of chromium (VI) from water using nanoscale zerovalent iron particles supported on herb-residue biochar. Journal of Environmental Management 197, 331–337.

Shen, Z., Hou, D., Xu, W., Zhang, J., Jin, F., Zhao, B., Pan, S., Peng, T., Alessi, D.S.J.S.o.t.t.e., 2018a. Assessing long-term stability of cadmium and lead in a soil washing residue amended with MgO-based binders using quantitative accelerated ageing. The Science of the Total Environment 643, 1571–1578.

Shen, Z., Hou, D., Zhao, B., Xu, W., Ok, Y.S., Bolan, N.S., Alessi, D.S., 2018b. Stability of heavy metals in soil washing residue with and without biochar addition under accelerated ageing. The Science of the Total Environment 619–620, 185–193.

Shen, Z., Jin, F., Wang, F., McMillan, O., Al-Tabbaa, A., 2015. Sorption of lead by Salisbury biochar produced from British broadleaf hardwood. Bioresource Technology 193, 553–556.

Shen, Z., Li, Z., Alessi, D.S., 2018c. Stabilization-based soil remediation should consider long-term challenges. Frontiers of Environmental Science & Engineering 12 (2).

Shen, Z., Som, A.M., Wang, F., Jin, F., McMillan, O., Al-Tabbaa, A., 2016a. Long-term impact of biochar on the immobilisation of nickel (II) and zinc (II) and the revegetation of a contaminated site. The Science of the Total Environment 542 (Pt A), 771–776.

Shen, Z., Tian, D., Zhang, X., Tang, L., Su, M., Zhang, L., Li, Z., Hu, S., Hou, D., 2018d. Mechanisms of biochar assisted immobilization of Pb(2+) by bioapatite in aqueous solution. Chemosphere 190, 260–266.

Shen, Z., Zhang, Y., Jin, F., McMillan, O., Al-Tabbaa, A., 2017a. Qualitative and quantitative characterisation of adsorption mechanisms of lead on four biochars. The Science of the Total Environment 609, 1401–1410.

Shen, Z., Zhang, Y., McMillan, O., Jin, F., Al-Tabbaa, A., 2017b. Characteristics and mechanisms of nickel adsorption on biochars produced from wheat straw pellets and rice husk. Environmental Science and Pollution Research International 24 (14), 12809–12819.

Shen, Z.T., Hou, D.Y., Jin, F., Shi, J.X., Fan, X.L., Tsang, D.C.W., Alessi, D.S., 2019a. Effect of production temperature on lead removal mechanisms by rice straw biochars. The Science of the Total Environment 655, 751–758.

Shen, Z.T., McMillan, O., Jin, F., Al-Tabbaa, A., 2016b. Salisbury biochar did not affect the mobility or speciation of lead in kaolin in a short-term laboratory study. Journal of Hazardous Materials 316, 214–220.

Shen, Z.T., Zhang, J.Z., Hou, D.Y., Tsang, D.C.W., Ok, Y.S., Alessi, D.S., 2019b. Synthesis of MgO-coated corncob biochar and its application in lead stabilization in a soil washing residue. Environment International 122, 357–362.

Shen, Z.T., Zhang, Y.H., Jin, F., Alessi, D.S., Zhang, Y.Y., Wang, F., McMillan, O., Al-Tabbaa, A., 2018e. Comparison of nickel adsorption on biochars produced from mixed softwood and Miscanthus straw. Environmental Science and Pollution Research 25 (15), 14626–14635.

Singh, B., Camps-Arbestain, M., Lehmann, J., 2017. Biochar: A Guide to Analytical Methods. Csiro Publishing.

Smith, S.R., 2009. A critical review of the bioavailability and impacts of heavy metals in municipal solid waste composts compared to sewage sludge. Environment International 35 (1), 142–156.

Sohi, S.P., 2012. Agriculture. Carbon storage with benefits. Science 338 (6110), 1034–1035.

Som, A.M., Wang, Z., Al-Tabbaa, A., 2013. Palm frond biochar production and characterisation. Earth and Environmental Science Transactions of the Royal Society of Edinburgh 103 (01), 39–50.

Suliman, W., Harsh, J.B., Abu-Lail, N.I., Fortuna, A.-M., Dallmeyer, I., Garcia-Perez, M., 2016. Modification of biochar surface by air oxidation: role of pyrolysis temperature. Biomass and Bioenergy 85, 1–11.

Sun, H., Zhou, Z.J.C., 2008. Impacts of charcoal characteristics on sorption of polycyclic aromatic hydrocarbons. Chemosphere 71 (11), 2113–2120.

Tag, A.T., Duman, G., Ucar, S., Yanik, J., 2016. Effects of feedstock type and pyrolysis temperature on potential applications of biochar. Journal of Analytical and Applied Pyrolysis 120, 200–206.

Thiruvenkatachari, R., Vigneswaran, S., Naidu, R.J.J.o.I., Chemistry, E., 2008. Permeable reactive barrier for groundwater remediation. Journal of Industrial and Engineering Chemistry 14 (2), 145–156.

Thurkettle, V., 1997. The marketing of British hardwoods. Forestry: An International Journal of Forest Research 70 (4), 319–325.

Tiessen, H., Cuevas, E., Chacon, P., 1994. The role of soil organic matter in sustaining soil fertility. Nature 371 (6500), 783.

Tong, H., Hu, M., Li, F.B., Liu, C.S., Chen, M.J., 2014. Biochar enhances the microbial and chemical transformation of pentachlorophenol in paddy soil. Soil Biology and Biochemistry 70, 142–150.

Trakal, L., Komárek, M., Száková, J., Zemanová, V., Tlustoš, P., 2011. Biochar application to metal-contaminated soil: evaluating of Cd, Cu, Pb and Zn sorption behavior using single-and multi-element sorption experiment. Plant Soil and Environment 57 (8), 372–380.

Uchimiya, M., Bannon, D.I., Wartelle, L.H., 2012a. Retention of heavy metals by carboxyl functional groups of biochars in small arms range soil. Journal of Agricultural and Food Chemistry 60 (7), 1798–1809.

Uchimiya, M., Bannon, D.I., Wartelle, L.H., 2012b. Retention of heavy metals by carboxyl functional groups of biochars in small arms range soil. Journal of Agricultural and Food Chemistry 60 (7), 1798–1809.

Uchimiya, M., Klasson, K.T., Wartelle, L.H., Lima, I.M., 2011. Influence of soil properties on heavy metal sequestration by biochar amendment: 1. Copper sorption isotherms and the release of cations. Chemosphere 82 (10), 1431–1437.

Violante, A., Cozzolino, V., Perelomov, L., Caporale, A., Pigna, M., 2010. Mobility and bioavailability of heavy metals and metalloids in soil environments. Journal of Soil Science and Plant Nutrition 10 (3), 268–292.

Wang, F., Wang, H., Al-Tabbaa, A., 2014. Leachability and heavy metal speciation of 17-year old stabilised/solidified contaminated site soils. Journal of Hazardous Materials 278, 144–151.

Wang, H., Lin, K., Hou, Z., Richardson, B., Gan, J.J., 2010. Sorption of the herbicide terbuthylazine in two New Zealand forest soils amended with biosolids and biochars. Journal of Soils and Sediments 10 (2), 283–289.

Xiao, F., Pignatello, J.J., 2015. π(+)-π interactions between (hetero)aromatic amine cations and the graphitic surfaces of pyrogenic carbonaceous materials. Environmental Science & Technology 49 (2), 906.

Xin, J., Liu, X., Liu, W., Zheng, X., 2014. Effects of biochar–BDE-47 interactions on BDE-47 bioaccessibility and biodegradation by Pseudomonas putida TZ-1. Ecotoxicology and Environmental Safety 106, 27–32.

Xing, B., Pignatello, J.J.J.E.s., technology, 1997. Dual-mode sorption of low-polarity compounds in glassy poly (vinyl chloride) and soil organic matter. Environmental Science and Technology 31 (3), 792–799.

Yang, H., Yan, R., Chen, H., Lee, D.H., Liang, D.T., Zheng, C., 2006. Pyrolysis of palm oil wastes for enhanced production of hydrogen rich gases. Fuel Processing Technology 87 (10), 935–942.

Yang, K., Yang, J., Jiang, Y., Wu, W., Lin, D.J.E.P., 2016a. Correlations and adsorption mechanisms of aromatic compounds on a high heat temperature treated bamboo biochar. Environmental Pollution 210, 57–64.

Yang, X., Liu, J., McGrouther, K., Huang, H., Lu, K., Guo, X., He, L., Lin, X., Che, L., Ye, Z., 2016b. Effect of biochar on the extractability of heavy metals (Cd, Cu, Pb, and Zn) and enzyme activity in soil. Environmental Science and Pollution Research 23 (2), 974–984.

Yang, Y., Wei, Z., Zhang, X., Chen, X., Yue, D., Yin, Q., Xiao, L., Yang, L., 2014. Biochar from Alternanthera philoxeroides could remove Pb(II) efficiently. Bioresource Technology 171, 227–232.

Yaning, Y., Guangyao, S., 2003. Enhanced pesticide sorption by soils containing particulate matter from crop residue burns. Environmental Science and Technology 37 (16), 3635–3639.

Yoon, J.-H., Kim, Y.-N., Shin, D.-C., Kim, K.-R., Kim, K.-H.J.S., 2017. Management of animal carcass disposal sites using a biochar permeable reactive barrier and fast growth tree (Populus euramericana). A Field Study in Korea 9 (3), 457.

Zeng, G., Wu, H., Liang, J., Guo, S., Huang, L., Xu, P., Liu, Y., Yuan, Y., He, X., He, Y., 2015. Efficiency of biochar and compost (or composting) combined amendments for reducing Cd, Cu, Zn and Pb bioavailability, mobility and ecological risk in wetland soil. Rsc Advances 5 (44), 34541–34548.

Zeng, Z., Zhang, S.D., Ting-Qiang, L.I., Zhao, F.L., Zhen-Li, H.E., Zhao, H.P., Yang, X.E., Wang, H.L., Zhao, J., Rafiq, M.T., Biotechnology, 2013. Sorption of ammonium and phosphate from aqueous solution by biochar derived from phytoremediation plants. Journal of Zhejiang University Science B 14 (12), 1152–1161.

Zhang, A., Cui, L., Pan, G., Li, L., Hussain, Q., Zhang, X., Zheng, J., Crowley, D., 2010. Effect of biochar amendment on yield and methane and nitrous oxide emissions from a rice paddy from Tai Lake plain, China. Agriculture, Ecosystems & Environment 139 (4), 469–475.

Zhang, G., Zhang, Q., Sun, K., Liu, X., Zheng, W., Zhao, Y., 2011. Sorption of simazine to corn straw biochars prepared at different pyrolytic temperatures. Environmental Pollution 159 (10), 2594–2601.

Zhang, P., Sun, H., Yu, L., Sun, T., 2013a. Adsorption and catalytic hydrolysis of carbaryl and atrazine on pig manure-derived biochars: impact of structural properties of biochars. Journal of Hazardous Materials 244–245 (3), 217–224.

Zhang, X., Wang, H., He, L., Lu, K., Sarmah, A., Li, J., Bolan, N.S., Pei, J., Huang, H., 2013b. Using biochar for remediation of soils contaminated with heavy metals and organic pollutants. Environmental Science and Pollution Research International 20 (12), 8472–8483.

Zhang, X., Wu, Y., 2017. Application of coupled zero-valent iron/biochar system for degradation of chlorobenzene-contaminated groundwater. Water Science & Technology 75 (3), 571–580.

Zhang, Y., Jin, F., Shen, Z., Lynch, R., Al-Tabbaa, A.J.J.o.H.M., 2018. Kinetic and equilibrium modelling of MTBE (methyl tert-butyl ether) adsorption on ZSM-5 zeolite. Batch and column studies 347, 461.

Zheng, H., Wang, Z., Zhao, J., Herbert, S., Xing, B., 2013. Sorption of antibiotic sulfamethoxazole varies with biochars produced at different temperatures. Environmental Pollution 181 (181C), 60–67.

Zhou, D., Li, Y., Zhang, Y., Zhang, C., Li, X., Chen, Z., Huang, J., Li, X., Flores, G., Kamon, M., 2014a. Column test-based optimization of the permeable reactive barrier (PRB) technique for remediating groundwater contaminated by landfill leachates. Elsevier logo Journals & Books 168, 1–16.

Zhou, Y., Gao, B., Zimmerman, A.R., Cao, X., 2014b. Biochar-supported zerovalent iron reclaims silver from aqueous solution to form antimicrobial nanocomposite. Chemosphere 117, 801–805.

Zhu, D., Kwon, S., Pignatello, J.J., 2005. Adsorption of single-ring organic compounds to wood charcoals prepared under different thermochemical conditions. Environmental Science & Technology 39 (11), 3990–3998.

Zhu, D., Pignatello, J.J., 2005. Characterization of aromatic compound sorptive interactions with black carbon (charcoal) assisted by graphite as a model. Environmental Science & Technology 39 (7), 2033–2041.

Zhu, Q., Wu, J., Wang, L., Yang, G., Zhang, X., 2015. Effect of biochar on heavy metal speciation of paddy soil. Water, Air, and Soil Pollution 226 (12).

Chapter 7

Application of slow-release materials for in situ and passive remediation of contaminated groundwater

Chih-Ming Kao, Yih-Terng Sheu, Jiun-Hau Ou, Wei-Han Lin
Institute of Environmental Engineering, National Sun Yat-Sen University, Kaohsiung, Taiwan

1. Background

Groundwater contaminated by organic chemicals (e.g., petroleum hydrocarbons, halogenated organic solvents) is a worldwide environmental problem. At many organic-chemical-contaminated sites, the residual amounts of organic pollutants remain in the subsurface and form pure liquid phase [commonly referred to as nonaqueous-phase liquids (NAPLs)] within soil pores. The slowly dissolved and dispersed free-phase or residual NAPLs result in continuous source of groundwater contamination and organic contaminant plumes. In many cases, it is very difficult to locate and remove the NAPLs, and thus, organic-chemical-contaminated groundwater remediation becomes a lengthy and costly process. From an engineering and practical point of view, remediation of NAPL-contained groundwater sites can be focused on preventing further migration of the dissolved phase of contamination. Moreover, application of slow-release materials (SRMs) for in situ and passive groundwater remediation can significantly lower the operational and maintenance cost during the long-term groundwater remediation. Therefore, the concept of SRM application is a more economic approach that is desirable for organic-chemical-contaminated groundwater remediation and also provides a long-term technique for plume migration control (O'Connor et al., 2018; Zhang et al., 2018a).

1.1 Groundwater contamination and plume control

Groundwater is an important water resource on earth. However, groundwater is not visible because it exists in the subsurface below the land surface. As a

Sustainable Remediation of Contaminated Soil and Groundwater. https://doi.org/10.1016/B978-0-12-817982-6.00007-0

result, groundwater usually could not receive any attention until its pollution is discovered. Currently, groundwater at many gas station and storage tank sites, surface spill sites, existing and former industrial sites, and disposal areas is contaminated by a variety of organic compounds that were released into the environment intentionally or accidently. Groundwater contamination causes a widespread and significant ecosystem impact, and the sources of groundwater contamination include leakages from land application sites, surface waste ponds, landfills, accidental spills, above-ground tanks (AST), underground storage tanks (UST), injection wells, waste disposal sites, septic tanks, pipelines, and acid mine drainage (Usman et al., 2017; Fetter et al., 2017; Sharma et al., 2018).

The factors that affect the fate and transport of organic contaminants in the subsurface include the following: groundwater advection, dispersion, sorption, dissolution, dilution, and biodegradation (ITRC, 2004). Hence, it is a challenge to groundwater hydrogeologists and environmental engineers that must be able to understand those mechanisms influencing the contaminant fate and transport to effectively develop remediation schemes for plume migration control.

On the road to becoming an industrialized country, many developing countries cannot immunize from such soil and groundwater contamination problems. Many of these developing countries fail to effectively manage the soil and groundwater contamination problems because of lacking adequate regulations for subsurface environment protection. Without a complete legal foundation, remedial measures are not legally binding should contamination occur. Moreover, without adequate environmental protection laws, the existing regulatory system would not be able to provide incentives for the development of innovative and novel cleanup technologies for contaminated site remediation. Therefore, the US (or European countries) soil/groundwater protection laws are adopted by many developing countries to establish their own soil/groundwater-related regulation (Grimski et al., 2018). However, the introduction of the new soil and groundwater laws would also become a challenge for the industries because of the required expenses on site investigation and remediation. To remediate soil and groundwater contaminated sites, more costly and applicable remedial technologies need to be developed to reduce the burdens to the governmental agencies, site owners, and responsible parties of the contaminated sites. Moreover, the site remediation market will grow rapidly in the developing countries due to the discovery of numerous priority sites continuously.

Because contaminated groundwater is mobile and it would migrate to farther downgradient areas causing the spreading of the contaminants and deterioration of the downgradient environments, it becomes the driving force for contaminant removal and transport control. The necessity for contaminant plume containment led to the application of pump and treat technology, which has become a common and popular remedial option for both containment and mass removal since late 1970s. More than two decades of experience has taught us that pump and treat technology can be only used for interim plume migration control, but it cannot complete the site cleanup task and solve the contamination problem. Its high operational and maintenance cost also makes it become an inappropriate option for a long-term plume control (Besha et al., 2017; Huang et al., 2017; Su et al., 2017).

1.2 Nonaqueous-phase liquids

Nonaqueous-phase liquids (NAPLs) are hydrocarbons, which exist as an immiscible and separate phase, when in contact with water. Due to the different physical and chemical properties of NAPL and water, a physical interface between the two liquids is formed, and thus, their mixing is prevented. NAPLs can be divided into light NAPLs (LNAPLs) and dense NAPLs (DNAPLs). LNAPL has a density lower than that of water, and DNAPL has a density higher than that of water (Tick et al., 2015; Besha et al., 2017; Field, 2018).

In most cases, NAPLs would migrate downward through the gravity force after release. Part of the NAPLs would be retained in the soil pores and become residual globules, which would become the long-term sources of groundwater contamination, which cause significant adverse impacts on subsurface environments and ecosystems (Jacobson, 2017; Tomlinson et al., 2017; Rice and Stolzenburg, 2018). The most common LNAPL-related groundwater contamination problems result from the release of petroleum products. These products are typically multicomponent organic mixtures composed of chemicals with varying degrees of water solubility. Some additives such as alcohols and methyl tertiary-butyl ether (MTBE) are very soluble, some components such as benzene, toluene, ethylbenzene, and xylenes (BTEX), main components of gasoline, are slightly soluble, and some components such as n-dodecane and n-heptane are only slightly soluble. Thus, when LNAPL is leaked into the subsurface, it will move downward until it reaches the water table (or capillary fringe) or a strata with low permeability (e.g., clay layer). Then the LNAPL will migrate laterally as a free-phase (free product) and continuous layer along the water table due to dispersion mechanisms (Schwarzenbach and Gschwend, 2016; Lenhard et al., 2017). Moreover, volatilization can result in further spreading of contamination. Groundwater in contact with LNAPL will result in the dissolution of chemical components from LNAPL into the water phase and form a dissolved-phase contaminant plume (Frollini et al., 2016; Chiu et al., 2017).

The DNAPLs are present at many industrial and hazardous waste sites. Due to the numerous variables influencing DNAPL fate and transport in the subsurface environments, DNAPLs are largely undetected, which makes the DMAPL site remediation become more difficult. The most general chemical categories found in US Superfund sites are halogenated volatiles and semivolatiles halogenated/nonhalogenated organic compounds. They are commonly found in the following waste-producing industries and processes: coal tars, solvents, pesticides, and wood preserving wastes (creosote, pentachlorophenol). The most commonly found DNAPL groups to date are chlorinated solvents (Widdowson, 2004; Sale and Newell, 2010; Sheu et al., 2018).

When a DNAPL spill occurs, DNAPL is initially trapped in the form of immobile blobs by capillary forces in the subsurface environments (Sale and Newell, 2010). Because DNAPL chemicals usually have densities higher than that of water, they tend to migrate to the deeper zone of the aquifer and usually form DNAPL pools on top of the impermeable layers (e.g., clay layer, bedrock). DNAPLs can slowly dissolve and release part of the components to the groundwater and pollute the aqueous phase for a long period of time (Wu et al., 2018; Guo et al., 2019).

Commonly found DNAPLs in contaminated sites are chlorinated solvents, particularly chlorinated ethylenes including perchloroethylene (PCE), trichloroethylene (TCE), dichloroethylene (DCE), and vinyl chloride (VC). Other DNAPL pollutants include polycyclic aromatic hydrocarbons (PAHs) and polychlorinated biphenyls (PCBs). Creosote and coal tar are commonly found PAH-dominated DNAPLs in contaminated sites. Chlorinated solvents usually have lower viscosities and higher densities compared with creosote and coal tar, and this makes them more mobile in the subsurface environments (Baciocchi et al., 2014). Given that it is often not easy to locate and remove the DNAPLs in the subsurface, groundwater remediation needs to focus on preventing further migration of the dissolved contamination. Because this plume control needs to be operated for a long period of time, more costly remedial options are desirable for DNAPL-contaminated groundwater remediation. Thus, the passive remediation technology can be applied to reduce the operational and maintenance cost (ITRC, 2003, 2004; Faisal et al., 2018; Martínez et al., 2019).

1.3 Remediation technologies

The difficulties of NAPL-contaminated groundwater remediation is attributable to the following causes: (1) the heterogeneity and complexity of the subsurface environments; (2) incomplete DNAPL delineation; (3) existence of varied NAPL phases; (4) different behavior of each NAPL phase (Lien et al., 2016; Sheng et al., 2018). The slow dissolution of residual NAPL results in a contaminant plume in groundwater. Given that it is often not possible to locate and remove the residual NAPL, remediation must focus on preventing further migration of the dissolved contamination. The plume control must be maintained for a long period of time. Therefore, some more economic approaches such as in situ and passive technologies are desirable for groundwater remediation to provide for long-term control of contaminated groundwater (Huang et al., 2014; Sheu et al., 2015; Guo et al., 2019).

1.3.1 Pump and treat technology

Pump and treat technology is the first remedial system used for contaminated groundwater remediation. It is a remedial option using physical process for contaminant removal. Physical processes involve the application of the physical properties of contaminants to affect their transfer from soil and groundwater to other media where they can be treated or destroyed. Pump and treat technology applies groundwater extraction followed by above-ground treatment typically using ex-situ wastewater treatment units for dissolved-phase contaminants removal.

To remove polluted groundwater from the subsurface, a pumping or extraction system is installed. Pumps are used to extract polluted groundwater to the ground surface where it can be treated more easily for pollutant removal. This system consists of different designs of a pumping system consisting of extraction wells equipped with pumps. Pumps can pull the polluted groundwater into the wells and result in a cone of groundwater depression. The extracted water is stored in a holding tank and then transferred to a treatment system for contaminant removal (Pennell and Abriola, 2017).

There are a number of commonly used treatment methods (e.g., sorption, air stripping, biological treatment, chemical oxidation, chemical precipitation), which can be used for groundwater cleanup to remove or destroy contaminants. The treated (cleaned) water can then be pumped back into the ground, into a public sewer, or surface water bodies. To improve its effectiveness, the source of the pollution can be taken away so that it will not continue to seep into the groundwater. For example, leaking underground storage tanks need to be pulled and removed from the subsurface, and the surrounding polluted soils with residual oils also need to be cleaned up to remove the source of the contamination (Boal et al., 2015). However, aquifer restoration for NAPL sites usually requires much more groundwater pumping, and complete groundwater restoration may not be possible (Hoponick, 2006; Suthersan et al., 2015).

1.3.2 Soil vapor extraction and air sparging

Soil vapor extraction (SVE) can be used to remove volatile organic compounds (VOCs), in the form of vapors, from the soils in the unsaturated zone above the water table. The organic vapors are removed via gas extraction using a vacuum pump to extract (pull) out the VOC vapors. SVE requires drilling extraction wells within the polluted area. These wells are drilled into the soil, but not the groundwater (Burris et al., 2018).

Air sparging (AS) can be applied to remove VOC vapors from polluted soil and groundwater phases below the water table. When air bubbles are pumped to the subsurface via a blower, the VOCs are evaporated and removed. Air pumped into the wells disturbs the groundwater, and this can help the VOCs change into vapors. AS wells can be drilled into the ground, and SVE and AS can work together to remove VOCs from contaminated sites. The number of

AS and SVE wells can range from one to hundreds, depending on the size of the polluted area. Once the VOC vapors are extracted from the subsurface, they are treated by the above-ground air pollution treatment facilities via sorption, oxidation, or thermal mechanisms. The air flows used in SVE and AS can also increase the oxygen concentrations in soils and groundwater, and thus, the in situ bioremediation efficiency can be enhanced (O'Steen and Howard, 2014; Chao et al., 2018; Liao et al., 2018). Fig. 7.1 shows the schematic diagram of the SVE/AS system.

1.3.3 Chemical oxidation

Chemical oxidation can be applied for contaminant destruction to reduce the contaminant concentrations in contaminated groundwater. During the chemical oxidation processes, oxidants are used for pollutants oxidation (destruction) and oxidize chemicals to less hazardous by-products or end products (e.g., carbon dioxide, water). Many types of chemicals (e.g., fuels, solvents, pesticides) can be destroyed by the chemical oxidation processes.

When in situ chemical oxidation (ISCO) is applied for site remediation, polluted soils are not excavated and groundwater is not extracted. Instead, oxidants are injected into the subsurface via the remediation wells or injection wells around the hot zone or polluted area for contaminants oxidation. The injected oxidants react with the target chemicals and result in the breakdown of the chemicals (Ameta and Ameta, 2018).

To accelerate the site remediation via the ISCO process, groundwater circulation can be used to improve the oxidants mixing with the contaminants

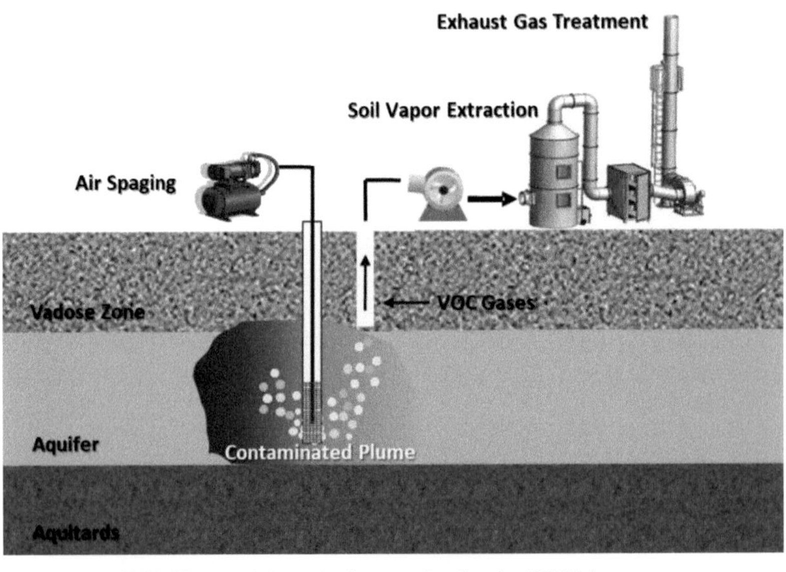

FIGURE 7.1 Schematic diagram showing the SVE/AS system.

in the subsurface. During the circulation stage, oxidants are pumped in via the upgradient wells and extracted from downgradient wells. The extracted groundwater is then pumped back (recirculated) down the upgradient injection wells. The injection, pumping, and mixing processes help the oxidation of the contaminants and accelerate the site cleanup. Detailed contaminant plume delineation and site investigation are required to effectively determine the right spots in the ground for remediation wells installation.

Chemical oxidation using hydrogen peroxide (H_2O_2) as the oxidant involves free radicals production when catalyst is added (Fenton reaction). During the oxidation process, H_2O_2 is decomposed to hydroxyl radical (•OH), which is a strong oxidant. Because of its nonselective characteristic, hydroxyl radical can react with most organic contaminants around the injection point and the influence area (Ameta and Ameta, 2018). The Fenton reaction is composed of two different processes: (1) Fenton reaction applies ferrous iron (Fe^{2+}) as the catalyst, and (2) Fenton-like reaction uses ferric iron (Fe^{3+}) or iron oxides [e.g., magnetite (Fe_3O_4), goethite (α-FeOOH), α-Fe_2O_3, iron chelates] as the catalyst (Nazari and Setayesh, 2018).

Potassium permanganate ($KMnO_4$) has been used as the oxidant during the ISCO process. Researchers demonstrated that $KMnO_4$ could be applied to degrade chlorinated organic compounds through the oxidative dechlorination process for groundwater remediation (Tsai et al., 2009; Liang et al., 2014). Due to its ability for chlorinated compound oxidation within a low-permeability media, $KMnO_4$ has been widely applied in ISCO schemes. Moreover, oxidant $KMnO_4$ has been well studied for the treatment of water and wastewater, and it has been investigated under laboratory and field conditions to determine its potential for in situ remediation of chlorinated solvents (Liang et al., 2014).

Persulfate oxidation has also been applied as an oxidant to remediate soils or groundwater contaminated by organic contaminants (Liu et al., 2017). Persulfate [peroxydisulfate ($S_2O_8^{2-}$)] is a strong oxidant, and it can be activated by ferrous iron for sulfate radicals ($SO_4^{-\cdot}$) ($E^0 = 2.6-3.1$ V) production (Zhou et al., 2017). Because of its strong oxidation potential, it is able to oxidize enormous organic compounds (Devi et al., 2016; Liu et al., 2017). Sulfate radical is more stable than other oxidants such as H_2O_2, and thus, it can exist in soils and groundwater for a longer reaction period after injection (Gong et al., 2016). This characteristic allows persulfate to have more opportunities to react with organic contaminants and result in the degradation of target compounds (Gong et al., 2016). Moreover, it has a lower affinity to natural organic matters (NOM), so it is suitable to be used to clean up contaminated sites containing higher NOM (Matzek and Carter, 2016, 2017).

Ozone is another oxidant; however, it is not easy to be applied for field application because it is a gas-phase chemical. Chemical oxidation using ozone can produce significant amount of heat, which can result in chemicals evaporation. The evaporated organic compounds can be captured by soil vapor extraction system for further treatment (Liu et al., 2014; Ji et al., 2017; Peluffo et al., 2018).

1.3.4 Bioremediation

Bioremediation is an environment-friendly approach, and it has been used for organic-compound-contaminated site remediation. Bioremediation applies microorganism for contaminant biodegradation, and it can achieve a complete mineralization of organic contaminant (Safdari et al., 2018). For biodegradable organic contaminants (e.g., petroleum hydrocarbons), bioremediation becomes a feasible remedial option for site cleanup. The efficiency of in situ bioremediation can be improved through biostimulation (via stimulation of indigenous bacteria by supplement of nutrients, oxygen, and other additives) and bioaugmentation (via inoculation of specific bacteria for enhanced contaminant biodegradation). Either biostimulation or bioaugmentation can be used to accelerate the site remediation efficiency when environmental conditions are not suitable for the occurrence of bioremediation (for example, lacking of electron acceptors, nutrients, and contaminant biodegrading bacteria) (Shahsavari et al., 2017).

Addition of biodegradable substrates (either solid or liquid substrates) (e.g., molasses, compost) for microbial proliferation has been used to enhance the contaminant biodegradation efficiencies (Kim et al., 2018). Frequent substrate supplement into the remediation zone can be applied to maintain a higher substrate concentrations during the bioremediation process (Gholami-Shiri et al., 2017; Dai et al., 2018). Addition of organic soluble substrates or emulsified substrates (electron donors) into the chlorinated-solvent-contaminated sites is able to enhance the reductive dechlorination rates of the chlorinated compounds (e.g., PCR, TCE) (Kanitkar et al., 2017; Sheu et al., 2018).

In situ bioremediation has the following advantages: relative simplicity, low cost, and higher removal efficiency for certain target contaminants (Sarkar et al., 2017). If indigenous microbial consortia and intrinsic biodegradation processes are used for site remediation, it is called intrinsic bioremediation. However, if substrates, nutrients, or specific microbial species are added into the subsurface to enhance the bioremediation efficiency, it is called enhanced bioremediation (Lin et al., 2017; Sheu et al., 2018; Yang et al., 2019).

1.4 Application of slow-release materials for a long-term remediation of NAPL-contaminated sites

The continuously released chemical contaminants from free or residual NAPLs usually cause a plume of contaminated groundwater. Thus, NAPLs become a long-term source of groundwater contamination, and thus, remediation of NAPL-contaminated sites is a lengthy and costly task. To reduce the operational and maintenance cost of the groundwater remedial system, one cost-effective approach that attracts increased attention is the application of SRMs within the plume or at plume boundary for in situ, passive, and long-term site cleanup and plume migration control.

A series of innovative SRMs have been developed to enhance the remediation of DNAPL (mainly chlorinated solvents), LNAPL, and heavy-metal-contaminated sites. The developed SRM products can be applied at different sites with different environmental and hydrogeological conditions to reduce the contaminant concentrations and environmental risks. From the sustainable development perspective, the developed SRM system is an environmental-friendly technology, which meets some of the requirements of the green remediation (in situ and passive system with the reduction of carbon emissions during site remediation), and thus, it is a cleaner and cutting-edge remediation technology (Cheng et al., 2019; O'Connor et al., 2018). This system will also have an impact on remediation markets and practices. With the application of the developed SRM technologies, the remediation cost can be notably reduced with an acceptable remediation efficiency.

1.4.1 Passive permeable reactive barrier system

The installation of permeable reactive barrier (PRB) or zones within aquifers for passive remediation of contaminated groundwater has been applied since the early 1990s (ITRC, 2004; Obiri-Nyarko et al., 2014; Sheu et al., 2016). When polluted groundwater flows via the hydraulic gradients to the PRB systems containing specially designed reactive materials or substrates, the contaminants are degraded, destroyed, or removed inside or around the downgradient area of the PRBs. Thus, groundwater emerging from the other side of the PRB system is remediated (ITRC, 2004; Obiri-Nyarko et al., 2014; Faisal et al., 2018). Compared to the currently used groundwater remediation technologies including pump and treat system and air sparging, the passive PRB system has the following advantages: treatment in situ, no groundwater extraction and reinjection, no above-ground water treatment systems, lower maintenance costs.

Reactive materials or substrates may be designed to remove contaminants from groundwater by (1) sorption or precipitation, (2) chemical oxidation, and (3) biodegradation. For example, different kinds of reactive metal media (e.g., zerovalent iron, zerovalent copper, ferrous-iron-contained composite) can be installed inside the permeable reactive zones or barriers to remediate chlorinated-solvent-contaminated aquifers via the chemical reduction mechanisms (Fu et al., 2014; He et al., 2015; Sheu et al., 2016; Su et al., 2017; Zhang et al., 2018b; Santisukkasaem and Das, 2018).

Supplement of alkaline materials [e.g., lime [$Ca(OH_2)$], sodium sulfide (Na_2S)] into heavy-metal-contaminated groundwater can create a reactive zone for heavy metal precipitation via the formation of heavy metal and sulfide complex so the heavy metals can be removed from the groundwater. The PRB system can be filled with limestone or crushed shells to neutralize acid mine drainage for pH control (Li et al., 2011; Huang et al., 2017). Sulfate-reducing bacteria can be applied to enhance the sulfate reduction process for sulfate and heavy metal removal from groundwater (Ayala-Parra et al., 2016).

During the practical application, sands can be used to mix the developed reactive materials to increase the permeability of the PRB system, which can improve the reactivity of the materials with groundwater. The funnel-and-gate PRBN system has been developed to divert the contaminated groundwater to the "gate" zone filled reactive materials, and thus, the contaminant removal efficiency can be improved. In most cases, the PRBs are filled with soils so they cannot be seen above ground (ITRC, 2004).

PRBs via the natural hydraulic gradient without groundwater pumping. PRBs can be designed and installed for the protection of downgradient receptors and for source zone remediation and containment to meet specific numerical objectives (ITRC, 2004). Figures 7.2 to 7.4 present the examples of PRB configurations.

To reduce the pollutant loadings to the PRB system, the major contaminant sources need to be removed so the remediation time can be minimized. Moreover, the polluted soils around the PRB zone also need to be excavated or cleaned to extend the life of the system. During the operational period of the PRB system, the performance evaluation needs to be conducted periodically to determine the effectiveness of the reactive materials (Yang et al., 2015; Gholami et al., 2018; Gibert et al., 2018; Cheng et al., 2019) (Figs. 7.2–7.4).

1.4.2 Biobarrier system

Many groundwater pollutants are organic compounds (e.g., petroleum hydrocarbons), and they are reduced chemicals. Therefore, these contaminants tend to proceed the oxidation process to innocuous compounds thermodynamically. As for the oxidized organic pollutants (e.g., nitrate, chlorinated solvents), they tend to undergo the reduction reactions. In the subsurface environments, different microorganisms can mediate these metabolic reactions to obtain energy and synthesize new cells when electron donors (e.g., petroleum hydrocarbon) and electron acceptors (e.g., oxygen, nitrate, ferric iron, sulfate) exist.

The biobarrier system usually include a porous wall (or barrier), which is installed perpendicular to the groundwater flow direction. It has a cross-sectional area covering the contaminant plume. The biobarrier contains filled materials, which are used to enhance the in situ bioremediation. The biobarrier can be used as a stand-alone system, or it can be used as a unit (or a process) of a series of treatment technologies (treatment train system). To design an effective and feasible biobarrier system, the contaminated site needs to be well characterized in terms of soil formation (soil permeability), groundwater hydrogeology (groundwater flow velocity, groundwater flow direction, hydraulic conductivity), and contaminant characteristics (contaminant

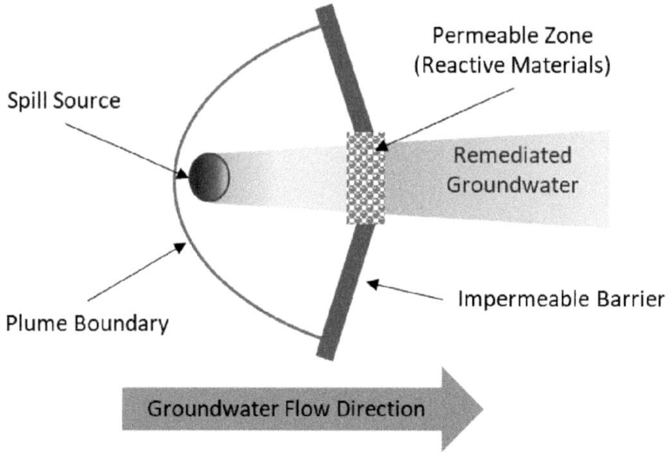

FIGURE 7.2 Schematic diagram showing the reactive materials in the funnel-and-gate PRB system.

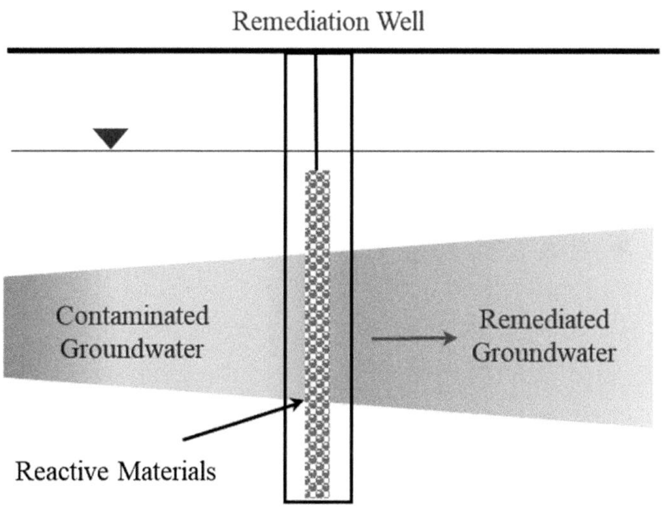

FIGURE 7.3 Schematic diagram showing the reactive materials in the PRB (remediation well) system.

concentrations, contaminant physical/chemical properties) (Daghio et al., 2015; Lee et al., 2018; Talley, 2016; Peng et al., 2018). Fig. 7.5 is the schematic diagram showing the biobarrier system.

Oxidative biodegradation process needs the supplement of electron acceptors, which can be applied by microorganisms. Microorganisms can respire the electrons removed from the contaminants (electron donors). This electron

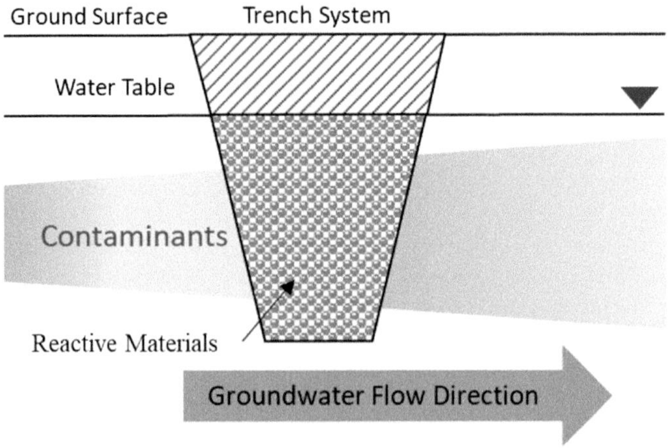

FIGURE 7.4 Schematic diagram showing the reactive materials in PRB (trench) system.

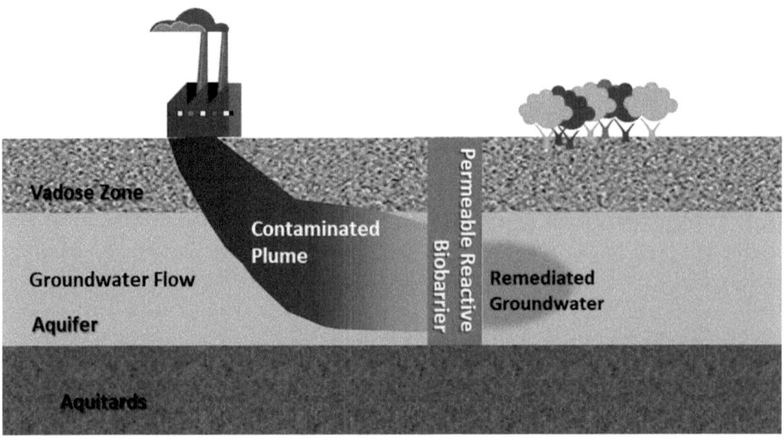

FIGURE 7.5 Schematic diagram showing the biobarrier system.

transfer can release energy to activate and drive the life functions of microorganisms. Oxygen is used as the electron acceptor under aerobic conditions. Nitrate, ferric iron, sulfate, and carbon dioxide can serve as electron acceptors under anaerobic conditions. Microbial consortia can perform a sequential utilization of different kind of electron acceptor at contaminated sites under a preferential order of redox potential. Oxygen is utilized preferentially over other electron acceptors because this reaction produces more energy to the microorganisms and results in a higher contaminant oxidation (biodegradation) rate. Because the amount of available oxygen is usually exceeded by the organic contaminants in site groundwater, anaerobic conditions are developed after the consumption of available oxygen especially in the highly contaminated areas. Oxygen supplement is usually required at organic-compound-contaminated sites to enhance the aerobic bioremediation process. Biobarrier systems containing oxygen-releasing compounds can be established for oxygen supplement to maintenance an acceptable dissolved oxygen concentrations in the subsurface to enhance the aerobic biodegradation of organic contaminants with a lower operational cost (Corseuil et al., 2015; Müller et al., 2017; Safdari et al., 2018).

However, oxygen supplement can also cause accidental oxidation of metals and minerals in their reduced forms. For example, reduced iron (ferrous iron) in the subsurface can be oxidized to ferric iron (oxidized form of iron), which can result in the reduced permeability of soil formation and clogging of the soil pores. Therefore, controlled release of oxygen (SRM technology for oxygen release) can be applied to minimize the amount of oxygen supplement and also minimize the clogging problems caused by the metal precipitation. Such problems can be also partially solved by the supplement of nitrate as the alternative electron acceptor to replace oxygen. Nitrate-releasing materials have been developed for controlled nitrate lease and nitrate supplement (Zhao et al., 2015). The nitrate-releasing materials have also been applied to enhance the bioremediation of petroleum-hydrocarbon-contaminated groundwater under denitrifying conditions (Müller et al., 2017; Chen et al., 2018).

Reductive biodegradation needs the supplement of electrons, which are transferred to pollutants by microorganisms. Appropriate electron donors are usually sufficient in the anaerobic subsurface environments. Therefore, reducible pollutants including hexavalent chromium, nitrate, sulfate, and trichloroethylene can be transformed under reduced conditions by anaerobic microorganisms. These pollutants usually persist under aerobic conditions with low reduction capacities. When applied in situ, reduced biobarriers can be filled with reducing equivalents such as organic substrates (e.g., composts, emulsified oil, molasses), which have been applied to stimulate the anaerobic microorganisms and enhance the anaerobic biodegradation of contaminants in oxidized state.

1.4.3 Slow-release liquid substrates

Slow-release liquid substrates (one kind of SRM) are usually produced by biodegradable liquid substrates such as edible oils (e.g., vegetable oil), which have been developed for continuous organic substrate release. Biodegradable carbon substrates released from the SRMs can consume dissolved oxygen in the subsurface and shift the redox conditions from oxidized stage to reduced stage, and thus, the anaerobic biodegradation of contaminants (e.g., chlorinated solvents, nitrate) can be enhanced. The developed slow-releasing liquid substrates can be injected into a line of remediation wells (perpendicular to the groundwater flow direction) or a trench system to form a biobarrier. The released substrates from the biobarrier can stimulate indigenous microbial consortia for the in situ site bioremediation.

Because edible oils have large globule diameters, higher viscosity, and molecular weights, using edible oils alone will result in the soil pore clogging problems (Liang et al., 2013). Therefore, edible oils can be emulsified to reduce their globule diameters with a range from micro- or nanoscale size. This can allow the liquids with small globule diameters to migrate to a farther downgradient area and form a wider biobarrier zone. Moreover, edible oils contain long-chain fatty acids, which are not easily degradable. During the emulsified oil production, some easily biodegradable substrates can be applied and mixed with the emulsified oil to improve its biodegradability and applicability for practical applications.

When applied in situ, substrates biodegradation under anaerobic conditions can result in groundwater acidification problem due to the production of fatty acids. The acidified groundwater will inhibit the biodegradation efficiency of organic contaminants. For example, the reductive dechlorinating of chlorinated solvents will be inhibited under low pH conditions ($pH < 5.5$) (Lien et al., 2016; Sheu et al., 2018). Thus, pH control and pH adjustment need to be conducted in situ when liquid substrates are injected into the subsurface to maintain an optimal pH range and an acceptable contaminant biodegradation efficiency.

The nanoscale zerovalent iron (nZVI) is an efficient and commonly used reducing agent for the reduction of chlorinated organic compounds via the chemical reaction (reduction) process (Sheu et al., 2016). The reactive wall (barrier) containing nZVI has been used for the remediation of chlorinated-solvent-contaminated groundwater. When nZVI is injected into the subsurface, nZVI can be oxidized to ferrous iron [Fe(II)] and the products include H_2 and OH^-. The produced hydrogen can serve as the electron donor and enhance the reductive dechlorination of chlorinated solvents under anaerobic conditions.

A long-lasting emulsified colloidal substrate (LECS) was developed for a long-term and continuous biodegradable carbon and nZVI releases to enhance the remediation of TCE-contaminated groundwater under biological dechlorinating and chemical reduction processes (Sheu et al., 2015). The LECS contained vegetable oil, molasses, lactate, surfactants (Simple Green and lecithin), minerals, and nZVI with a uniform droplet diameter of 0.7 μm. The emulsified solution can be also used as a dispensing agent, and nZVI particles (with diameter 100–200 nm) can distribute in the emulsion evenly without aggregation. Microcosm results showed that the LECS caused a rapid increase in the total organic carbon concentration (up to 488 mg/L), and reductive dechlorination of TCE was significantly enhanced. Acidification problem can be prevented due to the production of hydroxide ion after nZVI oxidation. The odor problem caused by the produced hydrogen sulfide can be reduced due to the formation of iron sulfide (Sheu et al., 2015).

2. Controlled-release materials for groundwater remediation

Because contaminated groundwater remediation is a lengthy and costly process, in situ and passive remedial systems have been developed and applied for site remediation to reduce the remediation cost. PRB (or barrier) systems have received a great deal of attention as a cost-effective method for in situ remediation of contaminated groundwater. The operational and maintenance cost of the PRB (or barrier) system can be significantly reduced if controlled-release materials are used in the PRB (or barrier) systems for groundwater remediation.

Controlled-release techniques have attracted increased attention in different fields including pharmaceutical, environmental, and agrochemical industries (Kamaly et al., 2016; Liang, and Chen, 2017; Cardarelli, 2018). Researchers applied paraffin, polyester mesh bags, Boler wax, Piccolyte resin S115, Epolene C-16, and stearic acid to produce the encapsulated or wrapping materials (Liu et al., 2014; Liang, and Chen, 2017; O'Connor et al., 2018; Tian et al., 2019). Some of these materials are limited for their applicability because they might be harmful to ecosystems, and they are not easily biodegradable.

Biodegradable polymers are commonly used materials used for controlled-release materials production. Among the commercialized polymers, polycaprolactone (PCL) $[(C_6H_{10}O_2)n]$ is one of the popular polymers, which has drawn increased attention due to its high biodegradability, high flexibility, and hydrophobic nature (Carr, 2016; O'Connor et al., 2018). However, the commercial application of PCL is limited due to its high production cost. To reduce the production cost and minimize the usage of PCL, some biodegradable natural biopolymers (e.g., starch, cellulose, chitin) with low cost have been applied to mix and blend with PCL to make some new materials with similar functions and properties (Rahim et al., 2016; Chen et al., 2018). Because starch is a biodegradable, abundant, renewable, inexpensive material (Wilpiszewska et al., 2016; Flores-Céspedes et al., 2018), it has been used with PCL for new controlled-release material production. A wide variety of controlled-release materials have been developed and successfully used in PRB and barrier systems for the removal of groundwater contaminants. Although the remedial goals vary from site to site, one of the main objectives for site remediation is to prevent the off-site contaminants migration (Liang et al., 2014; Bo et al., 2015; Xie et al., 2017).

2.1 Release of chemical oxidants

Researchers have developed various controlled-oxidant-release materials for a long-term oxidant release to enhance the ISCO for groundwater remediation (Yang et al., 2015; Liang et al., 2015; Yang et al., 2016; O'Connor et al., 2018). A well-based reactive barrier using potassium permanganate ($KMnO_4$)-

release materials has been developed as an in situ and passive system for the remediation of contaminated groundwater (Yao et al., 2016; Yang et al., 2018). The slowly released potassium permanganate materials can oxidize organic contaminants in situ and clean up contaminated groundwater.

Liang et al. (2015) developed an oxidant-release technology combining ISCO and PRB concepts with a controlled oxidant release rate to clean up groundwater polluted by TCE. In this study, $KMnO_4$ was released from the $KMnO_4$-releasing composite (PRC) for TCE oxidation. The PRC contained $KMnO_4$, PCL, and starch with a weight ratio of 2:1.14:0.96. Starch was blended with PCL to modify the physical properties of the PRC materials. Moreover, researchers developed persulfate-releasing materials by blending PCL, starch, and persulfate together for continuous persulfate (oxidant) release for petroleum-hydrocarbon-contaminated groundwater remediation with a persulfate release rate of 31 to 8 mg persulfate per day per g of material (Liang et al., 2014, 2015; Yang et al., 2018).

2.2 Release of dissolved oxygen

Oxygen-releasing material (ORM) has been produced and widely used to enhance the aerobic bioremediation of organic-compound-contaminated sites (Lin et al., 2017; Zhang et al., 2018c). Compatibility of the ORM with the environment and oxygen-releasing characteristics (Huang et al., 2015; Lin et al., 2017) usually affect the effectiveness of ORM application on in situ bioremediation.

Binding materials (e.g., cement, gypsum, starch, CaO_2) have been applied to mix with oxygen-releasing chemicals (e.g., CaO_2, MgO_2) to make ORMs for continuous oxygen release (Ahmad et al., 2007). However, some materials (e.g., cement, gypsum) are not biodegradable and might also cause clogging problem after site application (Shabalala et al., 2017; Fenyvesi et al., 2019). Lin et al. (2017) used the green materials biochar, alginate, polyvinyl alcohol (PVA), and citrate to prepare ORMs (Liang et al., 2015), and they found that the produced ORMs can provide dissolved oxygen effectively and enhance the growth of indigenous degraders for contaminant biodegradation in groundwater (O'Connor et al., 2018).

Lin et al. (2017) used a novel method to develop the encapsulated oxygen-releasing beads (encap-ORBs) to enhance the BTEX biodegradation in groundwater. Polyvinyl alcohol (PVA) was used to increase the integrality of the encap-ORBs. Results from their study show that the produced encap-ORBs had a high integrality and oxygen-releasing capacity for site remediation. Their results indicate that encap-ORBs can effectively release oxygen for more than 120 days.

Researchers have also reported that higher contaminant concentrations can be toxic to indigenous microorganisms, and thus, the growth of microorganism is inhibited (Chen et al., 2015). Encapsulation technique can be applied to reduce and alleviate the contaminant toxicity to indigenous microorganisms (Tsai et al., 2014; Gholami et al., 2019). Therefore, the effectiveness of contaminant biodegradation can be improved due to the increased microbial activity and population (Mosmeri et al., 2017).

The peroxide chemicals including CaO_2 and MgO_2 are used as the oxygen-releasing compound (ORC) to prepare ORMs for in situ groundwater bioremediation (Huang et al., 2015; Gholami et al., 2019). Oxygen can be produced when ORC reacts with water, and the produced oxygen can increase the dissolved oxygen concentration in groundwater. Moreover, CaO_2 nanoparticles with high surface areas have been prepared for efficient oxygen production and utilization, and thus, effective contaminant biodegradation can be obtained (Huang et al., 2015; Mosmeri et al., 2018).

Results from field application show that increased dissolved oxygen concentrations and decreased organic contaminant concentrations (e.g., petroleum hydrocarbons) were observed in groundwater after ORCs injection (Lu et al., 2017; Mosmeri et al., 2018). The results suggest that ORC application can enhance the aerobic biodegradation efficiency of contaminants. Reduction of operational and maintenance cost can be obtained with the application of the in situ and passive ORMs for groundwater remediation (Zhou et al., 2017; Zhang et al., 2018c; Gholami et al., 2019).

2.3 Release of carbon substrates for chlorinated-solvent-contaminated groundwater remediation

Chlorinated solvents are commonly found organic pollutants in groundwater (Kao et al., 2016; Su et al., 2017). Chlorinated solvents usually form DNAPLs, which result in a continuous and long-term groundwater contamination source (Lee et al., 2015; Cápiro et al., 2015; Mao et al., 2015). The dissolution of DNAPL causes a contaminant plume in the subsurface (Chiu et al., 2013; Wilkin et al., 2014; Frascari et al., 2015).

Researchers reported that the anaerobic reductive dechlorination is a feasible and cost-effective remedial option for the remediation chlorinated-solvent-contaminated groundwater (Liang et al., 2015; Lin et al., 2015; Sheu et al., 2016). To activate the reductive dechlorination process, alternate carbon sources (or primary substrates) need to be supplied (Chen et al., 2014). Under sequential reductive dechlorinating conditions, chlorinated solvents (e.g., PCE) can be microbiologically transformed to less-chlorinated compounds (e.g., TCE, DCE, VC) (Cobas et al., 2013; Lin et al., 2018). A chlorine atom can be replaced by a hydrogen atom at each reductive dechlorination step, and the dechlorination efficiency for each step decreases as chlorine is removed (Sheu et al., 2018; Hermon et al., 2019). Dehalorespiring bacteria including *Dehalococcoides*, *Trichlorobacter*, *Anaeromyxobacter*, *Dehalobacter*, and *Desulfitobacterium* can be used to activate and enhance the reduction of chlorinated compounds (Kwon et al., 2016; Kao et al., 2016).

Soluble substrates (electron donors), such as lactate and acetate, or immobile substrates, such as bark mulch, compost, and peanut shell, can be supplied as the primary substrates. However, soluble substrates must be supplied frequently due to the higher solubility characteristics of the substrates. This will cause the increase in capital and operational costs of the remediation process. Although solid substrates can be sued for continuous substrate supplement, their application generally results in the decrease in soil permeability and clogging problems (Löffler et al., 2013; Cápiro et al., 2015; Tang et al., 2016; Heyob et al., 2017).

Slow-release liquid substrates produced by edible oils have been developed for a long-term organic substrate supplement. The slowly and continuously released carbon substrates could enhance anaerobic biodegradation of chlorinated solvents (Yu, 2015; Sheu et al., 2016; Wu et al., 2018). The slow-release liquid substrates are injected into the subsurface to form a biobarrier system, and the released substrates can stimulate aquifer microbial activity for chlorinated-solvent dechlorination (Obiri-Nyarko et al., 2014; Thornton et al., 2016; Sheu et al., 2016). Application of the anaerobic biodegradation and PRB (or permeable reactive zone) for chlorinated-solvent-contaminated aquifer cleanup would be a feasible and cost-effective remedial option (Kuo et al., 2012; Faisal and Abd Ali, 2017; Liu et al., 2017).

2.4 Release of carbon substrates and sulfate for heavy-metal-contaminated groundwater remediation

Dissolved oxygen (DO) can be used as the electron acceptor and applied to improve the aerobic biodegradation efficiency. Application of ORCs is one of the options for continuous DO supplement to the contaminated sites to enhance the aerobic biodegradation mechanisms (Huang et al., 2017). Anaerobic microorganisms can use other electron acceptors such as NO_3^-, Mn^{4+}, Fe^{3+}, SO_4^{2-} , and CO_2 to enhance the anaerobic bioremediation efficiency under anaerobic conditions (Thornton et al., 2016).

Most subsurface environments are under reduced conditions if they are contaminated by organic contaminants such as petroleum hydrocarbons (Kao et al., 2016). Because sulfate is more water-soluble than oxygen [the solubility of sulfate salts varied from 2 g/L ($CaSO_4$) to 500 g/L ($ZnSO_4$)] (Huang et al., 2017), sulfate reduction becomes a more practical and cost-effective remedial option to clean up petroleum-hydrocarbon-polluted sites. Sulfate-reducing bacteria (SRB) apply sulfate as the electron acceptor and use organic compounds such as petroleum hydrocarbons as the electron donors under the sulfate reduction processes. Advantages of using sulfate as the electron acceptor include the following: (1) high water solubility, (2) low chemical cost, and (3) wider pH range (pH 5–9.5) during application (Dixit et al., 2015; Zhang et al., 2016a; Huang et al., 2017).

Sulfide is the end product of sulfate reduction process. Sulfide can react with cations and form insoluble precipitates. Supplement of carbon sources (electron donors) to enhance the sulfate reduction process can result in metal precipitates. This process is called bioprecipitation (formation of metal sulfide), which is applied for metal removal (Kerkar and Das, 2017; Huang et al., 2017). This process can stabilize the metals in the solid phase (Kumar et al., 2015; Kiran et al., 2016; Peng et al., 2018). It can be used as an alternative method for metal removal from groundwater (Xiong et al., 2012). Compared to the conventional heavy-metal removal techniques (e.g., adsorption, ion-exchange, chemical precipitation) (Xiong et al., 2012; Kiran et al., 2016), bioprecipitation via the sulfate reduction is a more cost-effective and feasible in situ remedial option for the remediation of heavy-metal-contaminated groundwater (Hsu et al., 2010; Zhang and wang, 2016; Huang et al., 2017).

When toluene is used as the target contaminant (carbon source), the stoichiometric equation for the sulfate reduction is described in the following (Eq. 1) (Hoa et al., 2007; Mal et al., 2016):

$$C_7H_8 + 4.5\ SO_4^{2-} + 9\ H^+ \rightarrow 2.25\ H_2S + 2.25\ HS^- + 7\ CO_2 + 4\ H_2O \qquad (7.1)$$

The produced hydrogen sulfide can react with metal ions to form metal sulfide precipitates (Cabrera et al., 2006) (Eq. 2).

$$M + H_2S \rightarrow MS_{(S)} + 2H^+ \qquad (7.2)$$

The concept of PRBs, slow organic compound release, slow sulfate release, and sulfate reduction can be applied and combined for the development of sulfate-reducing biobarrier system containing sulfate-releasing materials and OCR materials for continuous and passive sulfate and biodegradable organic carbon supplement to reduce the remediation cost of active sulfate injection system.

Huang et al. (2015) developed sulfate-releasing materials for continuous sulfate release for the enhancement of in situ sulfate reduction.

The sulfate-releasing materials were prepared by mixing polylactic acid (PLA), magnesium sulfate, starch or rice husk, and water together via a kneader. PLA was used as a binder and magnesium sulfate was used as the sulfate supplement for in situ sulfate reduction. Starch or rice husk powder was used to increase the permeability of the mixture. The sulfate-releasing rate varied from 1.5 to 3 mg sulfate per day per g of material for 180 days with a release rate of 2×10^{-2} 1/d.

Kuo et al. (2014) and Sheu et al. (2016) developed emulsified substrate for continuous carbon substrate supplement in groundwater. The emulsified substrate contained soybean oil, lactate, biodegradable surfactant, and nutrients, and thus, it was able to provide carbon sources continuously. In their practical field-scale study, emulsified substrate solution was injected into a chlorinated-solvent-contaminated site to form a biobarrier for the containment and bioremediation of the chlorinated-solvent plume (Kuo et al., 2014; Sheu et al.,

2016). Field results show that the total organic carbon (TOC) went up from 12 mg/L to 9,800 mg/L after 45 days of emulsified substrate injection and dropped to 1,500 mg/L after 230 days of substrate injection. The increased TOC resulted in the depletion of groundwater DO in the injection wells after 5 days of the substrate supplement.

Sulfate-releasing materials and OCR materials (e.g., emulsified substrate) can be used to form the downgradient bioremediation zone to biodegrade heavy-metal-contaminated groundwater migrating into the biobarrier zone. The OCR materials could release biodegradable organic compounds continuously, and the sulfate-releasing materials would release sulfate slowly and continuously when they contact with groundwater. The released sulfate would then enhance in situ organic compounds biodegradation under sulfate-reducing conditions. The produced sulfide would react with heavy metal ions and from metal-sulfide precipitates, which would reduce the metal concentration in groundwater and also reduce the risks to human health and ecosystem caused by the high metal concentrations in groundwater.

3. Practical application

The occurrence of groundwater contamination and the quality of groundwater have become a major issue since the discovery of numerous hazardous waste sites in the 1970s. Alarmingly high concentrations of various organic/inorganic contaminants have been detected in groundwater at many locations in developing and developed countries worldwide. Because it is often not possible to locate and remove the free-phase and residual NAPLs, site remediation must focus on preventing further migration of the dissolved contamination. This contaminant plume control must be maintained for a long period of time. Therefore, some more economic approaches are desirable for groundwater remediation to provide for a long-term control of contaminated groundwater.

One cost-effective and practical approach to clean up contaminated aquifers that is attracting increased attention is the application of SRMs for a long-term contaminant removal and transport control. Because SRM application for site remediation is an in situ and passive technology, the operational and maintenance cost can be significantly reduced. To improve the effectiveness of the SRM system on plume control, a more aggressive remedial option can be taken for source removal or control. For example, hot-spot soil excavation, free product extraction, and surfactant flushing and extraction can be applied at the contaminant source zone treatment. After the aggressive source removal, the passive SRM system can be used for the low-level contaminant treatment to clean up dissolved phase contaminants and also to control plume migration.

Thus, a treatment train remedial system containing two or several different treatment technologies can be designed and applied for contaminated groundwater remediation. The integrated processes or a series of treatments

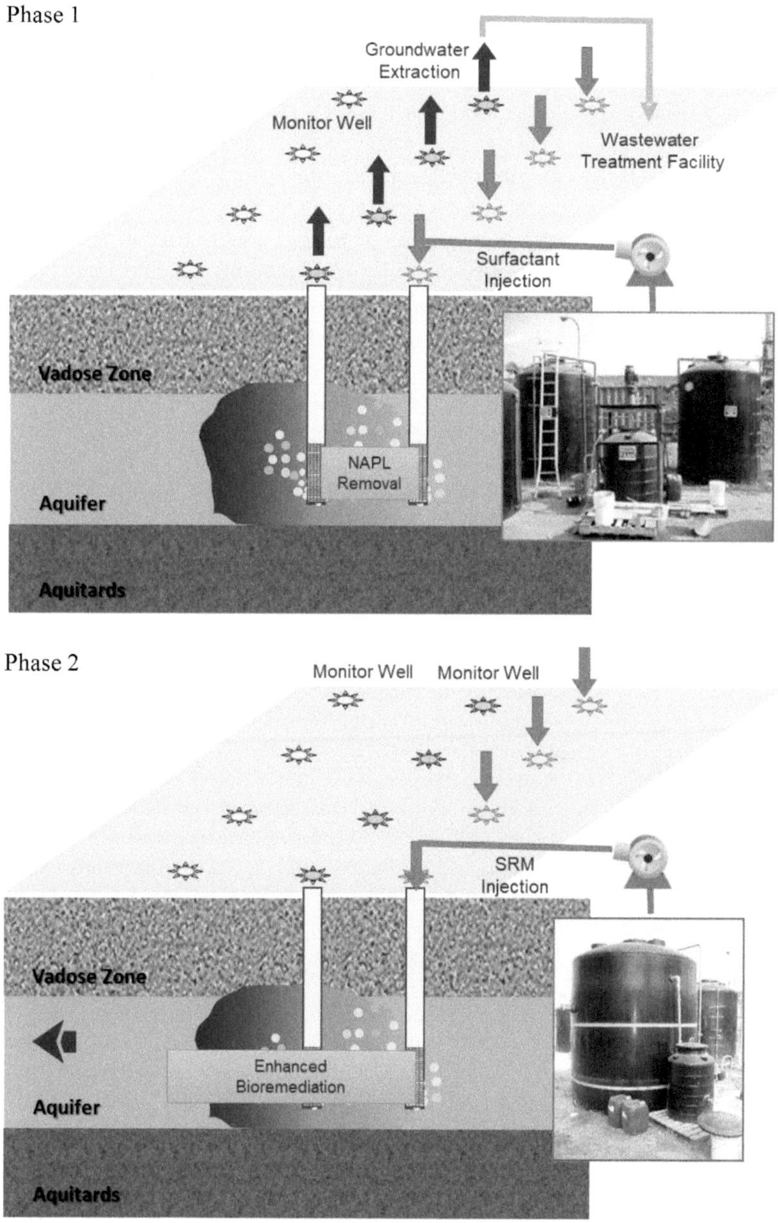

FIGURE 7.6 Schematic diagram of the proposed treatment train system: Phase 1, Phase 2.

are combined in sequence to provide a more realistic, practical, and less costly treatment option. In the treatment train design, SRM system can be used as the final stage following the aggressive technologies for a long-term control.

Fig. 7.6 shows the schematic diagram of the proposed treatment train system containing the multiphase extraction system or surfactant flushing/extraction installed at the source zone for DNAPL (free product) (Phase 1) removal followed by in situ and passive SRM system for residual DNAPL removal and plume control via the enhanced bioremediation mechanism at a DNAPL (e.g., TCE)-contaminated groundwater site (Phase 2). During the Phase 2 stage, emulsified oil, which is served as SRM, can be injected into the aquifer for the continuous supplement of biodegradable carbon substrate to enhance the reductive dechlorination (biodegradation) of chlorinated compounds. The treatment train system using SRM in Phase 2 stage would be expected to provide a more efficient and cost-effective alternative to remediate DNAPL-contaminated sites.

References

Ameta, S., Ameta, R. (Eds.), 2018. Advanced Oxidation Processes for Wastewater Treatment: Emerging Green Chemical Technology. Academic Press.

Ahmad, F., Schnitker, S.P., Newell, C.J., 2007. Remediation of RDX- and HMX-contaminated groundwater using organic mulch permeable reactive barriers. Journal of Contaminant Hydrology 90 (1-2), 1–20.

Ayala-Parra, P., Sierra-Alvarez, R., Field, J.A., 2016. Treatment of acid rock drainage using a sulfate-reducing bioreactor with zero-valent iron. Journal of Hazardous Materials 308, 97–105.

Baciocchi, R., D'Aprile, L., Innocenti, I., Massetti, F., Verginelli, I., 2014. Development of technical guidelines for the application of in-situ chemical oxidation to groundwater remediation. Journal of Cleaner Production 77, 47–55.

Besha, A.T., Bekele, D.N., Naidu, R., Chadalavada, S., 2017. Recent advances in surfactant-enhanced In-Situ Chemical Oxidation for the remediation of non-aqueous phase liquid contaminated soils and aquifers. Environmental Technology and Innovation 9, 303–322.

Bo, L., Shengen, Z., Chang, C.C., Zhanfeng, D., Hongxiang, L., 2015. Emerging pollutants-Part II: treatment. Water Environment Research 87 (10), 1873–1900.

Boal, A.K., Rhodes, C., Garcia, S., 2015. Pump-and-Treat groundwater remediation using chlorine/ultraviolet advanced oxidation processes. Groundwater Monitoring and Remediation 35 (2), 93–100.

Burris, D.R., Johnson, P.C., Hinchee, R.E., Dahlen, P.R., 2018. 1, 4-Dioxane soil remediation using enhanced soil vapor extraction (XSVE): II. Modeling. Groundwater Monitoring and Remediation 38 (2), 49–56.

Cabrera, G., Pérez, R., Gomez, J.M., Abalos, A., Cantero, D., 2006. Toxic effects of dissolved heavy metals on Desulfovibrio vulgaris and Desulfovibrio sp. Strains. Journal of Hazardous Materials 135, 40–46.

Cápiro, N.L., Löffler, F.E., Pennell, K.D., 2015. Spatial and temporal dynamics of organohalide-respiring bacteria in a heterogeneous PCE–DNAPL source zone. Journal of Contaminant Hydrology 182, 78–90.

Cardarelli, N.F., 2018. Controlled Release Pesticides Formulations: 0. CRC press.

Carr, S.A., 2016. The Use of Poly-lactic Acid and Poly-(ε-Caprolactone) Biodegradable Polymers to Design Controlled-Release KMnO 4 Structures for Water Remediation. Doctoral Dissertation. North Carolina Agricultural and Technical State University.

Chao, H.P., Hsieh, L.H.C., Tran, H.N., 2018. Increase in volatilization of organic compounds using air sparging through addition in alcohol in a soil-water system. Journal of Hazardous Materials 344, 942–949.

Chen, J., Lü, S., Zhang, Z., Zhao, X., Li, X., Ning, P., Liu, M., 2018. Environmentally friendly fertilizers: a review of materials used and their effects on the environment. The Science of the Total Environment 613, 829–839.

Chen, L., Liu, Y., Xia, F., Liu, F., Li, B., 2018. Influences of benzene or toluene on dechlorination of perchloroethene and trichloroethene in a simulated zero-valent iron permeable reactive barrier. Water Quality Research Journal 53 (2), 61–71.

Chen, M., Xu, P., Zeng, G., Yang, C., Huang, D., Zhang, J., 2015. Bioremediation of soils contaminated with polycyclic aromatic hydrocarbons, petroleum, pesticides, chlorophenols and heavy metals by composting: applications, microbes and future research needs. Biotechnology Advances 33 (6), 745–755.

Chen, M.Y., Su, Y.F., Shih, Y.H., 2014. Effect of geochemical properties on degradation of trichloroethylene by stabilized zerovalent iron nanoparticle with Na-acrylic copolymer. Journal of Environmental Management 144, 88–92.

Cheng, Z., Feng, K., Xu, D., Kennes, C., Chen, J., Chen, D., Dionysiou, D.D., 2019. An innovative nutritional slow-release packing material with functional microorganisms for biofiltration: characterization and performance evaluation. Journal of Hazardous Materials 366, 16–26.

Chiu, H.Y., Verpoort, F., Liu, J.K., Chang, Y.M., Kao, C.M., 2017. Using intrinsic bioremediation for petroleum–hydrocarbon contaminated groundwater cleanup and migration containment: effectiveness and mechanism evaluation. Journal of the Taiwan Institute of Chemical Engineers 72, 53–61.

Chiu, H.Y., Hong, A., Lin, S.L., Surampalli, R.Y., Kao, C.M., 2013. Application of natural attenuation for the control of petroleum hydrocarbon plume: mechanisms and effectiveness evaluation. Journal of Hydrology 505, 126–137.

Cobas, M., Ferreira, L., Tavares, T., Sanroman, M.A., Pazos, M., 2013. Development of permeable reactive biobarrier for the removal of PAHs by Trichoderma longibrachiatum. Chemosphere 91, 711–716.

Corseuil, H.X., Gomez, D.E., Schambeck, C.M., Ramos, D.T., Alvarez, P.J., 2015. Nitrate addition to groundwater impacted by ethanol-blended fuel accelerates ethanol removal and mitigates the associated metabolic flux dilution and inhibition of BTEX biodegradation. Journal of Contaminant Hydrology 174, 1–9.

Daghio, M., Tatangelo, V., Franzetti, A., Gandolfi, I., Papacchini, M., Careghini, A., Bestetti, G., 2015. Hydrocarbon degrading microbial communities in bench scale aerobic biobarriers for gasoline contaminated groundwater treatment. Chemosphere 130, 34–39.

Dai, Y., Liang, Y., Xu, X., Zhao, L., Cao, X., 2018. An integrated approach for simultaneous immobilization of lead in both contaminated soil and groundwater: laboratory test and numerical modeling. Journal of Hazardous Materials 342, 107–113.

Devi, E.S., Alanthadka, A., Tamilselvi, A., Nagarajan, S., Sridharan, V., Maheswari, C.U., 2016. Metal-free oxidative amidation of aldehydes with aminopyridines employing aqueous hydrogen peroxide. Organic and Biomolecular Chemistry 14 (35), 8228–8231.

Dixit, R., Malaviya, D., Pandiyan, K., Singh, U., Sahu, A., Shukla, R., Paul, D., 2015. Bioremediation of heavy metals from soil and aquatic environment: an overview of principles and criteria of fundamental processes. Sustainability 7 (2), 2189–2212.

Faisal, A.A., Abd Ali, Z.T., 2017. Remediation of groundwater contaminated with the lead–phenol binary system by granular dead anaerobic sludge-permeable reactive barrier. Environmental Technology 38 (20), 2534–2542.

Faisal, A.A.H., Sulaymon, A.H., Khaliefa, Q.M., 2018. A review of permeable reactive barrier as passive sustainable technology for groundwater remediation. International Journal of Environmental Science and Technology 15 (5), 1123–1138.

Fenyvesi, É., Gruiz, K., Morillo, E., Villaverde, J., 2019. Traditional and innovative methods for physical and chemical remediation of soil contaminated with organic contaminants. In: Engineering Tools for Environmental Risk Management: 4. Risk Reduction Technologies and Case Studies, p. 285.

Fetter, C.W., Boving, T., Kreamer, D., 2017. Contaminant Hydrogeology. Waveland Press.

Field, M.S., 2018. Investigating and remediating contaminated karst aquifers. In: Karst Groundwater Contamination and Public Health. Springer, Cham, pp. 101–115.

Flores-Céspedes, F., Daza-Fernández, I., Villafranca-Sánchez, M., Fernández-Pérez, M., Morillo, E., Undabeytia, T., 2018. Lignin and ethylcellulose in controlled release formulations to reduce leaching of chloridazon and metribuzin in light-textured soils. Journal of Hazardous Materials 343, 227–234.

Frascari, D., Zanaroli, G., Danko, A.S., 2015. In situ aerobic cometabolism of chlorinated solvents: a review. Journal of Hazardous Materials 283, 382–399.

Frollini, E., Piscitelli, D., Verginelli, I., Baciocchi, R., Petitta, M., 2016. A methodological approach to assess the dissolution of residual LNAPL in saturated porous media and its effect on groundwater quality: preliminary experimental results. Water, Air, and Soil Pollution 227 (10), 379.

Fu, F., Dionysiou, D.D., Liu, H., 2014. The use of zero-valent iron for groundwater remediation and wastewater treatment: a review. Journal of Hazardous Materials 267, 194–205.

Gholami, F., Mosmeri, H., Shavandi, M., Dastgheib, S.M.M., Amoozegar, M.A., 2019. Application of encapsulated magnesium peroxide (MgO_2) nanoparticles in permeable reactive barrier (PRB) for naphthalene and toluene bioremediation from groundwater. The Science of the Total Environment 655, 633–640.

Gholami, F., Shavandi, M., Dastgheib, S.M.M., Amoozegar, M.A., 2018. Naphthalene remediation form groundwater by calcium peroxide (CaO_2) nanoparticles in permeable reactive barrier (PRB). Chemosphere 212, 105–113.

Gholami-Shiri, J., Mowla, D., Dehghani, S., Setoodeh, P., 2017. Exploitation of novel synthetic bacterial consortia for biodegradation of oily-sludge TPH of Iran gas and oil refineries. Journal of environmental chemical engineering 5 (3), 2964–2975.

Gibert, O., Assal, A., Devlin, H., Elliot, T., Kalin, R.M., 2018. Performance of a field-scale biological permeable reactive barrier for in-situ remediation of nitrate-contaminated groundwater. The Science of the Total Environment.

Gong, Y., Li, J., Zhang, Y., Zhang, M., Tian, X., Wang, A., 2016. Partial degradation of levofloxacin for biodegradability improvement by electro-Fenton process using an activated carbon fiber felt cathode. Journal of Hazardous Materials 304, 320–328.

Grimski, D., Makeschin, F., Glante, F., Bartke, S., 2018. INSPIRATION: stakeholder perspectives on future research needs in soil, land use, and land management—towards a strategic research agenda for europe. In: International Yearbook of Soil Law and Policy 2017. Springer, Cham, pp. 475–497.

Guo, Z., Brusseau, M.L., Fogg, G.E., 2019. Determining the long-term operational performance of pump and treat and the possibility of closure for a large TCE plume. Journal of Hazardous Materials 365, 796–803.

He, Y.T., Wilson, J.T., Su, C., Wilkin, R.T., 2015. Review of abiotic degradation of chlorinated solvents by reactive iron minerals in aquifers. Groundwater Monitoring and Remediation 35 (3), 57–75.

Hermon, L., Hellal, J., Denonfoux, J., Vuilleumier, S., Imfeld, G., Ferreira, S., Joulian, C., 2019. Functional genes and bacterial communities during organohalide respiration of chloroethenes in microcosms of multi-contaminated groundwater. Frontiers in Microbiology 10, 89.

Heyob, K.M., Blotevogel, J., Brooker, M., Evans, M.V., Lenhart, J.J., Wright, J., Mouser, P.J., 2017. Natural attenuation of nonionic surfactants used in hydraulic fracturing fluids: degradation rates, pathways, and mechanisms. Environmental science and technology 51 (23), 13985–13994.

Hoa, T.T.H., Liamleam, W., Annachhatre, A.P., 2007. Lead removal through biologicalsulfate reduction process. Bioresource Technology 98, 2538–2548.

Hoponick, J.R., 2006. Status Report on Innovative In Situ Remediation Technologies Available to Treat Perchlorate-Contaminated Groundwater. U.S. EPA, Office of Superfund Remediation and Technology Innovation Technology Innovation and Field Services Division, Washington, D.C.

Hsu, H.F., Jhuo, Y.S., Kumar, M., Ma, Y.S., Lin, J.G., 2010. Simultaneous sulfate reduction and copper removal by a PVA-immobilized sulfate reducing bacterial culture. Bioresource Technology 101, 4354–4361.

Huang, G., Liu, F., Yang, Y., Deng, W., Li, S., Huang, Y., Kong, X., 2015. Removal of ammonium-nitrogen from groundwater using a fully passive permeable reactive barrier with oxygen-releasing compound and clinoptilolite. Journal of Environmental Management 154, 1–7.

Huang, S., Yang, Y., Li, Q., Su, Z., Yuan, C., Ouyang, K., 2017. Evaluation of the effects of lime–bassanite–charcoal amendment on the immobilization of cadmium in contaminated soil. Bulletin of Environmental Contamination and Toxicology 98 (3), 433–438.

Huang, W.H., Dong, C.D., Chen, C.W., Surampalli, R.Y., Kao, C.M., 2017. Application of sulfate reduction mechanisms for the simultaneous bioremediation of toluene and copper contaminated groundwater. International Biodeterioration and Biodegradation 124, 215–222.

Huang, W.H., Sheu, Y.T., Lien, P.J., Hsiao, Y.S., Kao, C.M., 2014. Investigation and remedial approach development for a TCE spill site: a case study. In: Advanced Materials Research, vol. 912. Trans Tech Publications, pp. 1884–1887.

Interstate Technology and Regulatory Council (ITRC), 2003. An Introduction to Characterizing Sites Contaminated with DNAPLs. Technical/Regulatory Guidance.

Interstate Technology and Regulatory Council (ITRC), 2004. Strategies for Monitoring the Performance of DNAPL Source Zone Remedies. Technical/Regulatory Guidance.

Jacobson, S.N., 2017. Non-Aqueous Phase Liquid Transport and Dynamics in Simulated Karst Conduits.

Ji, Y., Wang, L., Jiang, M., Lu, J., Ferronato, C., Chovelon, J.M., 2017. The role of nitrite in sulfate radical-based degradation of phenolic compounds: an unexpected nitration process relevant to groundwater remediation by in-situ chemical oxidation (ISCO). Water Research 123, 249–257.

Kamaly, N., Yameen, B., Wu, J., Farokhzad, O.C., 2016. Degradable controlled-release polymers and polymeric nanoparticles: mechanisms of controlling drug release. Chemical Reviews 116 (4), 2602–2663.

Kanitkar, Y.H., Stedtfeld, R.D., Hatzinger, P.B., Hashsham, S.A., Cupples, A.M., 2017. Development and application of a rapid, user-friendly, and inexpensive method to detect Dehalococcoides sp. reductive dehalogenase genes from groundwater. Applied Microbiology and Biotechnology 101 (11), 4827–4835.

Kao, C.M., Liao, H.Y., Chien, C.C., Tseng, Y.K., Tang, P., Lin, C.E., Chen, S.C., 2016. The change of microbial community from chlorinated solvent-contaminated groundwater after biostimulation using the metagenome analysis. Journal of Hazardous Materials 302, 144–150.

Kerkar, S., Das, K.R., 2017. Bioremediation of heavy metals from saline water using hypersaline dissimilatory sulfate-reducing bacteria. In: Marine Pollution and Microbial Remediation. Springer, Singapore, pp. 15–28.

Kim, J., Lee, A.H., Chang, W., 2018. Enhanced bioremediation of nutrient-amended, petroleum hydrocarbon-contaminated soils over a cold-climate winter: the rate and extent of hydrocarbon biodegradation and microbial response in a pilot-scale biopile subjected to natural seasonal freeze-thaw temperatures. The Science of the Total Environment 612, 903–913.

Kiran, M.G., Pakshirajan, K., Das, G., 2016. Heavy metal removal from multicomponent system by sulfate reducing bacteria: mechanism and cell surface characterization. Journal of Hazardous Materials 324, 62–70.

Kumar, N., Chaurand, P., Rose, J., Diels, L., Bastiaens, L., 2015. Synergistic effects of sulfate reducing bacteria and zero valent iron on zinc removal and stability in aquifer sediment. Chemical Engineering Journal 260, 83–89.

Kuo, Y.C., Liu, J.K., Chien, H.Y., Chen, C.C., Kao, C.M., 2012. Remediation of TCE-contaminated groundwater using integrated biosparging and enhanced bioremediation system. Research Journal of Chemistry and Environment 16, 37.

Kuo, Y.C., Liang, S.H., Wang, S.Y., Chen, S.H., Kao, C.M., 2014. Application of emulsified substrate biobarrier to remediate TCE-contaminated groundwater: a pilot-scale study. Journal of Hazardous, Toxic, and Radioactive Waste 18 (2).

Kwon, K., Shim, H., Bae, W., Oh, J., Bae, J., 2016. Simultaneous biodegradation of carbon tetrachloride and trichloroethylene in a coupled anaerobic/aerobic biobarrier. Journal of Hazardous Materials 313, 60–67.

Lee, S.S., Kaown, D., Lee, K.K., 2015. Evaluation of the fate and transport of chlorinated ethenes in a complex groundwater system discharging to a stream in Wonju, Korea. Journal of Contaminant Hydrology 182, 231–243.

Lee, T.H., Tsang, D.C.W., Chen, W.H., Verpoort, F., Sheu, Y.T., Kao, C.M., 2018. Application of an emulsified polycolloid substrate biobarrier to remediate petroleum-hydrocarbon contaminated groundwater. Chemosphere 219, 444–455.

Lenhard, R.J., Rayner, J.L., Davis, G.B., 2017. A practical tool for estimating subsurface LNAPL distributions and transmissivity using current and historical fluid levels in groundwater wells: effects of entrapped and residual LNAPL. Journal of Contaminant Hydrology 205, 1–11.

Li, Y., Liu, Z., Li, Q., Zhao, Z., Liu, Z., Zeng, L., 2011. Removal of arsenic from Waelz zinc oxide using a mixed NaOH–Na2S leach. Hydrometallurgy 108 (3–4), 165–170.

Liang, C., Chen, C.Y., 2017. Characterization of a sodium persulfate sustained release rod for in situ chemical oxidation groundwater remediation. Industrial and Engineering Chemistry Research 56 (18), 5271–5276.

Liang, S.H., Chen, K.F., Wu, C.S., Lin, Y.H., Kao, C.M., 2014. Development of KMn04-releasing composites for in situ chemical oxidation of TCE-contaminated groundwater. Water Research 54, 149–158.

Liang, S.H., Wang, S.Y., Chang, Y.M., Kao, C.M., 2015. Treatment of TCE-contaminated groundwater using in situ potassium permanganate oxidation: effects and kinetics evaluation. Research Journal of BioTechnology 10, 20–24.

Liang, S.H., Kuo, Y.C., Chen, S.H., Chen, C.Y., Kao, C.M., 2013. Development of a slow polycolloid-releasing substrate (SPRS) biobarrier to remediate TCE-contaminated aquifers. Journal of Hazardous Materials 254–255 (15), 107–115.

Liao, X., Yan, X., Ma, D., Zhao, D., Sun, L., Li, Y., Tao, H., 2018. The research and development of technology for contaminated site remediation. In: Twenty Years of Research and Development on Soil Pollution and Remediation in China. Springer, Singapore, pp. 785–798.

Lien, P.J., Yang, Z.H., Chang, Y.M., Tu, Y.T., Kao, C.M., 2016. Enhanced bioremediation of TCE-contaminated groundwater with coexistence of fuel oil: effectiveness and mechanism study. Chemical Engineering Journal 289, 525–536.

Lin, C.W., Wu, C.H., Guo, P.Y., Chang, S.H., 2017. Innovative encapsulated oxygen-releasing beads for bioremediation of BTEX at high concentration in groundwater. Journal of Environmental Management 204, 12–16.

Lin, C.H., Shih, Y.H., MacFarlane, J., 2015. Amphiphilic compounds enhance the dechlorination of pentachlorophenol with Ni/Fe bimetallic nanoparticles. Chemical Engineering Journal 262, 59–67.

Lin, J.L., Dong, C.D., Chen, C.W., Chen, S.H., Hsieh, T.E., Kao, C.M., 2017. Development of a two-stage washing and biodegradation system to remediate octachlorinated dibenzo-p-dioxin-contaminated soils. International Journal of Environmental Science and Technology 14 (9), 1919–1930.

Lin, K.S., Mdlovu, N.V., Chen, C.Y., Chiang, C.L., Dehvari, K., 2018. Degradation of TCE, PCE, and 1, 2–DCE DNAPLs in contaminated groundwater using polyethylenimine-modified zero-valent iron nanoparticles. Journal of Cleaner Production 175, 456–466.

Liu, H., Bruton, T.A., Doyle, F.M., Sedlak, D.L., 2014. In situ chemical oxidation of contaminated groundwater by persulfate: decomposition by Fe (III)-and Mn (IV)-containing oxides and aquifer materials. Environmental science and technology 48 (17), 10330–10336.

Liu, S., Yang, Q., Yang, Y., Ding, H., Qi, Y., 2017. In situ remediation of tetrachloroethylene and its intermediates in groundwater using an anaerobic/aerobic permeable reactive barrier. Environmental Science and Pollution Research 24 (34), 26615–26622.

Löffler, F.E., Yan, J., Ritalahti, K.M., Adrian, L., Edwards, E.A., Konstantinidis, K.T., Müller, J.A., Fullerton, H., Zinder, S.H., Spormann, A.M., 2013. Dehalococcoides mccartyi gen. nov., sp. nov., obligately organohalide-respiring anaerobic bacteria relevant to halogen cycling and bioremediation, belong to a novel bacterial class, Dehalococcoidia classis nov., order Dehalococcoidales ord. nov. and family Dehalococcoidaceae fam. nov., within the phylum Chloroflexi. International Journal of Systematic and Evolutionary Microbiology 63, 625–635.

Lu, S., Zhang, X., Xue, Y., 2017. Application of calcium peroxide in water and soil treatment: a review. Journal of Hazardous Materials 337, 163–177.

Mal, J., Nancharaiah, Y.V., van Hullebusch, E.D., Lens, P.N., 2016. Effect of heavy metal co-contaminants on selenite bioreduction by anaerobic granular sludge. Bioresource Technology 206, 1–8.

Mao, D., Revil, A., Hort, R.D., Munakata-Marr, J., Atekwana, E.A., Kulessa, B., 2015. Resistivity and self-potential tomography applied to groundwater remediation and contaminant plumes: sandbox and field experiments. Journal of Hydrology 530, 1–14.

Martínez, N.M., Basallote, M.D., Meyer, A., Cánovas, C.R., Macías, F., Schneider, P., 2019. Life cycle assessment of a passive remediation system for acid mine drainage: towards more sustainable mining activity. Journal of Cleaner Production 211, 1100–1111.

Matzek, L.W., Carter, K.E., 2016. Activated persulfate for organic chemical degradation: a review. Chemosphere 151, 178–188.

Matzek, L.W., Carter, K.E., 2017. Sustained persulfate activation using solid iron: kinetics and application to ciprofloxacin degradation. Chemical Engineering Journal 307, 650–660.

Mosmeri, H., Alaie, E., Shavandi, M., Dastgheib, S.M.M., Tasharrofi, S., 2017. Bioremediation of benzene from groundwater by calcium peroxide (CaO_2) nanoparticles encapsulated in sodium alginate. Journal of the Taiwan Institute of Chemical Engineers 78, 299–306.

Mosmeri, H., Gholami, F., Shavandi, M., Alaie, E., Dastgheib, S.M.M., 2018. Application of magnesium peroxide (MgO_2) nanoparticles for toluene remediation from groundwater: batch and column studies. Environmental Science and Pollution Research 25 (31), 31051–31061.

Müller, J.B., Ramos, D.T., Larose, C., Fernandes, M., Lazzarin, H.S., Vogel, T.M., Corseuil, H.X., 2017. Combined iron and sulfate reduction biostimulation as a novel approach to enhance BTEX and PAH source-zone biodegradation in biodiesel blend-contaminated groundwater. Journal of Hazardous Materials 326, 229–236.

Nazari, P., Setayesh, S.R., 2018. Effective degradation of Reactive Red 195 via heterogeneous electro-Fenton treatment: theoretical study and optimization. International Journal of Environmental Science and Technology 1–18.

O'Steen, W.N., Howard Jr., R.O., May 2014. Challenges in planning for groundwater remedy transition at a complex site. In: 2014 NGWA Groundwater Summit. Ngwa.

Obiri-Nyarko, F., Grajales-Mesa, S.J., Malina, G., 2014. An overview of permeable reactive barriers for in situ sustainable groundwater remediation. Chemosphere 111, 243–259.

O'Connor, D., Hou, D., Ok, Y.S., Song, Y., Sarmah, A., Li, X., Tack, F.M., 2018. Sustainable in situ remediation of recalcitrant organic pollutants in groundwater with controlled release materials: a review. Journal of Controlled Release 283, 200–213.

Peluffo, M., Rosso, J.A., Morelli, I.S., Mora, V.C., 2018. Strategies for oxidation of PAHs in aged contaminated soil by batch reactors. Ecotoxicology and Environmental Safety 151, 76–82.

Peng, W., Li, X., Xiao, S., Fan, W., 2018. Review of remediation technologies for sediments contaminated by heavy metals. Journal of Soils and Sediments 18 (4), 1701–1719.

Pennell, K.D., Abriola, L.M., 2017. Surfactant-enhanced aquifer remediation: fundamental processes and practical applications. In: Fundamentals and Applications of Bioremediation. Routledge, pp. 693–750.

Rahim, M., Abu, N., Mas Haris, M.R.H., 2016. The effect of pH on the slow-release behaviour of 1-and 2-naphthol from chitosan film. Cogent Chemistry 2 (1), 1234345.

Rice, J.M., Stolzenburg, T.R., 2018. Device and Method for Trapping Non-aqueous-phase Liquids Prior to a Permeable Reactive Barrier for the Treatment of Contaminated Groundwater. U.S. Patent No. 9,884,771. 6 Feb. 2018.

Safdari, M.S., Kariminia, H.R., Rahmati, M., Fazlollahi, F., Polasko, A., Mahendra, S., Fletcher, T.H., 2018. Development of bioreactors for comparative study of natural attenuation, biostimulation, and bioaugmentation of petroleum-hydrocarbon contaminated soil. Journal of Hazardous Materials 342, 270–278.

Sale, T., Newell, C.J., 2010. Impacts of source management on chlorinated solvent plumes. In Situ Remediation of Chlorinated Solvent Plumes. Springer, New York, NY, pp. 185–216.

Santisukkasaem, U., Das, D.B., 2018. A non-dimensional analysis of permeability loss in zero-valent iron permeable reactive barrier (PRB). Transport in Porous Media 1–21.

Sarkar, P., Roy, A., Pal, S., Mohapatra, B., Kazy, S.K., Maiti, M.K., Sar, P., 2017. Enrichment and characterization of hydrocarbon-degrading bacteria from petroleum refinery waste as potent bioaugmentation agent for in situ bioremediation. Bioresource Technology 242, 15–27.

Schwarzenbach, R.P., Gschwend, P.M., 2016. Environmental Organic Chemistry. John Wiley and Sons.

Shabalala, A.N., Ekolu, S.O., Diop, S., Solomon, F., 2017. Pervious concrete reactive barrier for removal of heavy metals from acid mine drainage– column study. Journal of Hazardous Materials 323, 641–653.

Shahsavari, E., Poi, G., Aburto-Medina, A., Haleyur, N., Ball, A.S., 2017. Bioremediation approaches for petroleum hydrocarbon-contaminated environments. In: Enhancing Cleanup of Environmental Pollutants. Springer, Cham, pp. 21–41.

Sharma, P.K., Mayank, M., Ojha, C.S.P., Shukla, S.K., 2018. A review on groundwater contaminant transport and remediation. ISH Journal of Hydraulic Engineering 1–10.

Sheng, Y., Zhang, X., Zhai, X., Zhang, F., Li, G., Zhang, D., 2018. A mobile, modular and rapidly-acting treatment system for optimizing and improving the removal of non-aqueous phase liquids (NAPLs) in groundwater. Journal of Hazardous Materials 360, 639–650.

Sheu, Y.T., Chen, S.C., Chien, C.C., Chen, C.C., Kao, C.M., 2015. Application of a long-lasting colloidal substrate with pH and hydrogen sulfide control capabilities to remediate TCE-contaminated groundwater. Journal of Hazardous Materials 284, 222–232.

Sheu, Y.T., Lien, P.J., Chen, K.F., Ou, J.H., Kao, C.M., 2016. Application of NZVI-contained emulsified substrate to bioremediate PCE-contaminated groundwater—A pilot-scale study. Chemical Engineering Journal 304, 714–727.

Sheu, Y.T., Tsang, D.C., Dong, C.D., Chen, C.W., Luo, S.G., Kao, C.M., 2018. Enhanced bioremediation of TCE-contaminated groundwater using gamma poly-glutamic acid as the primary substrate. Journal of Cleaner Production 178, 108–118.

Su, C., Puls, R.W., Krug, T.A., Watling, M.T., O'Hara, S.K., Quinn, J.W., Ruiz, N.E., 2017. Long-term performance evaluation of groundwater chlorinated solvents remediation using nanoscale emulsified zerovalent iron at a Superfund site. In: Applying Nanotechnology for Environmental Sustainability. IGI Global, pp. 92–111.

Suthersan, S., Killenbeck, E., Potter, S., Divine, C., LeFrancois, M., 2015. Resurgence of pump and treat solutions: directed groundwater recirculation. Groundwater Monitoring and Remediation 35 (2), 23–29.

Talley, J.W., 2016. Roadblocks to the implementation of biotreatment strategies. In: Bioremediation of Recalcitrant Compounds. CRC Press, pp. 41–58.

Tang, J., Li, X., Luo, Y., Li, G., Khan, S., 2016. Spectroscopic characterization of dissolved organic matter derived from different biochars and their polycylic aromatic hydrocarbons (PAHs) binding affinity. Chemosphere 152, 399–406.

Thornton, S.F., Morgan, P., Rolfe, S.A., 2016. Bioremediation of hydrocarbons and chlorinated solvents in groundwater: characterisation, design and performance assessment. In: Hydrocarbon and Lipid Microbiology Protocols. Springer, Berlin, Heidelberg, pp. 11–64.

Tian, H., Liu, Z., Zhang, M., Guo, Y., Zheng, L., Li, Y.C., 2019. Bio-based polyurethane, epoxy resin and polyolefin wax composite coating for controlled-release fertilizer. ACS Applied Materials and Interfaces 11, 5380–5392.

Tick, G.R., Harvell, J.R., Murgulet, D., 2015. Intermediate-scale investigation of enhanced-solubilization agents on the dissolution and removal of a multicomponent dense nonaqueous phase liquid (DNAPL) source. Water, Air, and Soil Pollution 226 (11), 371.

Tomlinson, D.W., Rivett, M.O., Wealthall, G.P., Sweeney, R.E., 2017. Understanding complex LNAPL sites: illustrated handbook of LNAPL transport and fate in the subsurface. Journal of Environmental Management 204, 748–756.

Tsai, T.T., Kao, C.M., Yeh, T.Y., Liang, S.H., Chien, H.Y., 2009. Application of surfactant enhanced permanganate oxidation and biodegradation of trichloroethylene in groundwater. Journal of Hazardous Materials 161, 111–119.

Tsai, T.T., Liu, J.K., Chang, Y.M., Chen, K.F., Kao, C.M., 2014. Application of polycolloid-releasing substrate to remediate trichloroethylene-contaminated groundwater: a pilot-scale study. Journal of Hazardous Materials 268, 92–101.

Usman, M., Yasin, H., Nasir, A., Mehmood, W., 2017. A case study of groundwater contamination due to open dumping of municipal solid waste in Faisalabad, Pakistan. Earth Sciences Pakistan 1 (2), 15–16.

Widdowson, M.A., 2004. Modeling natural attenuation of chlorinated ethenes under spatially varying redox conditions. Biodegradation 15 (6), 435–451.

Wilkin, R.T., Acree, S.D., Ross, R.R., Puls, R.W., Lee, T.R., Woods, L.L., 2014. Fifteen-year assessment of a permeable reactive barrier for treatment of chromate and trichloroethylene in groundwater. The Science of the Total Environment 468–469 (15), 186–194.

Wilpiszewska, K., Spychaj, T., Paździoch, W., 2016. Carboxymethyl starch/montmorillonite composite microparticles: properties and controlled release of isoproturon. Carbohydrate Polymers 136, 101–106.

Wu, M., Cheng, Z., Qin, G., Lei, M., Wu, J., Wu, J., Lin, J., 2018. The change of representative elementary volume of DNAPL influenced by surface active agents during long-term remediation period in heterogeneous porous media. The Science of the Total Environment 625, 1175–1190.

Xie, Y., Zhang, D., Lou, S., He, F., Yin, L., 2017. Slowly released carbon source from composite materials system for removing nitrate pollution in groundwater. RSC Advances 7 (17), 10215–10220.

Xiong, W., Mathies, C., Bradshaw, K., Carlson, T., Tang, K., Wang, Y., 2012. Benzene removal by a novel modification of enhanced anaerobic biostimulation. Water Research 46, 4721–4731.

Yang, C.F., Liu, S.H., Su, Y.M., Chen, Y.R., Lin, C.W., Lin, K.L., 2019. Bioremediation capability evaluation of benzene and sulfolane contaminated groundwater: determination of bioremediation parameters. The Science of the Total Environment 648, 811–818.

Yang, S., Oostrom, M., Truex, M.J., Li, G., Zhong, L., 2016. Injectable silica–permanganate gel as a slow-release MnO_4^- source for groundwater remediation: rheological properties and release dynamics. Environmental Science: Processes and Impacts 18 (2), 256–264.

Yang, Y., Yang, J., Wu, W.M., Zhao, J., Song, Y., Gao, L., Jiang, L., 2015. Biodegradation and mineralization of polystyrene by plastic-eating mealworms: Part 2. Role of gut microorganisms. Environmental Science and Technology 49 (20), 12087–12093.

Yang, Z.H., Ou, J.H., Dong, C.D., Chen, C.W., Lin, W.H., Kao, C.M., 2018. Remediation of TCE-contaminated groundwater using $KMnO_4$ oxidation: laboratory and field-scale studies. Environmental Science and Pollution Research 1–12.

Yao, G., Bliss, K.M., Crimi, M., Fowler, K.R., Clark-Stone, J., Li, W., Evans, P.J., 2016. Radial basis function simulation of slow-release permanganate for groundwater remediation via oxidation. Journal of Computational and Applied Mathematics 307, 235–247.

Yu, Y., 2015. Research on soybean protein wastewater treatment by the integrated two-phase anaerobic reactor. Saudi Journal of Biological Sciences 22 (5), 526–531.

Zhang, F., Zhang, M., Su, X., Zhang, S., Zhang, Q., Zhong, S., 2018. Preparation and application of oxygen slow-releasing materials for in situ manganese removal from groundwater. International Journal of Environmental Science and Technology 1–10.

Zhang, M., Wang, H., 2016. Preparation of immobilized sulfate reducing bacteria (SRB) granules for effective bioremediation of acid mine drainage and bacterial community analysis. Minerals Engineering 92, 63–71.

Zhang, M., Wang, H., Han, X., 2016. Preparation of metal-resistant immobilized sulfate reducing bacteria beads for acid mine drainage treatment. Chemosphere 154, 215–223.

Zhang, Q., Zhou, W., Cui, Y., Lyu, C., Liu, T., Zhang, R., Guo, L., 2018. Iron triad nanomaterials and their sustainable application in the environment. Environmental Science: Nano 5 (2), 246–256.

Zhang, W., Ruan, X., Bai, Y., Yin, L., 2018. The characteristics and performance of sustainable-releasing compound carbon source material applied on groundwater nitrate in-situ remediation. Chemosphere 205, 635–642.

Zhao, Y., Qu, D., Hou, Z., Zhou, R., 2015. Enhanced natural attenuation of BTEX in the nitrate-reducing environment by different electron acceptors. Environmental Technology 36 (5), 615–621.

Zhou, Y., Fang, X., Zhang, Z., Hu, Y., Lu, J., 2017. An oxygen slow-releasing material and its application in water remediation as oxygen supplier. Environmental Technology 38 (22), 2793–2799.

Chapter 8

Controlling secondary pollution impacts during enhanced in situ anaerobic bioremediation

Daniel R. Griffiths
Parsons Corporation, Denver, CO, United States

1. Introduction

The in situ bioremediation remedial technology offers an excellent balance between cost, effectiveness, and maintaining site buildability, but there are many potential pitfalls that may occur through lack of sufficient site characterization, inadequate application design, and/or improper installation. In situ biological solutions depend upon the interactions between site soils, groundwater, microbiology, contaminants, and emplaced reactants to degrade site constituents. Understanding site conditions and how they relate under natural conditions and how they will relate under modified conditions is crucial to the performance of any in situ remedial design. Thus, practitioners of in situ remedial technologies, particularly those involving existing or modified microbiology as the desired contaminant degradation mechanism, must have a firm understanding of the site conceptual model and all associated aspects of their site. Failure to collect sufficient site data and develop a sufficient understanding of site conditions invites failure in achieving remedy performance goals and potentially the creation of site conditions that are more harmful than baseline.

Site characterization and the associated level of effort needed for remedy design are generally more extensive than those required for site delineation and are a balance between the cost to collect investigation data, remedy performance requirements, and remedy implementation cost (Simkin and Norris, 2010). Predesign investigation needs are also governed by site complexity, potential receptors that could be impacted by remedial efforts, and the selected remedial technology.

Sustainable Remediation of Contaminated Soil and Groundwater. https://doi.org/10.1016/B978-0-12-817982-6.00008-2

In situ bioremediation is a technology that can be applied in both oxic and anoxic geochemical conditions depending on the contaminants to be degraded and the desired degradation mechanism. This chapter provides an overview of common problems that may be encountered during the application of enhanced anaerobic in situ bioremediation. Many of the challenges presented are driven by failures during the site characterization and/or remedy design steps and are very difficult to remedy quickly once they occur. The objective of this chapter is not to provide an exhaustive compendium of problems that may occur during in situ bioremediation application implementation and performance together with fixes for each problem, but to alert the reader to the importance of sufficient site characterization and proper application design such that problems can be avoided from the outset. In some cases, where available, fixes to problems that may occur are presented.

2. Anaerobic enhanced bioremediation overview

Enhanced anaerobic bioremediation is an effective method of degrading various chlorinated solvents and other contaminants dissolved in groundwater. One of the major advantages of this technology is that it can lead to complete destruction of contaminant mass in situ with little impact on site infrastructure or need for secondary treatment trains. A second major advantage of enhanced in situ anaerobic bioremediation is that this technology can generally be implemented at a cost that is low relative to more expensive remedial systems (e.g., groundwater extraction, air sparging, permeable reactive iron barriers, or chemical oxidation).

When anaerobic degradation of chlorinated aliphatic hydrocarbons (CAHs) occurs naturally, it is considered to be a component of natural attenuation (USEPA, 1998). Unfortunately, monitored natural attenuation (MNA) alone may not be sufficient to achieve remedial objectives in a timely manner at many sites that are contaminated with chlorinated solvents. Engineering this approach involves the acceleration of natural destructive mechanisms by adding organic substrates and potentially other geochemical modifiers to drive site groundwater geochemistry into pH neutral anoxic conditions and to provide a supply of electron donor to the microbial population. The addition of an organic substrate to an aquifer stimulates microbial activity and growth, resulting in the formation of an anoxic environment that is conducive to anaerobic microbial processes including the reductive dechlorination of chlorinated solvent mass. The induction of anoxic geochemical conditions also allows for less desirable processes to occur including the production of undesirable by-products (e.g., methane, hydrogen sulfide, etc.) and the mobilization of metal species present in the soil column (e.g., iron, manganese, arsenic, etc.).

A variety of organic substrates have been applied to the subsurface to promote anaerobic degradation of chlorinated solvents to innocuous end products over the last two decades as this technology has been developed and implemented. Examples of easily fermentable organic substrates include carbohydrates and sugars, vegetable oils, bark mulch, and low-molecular-weight fatty acids (AFCEE, 2007). Organic substrate injections enhance chlorinated solvent degradation by depleting the aquifer of dissolved oxygen and lowering the oxidation–reduction potential, which results in stimulation of anoxic conditions that are conducive to anaerobic degradation processes. After the dissolved oxygen is consumed, anaerobic microorganisms use available native electron acceptors in the following order of preference: nitrate, manganese, and ferric iron hydroxides or oxyhydroxides, sulfate, and finally carbon dioxide (methanogenesis) (Wiedemeier et al., 1999). Differing microbial and geochemical processes occur under each geochemical regime.

The biodegradation of organic substrate also leads to the production of metabolic acids, including lactic, propionic, butyric, and acetic acids. Fermentation of these low-molecular-weight acids produces molecular hydrogen (H_2), which is the primary electron donor used for anaerobic reductive dechlorination of CAHs. Although hydrogen appears to be the most important electron donor for reductive dechlorination (Maymo-Gatell et al., 1997; Fennell et.al., 1997), other fermentation products (e.g., acetate or formate) have been identified as direct electron donors for reductive dechlorination.

Under anoxic conditions, chlorinated solvents are susceptible to degradation through one or more of the following reactions (AFCEE, 2007):

- **Direct anaerobic reductive dechlorination** is a biologically mediated reaction that allows microorganisms to gain energy as one or more chlorine atoms on a chlorinated solvent molecule are replaced with hydrogen atoms in an anaerobic environment. In this reaction, the chlorinated solvent mass serves as the electron acceptor and molecular hydrogen typically serves as the electron donor. During enhanced bioremediation applications, hydrogen used in this reaction is supplied by fermentation of organic substrates. In general, direct anaerobic reductive dechlorination occurs by sequential removal of chlorine ions. For example, the chlorinated ethene solvents are transformed sequentially from the most chlorinated compound tetrachloroethene through cis-1,2-dichloroethene and vinyl chloride to ethene. This reaction has also been referred to as halorespiration or dehalorespiration by the United States Environmental Protection Agency (USEPA, 2000). A reaction in which two chlorine atoms are removed and replaced by two hydrogen atoms in one step is termed dichloroelimination.
- **Cometabolic anaerobic reductive dechlorination** is a reaction in which a chlorinated solvent is reduced by an enzyme or cofactor that is produced during microbial metabolism of another compound in an anaerobic environment. In such a case, biodegradation of the cometabilized chlorinated solvent does not yield any energy or growth benefit for the microbe mediating the reaction (USEPA, 2000).
- **Anaerobic oxidation** is a biologically mediated reaction in which less chlorinated solvents, such as *cis*-1,2-dichloroethene and vinyl chloride, are directly oxidized (i.e., used as an electron donor) to carbon dioxide,

hydrogen, and chloride. This reaction has been documented to occur most prevalently under iron- and manganese-reducing conditions (Bradley and Chappelle, 1997; Bradley et al., 1998a, 1998b). This degradation pathway has been successfully documented in the laboratory and in the field for chlorinated ethene compounds. Chlorinated ethane compounds can theoretically be degraded through anaerobic oxidation. However, this process has yet not been documented for chlorinated ethane compounds.

- **Abiotic degradation** is a chemical reaction where a chlorinated solvent molecule is mineralized through contact with a reduced compound or element. An example of this degradation mechanism is the reaction that is targeted when zerovalent iron (Fe^0) is placed in the subsurface as a permeable reactive barrier for treatment of CAH-impacted groundwater. Note that the addition of an organic substrate may also create an anaerobic environment that forms reactive compounds (e.g., metal sulfides) that can degrade CAHs (Butler and Hayes, 1999). In this way, organic substrate addition may indirectly stimulate abiotic degradation by producing geochemical conditions that favor the formation of compounds that can chemically react with CAHs in groundwater.

3. Potential secondary impacts and optimization strategies

In situ anaerobic bioremediation is a well-proven remedial technology that has been in use for more than two decades to degrade a host of contaminants including chlorinated solvents, organometals, refrigerants, petroleum constituents, and some pesticides. More contaminants are added to the list of compounds that are biodegradable every year. However, adequate understanding of site conditions and proper application design are critical to achieve objectives regardless of which contaminant class is present and targeted for biological degradation.

The induction and maintenance of anoxic conditions in order to induce anaerobic degradation processes are the basis for applying this remedial technology but are also the root of common problems that can occur. The potential for problems is exacerbated on sites where natural geochemistry is aerobic because the induction of anoxic conditions comes as a geochemical shock to site geochemistry and can cause large changes in mineral and accompanying metals' solubility among other potential problems discussed in more detail in following subsections. The following subsections describe several potential secondary impact issues associated with this technology.

4. Groundwater mounding, daylighting, and poor distribution

The injection of organic substrates (e.g., soybean oil, lactate, molasses, etc.) typically involves the injection of large volumes of water containing low concentrations of organic substrate to achieve desired substrate distribution.

Liquid substrates and water are typically injected into the subsurface using injection wells, direct push points, or a combination of the two. The spacing and placement of injection locations (injection wells or direct push points) are critical to the achievement of desired substrate distribution in that at each injection location, the volume of injected materials must be matched to desired substrate radius of influence (e.g., how far from the injection location substrate travels radially) given site soil conditions (stratigraphy, soil porosity within the injection interval, etc.).

During the injection of substrate and water into the subsurface, pressure is applied to the fluid stream to increase injection rates and to force the injection fluid out into the subsurface hydrogeologic formation. In many cases, practitioners apply excessive injection pressure to force more injection fluid into the ground faster and thereby reduce their time in the field and their cost. However, this practice often leads to problems when the injection pressure in use is excessive but not high enough to induce temporary soil fractures. Under typical site conditions, where the depth of injection is less than approximately 30 m below the water table, injection pressures should be maintained at slightly above the sum of the hydrostatic pressure and pressure loss due to friction within the injection system in use. This "ideal" injection pressure is generally in the 30–50 pounds per square inch (PSI) range with lower pressures used for shallow injection projects and higher pressures used for deeper applications. Alternatively, substrate injection can be achieved through the induction of temporary soil fractures wherein substrate is injected at very high pressures (150–300 psi or more) such that soil fractures are induced and substrate is distributed through the interconnected fractures.

The injection of substrates at intermediate pressures, pressures in excess of a "low pressure injection" but below pressures associated with soil fracturing, generally leads to problems during application installation including excessive hydraulic mounding, substrate daylighting on the ground surface, and unintended or nonoptimal substrate distribution. These problems are particularly prevalent on sites with low-permeability soils (fine sands, silts, and clays) and where subsurface structures like utilities are present. In these cases where substrate is injected at pressure and fluid is added to the subsurface system faster than it can disperse away from the injection locations, hydraulic mounding, where the water table rises due to the extra fluid being injected, results. In some cases, the hydraulic mounding goes unnoticed, and in others, mounding is severe such that the water table rises to above the ground surface and daylighting of substrate onto the ground surface results. Daylighting of significant quantities of substrate along with water that likely contains contaminants causes application delay and costly cleanup. The unintended distribution of substrate into utility corridors is also common, which results in a total loss of control in substrate distribution. In these cases, it is not uncommon to find organic substrate distributed hundreds of meters along utility corridors, well beyond the intended treatment area footprint, which can cause significant future risk to previously unidentified potential receptors.

The key to avoiding or at least minimizing problems related to excessive mounding and unintended substrate distribution is to develop a strong understanding of subsurface conditions during the design phase. Sufficient geologic, hydraulic, and site use data must be collected prior to implementation such that a site conceptual model, including the interactions between man-made structures and site conditions, can be developed and understood. It is also important to understand where contaminant mass remains in the subsurface (e.g., are remaining contaminants largely bound up in low-permeability soils? Can a source area be identified?), such that these areas can be targeted for substrate injection. Once site conditions are properly understood, an injection plan can be developed to deliver substrate to the right areas effectively and efficiently. There are many guides available within the industry that are designed to help the site professional develop a robust site conceptual model (e.g., USACE, 2012; NJDEP, 2011).

During the installation of an in situ bioremediation system, adequate monitoring wells and piezometers should be installed in and around the injection area such that the depth to the water table can be monitored during injection to identify signs of excessive hydraulic mounding before unacceptable impacts occur. Subsurface utility corridors and other underground structures should also be monitored to the degree possible to watch for signs of substrate appearance (e.g., monitor stormwater outfalls in the vicinity of the injection area for signs of substrate appearance). It is critical to have a sufficient monitoring plan in place such that signs of unintended substrate distribution (e.g., excessive mounding, etc.) can be identified early and managed aggressively.

On many sites, site geologic, hydraulic, and/or infrastructure conditions make substrate injection very difficult and high risk even with a strong understanding of site conditions and a robust in situ bioremediation design. However, practitioners do have some options available to them that may allow for the successful distribution of substrates under difficult conditions. Livadas (2003), Pac et al. (2005), and NAVFAC (2013) provide extensive discussions of injection and substrate delivery methods. The first two technologies outlined next can be implemented during the planning phase or adopted as a change in the field during implementation if excessive mounding is observed. However, recirculation and soil fracturing will not remedy unintended substrate distribution that has already occurred.

4.1 Recirculation

Recirculation is an injection technique wherein groundwater is extracted at the same time substrate is injected. This process of removing groundwater while new fluid is added in the form of substrate mixed with water effectively reduces mounding and at the same time improves substrate distribution in the subsurface. Typically the extracted groundwater stream is filtered to remove silt/clay particles, amended with organic substrates, and reinjected without any additional treatment to further reduce the volume of new fluid being added to the site hydraulic system and to further reduce the potential for unacceptable mounding and unintended substrate distribution. Recirculating in situ bioremediation applications have been implemented by the author in California, Texas, Ohio, Mississippi, New York, and New Jersey (e.g., Griffiths et al., 2018; Diller et al., 2018).

Ideally recirculation is incorporated into an in situ bioremediation application during the design phase such that the design can be optimized and recirculation can be fully integrated to achieve desired results and the lowest possible cost. However, in many cases, recirculation is added "after the fact" when poor or intended substrate distribution is identified in the field during implementation. In cases where recirculation is adopted during implementation, it is likely that the addition of recirculation will require a temporary halt to injection activities, client and regulatory agency interaction, and renegotiation of cost and schedule. The addition of recirculation to the injection paradigm will also likely require installation of additional wells for groundwater extraction as well as changes to the equipment plan to include groundwater extraction pumps, additional conveyance line, and water storage tanks.

Recirculation can be applied to in situ bioremediation applications by adding groundwater extraction wells within the intended treatment area and screened within the same vertical interval where substrate injection is planned. The number of groundwater extraction wells and proximity of extraction wells to injection locations will depend on site stratigraphy. In general, sites with lower-permeability soils (e.g., fine sands and silts) will require more closely spaced injection and extraction wells to achieve intended substrate distribution and control hydraulic mounding during injection.

Fig. 8.1 shows a cross-sectional view of a permeable reactive barrier that was implemented using recirculation at a site in New York. During this application, wells were installed in a linear transect perpendicular to groundwater flow and cutting across a chlorinated solvent plume. After the wells were in place, groundwater extraction pumps were installed in every other well and plumbed to a series of filters for sediment removal and tanks for water storage. The stored water was then pumped out of the tanks, amended with organic substrates (soy bean oil and pH buffer), and injected into the intervening wells. This process of groundwater extraction, substrate addition, and injection was conducted at the same time and in one continuous process. The result, as shown in the diagram, was induced flow between the injection and extraction wells, improved distribution between well pairs, and ultimately tighter control over substrate distribution such that a continuous permeable reactive barrier was implemented.

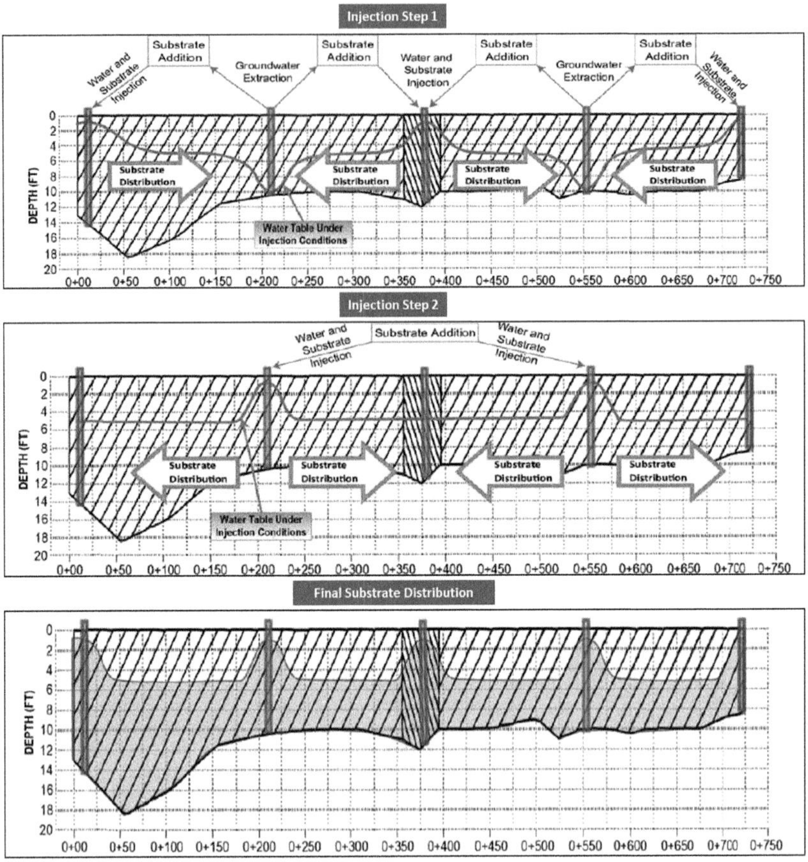

FIGURE 8.1 Cross-sectional view of a recirculation system.

4.2 Soil fracturing

Soil fracturing, like recirculation, is an injection methodology that can be incorporated into an in situ bioremediation design during the design phase or during implementation if injection is not proceeding as expected. Adopting soil fracturing after the design phase will incur some inefficiencies in terms of time and dollars spent, likely in the form of design rework, multiple mobilizations, and injection equipment changeout. However, identifying distribution problems early in the application phase and fixing them, even at higher cost, is preferable to installing a flawed design and waiting for the performance monitoring phase of the project to identify those failures through poor remedy performance.

Soil fracturing is most commonly applied on sites where site soils consist of low-permeability fine sands, silts, and clays to induce temporary fractures in subsurface soils such that injectants (e.g., organic substrates, pH buffer, bioaugmentation culture, etc.) can be distributed more effectively. This technology is applied most frequently using direct push drilling methods and equipment, though in special circumstances other drilling methods (e.g., rotosonic) can be used where the application depth is excessive or site soils preclude the use of direct push. Soil fracturing can be achieved using a number of fracturing methods including blast fracturing, pneumatic fracturing, and hydraulic fracturing (FRTR, 2019). Hydraulic fracturing is most commonly used in an environmental restoration setting where the goal is to deliver substrates or other reactants to the subsurface. Applying soil fracturing to sites with low-permeability soils allows for more rapid and effective substrate distribution and has the added benefit of establishing networks of preferential pathways within the impacted site media that allows for the treatment of areas that were not directly accessible before fracturing took place. Fracturing may also be applicable to some sites where back diffusion of contaminants trapped in the aquifer matrix is likely to drive long-term plume presence (USEPA, 2018).

Soil fracturing uses water or the bioremediation substrates themselves to induce temporary fractures in the soil matrix. In unconsolidated subsurface materials (fine sands, silts, and clays), boreholes are advanced to the depth where the fracturing will occur. Generally the borehole is cased to this depth (usually accomplished using a rotosonic, hollow stem auger, or dual-tube direct push casing), and a short screen injection probe is inserted at the bottom of the desired injection interval. Water or substrates are then injected at high pressure (typically 200 to 300 psi), temporary soil fractures are induced, and substrate is distributed along the induced fractures. The pressure needed for the initiation is typically low (less than 100 psi), but applying increased pressure at least initially results in longer fracture propagation and improved substrate distribution. At start, the injection pressure is relatively high and then falls as the fractures propagate (Suthersan, 1999 and USEPA, 1994). This fracture injection process is illustrated in Fig. 8.2.

Soil fracture as a delivery method for substrates should be applied with caution as the fractures that are induced are difficult to control in terms of three-dimensional propagation direction and lateral extent. In the example depicted in Fig. 8.2, there is significant potential for fractures to propagate downward into the B-zone to potentially transport contaminant mass and substrates into the previously unimpacted B-zone aquifer unit causing spreading of contaminants and impact of additional water resources. In addition, injecting substrate using soil fracture at shallow depths carries with it some potential for disturbing nearby infrastructure and potentially distributing substrates in unintended areas, though these risks can be mitigated with careful planning. Soil fracturing should also be avoided in close proximity to sensitive receptors such as surface water bodies, particularly if the intended injection depth is shallow with respect to the depth of the nearby surface water.

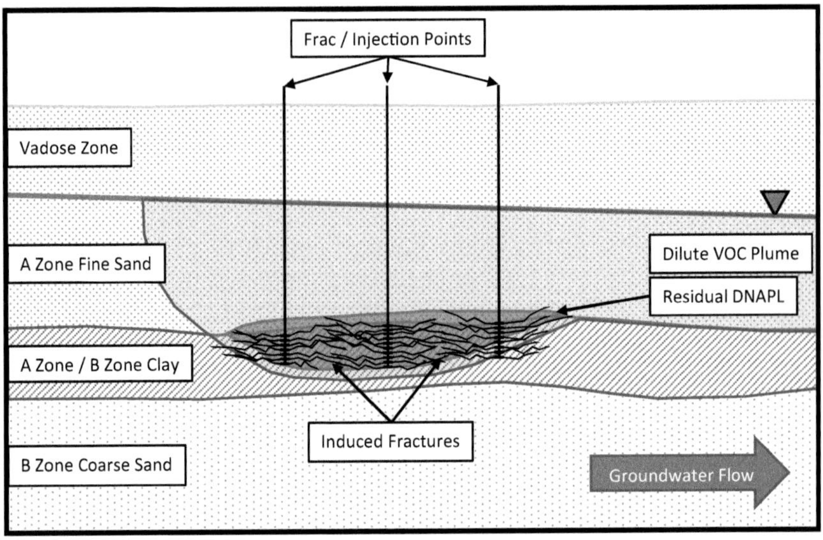

FIGURE 8.2 Conceptual soil fracture application.

4.3 Chemical oxidant injection

In the event that unintended substrate distribution does occur and previously unidentified receptors are at risk, the organic substrates that migrated beyond the intended treatment area and that are driving this newly identified risk can be degraded through the application of chemical oxidant materials (e.g., persulfate, hydrogen peroxide, etc.). Though careful consideration should be given before chemical oxidation is applied in close proximity to receptors. Using chemical oxidants to rapidly degrade organic materials in situ is a commonly applied technology used to degrade other organic materials such as petroleum constituents. Using chemical oxidants to rapidly degrade organic substrates in situ works in exactly the same way. The area of unintended substrate impact is identified through drilling and sampling, the quantity of organics, expressed as the sum of substrate mass plus naturally occurring organics in the effected soil matrix, and the appropriate mass of oxidant is calculated. This chemical oxidation design process has been documented and presented numerous times by various authors (e.g., Shammas et al., 2005; ITRC, 2005; Siegrist et al., 2011). The reader is invited to refer to detailed chemical oxidation design guidance provided by others for their chemical oxidation design and implementation needs.

4.4 Air sparging

Air sparging is a similar technology to chemical oxidant injection, presented earlier, in that this technology is a means to introducing oxygen to the subsurface to promote oxic conditions and drive aerobic processes. In this case, the goal is to maintain subsurface geochemical conditions in the oxic range to prevent the anaerobic production of methane, hydrogen sulfide, carbon dioxide, and other potentially risk-driving gases while promoting the aerobic degradation of substrates present in unintended areas. As in the case of chemical oxidation, air sparging could be implemented after unintended organic substrate distribution is discovered and could thus be applied as a "recovery" application to protect newly impacted or newly identified receptors.

Air sparging in its simplest form is implemented by installing a series of injection wells with screens situated below the water table and ideally below the zone where substrate was emplaced. Atmospheric air is then injected to force oxygen into the subsurface system to induce and maintain aerobic conditions. Air sparging can also be implemented using horizontal wells screened below the water table and engineered trenches to deliver air or other gaseous reactants as site conditions require (Johnson et al., 2010).

Air sparging could be applied to aerobically degrade organic substrates and prevent the production of potentially harmful gases by installing a series of air sparge wells in the area of interest and then installing positive displacement air pumps to force atmospheric air (or pure oxygen) into the subsurface. The design and implementation of air sparge systems have been published many times in the past by other authors (e.g., NAVFAC, 2001; Johnson et al., 2010; USACE, 2013; USEPA, 2017).

Air sparging may be a more attractive option to chemical oxidation for a variety of reasons (e.g., cheaper, less potential for negative consequences, more acceptable to regulatory agencies), but the practitioner must be aware that air sparging and "natural" degradation rates under these conditions are likely to be much slower than degradation rates achieved by the injection of chemical oxidants because oxygen is much less reactive than chemical oxidants. Thus, while air sparging may be less impactful, it is also much slower and by extension must be maintained and operated over a longer period of time. To date, the author is not aware of any air sparge applications that have been installed specifically to remediate unintended organic substrate distribution, though there are many air sparge applications that have been published in literature for the remediation of petroleum-related organics, to protect potential receptors, and to induce and maintain oxic geochemical conditions in the subsurface.

5. Unintended impacts on downgradient secondary groundwater quality

The basis for the in situ anaerobic bioremediation technology for the treatment of chlorinated solvents is to inject organic substrates to induce and maintain anoxic geochemical conditions to drive anaerobic processes including biodegradation and remediation of the chlorinated solvents of interest. Inducing anoxic conditions also has less desirable impacts on secondary water quality conditions within the treatment area (e.g., color, odor, dissolved iron, manganese, turbidity, total dissolved solids (TDS), sulfides, suppressed pH, biofouling, etc.) (ESTCP, 2005 and AFCEE, 2007). Elevated levels of total organic carbon (TOC), lower redox geochemistry compared to background, and elevated levels of metabolic by-products (e.g., carbon dioxide, ferrous iron, methane) can migrate out of the treatment area with groundwater flow, impacting downgradient groundwater quality.

During the design phase of an in situ bioremediation application, a risk and receptor survey should be completed to identify potential receptors in the vicinity of the intended application. A carefully executed survey should identify potential downgradient receptors that may be impacted by secondary water quality issues caused by the bioremediation application such that mitigation measures can be identified and designed into the remedy to protect downgradient receptors from secondary water quality impacts while the primary constituents of interest are biologically degraded in the upgradient source area. The most common mitigation measure to control secondary water quality impacts is simply to provide sufficient buffer zone between the bioremediation area and the identified downgradient receptors such that within the buffer zone, natural oxic geochemical conditions are reestablished and secondary water quality impacts naturally dissipate before reaching the receptors. However, it is not uncommon for secondary water quality impacts to migrate further and/or faster than expected.

The secondary water quality effects associated with the induced anoxic geochemical conditions are temporary, lasting only as long as the injected substrate is present in the subsurface. After the injected substrates are consumed, natural geochemical conditions will be reestablished and groundwater quality will ultimately return. However, this process can take years to achieve, particularly on sites where groundwater seepage rates are low and where high volumes/concentrations of substrates were emplaced. In some cases where solid organic substrates were emplaced (e.g., bark mulch, chitin, cotton gin waste, etc.), baseline geochemistry could take decades to return. In addition, in most cases, anaerobic bioremediation applications are emplaced for the long term and involve refresh of additional organic substrate when the previously injected mass reaches depletion. Thus, if unacceptable secondary water quality impacts are currently threatening a receptor (e.g., a water supply well), then it's likely that this threat will remain for an extended period of time and should be addressed aggressively.

The degree of response to secondary water quality impacts, and by extension the cost of the response, is dependent on the severity of the impacts and the receptors under impact. Secondary water quality impacts vary widely in terms of risk (e.g., methane impact may represent a higher risk than ferrous iron) and impact importance (e.g., methane intruding into a building basement has far higher potential for negative results than dissolved methane discharging to a stream). Thus, the response to secondary water quality impacts also varies widely and should be applied on a case-by-case basis. There is no one-size-fits-all response to addressing secondary water quality impacts.

5.1 Air sparge permeable reactive barrier

One such mitigation measure could be the installation and operation of an air sparge permeable reactive barrier system between the bioremediation application footprint and the downgradient receptor(s) to introduce oxygen into the subsurface system and convert the anoxic geochemical conditions induced by the bioremediation application back to oxic conditions before groundwater migrates to the receptor(s). Reverting geochemical conditions to the oxic range will alleviate many of the secondary water quality impacts caused by the bioremediation application while allowing the bioremediation application to continue to degrade chlorinated solvents within the application footprint. Ideally the need for a secondary water quality mitigation system would have been identified during the design phase allowing for the incorporation of the air sparge component into the overall remedy. However, things don't always go according to plan, and an air sparge system is a good way to overcome unforeseen secondary water quality problems while still making progress on the primary contaminants through anaerobic bioremediation. The use of air sparge systems to treat dilute contaminant plumes under maintained oxic conditions is discussed in USACE (2013) in the context of petroleum constituents.

The air sparge technology, when applied as a permeable reactive barrier, is most often implemented by installing a series of wells in a linear configuration and oriented perpendicular to the direction of groundwater flow. A conceptual layout of an air sparge system designed to address secondary water quality impacts is depicted in Fig. 8.3. Performance monitoring wells are generally installed in linear transects parallel to groundwater flow with monitoring wells installed upgradient from the intended air sparge treatment area, within the treatment area, and downgradient from the treatment area. The air sparge wells are installed such that the screens are placed below the water table and extend as deep below the water table as required to intercept the groundwater plume of interest. It should be noted that if the vertical interval requiring treatment to address secondary water quality issues is extensive (e.g., greater than 10–15 m), then it is likely that multiple air sparge vertical intervals will be required. The lateral spacing between the sparge wells is dependent on site conditions (e.g., stratigraphy, hydrogeology, etc.) and the volume of air that will be injected into each sparge well. Well spacing should be less than twice the calculated sparge radius of influence to provide a safety factor and to ensure that a complete permeable reactive barrier is implemented. The air

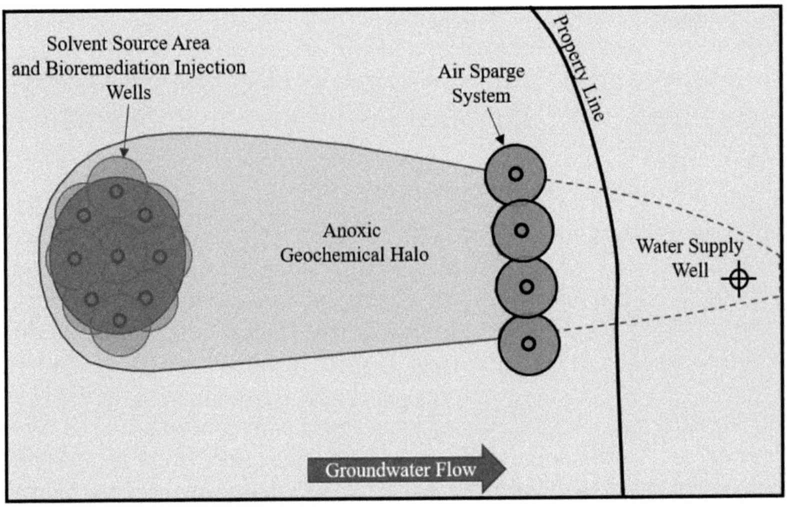

FIGURE 8.3 Conceptual air sparge application.

sparge design and installation process is presented in detail in NAVFAC, 2001, Johnson et al. (2010), USACE (2013), and USEPA (2017) among other works.

After all the air sparge and performance monitoring wells are installed, developed, and completed, a baseline round of groundwater samples should be collected from the performance monitoring well network to assess baseline, prestart conditions and to provide a comparison point for future performance sampling. The air sparge wells would then be plumbed to a positive displacement blower such that atmospheric air (or compressed oxygen) is injected below the water table at each sparge well location to induce and maintain oxic geochemical conditions. Following system start-up, performance data is generally collected relatively frequently (e.g., monthly) for the short term, becoming less frequent as the system continues to perform as expected.

There are some potential issues related to an air sparge system that is designed to induce and maintain oxic conditions. It is likely that significant precipitation of iron, manganese, and other materials will occur within the air sparge treatment area. These are the materials that the system is designed to address to protect downgradient receptors from secondary water quality impacts. However, long-term precipitation of these materials in the air sparge footprint will become problematic in that soil permeability will decline over time, and ultimately groundwater flow may divert around the air sparge footprint. Thus, precipitation impacts should be carefully considered against the desired system lifespan during the design phase. Geochemical precipitation within the air sparge treatment footprint can be addressed through the addition of acids (acidification) but at significant risk to the downgradient receptor(s).

5.2 Chemical oxidation

The worst-case scenario for an in situ anaerobic enhanced bioremediation application is likely the scenario wherein substrate is injected too close to a high-risk receptor mechanism such as a public water supply well, resulting in significant negative impacts to the quality of water produced by that well such that it is no longer useable even with ex situ treatment. In this extreme case, chemical oxidation could be applied to the area where substrate was distributed to rapidly degrade the emplaced substrates and more rapidly return groundwater geochemical conditions to baseline.

Chemical oxidation has the added benefit of being capable of degrading many chlorinated solvents in situ. Thus, a properly designed chemical oxidation application using a well-chosen oxidant could reduce secondary water quality impacts as well as degrade primary chlorinated solvent contaminant mass. However, applying chemical oxidation also brings its own risks, particularly in close proximity to a drinking water well or other sensitive receptor like the potential for the mobilization of naturally occurring metals as an example.

In this scenario, it is likely that predesign drilling and soil sampling would be required to define the lateral and vertical zone of organic substrate impact prior to the design and implementation of chemical oxidation to degrade the emplaced organic substrates. After the zone of organic substrate is defined in three dimensions, a carefully targeted oxidant injection plan would be prepared to deliver oxidant throughout the area of substrate impact. The mass of oxidant would be determined and designed to exceed the sum of naturally occurring oxidant demand (naturally occurring organic carbon in the soil matrix), the mass of organic substrate that was emplaced, and mass of other oxidant demands (e.g., contaminants) such that sufficient oxidant is injected to oxidize and degrade the emplaced substrate to meet objectives. The chemical oxidation design process as well as guidance on selection of oxidants is presented in detail in other documents (e.g., Shammas et. al., 2005; ITRC, 2005; Siegrist et al, 2011).

6. Inadequate pH control

A critical aspect to the successful performance of an anaerobic bioremediation application is the maintenance of pH conditions within the neutral range. The neutral range within which dechlorinating bacteria thrive is generally understood to be between pH 6 and pH 8 (Volkering and Pjils, 2004). When the bioremediation technology is applied and biologically mediated reductive dechlorination is induced, organic acids are formed through the fermentation of the injected substrates, and hydrochloric acid is generated through the production of molecular hydrogen and subsequent complexation with chloride (Robinson et. al., 2009). These acid-producing processes can overcome natural soil buffering capacity, leading to groundwater acidification and the creation of conditions that are inhibitory to dechlorinating microorganisms.

Low-pH conditions within the bioremediation application area result in low degradation rates and incomplete reductive dechlorination, which can lead to the accumulation of biodegradation intermediate products (e.g., cis-1,2-dichloroethene, vinyl chloride, etc.). Under highly acidic conditions, biodegradation can halt entirely resulting in application failure. Low-pH conditions can also result in the dissolution of naturally occurring metals (particularly arsenic) from the soil matrix and increased metal concentrations in groundwater.

Groundwater pH can be controlled and managed if addressed proactively. During the design phase, soil buffering capacity should be determined through sampling and analysis, which is balanced against acid production within the intended treatment area. In many cases, pH buffer products are injected with organic substrates to augment natural soil buffering capacity and provide a safety factor against groundwater acidification. Acid production and pH buffer calculation methods are available through previously published works (e.g., Robinson and Barry, 2009) and are relatively simple to execute as part of the design process.

In some cases, natural soil buffering capacity is low and/or groundwater is naturally acidic. The author has encountered conditions such as this particularly in Texas and along the gulf coast. In these areas, soils contain high concentrations of organic carbon and soil buffering capacity is already low or has been depleted and pH conditions are more difficult to control. At sites such as this, pH buffer addition is required and the buffer materials are likely to be depleted before the organic substrates are. Thus, careful analysis of pH data collected during performance monitoring sampling should be conducted to identify declining trends in pH before unacceptable acidic conditions develop.

7. Soil gas emissions

The primary goal of an enhanced bioremediation application is to induce and maintain anoxic geochemical conditions that are conducive to reductive dechlorination of chlorinated solvent mass. Dissolved gases are commonly produced under these anoxic geochemical conditions, and these produced gases may pose a risk to overlying structures and nearby underground utilities. Commonly produced gases include carbon dioxide, methane, and hydrogen sulfide among others. Production of these gases is unavoidable when applying anaerobic bioremediation to degrade chlorinated solvents, and in fact, generation of some dissolved gases (carbon dioxide and methane) is a positive line of evidence indicating that the bioremediation application has successfully induced proper conditions to support anaerobic degradation of chlorinated solvents (USEPA, 1998 and Wiedemeier et al., 1999).

Gas production and migration through the vadose zone are a significant risk associated with the application of bioremediation and must be taken seriously during the design phase to minimize this risk as the potential outcome of methane migrating into in-door air could be disastrous including development of atmospheres that are immediately dangerous to life and health (IDLH) or flammable atmospheres. Applying in situ bioremediation in close proximity to buildings and/or subsurface structures should be approached with caution, and safety measures should be included as design measures for the application or at least as contingency plans to be implemented in the event that unwanted application-derived gases migrate into structures.

Mitigation measures to address substrate that is distributed inadvertently to unwanted areas are presented in Sections 3.1.3 and 3.1.4 including options that could be used to address unwanted gas generation (air sparge). However, if bioremediation has already been applied and unwanted gases are being generated and subsequently are migrating into in-door air, air sparge is not likely to immediately address the problem, and impacted buildings may be unusable for some time. The only mitigation measures known and recognized as effective by this author are typical measures designed to improve air quality in a structure. These measures include subslab depressurization and venting the living space to immediately restore air quality. Continued operation of these safety measures would continue until gas generation halts or is mitigated through another in situ remedial measure.

8. Conclusions

In situ remedial technologies and enhanced bioremediation in particular offer practitioners attractive solutions that can be used to restore groundwater quality rapidly effectively and at low cost in comparison to more traditional competing technologies while minimizing impacts to site use. In situ remedial technologies also offer a very high degree of flexibility in implementation methodology such that they can be effectively applied at any depth that can be reached through drilling, a verity of geochemical conditions, and a wide range of soil conditions. However, in situ remedial technologies require a high degree of site characterization in order to be applied and a high degree of dependability. Improper or inadequate site characterization can and frequently does lead to suboptimal results, failure in achieving performance objectives, and in some cases, the formation of new site contaminants and conditions that are more toxic or higher risk than baseline. These negative impacts can range from the generation and migration of undesirable and hazardous gas generation to the mobilization of toxic metal species, to the production of more toxic intermediate products. To make matters worse, once undesirable conditions form, they can be very difficult and costly to remedy. Therefore, proper characterization prior to implementation is critical as is careful application and performance monitoring program design by a trained remediation professional with extensive in situ remediation experience. Finally, careful and precise application implementation is key to achieving remedial objectives.

References

Air Force Center for Environmental Excellence (AFCEE), 2007. Final Protocol for In-Situ Bioremediation of Chlorinated Solvents Using Edible Oil. October.

Bradley, P., Chapelle, F., 1997. Kinetics of DCE and VC mineralization under methanogenic and Fe(III)-reducing conditions. Environmental Science and Technology 31, 2692–2696.

Bradley, P., Landmeyer, J., Dinicola, R., 1998a. Anaerobic oxidation of 1,2-dichloroethene under Mn(IV)-Reducing conditions. Applied and Environmental Microbiology 64, 1560–1562.

Bradley, P., Chapelle, F., Wilson, J., 1998b. Anaerobic mineralization of vinyl chloride in Fe(III)-reducing, aquifer sediments. Journal of Contaminant Hydrology 31 (1–2), 111–127.

Butler, E., Hayes, K., 1999. Kinetics of the transformation of trichloroethylene and tetrachloroethylene by iron sulfide. Environmental Science and Technology 33, 2021–2027.

Diller, K., Griffiths, D., Ross, C., Mehta, M., 2018. Combined In Situ Chemical Reduction and. Situ Bioremediation Groundwater Treatment for a Chlorinated VOC Plume. Paper presented at the Eleventh International Conference on Remediation of Chlorinated and Recalcitrant Compounds, Palm Springs. CA. Battelle Press, Columbus, OH.

ESTCP (Environmental Security Technology Certification Program), 2005. A Review of Biofouling Controls for Enhanced In-Situ Bioremediation of Groundwater. ESTCP, Arlington, VA, USA. https://clu-in.org/download/contaminantfocus/dnapl/Treatment_Technologies/ER-0429-WhtPaper.pdf.

Fennel, D., Gossett, J., Zinder, S., 1997. Comparison of butyric acid, ethanol, lactic acid, and propionic acid as hydrogen donors for the reductive dechlorination of tetrachloroethene. Environmental Science and Technology 31 (3), 918–926.

Federal Remediation Technologies Roundtable (FRTR), 2019. Remediation Technologies Screening Matrix and Reference Guide Version 4, Section 4.6 Fracturing. https://frtr.gov/matrix2/section4/4-5.html.

Griffiths, D., Badik, B., Belanger, T., Bataglia, R., 2018. Performance and refresh of a full scale biowall system designed to treat chlorinated solvents in groundwater. In: Paper Presented at the Eleventh International Conference on Remediation of Chlorinated and Recalcitrant Compounds. Palm Springs, CA. Battelle Press, Columbus, OH. April.

Interstate Technology and Regulatory Council (ITRC), 2005. Technical and Regulatory Guidance for In-Situ Chemical Oxidation of Contaminated Soil and Groundwater, second ed. January. https://www.itrcweb.org/GuidanceDocuments/ISCO-2.pdf.

Johnson, P.C., Johnson, R.L., Bruce, C.L., 2010. Air sparging for the treatment of chlorianted solvent plumes. In: Stroo, H.F., Ward, C.H. (Eds.), In-Situ Remediation of Chlorinated Solvent Plumes. SERDP and ESTCP Remediation Technology Monograph Series. Springer Science and Business Media, New York, NY, USA, pp. 109–143.

Livadas, A., 2003. Injecting remediation compounds: how you inject is equally as important as what you inject. Remediation Journal 13, 105–110.

Maymo-Gatell, X., Chien, Y., Gossett, J., Zinder, S., 1997. Isolation of a bacterium that reductively dechlorinates tetrachloroethene to ethene. Science 276, 1568–1571.

Naval Facilities Engineering Command (NAVFAC), 2001. Final Air Sparging Guidance Document, NFESC Technical Report TR-2193-ENV. NAVAL FACILITIES ENGINEERING

COMMAND Washington, DC 20374-5065. August. https://clu-in.org/download/contaminantfocus/dnapl/Treatment_Technologies/Air_Sparg_TR-2193.pdf.

NAVFAC, 2013. Best Practices for Injection and Distribution of Amendments, TR-NAVFAC-EXWC-EV-1303. Naval Facilities Engineering Command, Engineering and Expeditionary Warfare Center, Port Hueneme, California, 93043-4370. March. https://clu-in.org/download/techfocus/chemox/Inject-amend-tr-navfac-exwc-ev-1303.pdf.

New Jersey Department of Environmental Protection (NJDEP), 2011. Technical Guidance for the Preparation and Submission of a Site Conceptual Model. December. https://www.nj.gov/dep/srp/guidance/srra/csm_tech_guidance.pdf.

Pac, T., Lewis, R., Frazier, J., 2005. Importance of delivery methods for remedial effectiveness. In: Proceedings, Fourth International Conference on Oxidation and Reduction Technologies for In-Situ Treatment of Soil and Groundwater, Chicago IL, USA, October 23–27, p. 41.

Robinson, C., Barry, D.A., 2009. Design tool for estimation of buffer requirement for enhanced reductive dechlorination of chlorinated solvents in groundwater. Environmental Modelling and Software 24 (11), 1332–1338. ISSN 1364-8152.

Robinson, C., Barry, D.A., McCarty, P.L., Gerhard, J.I., Kouznetsova, I., 2009. pH control for enhanced reductive bioremediation of chlorinated solvent source zones. The Science of the Total Environment 407 (16), 4560–4573. ISSN 0048-9697.

Shammas, N.K., Yang, J.Y., Yuan, P.C., Hung, Y.T., 2005. Chemical Oxidation. Chapter 7 of Physicochemical Treatment Processes. In: Handbook of Environmental Engineering Book Series, vol. 3. Springer Science and Business Media, New York, NY, USA, pp. 229–270.

Siegrist, R.L., Crimi, M., Timpkin, T.J. (Eds.), 2011. Situ Chemical Oxidation for Groundwater Remediation. SERDP and ESTCP Remediation Technology Monograph Series. Ward, C.H., Series Editor. Springer Science and Business Media, New York, NY, USA, p. 658.

Simkin, T., Norris, R., 2010. Engineering and implementation challenges for chlorinated solvent remediation. In: Stroo, H.F., Ward, C.H. (Eds.), In-Situ Remediation of Chlorinated Solvent Plumes. SERDP and ESTCP Remediation Technology Monograph Series. Springer Science and Business Media, New York, NY, USA, pp. 109–143.

Suthersan, S., 1999. In: Suthersan, S. (Ed.), Hydraulic and Pneumatic Fracturing in Remediation Engineering: Design Concepts. CRC Press, Boca raton, 1999. http://www2.bren.ucsb.edu/~keller/courses/esm223/SuthersanCh09HydraulicFracturing.pdf.

United States Environmental Protection Agency (USEPA), 1998. Technical Protocol for Evaluating Natural Attenuation of Chlorinated Solvents in Groundwater. National Risk Management Research Laboratory, Office of Research and Development, USEPA. EPA/600/R-98/128, Cincinnati, OH.

United States Environmental Protection Agency, 2000. Engineered Approaches to In Situ Bioremediation of Chlorinated Solvents: Fundamentals and Field Applications. Office of Solid Waste and Emergency Response, Division of Solid Waste and Emergency Response. EPA 542-R-00-008.

United States Army Corps of Engineers (USACE), 2012. Environmental quality, site conceptual models. An Engineering Manual. December. http://140.194.76.129/publications/.

USACE, 2013. Environmental Quality, In-Situ Air Sparging. An Engineering Manual. December. https://www.publications.usace.army.mil/Portals/76/Publications/EngineerManuals/EM_200-1-19.pdf?ver=2014-01-06-104233-540.

USEPA, 2017. How to Evaluate Alternative Cleanup Technologies for Underground Storage Tank Sites, A Guidance for Corrective Action Plan Reviewers. Office of Land and Emergency Management, USEPA. EPA/510/B-17/003, Cincinnati, OH. October.

USEPA, 2018. Environmental Fracturing – an Overview. Technology Innovation and Field Services Division, USEPA. September. https://clu-in.org/techfocus/default.focus/sec/Environmental_Fracturing/cat/Overview/.

USEPA, 1994. Manual: Alternative Methods for Fluid Delivery and Recovery EPA/625/R-94/003, 96 pp. https://clu-in.org/download/techfocus/fracturing/Frac-EPA-625-R-94-003.pdf. Accessed on April 14, 2019

Volkering, F., Pijls, C., 2004. Factors determining reductive dechlorination of cis-1,2-DCE at PCE contaminated sites. In: Proceedings of the Fourth International Conference on Remediation of Chlorinated and Recalcitrant Compounds (Monterey, CA; May 2004). Paper 3D-10. Battelle Press, Columbus, OH.

Wiedemeier, T., Rifai, H., Newell, C., Wilson, J., 1999. Natural Attenuation of Fuels and Chlorinated Solvents in the Subsurface. John Wiley and Sons, Inc., New York, NY, USA, p. 617.

Chapter 9

Star: a uniquely sustainable in situ and ex situ remediation process

Jason I. Gerhard[1], Gavin P. Grant[2], José L. Torero[3]
[1]*Department of Civil and Environmental Engineering, The University of Western Ontario, London, ON, Canada;* [2]*Savron, Toronto, ON, Canada;* [3]*Department of Civil, Environmental and Geomatic Engineering, University College London, London, United Kingdom*

1. Introduction

STAR (Self-sustaining Treatment for Active Remediation) is a rapidly growing thermal technology for treating contaminated soils, organic sludges, and wastes. STAR utilizes smouldering combustion to destroy organic contaminants. Smouldering has been traditionally regarded as a hazard to be avoided. However, Switzer et al. (2009) introduced the concept of applied, controlled smouldering to remediate contaminated soils. Smouldering is an exothermic oxidation reaction where the organic contaminant is the fuel. Oxidation of the fuel converts the contaminants into energy, leaving clean soil behind. Since this concept was established, numerous laboratory studies have extended our understanding of applied smouldering and pointed to its wide range of potential applications (Pirioni et al., 2011; Salman et al., 2015; Kinsman et al., 2016; Yerman et al., 2015, 2016; Yerman et al.,2016; Rashwan et al., 2016; Zanoni et al., 2017; Gianfelice et al., 2019). Numerical modeling studies have further expanded our knowledge of how STAR can be manipulated and optimized for environmental and economic benefits (Macphee et al., 2012; Hasan et al., 2015; Zanoni et al., 2018a, 2018b, 2019). Moreover, STAR has been successfully upscaled, emerging as a green remediation technology being commercially applied to contaminated soils in situ as well as contaminated soils and waste sludges ex situ (Switzer et al., 2014; Scholes et al., 2015; Grant et al., 2016). Numerous field pilot-tests and full-scale site remediation projects using STAR have been successfully completed to date (Savron, 2014, 2015, 2016a, 2016b, 2017a, 2017b, 2018a, 2018b, 2018c).

This chapter provides a summary of STAR technology for remediation and highlights the reasons it exhibits a low environmental footprint and rates

Sustainable Remediation of Contaminated Soil and Groundwater. https://doi.org/10.1016/B978-0-12-817982-6.00009-4

highly on sustainability assessments. The chapter begins with a summary of the science of smouldering and the basic principles behind its application to remediation. Highlighted within this summary is how energy is generated, transferred, stored, and accumulated during STAR since the positive energy balance is central to its sustainability. The chapter then describes the range of soils and contaminants amenable to STAR as well as some limiting factors in its application. Next, in situ and ex situ full-scale applications of STAR are described. Finally, STAR's sustainability aspects are described as well as two examples of quantitative sustainable remediation assessments on STAR projects.

2. Scientific principles

Combustion is an exothermic oxidation reaction:

$$\text{Carbon Compounds} + O_2 \rightarrow CO_2 + H_2O + \text{Energy} \tag{9.1}$$

It is activated by raising the fuel (i.e., the carbon compounds, in this case the contaminant) to its ignition temperature and ensuring the presence of oxygen. Smouldering is a flameless type of combustion; the reaction occurs within the surface of the fuel. Glowing red charcoal in a barbeque is a typical example of smouldering (Fig. 9.1).

Smouldering requires a porous fuel, such as charcoal, or an inert porous matrix embedded with fuel, such as soil containing coal tar. This is because the reaction requires sufficient oxygen (air) from the system's boundary to reach the burning fuel's surface to sustain the reaction. Once ignited, the reaction will continue indefinitely as long as sufficient air and fuel are available, since the reaction itself produces the energy required to initiate further reactions. In fact, under a wide range of conditions, the heat generated exceeds that required

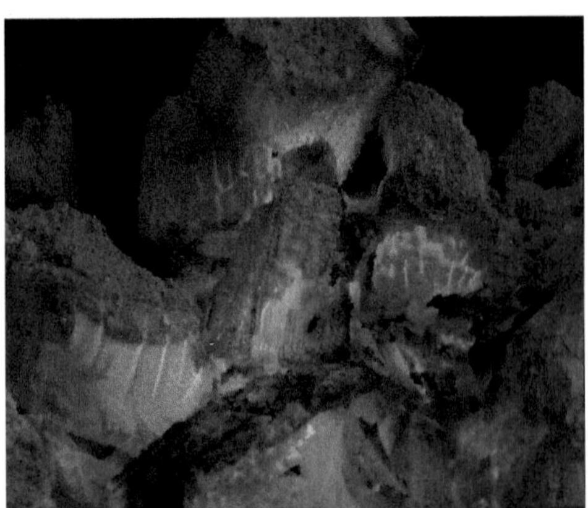

FIGURE 9.1 Smouldering charcoal example of a flameless, self-sustaining combustion reaction. *Photo: https://commons.wikimedia.org/wiki/File:Embers_01.JPG.*

to heat the fuel plus heat losses (e.g., boiling pore water). This positive energy balance underpins the fact that smouldering is self-sustaining under appropriate conditions (Zanoni et al., 2018a).

Consider a one-dimensional system containing smoulderable material, such as coal-tar-contaminated soil (Fig. 9.2). In the laboratory, a stainless-steel column is typically employed (e.g., Switzer et al., 2009). Heat is applied to the base of the system for a short time. There are numerous methods of applying heat that are functionally equivalent (convection, conduction, radiation); convection via hot air injection is illustrated. The soil rapidly absorbs the heat due, in particular, to the high heat capacity of the mineral grains. Therefore, only a small volume of soil becomes hot, and it exceeds the ignition temperature relatively quickly (e.g., minutes in the laboratory, hours in the field). The ignition temperature is dependent on contaminant type, but is typically approximately 200°C. Beyond this threshold, in the presence of continued air injection, the smouldering reaction ignites. After this time, provision of external energy is unnecessary, so the injected air is not heated. The burning contaminant is converted primarily to CO_2, H_2O, and heat (Eq. 9.1), leaving clean soil (i.e., no detectable carbon-based compounds). Once the contaminant is consumed, the reaction propagates in the direction of the flowing air (Fig. 9.2). This propagation of the reaction along the column is the manifestation of the self-sustaining process. The reaction travels upward at a steady velocity as long as the material ahead contains sufficient fuel and the airflow rate stays constant.

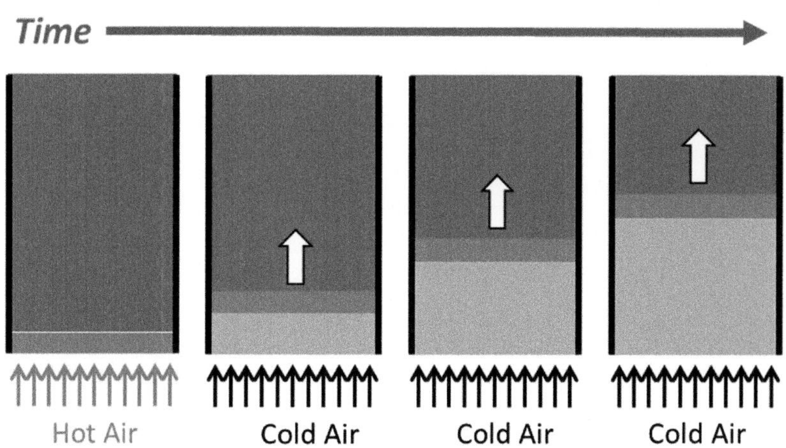

FIGURE 9.2 Time sequence illustrating the ignition and self-sustaining propagation of a smouldering reaction through contaminated soil. Dark Gray represents contaminated soil, red (gray in print version) is the smouldering reaction, and blue (light gray in print version) is remediated soil. The *white arrow* indicates the propagation direction of the self-sustaining smouldering reaction.

As illustrated, the smouldering reaction is a relatively thin (i.e., several centimeters) treatment zone. The temperature and energy characteristics surrounding this zone are explained in Fig. 9.3, which presents a snapshot of the reaction (one panel from Fig. 9.2) in more detail. The material above the reaction is contaminated while the material below the reaction is remediated. The temperature profile (right) reveals that the smouldering reaction coincides with the peak temperature, resulting from the release of energy associated with fuel oxidation (i.e., contaminant destruction). In fact, this process typically generates considerable excess energy. One of the keys to STAR's sustainability is that the porous medium retains that energy making heat transfer and reuse very efficient. Fig. 9.3 shows that the reaction zone is the center of a larger heated region comprising two additional zones: above is the preheating zone of the contaminated soil and below is the hot clean soil zone that is slowly cooling. Both above and below these heated zones the soil is at ambient temperatures. As Fig. 9.3 illustrates, heat is transferred—from within the reaction where energy is generated, and from below the reaction where heat is stored in clean soil—to the contaminated soil immediately above the reaction where it is deposited. Thus, smouldering is by its nature a highly efficient

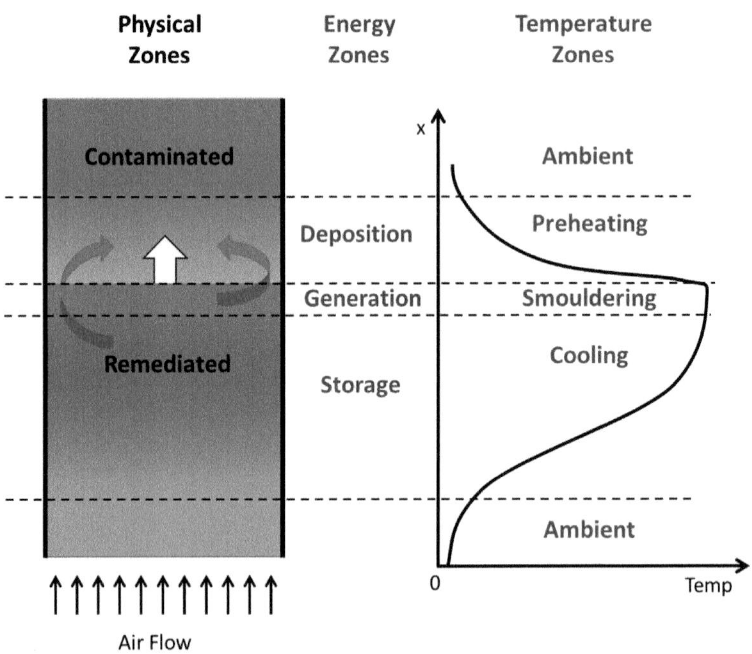

FIGURE 9.3 Profile of a self-sustaining smouldering reaction at a single time. Left: the physical distribution of heat, contaminated soil, and remediated soil. The *red arrows* (gray in print version) illustrate key heat transfer processes. Center: the zones that generate, store, and receive energy. Right: typical temperature profile with the temperature-related zones identified.

energy recycling process. It is because of this energy recycling that materials that would never burn via flaming combustion—such as wet logs, wet fabric, or wet contaminated soils and sludges—can successfully smoulder.

The propagation of a self-sustaining reaction is documented by observing a succession of temperature signals using thermocouples spaced along the reaction pathways in the contaminated material. Fig. 9.4 provides typical examples from the laboratory and from the field (Scholes et al., 2015). Both reveal that the self-sustaining reaction propagates in the direction of airflow (vertically upward for the laboratory example, horizontally outward in the in situ field example) with a specific velocity. The velocity is determined by the distance between the peak temperatures divided by the time between the peaks. Since the smouldering reaction separates clean from contaminated material, the velocity of the smouldering reaction (indicated by the white arrow in Fig. 9.2) corresponds to the contaminant mass destruction rate.

The velocity of the smouldering reaction in organic-contaminated soil can be described analytically by (MacPhee et al., 2012):

$$V^{f}_{i,j}(t) = \frac{\dot{m}^{AIR}_{i,j}(t)c_{pg}(T_s - T_{amb}) + A\dfrac{\Delta H_r}{v_O}\dot{m}^{AIR}_{i,j}(t)\cdot Y_{O,I}}{\left(\rho_{bs}c_{ps}T_s - \left(\rho_{bs}c_{ps} + S^{Org}_{i,j}(t)\varnothing\rho_{Org}c_{pOrg}\right)T_{amb}\right)} \quad i,j = x,y \quad (9.2)$$

where $\dot{m}^{AIR}_{i,j}(t)$ is the air mass flux, c_{ps}, c_{pOrg}, and c_{pg} are the specific heat

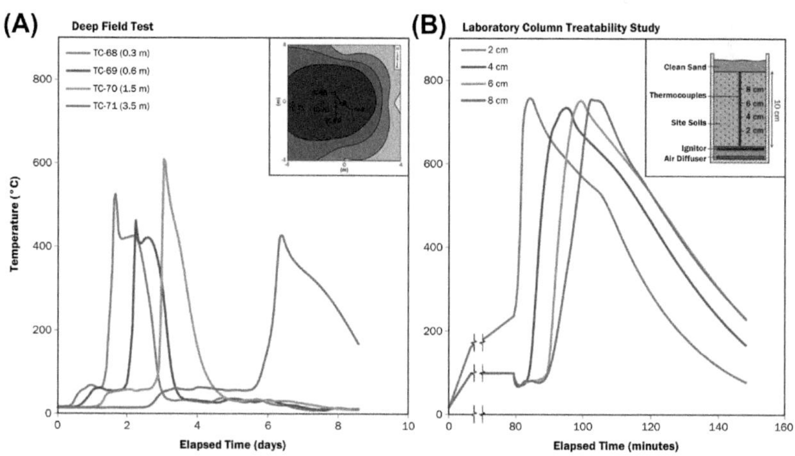

FIGURE 9.4 Temperature history trends for single line of thermocouples in the deep field test (A) and laboratory treatability study (B). Inset in (A) shows the thermocouple locations in plan view, while the inset in (B) is the laboratory column apparatus described fully elsewhere 6,14 with the vertically oriented thermocouple locations. *Reprinted with permission from Scholes, G., Gerhard, J., Grant, G.P., Major, D.W., Vidumsky, J.E., 2015. Smouldering remediation of coal tar contaminated soil: pilot field tests of STAR. Environmental Science and Technology, 1–9, doi:10.1021/acs.est.5b03177. Copyright (2015) American Chemical Society.*

constants for the soil, organic, and air phases, respectively, T_s and T_{amb} are the peak smouldering reaction temperature and the ambient temperature, respectively, ΔH_r represents the effective heat of smouldering of the organic, v_O is the oxygen/fuel overall stoichiometric coefficient, $Y_{O,I}$ is the initial mass fraction of oxygen present in the gas phase, ρ_{bs} is the bulk density of the soil, ρ_{Org} is the density of the organic phase, $S_{i,j}^{Org}(t)$ is the organic saturation, $\varnothing$ is the soil porosity, t is time, and A is a calibration parameter specific to each contaminant type. The first term in the numerator describes the energy transferred ahead via the convection of hot air. The second term in the numerator describes the energy generated by the smouldering of the organic contaminant. The term in the denominator describes the energy absorbed by the soil. Eq. (9.2) helps explain why the air mass flux delivered to the system is the dominant factor in controlling the velocity of the smouldering reaction: injecting air at a higher rate increases both terms in the numerator.

Indeed, laboratory and field measurements have confirmed that the smouldering velocity increases linearly with the injected air flux. Since the system operator sets the injected airflow rate, STAR provides considerable control on the rate of mass destruction. In a typical column experiment, the reaction propagates around 0.3 cm/min for an air flux of 4 cm/s (reported as Darcy flux: volume of air per unit area per unit time) to around 1.2 cm/min for an air flux of 25 cm/s (Pironi et al., 2011). In the field, practical considerations of air delivery means that the smouldering reaction typically propagates at approximately 0.5−1.0 m/d (Scholes et al., 2015; Grant et al., 2016). The contaminant mass destruction rate that corresponds to this propagation rate increases with the scale of the application: a larger smouldering reaction surface results in a proportionally larger mass destruction rate. The reaction surface's configuration in full-scale applications is discussed further in Section 5.

Smouldering in contaminated soil can never be a runaway or uncontrolled reaction, unlike in underground coal fires. Terminating the injected airflow causes the reaction to die quickly because the soil is such a significant heat sink (Switzer et al., 2009). If the airflow is continued, a self-sustaining smouldering reaction terminates naturally when no organic contaminant remains. After the process is complete, it is typical to observe dry, organic-free soil (Fig. 9.5). Typical reductions in soil of total petroleum hydrocarbons are greater than 98% (Switzer et al., 2009; Scholes et al., 2015; Grant et al., 2016). Many STAR applications, both in the laboratory and in the field, return "non-detect" levels of Total Petroleum Hydrocarbons (TPH) for soils that experience smouldering.

The reaction zone temperature depends primarily on contaminant type, typically ranging from 500 (e.g., for diesel) to 1000°C (e.g., for coal tar). Water-filled porosity tends to have little effect, since there is a boiling front that travels ahead of the reaction. In other words, smouldering is so energy efficient that there is typically sufficient excess energy to boil of the water and

FIGURE 9.5 Comparison of oil-contaminated soil before and after smouldering treatment in a field pilot test in Taiwan. *Reprinted with permission from Savron, 2017b. STARx (ex situ) HottpadTM pilot test for the treatment of petroleum hydrocarbon-impacted soils. Project Description. http://www.savronsolutions.com/case-studies.*

still have a net positive energy balance to support self-sustaining propagation. For this reason, smouldering remediation below the water table is common (Scholes et al., 2015; Grant et al., 2016).

3. Energy efficiency

A key aspect of STAR with respect to sustainability is the net positive energy generated by the process. This means that the energy generated by smouldering exceeds the sum of the heat sinks and losses: igniting the reaction, heating the soil up to smouldering temperatures, boiling off water and some volatiles, and radial heat losses. The net energy surplus increases with increasing scale of the STAR application. In other words, full field applications are more energy efficient than laboratory experiments and small pilot tests. This occurs for two main reasons. First, larger systems result in a larger thickness of contaminated soil along the treatment path, which generates more energy. Second, larger systems have a larger distance perpendicular to the smouldering travel path, which reduces heat losses. These two important aspects of the energy balance are discussed in turn next.

More total energy is generated as the propagation pathway for the STAR reaction increases. Referring to Fig. 9.2, this means a thicker contaminated soil layer in a taller column. Since at each step of Fig. 9.2, some net positive energy is generated, the total energy generated by STAR systems increases with increasing amounts of contaminant destroyed. Moreover, the majority of that energy is stored in the clean material behind the front (Fig. 9.3). Therefore, the energy present in the system accumulates with time as the treated pathway length grows.

Fig. 9.6 demonstrates this result with data from three column smouldering experiments that differ only in the height of the column. The total energy in the column at each time is calculated based on the distribution of temperature along the column (i.e., integration of the temperature profile as in Fig. 9.3 while accounting for the mass of heated soil and the soil's heat capacity). The increasing amount of energy with time occurs because of the increasing length

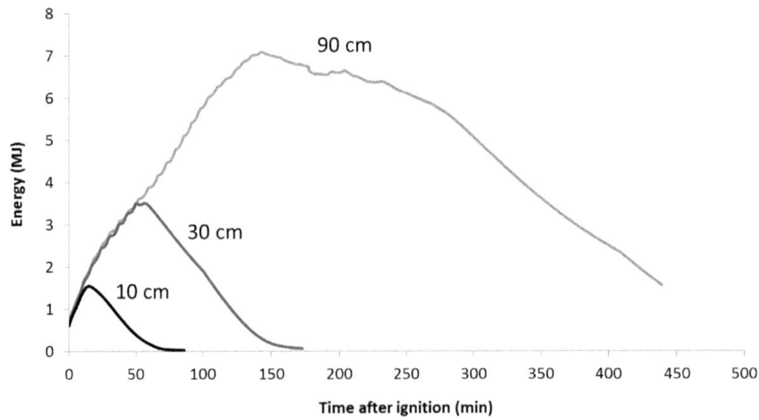

FIGURE 9.6 Total energy stored in the treated material over time for one-dimensional column experiments with three different lengths. Column experiments all used the same contaminated material, column diameter, and airflow rate. *Data courtesy of Laura Kinsman, Western University STAR Laboratory.*

of the cooling zone (Fig. 9.3) as STAR progresses; in other words, an increasing amount of the excess energy is stored in the system as hot, clean soil. The peak total stored energy for each experiment in the figure occurs when the smouldering reaction reaches the top of the contaminant pack. The subsequent energy decrease corresponds to the cooling of the treated material when smouldering has terminated (due to no more contamination) and injected air is removing stored heat by convection. All three experiments required approximately 2.8 MJ to ignite the smouldering reaction. The figure reveals that the 30 cm tall column produced 3.5 MJ by the end of treatment (~1.3 times the inputted energy) while the 90 cm column produced 7.1 MJ (~2.6 times the inputted energy). For field-scale treatment systems, the treated path length for the reaction is typically 1.5–4 m depending on the application and site characteristics. Therefore, typically there is considerable net energy generated and stored in the clean soil, reducing the costs of treatment and increasing the potential for energy recovery and reuse.

The second reason larger STAR treatment systems are more energy efficient than smaller systems is that the heat loss component of the energy balance is reduced as scale increases. Heat losses occur in the direction(s) perpendicular to that in which the smouldering reaction is propagating. In Fig. 9.3, these heat losses occur in the radial direction (toward the outside of the system, through the column walls) along the thickness of the heated zone. Because the heated zone thickness increases with time, as described earlier, the heat losses of the system also increase with time (Zanoni et al., 2018a). However, heat losses are a function of the surface area-to-volume ratio of the system. Therefore, as system applications increase in scale, and the volume of material increases greatly relative to the surface area of the outside boundary, the fraction of heat lost decreases substantially. In other words, since soil is such an excellent insulator, the more soil that is treated, the larger the fraction of energy that is stored and not lost to the environment. More energy stored translates into more energy that can be reused in smouldering and recycled (Fig. 9.3) and thus a more efficient treatment system.

The practical implications of this are that laboratory tests, while useful for determining if a contaminated soil will smoulder, are not ideal for establishing the limits of the process (e.g., the lowest contaminant concentration that can be smouldered, the lowest airflow rate that can support smouldering). That is because energy losses at the column scale are approximately 40%–50% (Martins et al., 2010) due to the high surface area and low volumes of these systems. Meanwhile, it means that field-scale applications are much more energy efficient and can therefore be expected to smoulder contaminants that are less energetic (i.e., lower concentration, poorer fuels) or are wetter or the operator can use lower air fluxes and achieve the same results. It also means that the cost of treatment decreases dramatically with increasing scale. Fig. 9.7 reveals how the ratio of input energy for ignition to the mass of soil treated by STAR drops dramatically as the scale of the system increases over four orders of magnitude (Scholes et al., 2015). Ultimately, this suggests that large-scale treatment systems are the most cost and energy efficient for STAR, which is not the case for many other technologies.

4. Contaminants treated and process limits

Smouldering remediation is most readily applied to complex, long-chain hydrocarbons such as coal tar, fuel oils, and unrefined or heavy petroleum hydrocarbons. These organic industrial contaminants contain significant embedded energy; for example, coal tar has a heat of combustion of ~35 MJ/kg, which is more than double that of wood. However, the suitability of a contaminant for smouldering cannot be evaluated by its heat of combustion; other characteristics of the compounds in the organic liquid/sludge/solid contaminant and in the soil play an important role. For example, compound volatility and the composition of mixed (multicomponent) organic liquids are important. Coal tar and heavy hydrocarbons exhibit low volatility in the majority of their hundreds of compounds, making their smouldering reaction highly exothermic and therefore self-sustaining. This means that their removal during smouldering is dominated by in situ oxidation rather than volatilization. For example, less than 2% of coal tar mass treated by smouldering was volatilized with greater than 98% destroyed during an in situ STAR application (Scholes et al., 2015). The stability of heavy hydrocarbons, which make them generally resistant to standard chemical, thermal, and biological treatments, is also the reason they are suitable for smouldering.

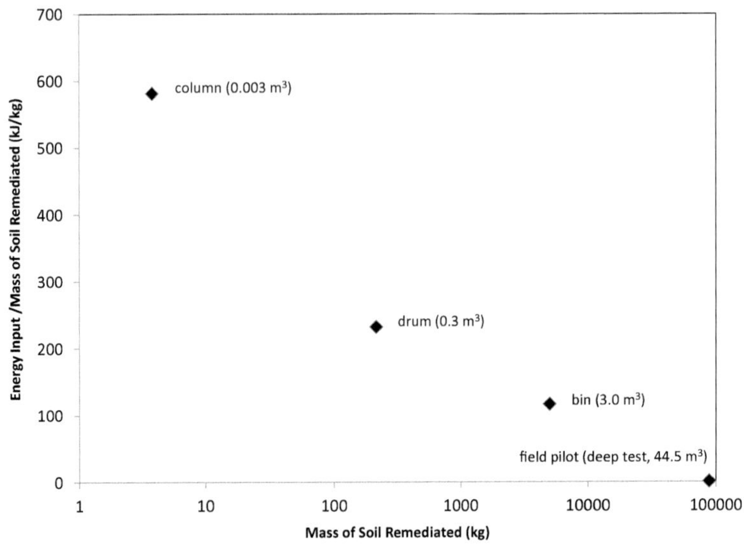

FIGURE 9.7 Ratio of ignition energy to total mass of soil remediated for increasing scales of coal tar smouldering tests demonstrating the increasing energy efficiency of the smouldering process with scale of application. *Reprinted with permission from Scholes, G., Gerhard, J., Grant, G.P., Major, D.W., Vidumsky, J.E., 2015. Smouldering remediation of coal tar contaminated soil: pilot field tests of STAR. Environmental Science and Technology, 1–9, doi:10.1021/acs.est.5b03177. Copyright (2015) American Chemical Society.*

More volatile contaminants, or the more volatile components of mixed organic liquids, can exhibit higher mass fractions volatilized due to the heating front (i.e., preheating zone, Fig. 9.3) that moves ahead of the smouldering reaction. If a contaminant is too volatile, such as gasoline, it may be difficult to ignite. Diesel has been treated with self-sustained smouldering in field pilot tests (Savron, 2016a); however, lighter compounds, such as gasoline and chlorinated solvents, tend to be too volatile to be smouldered directly. In these cases, neat or emulsified vegetable oil (EVO) can be injected into the soil as a nontoxic fuel to support smouldering. This has been demonstrated for smouldering trichloroethylene (TCE)-contaminated soil in laboratory studies, producing soil that had no detectable chlorinated solvents (or vegetable oil) remaining (Salman et al., 2015). That work demonstrated that of the chlorinated solvent mass removed from the soil during smouldering, ~25% was destroyed in situ and ~75% was volatilized and captured in the emissions. A similar ratio of destroyed to volatilized mass was observed in a field pilot test for EVO smouldering to remediate gasoline (Savron, 2016a). In such cases, EVO smouldering may be an inexpensive and green approach to accomplish thermal desorption, the goal of which is to volatilize most of the contaminant for recovery and treatment at the surface.

Other gaseous compounds can be produced during STAR in minor quantities that require careful attention for capture and treatment. Carbon monoxide is typically produced as a by-product in smouldering; although it is typically a fraction of the amount of carbon dioxide (i.e., most combustion is complete, not incomplete), it is a regulated compound that must be carefully monitored, captured, and handled. Typically, for in situ applications, a small fraction ($<1\%$) of remediated hydrocarbon mass is released as volatile components for subsequent capture and treatment (primarily benzene, toluene, ethylbenzene, xylenes, and naphthalene) (Grant et al., 2016). Incineration of chlorinated compounds and vegetable oil may, under certain conditions, produce gaseous by-products including volatile organic compounds, phosgene, and dioxins/furans (D/F). Laboratory studies of smouldering TCE solvent found PCE (perchloroethylene) generated at concentrations two orders of magnitude less than TCE emissions, and phosgene concentrations an order of magnitude less than strict regulatory limits (Salman et al., 2015). The same study found that aliphatic hydrocarbons in trace amounts were observed due to smouldering the injected vegetable oil.

D/F formation is complex and highly dependent on the type and mass fraction of chlorinated compounds in the organic liquid, the smouldering temperature, the smouldering rate, and numerous other factors. A study was conducted by the U.S. Environmental Protection Agency (USEPA) and the Battelle Memorial Institute involving the smouldering combustion of a creosote-contaminated soil containing pentachlorophenol (PCP) to consider potential D/F formation. The study concluded that STAR did not generate D/F, and that STAR resulted in D/F destruction since the D/F mass in the pre-treatment soils was more than 300,000 times the mass detected in the vapor sample (Battelle, 2012). However, the potential for its formation should be evaluated on a site-specific basis using laboratory treatability studies and confirmed with an on-site pilot test. In all cases where minor gaseous compounds of concern are expected, standard surface vapor treatment techniques are implemented (e.g., adsorption on activated carbon, destruction via thermal oxidation).

Bench-scale treatability studies are a simple and inexpensive way to evaluate whether a particular contaminated soil or liquid waste is a good candidate for smouldering remediation. Standard laboratory tests can determine whether the contaminated soil sample is self-sustaining alone or self-sustaining with the addition of EVO. In the case of liquid waste, a homogeneous mixture with sand is typically created first. The treatability study can provide the peak smouldering temperature, before/after contaminant concentrations, and identifies the presence/absence of compounds in the emissions. A treatability study can also reveal if the soil permeability is too low to permit sufficient airflow to sustain the reaction; in such cases where ex situ treatment is an option, a bulking agent (e.g., sand) can be added first. As discussed in Section 3, smouldering reactions at the bench scale are less robust (due to shorter propagation lengths and higher heat losses) than at the field scale. Thus, bench treatability studies are conservative: if a sample smoulders in the laboratory, it is very likely to smoulder in the field, and may do so at lower contaminant concentrations and/or with lower airflow rates.

Current research is pushing the boundaries of the material types that can be remediated with smouldering. Wastes that are typically considered too wet to be combusted in a conventional manner have been shown to be amenable to smouldering due to its energy efficiency. For example, smouldering is a promising alternative for converting wastewater treatment plant sludge into heat and inorganic ash (Rashwan et al., 2016). Moreover, feces mixed with sand can be smouldered, potentially offering a low-energy sanitation solution for the developing world (Yermán et al., 2015). Toward that end, a continuous smouldering reactor that receives a steady supply of feces/sand mixture, destroys the feces, and expels hot sand with recoverable energy has been developed (Favara et al., 2016).

Furthermore, emerging contaminants such as per- and polyfluoroalkyl substances (PFAS) have been demonstrated to be successfully eliminated from contaminated soil and sorptive media by STAR (Major, 2019). Such compounds are typically present in small absolute quantities but orders of magnitude above regulatory limits; thus, they cannot form the fuel for STAR. In this case, granular activated carbon (GAC) is a valuable surrogate fuel that also is typically loaded with PFAS from water treatment activities. Proof of concept tests have shown that PFAS-contaminated GAC and also PFAS-contaminated soil mixed with GAC have smouldered at above 900°C, corresponding to temperatures expected to thermally destroy PFAS compounds. Pretreatment samples, exhibiting in excess of 1,000,000 ug/kg PFAS were found to contain nondetect concentrations of PFAS after STAR (with a detection limit of 0.4 ug/kg) (Major, 2019). Since emerging contaminants represent a class of contaminants that are present at low but environmentally significant concentrations in very large volumes of soil and sorptive media to be treated, a sustainable remediation method for addressing them such as STAR is needed.

There are some limits on the STAR process. In some cases, the amount of energy generated is not enough to overcome heat sinks/losses and the reaction is not self-sustaining. This can occur if not enough organic contaminant or supplemental fuel is present in the soil. This threshold is typically ~3000–5000 mg/kg, with the exact value being soil and contaminant dependent (Grant et al., 2016). However, there is no known upper limit on the contaminant concentration within a soil, meaning the process is attractive as a mass reduction technology for highly contaminated sites. Indeed, more organic contamination results in more net positive energy produced and is thus beneficial with respect to maximizing both energy efficiency and the rate of mass destruction. Organic saturation of soil up to unity (i.e., 100% of porosity filled by organic contaminant) has been successfully treated by smouldering (Switzer et al., 2009).

The process can also fail to generate positive net energy if sufficient oxygen is unable to reach the reaction. This can occur for two main reasons. First, the injected airflow rate is too low. Modeling and experimental studies have both indicated that approximately 0.5 cm/s (Darcy flux) of injected air is a minimum threshold for self-sustained smouldering in laboratory- and pilot-scale tests (MacPhee et al., 2012; Hasan et al., 2015). Theoretically, there is an upper bound on airflow rates, since very high airflows can cool the reaction faster than energy is being accumulated. However, laboratory experiments suggest that this only occurs above approximately 30 cm/s (Yerman et al., 2016) while the practical upper limit of field equipment is typically around 2–5 cm/s, considering the air injection footprint of the typical application. Thus, practical applications of STAR need to ensure that the airflow rate is not too low, but can otherwise choose an airflow rate of any practical magnitude to set the mass destruction rate as desired.

The second reason the smouldering reaction may die due to oxygen starvation is excessively low soil permeability. Treatability studies have shown smouldering remediation to be successful in a wide range of soil types from gravel and fill to 70%–80% silt with clay. However, extremely low permeability materials such as thick clay and rock are not able to conduct sufficient airflow to support self-sustaining smouldering. An advantage of ex situ applications is that contaminated clays can be mixed with more permeable materials, such as sands, to create a sufficiently permeable material for smouldering.

5. Field applications

5.1 In situ STAR

STAR has been demonstrated as a successful in situ remediation technology in numerous pilot tests and also at full scale (Scholes et al., 2015; Grant et al., 2016; Savron, 2014, 2015, 2016a, 2016b, 2018a). The typical in situ configuration and associated equipment are illustrated in Fig. 9.8. [Note that a full animation of the in situ process from installation to completion is available at https://youtu.be/656B57-yxpo]. It illustrates how smouldering is ignited at a central ignition well, then the external heat is turned off, and the self-sustaining reaction is propagated outward (both laterally and upward in a growing semitorus or horizontally sliced donut shape) by air injected through the same well. Gases are captured at the surface and treated using standard equipment.

As an example, in situ smouldering was applied to a 2 m thick coal tar layer present at 8 m depth in an alluvial aquifer, 7.5 m below the water table (Scholes et al., 2015). The 15-hectare site was a former industrial facility in

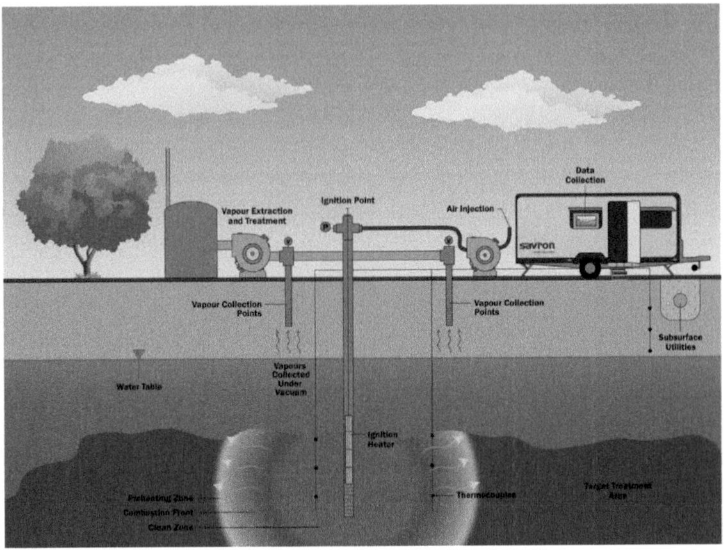

FIGURE 9.8 Standard configuration of in situ STAR smouldering treatment below the water table including typical equipment employed. *Courtesy of Savron; animated version available at https://youtu.be/656B57-yxpo.*

New Jersey, USA, where coal tar was disposed in lagoons that were backfilled in 1965. Ignition of STAR was achieved after approximately 3 hours of heating a region immediately around the well screen. Thereafter, the self-sustaining reaction traveled outward over the following 11 days in an approximately uniform manner in all four compass directions, reaching a radius of influence (ROI) of 3.7 m. During that time, coal tar destruction rates were measured at approximately 80 kg/day, as determined from combustion gases collected at the surface. An estimated total of 860 kg was destroyed with this single treatment event, with less than 2% recovered as organic compounds by the vapor collection system (Scholes et al., 2015). Pre- and post-treatment coring provided evidence of coal tar elimination in the treatment zone (Fig. 9.9). TPH concentrations, based on subsampling of the cores, were reduced from a pre-treatment average of 18,500 mg/kg to a post-treatment average of 450 mg/kg (Fig. 9.10). The fact that the zone being treated was below the water table was no barrier to in situ smouldering. The test also demonstrated that the smouldering reaction rapidly terminates when air injection is stopped. It was also demonstrated that heterogeneity of in situ permeability, which dictates the spatial distribution of airflow, can affect the distribution of smouldering propagation rates (Scholes et al., 2015). Further details about smouldering pilot tests, full-scale applications, and challenges associated with in situ applications are described in Scholes et al. (2015) and Grant et al. (2016) while additional in situ applications are described in Savron (2014, 2015, 2016a, 2016b, 2018a).

FIGURE 9.9 Soil cores from (A) before and (B) after an in situ STAR pilot test at a coal-tar-contaminated site. *Reprinted with permission from Scholes, G., Gerhard, J., Grant, G.P., Major, D.W., Vidumsky, J.E., 2015. Smouldering remediation of coal tar contaminated soil: pilot field tests of STAR. Environmental Science and Technology, 1–9, doi:10.1021/acs.est.5b03177. Copyright (2015) American Chemical Society.*

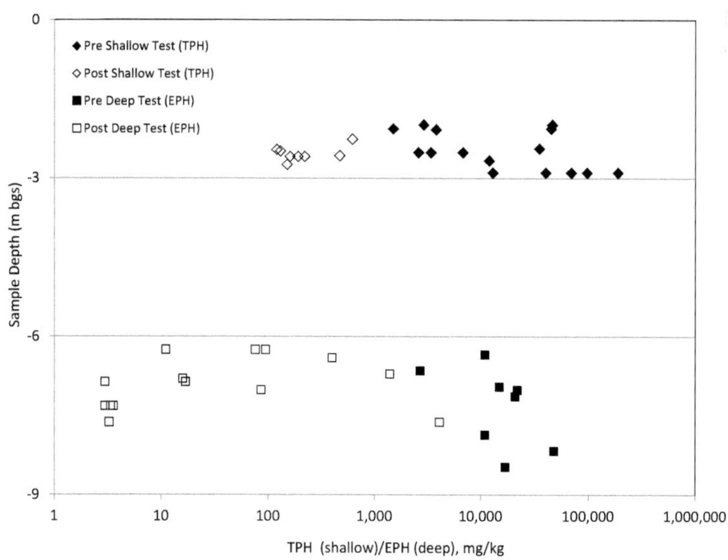

FIGURE 9.10 Pre- and post-TPH (shallow fill) and EPH (deep sand) soil concentrations by sample depth. All post test samples are collected from within inferred combustion zones. Note that the depth (vertical) axis is linear scale and the soil concentration (horizontal) axis is logarithmic. *Reprinted with permission from Scholes, G., GerhardJ., GrantG.P., MajorD.W., VidumskyJ.E., 2015. Smouldering remediation of coal tar contaminated soil: pilot field tests of STAR. Environmental Science and Technology, 1–9, doi:10.1021/acs.est.5b03177. Copyright (2015) American Chemical Society.*

The pilot test results described in Scholes et al. (2015) were used to design the full-scale site remediation that was subsequently applied at the New Jersey site. Extensive site investigation determined that approximately 40,000 cubic meters of soil was impacted with coal tar over an approximately 2 ha area within two main horizons: 2 and 8 m below ground surface (bgs). Fig. 9.11

FIGURE 9.11 Photograph of STAR full-scale in situ site remediation at a coal-tar-contaminated former industrial facility in New Jersey, USA. The black cable provides power to the heater in the ignition well, the black hose provides air to the same well, and the yellow (white in print version) hoses pull gases from the vapor extraction points to the treatment trailer. *Reprinted with permission from Grant, G., Major, D., Scholes, G.C., Horst, J., Hill, S., Klemmer, M.R., Couch, J. N., 2016. Smouldering combustion (STAR) for the treatment of contaminated soils: examining limitations and defining success, Remediation Journal 26, 27–51, Copyright (2016) John Wiley and Sons.*

shows a single treatment "cell." Each cell contained between 4 and 20 ignition/air injection wells operated simultaneously, a control trailer that dictated power delivery, air delivery, subsurface temperature monitoring, and vapor extraction, as well as a vapor treatment system. Three such systems were operated in parallel, progressing STAR remediation across the site in a continual process of treatment and post-treatment verification. During the 4 years of operation, nearly 2000 STAR ignitions were applied, destroying more than 50,000 kg of coal tar. Remedy verification within both horizons confirmed successful treatment. Soil remediation activities are now complete, allowing the site to proceed to redevelopment for valuable reuse.

5.2 Ex situ STAR

STAR has been applied as an ex situ treatment at the rapid assessment, pilot and full scales. A variety of ex situ STAR treatment systems have been constructed to apply the technology to a wide range of situations (Fig. 9.12). One

FIGURE 9.12 Photographs of STAR ex situ treatment systems. Upper Left: rapid field screening system ($\sim 1.5\ m^3$); Upper Right: pilot treatment reactor for small footprint and active refineries ($\sim 3\ m^3$); Lower Left: pilot scale, mobile Hottpad system ($\sim 10\ m^3$); Lower Right: Large-scale Hottpad system of six integrated modules ($\sim 200\ m^3$). *Courtesy of Savron and reprinted with permission from Savron, 2017a. STARx (ex situ) HottpadTM system to treat oil-water separator sludge and oil-impacted soils. Project Description. http://www.savronsolutions.com/case-studies; Savron, 2018c. STARx (ex situ) HottpadTM pilot test for the treatment of oil sludge. Project Description. http://www.savronsolutions.com/case-studies.*

full-scale approach is the Hottpad, a modular, low-cost, scalable, treatment system for contaminated soils and organic liquids and sludges (Fig. 9.13). Hottpad modules are fitted together to make a footprint of any desired size. For example, in one design, a six-pad system can treat 200 m^3 per batch (Fig. 9.12, bottom right). In another design, four 200 m^3 base units are integrated to treat 135 tonnes of soil per day (Savron, 2018b). Such a system presents opportunities for capturing energy from one pad and utilizing it to reduce the environmental footprint of the adjacent pads.

Fig. 9.13 illustrates that ignition of the pile is at the base, and the horizontal plane of the self-sustaining reaction travels upward fed by air injection through the base. As with in situ STAR, standard gas capture and treatment equipment is used. Ex situ treatment with smouldering can be more straightforward than in situ because issues associated with heterogeneity, airflow injection and distribution, and vapor management are more easily controlled and optimized. The smouldered material can be either (i) excavated contaminated soils or (ii)

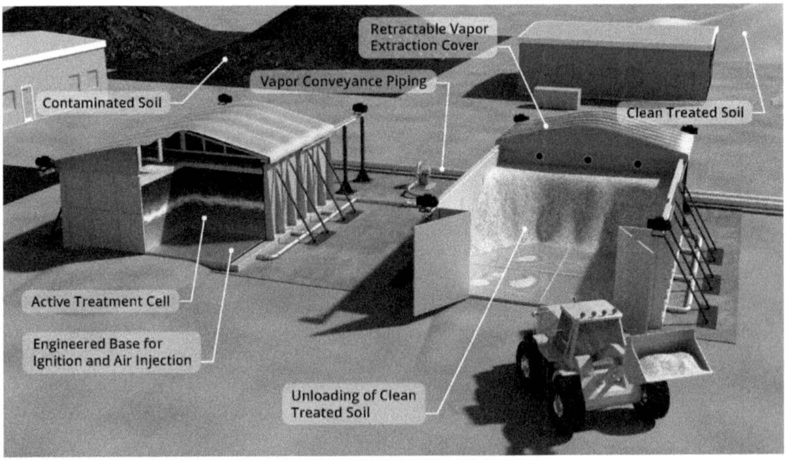

FIGURE 9.13 Sketch of Hottpad concept for applying STAR ex situ to contaminated soils and waste sludges. *Courtesy Savron.*

organic liquids/sludges from lagoons or industrial processes that are mixed intentionally with contaminated soils or locally sourced bulking material (e.g., sand). In the former case, cotreatment of organic liquids and contaminated soils offers significant benefits over disposal or remediation of either material separately. In the latter case, the inert bulking material can be reused in subsequent batches, since treatment renders it clean and relatively unaltered (Yermán et al., 2016). As explained in Section 3, when ex situ treatment was scaled up, the technology was shown to be more energy efficient and cost-effective (Switzer et al., 2014).

As an example, a pilot test of an ex situ treatment of coal-tar-contaminated soil, pretest concentrations averaging 31,000 mg/kg TPH were reduced to 10 mg/kg (Fig. 9.14) (Switzer et al., 2014). In another example, oil sludge from a lagoon was mixed with site soil in the Hottpad system shown in Fig. 9.12 (bottom right) with an ignition footprint of 78 m^2 and pile height of 2 m. Soil concentrations of up to 35,000 mg/kg TPH were reduced to non-detect (Savron, 2017a). This system was used to treat all the lagoon sludge on site in this remote SouthEast Asia location before being dismantled and shipped elsewhere. In another example, oil sludge was mixed with site soil at a sludge disposal and brick manufacturing facility in China. Two pilot units, designed within shipping containers similar to that shown in Fig. 9.12 (bottom left), were employed for STAR tests of eight contaminated mixtures ranging from 3300 to 66,300 mg/kg. All cases demonstrated self-sustained smouldering with posttreatment concentrations ranging from 167 mg/kg to a majority of nondetect (Savron, 2018c).

FIGURE 9.14 Pilot test of ex situ STAR treatment for coal-tar-contaminated sand. *Reprinted with permission from Switzer, C., Pironi, P., Gerhard, J.I., Rein, G., Torero, J.L., 2014. Volumetric scale-up of smouldering remediation of contaminated soils. Journal of Hazardous Materials 268, pp. 51–60. Copyright (2014) Elsevier.*

6. Sustainability

6.1 General considerations

As discussed in Chapter 2, green and sustainable remediation (GSR) is emerging as an important decision criterion in remedy selection. Moreover, the quantitative assessment tools to support GSR-based assessments of competing approaches for a site have matured and are now readily useable and reliable. GSR analyses emphasize the benefits of a technology with respect to environmental, social, and economic factors and provide an important complement to considerations of technical effectiveness, speed, and cost. Such a holistic approach to remedy selection is the latest evolution in evidence-based applications of remediation.

STAR is expected to rank highly in terms of GSR for a number of reasons. STAR is a source treatment technology, which is an approach that has a lower environmental footprint than plume treatment or dig and haul (D&H) (see Chapter 2). Also, STAR is primarily applied on-site; this means no off-site transport of soils/sludges for landfilling or incineration, with the associated hazards of truck mileage, including fuel consumption and accidents. Moreover, STAR destroys the majority of the organic contaminants; this is in contrast to other thermal and flushing technologies that aim to capture the contaminants above ground, requiring handling, storage, and off-site destruction or disposal with the associated risks and costs. It is also in contrast to approaches that do not clean up the site, such as permanent barriers, caps, and in situ stabilization/solidification. Any approach that contains the problem, or transfers it elsewhere such as D&H, leaves the potential of long-term liability if/when the contaminants migrate into the environment.

Furthermore, there are several aspects of STAR that make this remediation approach uniquely sustainable. First, since the reaction is self-sustaining, the contaminant-destroying smouldering reaction will continue for days to weeks without external energy input. This is in contrast to traditional in situ thermal technologies, which aim to volatilize, mobilize, or potentially pyrolize the contaminants and therefore require continuous and copious energy inputs. For ex situ STAR, its self-sustaining nature means it requires minimal energy input compared to incineration, which requires the continuous addition of a supplemental fuel such as diesel. Second, STAR continuously generates net gains in energy and stores it in the soil, meaning that the taller and wider the scale of the system, the more excess energy that is available for recovery and reuse. Third, in situ STAR can be applied surgically, akin to laser treatment of brain tumors, so that isolated contaminated pockets are strategically targeted with ignition points. This contrasts with numerous thermal and amendment technologies that are applied uniformly to the whole site, including relatively clean regions. It also contrasts with excavation, which inevitably removes and sometimes comingles clean and contaminated soil. Fourth, since the contaminant is the fuel, STAR generally works better for higher contaminant concentrations and indeed is ideal for heavily contaminated sites. This is in contrast to most biologic, chemical, and thermal remediation approaches that fail, perform worse, or are costlier at higher levels of contamination. Fifth, the only amendment STAR requires for injection is air, which is inexpensive and innocuous. This contrasts with remedies that require handling and injection of hazardous compounds, such as in situ oxidants and reductants, and those that require expensive amendments such as electricity, heat, steam, or nanoparticles. Sixth, STAR acts quickly—with the reaction propagating on the order of 0.5 m per day—while most in situ processes such as thermal, flooding, and in particular biological treatments require much longer time frames. In total, these advantages have led to STAR ranking highly on sustainability assessments against a wide range of alternative treatments.

6.2 Taiwan case study

This sustainability assessment is summarized from Chen et al. (2019). A pilot in situ STAR test was initiated by the Taiwan Environmental Protection Administration (TWEPA) at a site exhibiting high TPH concentrations (1030–18,900 mg/kg) at depths of 4.0–5.5 m bgs. Identified in the early 1990s, the site remained unremediated due to factors such as the depth of source area, the potential high cost of ex situ remediation, and the anticipated ineffectiveness of soil vapor extraction and a thermal desorption remedy. The pilot test area was 15 m by 15 m and equipped with two ignition points, 20 thermocouple bundles monitoring 6 depths each, and continuous monitoring of the emissions in the vapor capture system. The system operated for less than 72 hours and self-sustaining smouldering destroyed over 90% of the TPH within the treatment zone (97%–100% in the center of the treatment zone, less at the boundaries of the treatment area). The complete report with respect to the pilot test results and analysis can be found elsewhere (TWEPA, 2018).

The GSR tool of TWEPA was employed to compare the STAR remedy undertaken with two alternative technology options: in situ chemical oxidation (ISCO) and D&H. The methodology takes into account the entire life cycle of the project, including site investigation, remediation, and post-treatment verification phases. Within each phase, the Environment component involved computing the environmental footprint for the required materials and activities, reported as equivalent emissions (CO_2, NOx, SOx, and PM_{10}) and total energy, in the categories Raw Materials, Transportation, Equipment, Waste Disposal, Water Use and Wastewater Treatment, and Laboratory Analysis. The Economic component summed the capital and operation and maintenance costs throughout the project and considered the direct and indirect benefits to the local economy. The Social component considered two indicators: human health risk (workers and residents) and social justice and participation (noise, dust, and odor impacts on residents). All of the impacts were then normalized and ranked in accordance with the methodology.

STAR ranked as the most sustainable option overall. It ranked highest on Environment, principally because of the energy efficiency and short treatment time. It also ranked highest on the Economic component due to reduced cost of implementation associated with low energy requirements. D&H ranked first for the Social aspect; however, the analysis considered only excavating the footprint of the pilot test, which minimized the time required and the negative impact on the neighborhood. For a full-scale implementation, it is likely that the noise, dust, and odor of the D&H option would significantly reduce its Social score relative to the two in situ options (TWEPA, 2018).

6.3 United States case study

A sustainability analysis was completed for a planned full-scale STAR application at the New Jersey, USA, site (Geosyntec, 2013); the implementation of this remedy was presented in Section 5.1. The study followed the United States Sustainable Remediation Forum's (SURF) nine-step process for evaluating remediation sustainability (Favara et al., 2011). The remedy alternatives considered in addition to STAR were: In Situ Thermal Desorption (ISTD)/In Situ Thermal Stabilization (ISTS), In Situ Stabilization (ISS), Combination of Hydraulic (Pump and Treat, P&T) and Physical (Impermeable Barrier Wall), Containment (P&T + Wall), Surfactant–In Situ Chemical Oxidation (S–ISCO), Bio-Venting and Air Sparging, Excavation and Off-Site Treatment (D&H) (Geosyntec, 2013), and Low-Temperature Thermal Desorption (LTTD) (Grant et al., 2016). The carbon footprint of each remedy was evaluated using SimaPro, a Life-Cycle Assessment (LCA) software platform. The LCAs attempt to capture all outputs to the environment (e.g., kilograms of carbon dioxide) considering all materials and activities for the remedy application. For example, the STAR LCA included transportation of personnel to the site, manufacture of well points and thermocouples, energy usage for the heaters during ignition, energy usage for air delivery, vapor collection and treatment, and direct CO_2 emissions from the combustion process (Grant et al., 2016). The software also translates the outputs into three main impact categories, Ecosystem Quality, Human Health, and Resources, with each the sum of its detailed impact categories (e.g., climate change, ecosystem toxicity, particulate matter formation, ozone depletion, water depletion, etc.). Moreover, 1000 Monte Carlo simulations were run for each remedy to assess the distribution of uncertainty in the rankings due to the uncertainty associated with the input parameters.

The results of the analyses showed that STAR was ranked as the most sustainable remedy out of the nine alternatives. Fig. 9.15 demonstrates, for example, that STAR was determined to have approximately 10 times lower carbon footprint than either LTTD or Excavation. Fig. 9.16 illustrates the major contributing factors to the carbon footprint of these three remedies as applied at this site. The STAR footprint is dominated by the gases produced by coal tar combustion (primarily CO_2 and CO), the energy required to run the heaters and blowers, and the activated carbon required to scrub the emissions. While the electricity associated with the extraction system for LTTD is on the same order as STAR, the energy required to heat the subsurface is an order of magnitude larger than any aspect of the STAR process. The Excavation

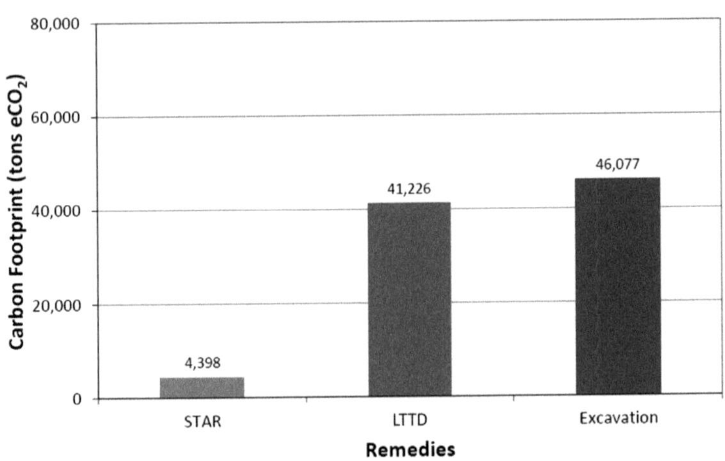

FIGURE 9.15 Carbon footprints calculated for STAR, Low-Temperature Thermal Desorption (LTTD), and Excavation (Dig and Haul), for the full-scale remediation of a coal-tar-contaminated site in New Jersey, USA. *Reprinted with permission from Grant, G., Major, D., Scholes, G.C., Horst, J., HillS., Klemmer, M.R., Couch, J.N., 2016. Smouldering combustion (STAR) for the treatment of contaminated soils: examining limitations and defining success, Remediation Journal 26, 27–51, Copyright (2016) John Wiley and Sons.*

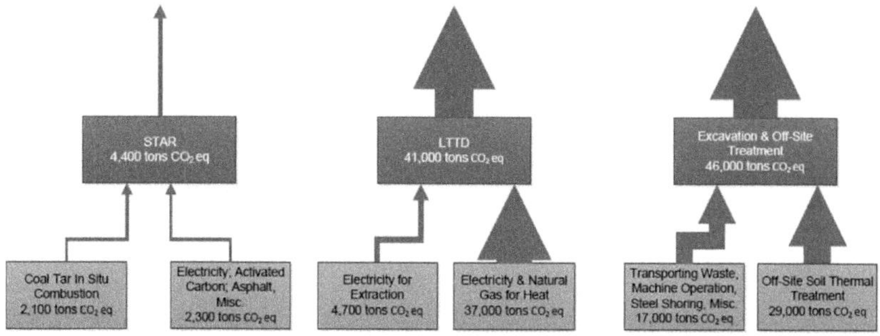

FIGURE 9.16 Primary contributing factors to the carbon footprints calculated for STAR, Low Temperature Thermal Desorption (LTTD), and Excavation and Off-Site Treatment, for the full-scale remediation of a coal-tar contaminated site in New Jersey, USA. *Reprinted with permission from Geosyntec, 2013. Sustainability evaluation of remedial alternatives report. Confidential Client Report. 124p.*

remedy footprint is shown to be dominated by machinery operation, transportation of soil, and in particular the off-site thermal treatment of the soil, which is an order of magnitude larger than any aspect of STAR. Moreover, the distribution of uncertainty in the STAR carbon footprint did not overlap with those of these other remedies. In other words, STAR was the most sustainable remedy regardless of variability in on-site implementation (Geosyntec, 2013).

These STAR analyses did not consider the option of capturing and reusing the net energy that STAR generates and stores in the soil. This is most straightforward in ex situ STAR applications. Including this option would even further reduce the carbon footprint of STAR by replacing some energy required from external sources with heat or electricity delivered by the STAR process itself.

7. Summary

STAR is a self-sustaining thermal process that destroys organic contaminants where they reside in the soil and on-site. STAR generates more energy from oxidizing the contaminant than is needed to ignite the process and the net energy gain grows with time and with the scale of application. A fraction of the energy is recycled naturally making smouldering a highly efficient process. Meanwhile the majority of the energy is stored in the soil and, particularly in an ex situ application, this energy can be captured and reused. Thus, STAR exhibits some unique characteristics that make it a highly sustainable remedy. This is supported by quantitative GSR analyses conducted for specific applications, where the carbon footprint and/or life-cycle impact was considerably lower than available alternate technologies.

In situ STAR is typically used as a source treatment technology where the goal is organic contaminant destruction to the extent practicable. Mass destruction rates on the order of hundreds of kilograms per day per ignition point are typical, with propagation rates on the order of feet per day. Thus, relative to most in situ technologies, STAR is rapid and capable of significant mass reductions.

Typical STAR sites are those characterized as being impacted by low-volatility recalcitrant compounds such as coal tar, creosote, or petroleum hydrocarbons at soil concentrations equal to or greater than approximately 3000 mg/kg TPH within silt, sand, or coarser geologic units. Soil concentrations below this level will be combusted and the process can tolerate "gaps" in contaminant distribution, but this level is required to maintain self-sustaining smouldering combustion (i.e., combustion without the input of external energy). Depth below the water table is an important design consideration but does not impact the suitability of the STAR process (i.e., STAR can be applied both above and below the water table).

STAR results in the near complete destruction of organic compounds wherever combustion occurs. Concentration reductions greater than 98% are common in treated areas, and post-treatment concentrations above nondetect or near nondetect levels are typically only associated with small areas where combustion did not occur within the target treatment zone.

Ex situ STAR is suited to the same contaminants as STAR, with the added benefit that materials can easily be manipulated, if necessary, to facilitate treatment. For example, the blending of high-concentration soils with low-concentration soils, or impermeable soils with permeable soils, can be carried out to create a mixture ideal for smouldering treatment and increase the range of soil types suitable for the process. In addition to soil treatment, STARx systems can be used as a waste disposal technology for liquid organic wastes and sludges from a wide variety of industries (e.g., crude oil tank bottoms, lagooned hydrocarbon residues, wastewater treatment sludge, pulp and paper residue, etc.). Cotreatment of contaminated soils with liquid organic wastes is the most efficient application, greatly reducing the costs and energy used for either application alone. The latest STAR research and pilot tests are proving that the technology is effective for a wider range of contaminants including light compounds (e.g., diesel, chlorinated solvents) and emerging contaminants (e.g., PFAS-contaminated soil and PFAS-loaded activated carbon).

Sustainability assessments combined with pilot tests and full-scale applications have shown that STAR has emerged as a remedy that is technically effective, cost-effective, and minimizes both energy consumption and impact on the environment.

References

Battelle Memorial Institute, 2012. Final Data Report for Lab-Scale Test for STAR, EPA Contract No. EP-C-10-001, Work Assignment 2–21. Author, Columbus, OH.

Chen, W.S.W., Lin, Y.H., Wu1, P.Y., Huang, C., 2019. Evaluation of sustainability on the sustainable thermal accelerated remediation (STAR) for soil remediation- a pilot case study on heavy end TPH. In: 24th Conference in Groundwater and Soil Contamination and Prevention Countermeasure, October 8, 2019, Osaka, Japan.

Favara, P.J., Krieger, T.M., Boughton, B., Fisher, A.S., Bhargava, M., 2011. Guidance for performing footprint analyses and life-cycle assessments for the remediation industry. Remediation Journal 21, 39–79.

Fabris, I., Cormier, D., Gerhard, J.I., Bartczak, T., Kortschot, M., Torero, J.L., Cheng, Y.L., 2016. Continuous, Self-Sustaining Smouldering Destruction of Simulated Faeces. Fuel 190, 58–66. https://doi.org/10.1016/j.fuel.2016.11.014.

Gianfelice, G., Della Zassa, M., Biasin, A., Canu, P., 2019. Onset and propagation of smouldering in pine bark controlled by addition of inert solids. Renewable Energy 32, 596–614. https://doi.org/10.1016/j.renene.2018.08.028.

Geosyntec, 2013. Sustainability evaluation of remedial alternatives report. Confidential client report 124.

Grant, G., Major, D., Scholes, G.C., Horst, J., Hill, S., Klemmer, M.R., Couch, J.N., 2016. Smouldering combustion (STAR) for the treatment of contaminated soils: examining limitations and defining success. Remediation Journal 26, 27–51.

Hasan, T., Gerhard, J.I., Hadden, R., Rein, G., 2015. Self-sustaining smouldering combustion of coal tar for the remediation of contaminated sand: two-Dimensional Experiments and Computational Simulations. Fuel 150, 288–297.

Kinsman, L., Gerhard, J.I., Torero, J.L., 2016. Smouldering remediation and non-aqueous phase liquid mobility. Journal of Hazardous Materials 325, 101–112. https://doi.org/10.1016/j.jhazmat.2016.11.049.

MacPhee, S.L., Gerhard, J.I., Rein, G., 2012. A novel method for simulating smouldering propagation and its application to STAR (Self-sustaining treatment for active remediation). Environmental Modelling and Software 31, 84–98.

Major, D., 2019. Demonstration of smouldering combustion treatment of PFAS-impacted investigation-derived waste. SERDP Project ER18-1593. https://www.serdp-estcp.org/content/download/48955/466822/file/ER18-1593%20Final%20Report.pdf.

Martins, M.F., Salvador, S., Thovert, J.F., Debenest, G., 2010. Co-current combustion of oil shale–Part 2: structure of the combustion. Front Fuel 89, 133–143.

Pironi, P., Rein, G., Switzer, C., Gerhard, J., Torero, J.L., 2011. Self-sustaining smouldering combustion for NAPL remediation: laboratory evaluation of process sensitivity to key parameters. Environmental Science and Technology 45 (7), 2980–2986.

Rashwan, T., Gerhard, J.I., Grant, G., 2016. Application of self-sustaining smouldering combustion for the destruction of wastewater biosolids. Waste Management, pp. 201–212. https://doi.org/10.1016/j.wasman.2016.01.037.

Salman, M., Gerhard, J.I., Major, D.W., Pironi, P., Hadden, R.M., 2015. Remediation of trichloroethylene-contaminated soils by STAR technology using vegetable oil smouldering. Journal of Hazardous Materials 285, 346–355.

Savron, 2014. Full-scale Application of STAR (in situ) to treat coal tar-impacted soils. Project Description. http://www.savronsolutions.com/case-studies.

Savron, 2015. STAR (in situ) pilot test to treat coal tar LNAPL and DNAPL in a fine sand aquifer. Project Description. http://www.savronsolutions.com/case-studies.

Savron, 2016a. STAR (in situ) pilot tests to treat refined petroleum hydrocarbons. Project Description. http://www.savronsolutions.com/case-studies.

Savron, 2016b. STAR (in situ) pilot test to treat Navy Special Fuel Oil (NSFO). Project Description. http://www.savronsolutions.com/case-studies.

Savron, 2017a. STARx (ex situ) HottpadTM system to treat oil-water separator sludge and oil-impacted soils. Project Description. http://www.savronsolutions.com/case-studies.

Savron, 2017b. STARx (ex situ) HottpadTM pilot test for the treatment of petroleum hydrocarbon-impacted soils. Project Description. http://www.savronsolutions.com/case-studies.

Savron, 2018a. STAR (in situ) pilot test for the treatment of petroleum hydrocarbon-impacted soils. Project Description. http://www.savronsolutions.com/case-studies.

Savron, 2018b. STARx: ex situ smouldering combustion for the treatment of contaminated soils and organic wastes. Project Description. http://www.savronsolutions.com/case-studies.

Savron, 2018c. STARx (ex situ) HottpadTM pilot test for the treatment of oil sludge. Project Description. http://www.savronsolutions.com/case-studies.

Scholes, G., Gerhard, J., Grant, G.P., Major, D.W., Vidumsky, J.E., 2015. Smouldering remediation of coal tar contaminated soil: pilot field tests of STAR. Environmental Science and Technology 1−9. https://doi.org/10.1021/acs.est.5b03177.

Switzer, C., Pironi, P., Gerhard, J.I., Rein, G., Torero, J.L., 2014. Volumetric scale-up of smouldering remediation of contaminated soils. Journal of Hazardous Materials 268, 51−60.

Switzer, C., Gerhard, J.I., Pironi, P., Rein, G., Torero, J.L., 2009. Self-sustaining smouldering combustion: a novel remediation process for non-aqueous phase liquids in porous media. Environmental Science and Technology 43, 5871−5877.

TWEPA, 2018. Screening Technologies of Priority in Soil and Groundwater Contamination Investigation and Remediation and Related Pilot Testing Study. EPA-106-GA13-03-A078. Environmental Protection Administration, Taipei, Taiwan (R.O.C.).

Yermán, L., Hadden, R.M., Carrascal, J., Fabris, I., Cormier, D., Torero, J.L., Gerhard, J.I., Krajcovic, M., Pironi, P., Cheng, Y.L., 2015. Smouldering combustion as a treatment technology for faeces: exploring the parameter space. Fuel 147, 108−116.

Yermán, L., Wall, H., Torero, J., Gerhard, J.I., Cheng, Y.-L., 2016. Smouldering combustion as a treatment technology for faeces: sensitivity to key parameters. Combustion Science and Technology 188 (6), 968−981. https://doi.org/10.1080/00102202.2015.1136299.

Yerman, L., Cormier, D., Fabris, I., Carrascal, J., Torero, J.L., Gerhard, J.I., Cheng, Y.L., 2016. Potential bio-oil production from smouldering. Waste and Biomass Valorization 8 (2), 1−10. https://doi.org/10.1007/s12649-016-9586-1.

Zanoni, M.A.B., Torero, J.L., Gerhard, J.I., 2019. Delineating and explaining the limits of self-sustained smouldering combustion. Combustion and Flame 201, 78−92. https://doi.org/10.1016/j.combustflame.2018.12.004.

Zanoni, M.A.B., Torero, J.L., Gerhard, J.I., 2018a. Determining the conditions that lead to self-sustained smouldering combustion by means of numerical modelling. Proceedings of the Combustion Institute. https://doi.org/10.1016/j.proci.2018.07.108.

Zanoni, M.A.B., Torero, J.L., Gerhard, J.I., 2018b. The role of local thermal non-equilibrium in modelling smouldering combustion of organic liquids. Proceedings of the Combustion Institute. https://doi.org/10.1016/j.proci.2018.05.177.

Zanoni, M.A.B., Torero, J.L., Gerhard, J.I., 2017. Determination of the interfacial heat transfer coefficient between forced air and sand at Reynold's numbers relevant to smouldering combustion. International Journal of Heat and Mass Transfer 114, 90−104. https://doi.org/10.1016/j.ijheatmasstransfer.2017.06.020.

Chapter 10

Long-term effectiveness of in situ solidification/ stabilization

Fei Jin
School of Engineering, University of Glasgow, Glasgow, United Kingdom

1. Introduction

Solidification/Stabilization (S/S) refers to the technology that: (i) encapsulates the waste in a monolithic solid of adequate structural integrity (solidification) and (ii) reduces the leaching potential of the contaminants by converting them into less mobile or less toxic forms via adsorption, precipitation, isomorphous substitution, and/or lattice incorporation (stabilization). It has been widely used for treating both inorganic and organic contaminants in various types of wastes (Conner and Hoeffner, 1998). For instance, in the UniteStates from 1982 to 2005, inorganic contaminants were almost exclusively treated by S/S technology ($\sim$80% of the projects) and a considerable number of sites with different organic contaminants (USEPA, 2007) were also treated by S/S. There have been so far many review articles on the history, employed reagent, binding mechanism, and evaluation procedures of this technology (Conner and Hoeffner, 1998) (Conner, 1990; Conner and Hoeffner, 2010; Paria and Yuet, 2006; USEPA, 1993; Pollard et al., 1991; Bone et al., 2004; Shi and Spence, 2004). In addition, various analytical techniques used to characterize the physical, molecular, bulk, and surface properties of S/S materials are reviewed in (Cocke, 1990). S/S can be applied both ex situ and in situ, and the latter is usually applied with deep soil mixing (DSM) technology. Due to the many advantages of in situ S/S compared to ex situ treatment, it has gained much attention in the last few decades. Nevertheless, there are still very limited field data on the evolution of the S/S materials over time under various environmental stresses and the long-term effectiveness of in situ S/S treatment. This chapter aims to critically review the literature associated with the long-term performance of in situ S/S and will mainly focus on: (i) elucidating the main degradation mechanisms for in situ S/S materials and the major

Sustainable Remediation of Contaminated Soil and Groundwater. https://doi.org/10.1016/B978-0-12-817982-6.00010-0

influencing factors; (ii) reviewing methodologies and techniques to evaluate the long-term durability of S/S materials in the laboratory; (iii) summarizing representative field data to showcase the long-term effectiveness of in situ S/S materials with different reagents and under various environmental conditions; and (iv) providing insights into the future prospects of improving the long-term effectiveness of in situ S/S and its assessment.

2. In situ solidification/stabilization

Compared to the ex situ S/S, in situ treatment has many advantages (Al-Tabbaaa et al., 2015): use of well-developed and versatile soil mixing equipment; reduction of the amount of soils to be excavated and thus saving disposal and landfill costs; reduction of the odors during remediation processes; lower material usage, potential ground improvement for immediate reuse (high strength and low permeability), and a lesser impact on the local community. In addition, the overall cost of in situ S/S is usually significantly lower than other remediation technologies such as in situ chemical oxidation and thermal desorption (Fleri and Whetstone, 2007). Nevertheless, the main concern for authorities and site owners regarding in situ S/S is that contaminants remain in the ground and the total concentration, in particular metals and metalloids, remains the same after successful treatment and the associated long-term risks are not clearly defined, which entails uncertainties in the future. It should be noted that in terms of risk assessment, the total concentration is insufficient to assess the contamination, and more detailed qualitative and quantitative determination of specific contaminant species and mobility should be provided in order to properly evaluate their toxicity to the potential environmental receptors such as microorganisms, plants, animals, water bodies, and humans (Guo et al., 2006).

The most commonly used binder in S/S is ordinary Portland cement (OPC) type I due to: (i) its relatively low cost and high availability across the world; (ii) its consistent compositions and properties with long-time track record; (iii) good modeling data on OPC hydration process and the environmental effects on the hydration products from cement and concrete science field; (iv) rich experimental and modeling data about leaching of contaminants from the S/S treated nuclear waste; and (v) interactions between various additives/adsorbents/soil minerals with OPC are studied extensively by the cement and concrete chemists. OPC type II (moderate sulfate resistance) and V (high sulfate resistance) are used when it is necessary to consider the sulfate resistance. Apart from OPC, commonly used additives include: carbonates, silicates, sulfides, and phosphates. In addition, activated carbon, modified clays, and a proprietary rubber particulate formulation are broadly useful for organic contaminants immobilization in S/S (Conner, 1997).

Currently the primary performance criteria of in situ S/S are similar to those of ex situ treatment including (Harbottle et al., 2007): strength, permeability, durability (wet—dry and freeze—thaw resistance), and leachability of the contaminants. The commonly used tests, specifications, and accepted values are summarized in (Perera et al., 2005). These criteria are specified at either 7 or 28 days, assuming that the changes of physical and chemical properties of the S/S materials over time will not undermine the inherent efficacy of S/S. The ultimate goal is to retain the physical and chemical immobilization of the contaminants in the long term and therefore the durability problem may be broadly classified into: (i) solidification problems, i.e., unable to maintain the required physical and mechanical and hydraulic properties of S/S materials and (ii) stabilization problems, i.e., inadequately immobilizing the contaminants in the long term. The two problems are not independent since the solidification problems will lead to the failure of physical encapsulation resulting in increased leachability in many cases. On the other hand, the continuous dissolution/leaching of contaminants and cement hydration products will eventually lead to the destruction of the microstructure of the S/S materials, undermining their physical and engineering properties.

S/S-treated wastes are very complex systems, and many factors contribute to their long-term durability: waste type, binder/additive type and dosage, curing time and temperature, and environmental stresses. Unlike ex situ S/S and subsequent landfilling, in situ treatment of contaminated soils will also be subject to high contamination heterogeneity, potentially poor mixing, and site-specific environmental conditions, which will have significant effects on the properties and assessment of the S/S materials (Malviya and Chaudhary, 2006; Fitch and Cheeseman, 2003). Even if the short-term performance of the in situ S/S is satisfactory, there is no guarantee that the efficacy will last forever. As indicated in (Klich et al., 1999a), even the best-quality concrete cannot be expected to last for centuries under extremely harsh conditions. Therefore, it is imperative to review the current in situ S/S practices and demonstrate its long-term effectiveness under the natural environment in order to enhance the public knowledge and ensure that the credentials for this in situ remediation technology are well received by stakeholders.

3. Immobilization and leaching mechanisms of contaminants in S/S materials

3.1 Immobilization mechanism

In order to assess the long-term durability of in situ S/S, it is important that the immobilization and leaching mechanisms are well understood so that binder compositions can be properly designed to achieve stronger binding between the S/S matrix and the contaminants and proper models can be built for the prediction of the leachability of the contaminants from the site to the surroundings. S/S can be used to immobilize both metal(loid)s and organics, but the mechanisms of immobilization are different.

Organic contaminants are usually sorbed/chelated or encapsulated in the pores of the final product of the S/S matrix. Since cement hydration is exothermic, some organic compounds may be destroyed or altered during the in situ S/S treatment (Conner and Hoeffner, 1998). The leachability of the immobilized organics depends on their solubility in water and diffusivity through the S/S matrix (Paria and Yuet, 2006). Most of the organic compounds significantly interfere with the cement hydration retarding the setting and hardening processes (Pollard et al., 1991). Another potential hazard is the possibility of gas generation resulting from the decomposition of some organic contaminants (Glasser, 1997). Activated carbon and modified clays are found to perform excellently in the sorption of various organic contaminants and hence help to reduce the retarding effect of them on the cement hydration. In these cases, immobilization was achieved by physical or chemical adsorption of the organics on the surface/interlayer and pore space of the additives (Pollard et al., 1991), which were further protected by the solidified cement matrix. However, it was reported that clay minerals are unstable in contact with cement pore water in the long term (Savage et al., 2007), which might lead to the release of the organics as well as properties change of the S/S materials. Currently there is little research on the evolution of the organics' immobilization mechanisms over time in the S/S matrix irrespective of whether adsorbents are added or not.

The immobilization mechanisms for metal(loid)s in S/S have been studied extensively, and the main processes involved depend on the employed binder/additive used and the type/dosage of the metal(loid)s. Apart from physical encapsulation, primary chemical immobilization mechanisms include (Cocke, 1990; Glasser, 1993): (i) adsorption: the main hydration product, C–S–H and AFt from OPC hydration have high specific surface area and are able to adsorb various inorganic ions (Gougar et al., 1996). Common additives such as pulverized fly ash (PFA) and zeolites can also provide additional adsorption sites for metal(loid)s in the S/S matrix (Alver and Metin, 2012; Panday et al., 1985; Sun et al., 2010; Motsi et al., 2009); (ii) precipitation: the high pH environment created by OPC hydration facilitates the precipitation of many inorganic ions in their hydroxide forms (Conner and Hoeffner, 1998). In addition, carbonates, silicates, sulfides, and phosphates can be added to precipitate inorganics in various forms depending on their solubility products (Conner); (iii) crystallochemical incorporation: both C–S–H and AFt phases are able to incorporate a wide range of ions (e.g., Pb, Sr, CrO_4^{2-}, Cr^{3+}) into their crystal structures (Rose et al., 2000; Chrysochoou and Dermatas, 2006). When ground granulated blastfurnace slag (GGBS) is used in the binder composition, hydrotalcite ($Mg_6Al_2CO_3(OH)_{16}\cdot 4(H_2O)$) may form, which not only has superior adsorptive capacity toward many inorganic ions (Cavani et al., 1991) but can also incorporate them into the crystal structure by isomorphous substitution (Jin and Al-Tabbaa, 2014); and (iv) oxidation/reduction: metals of multivalence such as As, Cr, Se, Ni, Hg, Mn will be strongly influenced by the redox potential within the S/S matrix, which can be regulated by dosing various additives. For instance, GGBS contains sulfide, which may help to reduce Cr^{6+} to Cr^{3+} followed by precipitation of its hydroxide form (Zhang et al., 2017). Hydrogen peroxide or zerovalent iron can be added to regulate the redox potential for better immobilization of specific metal(loid)s (Chrysochoou and Dermatas, 2006).

3.2 Methods of studying immobilization mechanism

Direct investigation on the immobilization mechanism is to study the element speciation by various microstructural, mineralogical, and spectroscopic analyses. Powder X-ray diffraction (PXRD) is used to identify crystalline phases, and thermogravimetric analysis (TGA) enables the quantification of phases with distinctive thermal properties at a high resolution (Mollah et al., 2004; Ortego et al., 1989). Fourier transform infrared spectroscopy (FTIR), Raman spectroscopy, and nuclear magnetic resonance (NMR) are proved useful to study the cement hydration process and the effect of carbonation on the structural change of the hydration products (Mollah et al., 2004; Yousuf et al., 1993). Scanning electron microscopy (SEM) equipped with energy-dispersive X-ray spectroscopy (EDS) is widely adopted to observe the morphology of discrete particles and determine the element abundance/distribution to deduce the newly formed phases (Jin and Al-Tabbaa, 2014; Klich et al., 1999b). X-ray photoelectron spectroscopy (XPS) provides information about chemical bonds and the relative abundance of difference species on particle surfaces (Yousuf et al., 1993; Briggs and Seah, 2003). X-ray absorption spectroscopy (XAS) enables the in situ identification and quantification of metals' redox states (Landrot et al., 2009). In addition, gas adsorption and mercury intrusion porosimetry are useful to determine the textural properties and pore profile of the S/S materials, which can indicate their physical integrity at nano- and micro-levels (Lowell et al., 2012; Du et al., 2014a).

It is usually required that a combination of different techniques be used in order to confirm the element speciation and hence the immobilization mechanism considering the complex nature of the waste/binder interactions. Extra caution should be taken during the sample preparation and in data interpretation due to the possibility of the change of element speciation and the representativeness of these analytical analyses.

Apart from the direct investigation, fraction analysis using sequential extraction methods has been used widely to study the speciation of metals in soils or sediments (Tessier et al., 1979; Ure et al., 1993; Rauret et al., 1999; Mossop and Davidson, 2003). The metals are classified into five fractions (Tessier et al., 1979): (i) exchangeable ions; (ii) carbonate-bound ions; (iii) ions bound to Fe—Mn oxides (iv) ions bound to organic matter and sulfides;

and (v) residual ions, and the first two fractions are combined in (Ure et al., 1993). It should be noted that the speciation analysis is only relevant to the natural soils or sediments, but when immobilization agents are added, particularly for cement and other additives such as silicates, the interactions between the extraction agents and the S/S materials and hence the dissociation mechanisms of the metals will be very complex. Although modifications have been made to compensate the alkalinity due to the cement addition (Li et al., 2001), the speciation cannot be reliable, e.g., strong acid may destruct the C−S−H structure and leach out the heavy metals incorporated, but they may be assigned as carbonate-bound ions. In addition, the sequential extraction test only provides information at the designated time without considering the evolution of the cementitious materials, which may over- or underestimate the S/S effectiveness related to the cement hydration and interaction with the soil and contaminants (see Section 4). Nevertheless, the concept may still be valid in evaluating the effectiveness of in situ S/S since the sequence of the extraction could indicate the strength of the binding of the metals within the S/S matrix and their potential to leach out.

3.3 Leaching mechanism

For design purpose, it is desirable that the long-term performance of the S/S system can be adequately modeled and predicted to ensure cost-effectiveness at the design stage and safety in the long run. Thus, it is imperative to understand the release mechanism of the contaminants from the S/S matrix and hence to model their behavior over time. The models should take into account the physicochemical characteristics of the S/S material and the main mass transfer phenomena occurring during the leaching process.

The leaching mechanism is commonly studied in a tank leaching test (see Section 4.3.1), e.g., ANS16.1, which was initially developed for the evaluation of solidified low-level radioactive wastes and has also been widely used for S/S-treated hazardous wastes (American National Standards Institute (ANSI), 1986). It is assumed that bulk diffusion is the main mechanism for contaminant release from the S/S matrix, which can be described by:

$$\frac{\partial C}{\partial t} = D_e \frac{\partial^2 C}{\partial^2 x}$$

where D_e is the effective diffusion coefficient (cm^2/S); C is the concentration of a contaminant (g/cm^3); t equals time (s); and x equals the distance (cm).

Some modifications are used to simulate various environments (mainly by changing the pH and chemical compositions of the leachants in ANS 16.1 (Liu et al., 2013; Du et al., 2014b; Guo et al., 2001)). More advanced models are developed to predict the long-term leachability of S/S-treated wastes based on the measurement of wastes' intrinsic physical and chemical properties (e.g., acid neutralization capacity, ANC) and contaminant mobility parameters (e.g.,

D_e) under various leachant characteristics, specimen forms, and hydraulic regimes (Côte et al., 1987; Côté and Bridle, 1987; Tiruta-Barna et al., 2004; De Windt and Badreddine, 2007; Catalan et al., 2002). Baker and Bishop (Baker and Bishop, 1997) indicated that this diffusion model does not consider the effect of acidity on the leaching process, and hence, they used the shrinking unreacted core (SUC) model (Hinsenveld, 1993) to better account for the acid infiltration into the S/S matrix. It was found that the leaching of the metals under acidic conditions can be well modeled compared to the experimental data from two site samples (Baker and Bishop, 1997).

However, they still cannot fully replicate the field conditions where multiple environmental stresses are exerted on the S/S materials simultaneously. In addition, the whole idea is based on the intrinsic properties of materials that were measured at a certain curing time, which neglects the evolution of these properties with regard to the environmental conditions (see Section 4). Apparently, porosity, ANC, and effective diffusion coefficients will have little meaning when the matrix is changing over time (Akhter et al., 1997). These results emphasize the importance of long-term testing to identify specific combinations of S/S binders and wastes that are prone to undergo respeciation and consequent leaching mechanism and kinetics.

Considerable progress has been made in geochemical and reactive transport models, which are based on batch and dynamic leaching tests as well as mineralogical investigations (De Windt and Badreddine, 2007; De Windt et al., 2007). Theoretically, not only the effect of physical changes of the materials (volume change, cracking, etc.) can be taken into account but also chemical changes (carbonation, sulfate attack, etc.) can be incorporated to deal with realistic exposure conditions to improve the prediction accuracy. This will surely increase the complexity of the computation but also necessitates the full understanding of the chemical and physical evolution of the S/S materials under various environmental conditions with the help of advanced techniques (Klich et al., 1999b). In this regard, the rich literature from the cement and concrete science and radioactive waste management could contribute significantly (Lothenbach et al., 2008; Kosakowski and Berner, 2013).

4. Evolution of the S/S materials over time and its degradation mechanisms under environmental stresses

The long-term performance of in situ S/S-treated contaminated soils does not only depend on the quality of the binder/additive used but also depends on the exposed environmental conditions. Since OPC is the main binder used in S/S, the solidified contaminated soils behave similarly to the concrete materials to some extent, and hence, when considering the long-term performance of in situ S/S treatment, the experience gained from concrete durability studies proved particularly relevant. Fortunately there is rich literature about the environmental impact on the durability of cement and concrete (Roy, 1999; Guettala

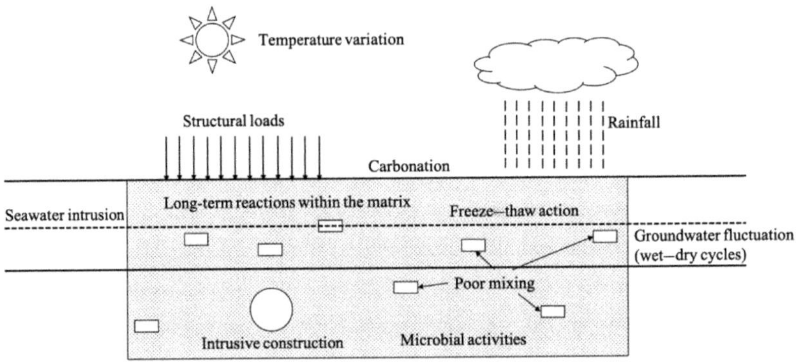

FIGURE 10.1 Influencing factors for the long-term effectiveness of in situ S/S. *Credit: Fei Jin.*

et al., 2006; Castellote et al., 2008; Wei et al., 2013) and ex situ S/S-treated wastes (Malviya and Chaudhary, 2006; Chen et al., 2009; Ludwig et al., 2000), which can shed light on the long-term effectiveness of in situ S/S soils. Regarding the influencing factors that affect the cement hydration and contaminant leachability in the short term, readers can refer to (Malviya and Chaudhary, 2006; Hills and Pollard, 1997). Fig. 10.1 shows a schematic diagram of the various processes that could occur during the service life of in situ S/S soils. It should be noted that there may be multiple actions involved in one process. For instance, the groundwater fluctuation not only brings about wet–dry cycling but also induces microorganisms and possibly sulfate sources into the ground. Among these processes, the poor mixing issue should hopefully be addressed at the design/operation stage and the potential hazards regarding intrusive construction (e.g., creating exposure pathway) and additional structural loads (e.g., undermining physical integrity) within or on top of the in situ S/S treated soils should be subject to additional risk assessment. The focus here is the effect of long-term reactions within the matrix and different types of environmental stresses on the long-term effectiveness of in situ S/S.

4.1 Continuing hydration of OPC and pozzolanic reaction

Cement hydration is an exothermal reaction, and its kinetics is related to many factors including its chemical composition, fineness, water/cement ratio, and curing temperature/pressure as well as the additives (Lin and Meyer, 2009). Modern OPC normal concrete hydrates fast and almost completes the process in 1 year under ambient conditions (Lothenbach, 2010). The pozzolanic reaction occurs much slower depending on the reactivity of the SiO_2 contained in the pozzolans. In addition, some contaminants significantly retard the cement hydration and hence the pozzolanic reaction, particularly the organics (Malviya and Chaudhary, 2006). It was reported that PFA particles are still present in the in situ PC-PFA-treated soils contaminated with high levels of metals and organics after 17 years (Wang et al., 2014). One method to accelerate the hydration and hence the aging of the cement stabilized waste is to increase the curing temperature. The effect of temperature on the cement hydration rate is described by the Arrhenius equation:

$$k = A \cdot e^{\frac{-E_a}{RT}} \tag{10.1}$$

where k is the rate of heat evolution; A equals the proportionality constant; T is the temperature in K; R is the universal gas constant (8.314 J/mol/K); and E_a is the activation energy (J/mol).

The concept of "equivalent age" is to use E_a to predict the hydration behavior of cement at different curing temperatures. The equation developed in (Hansen and Pedersen, 1977) is the most common expression to calculate the equivalent age:

$$t_e(T_r) = \sum_0^t e^{\frac{-E_a}{R}\left(\frac{1}{T_c} - \frac{1}{T_r}\right)} \cdot \Delta t \tag{10.2}$$

where $t_e(T_r)$ is the equivalent age at the reference temperature T_r; T_r and T_c are the reference temperature and true curing temperature, respectively; Δt is the curing time interval; E_a and R are the same as defined earlier; and here E_a is assumed to be independent of temperature in order to perform the calculation.

Curing the S/S materials at higher temperatures will thus accelerate the aging compared to the ambient temperature curing, and the equivalent curing age can be calculated using Eq. (10.2). Using this method, the properties of cement-stabilized wastes can be modeled for hundreds of years. However, there are many flaws of the aging test including: (i) the assumption of constant E_a may not hold true for S/S wastes; (ii) the temperature effect on the contaminant's respeciation and volatilization cannot be quantified; and (iii) cement hydration products and reaction products with the contaminants/additives formed at higher temperatures are likely to be different from the ones at lower temperatures, displaying different sorption behaviors and engineering properties (Kjellsen et al., 1991).

For normal concrete materials, as long as the 28-day performance is acceptable, the continuing hydration and pozzolanic reactions should improve the strength and impermeability over time. Nevertheless, S/S materials may behave differently. For instance, Kirt (Kirk, 1996) summarized the accelerated aging studies conducted by University of Cincinnati, where several physical and chemical tests were performed on the aged and unaged specimens. They found that the accelerated aging increased the porosity in the aged specimens leading to significant loss of unconfined compressive strength (45% reduction at 50 years and 80% reduction at 100 years, respectively, compared to 295 days). Meanwhile, permeability increased by a three- to fourfold accordingly.

In addition, it is potentially negative of the pozzolanic reaction between the modified clays and OPC. Modified bentonite added to the S/S binders reduces the permeability of the in situ S/S soils (Jin et al., 2016) and adsorbs a wide variety of contaminants depending on the modification method (Celis et al., 2000; Montgomery et al., 1991). However, the bentonite structure may be destructed over time by the high pH resulting from the cement hydration, leading to the formation of crystalline phases (Sánchez et al., 2006) and may lead to strength and impermeability loss (Al-Tabbaa and Boes, 2002) and reduction of adsorption capacity, which warrants further studies.

4.2 Binder–soil–contaminant interactions

The contaminated soil compositions may consist of soil particles, inorganic and organic contaminants, extraneous items (e.g., plastics, wood), and the contamination level are highly heterogeneous. The addition of various types of binders and additives makes the system even more complex. Although the immobilization mechanism has been briefly discussed in Section 3, the interactions among binder, soil, and contaminants will be temporally and spatially dependent. For instance, Mollah et al. (Mollah et al., 2004) studied PC type V treated As(V)-bearing waste at 10 days, 1 month, and 10 years with FTIR and XRD. They found that oxyanion-substituted ettringite, $Ca(AsO_4)_2$, $CaHAsO_4 \cdot H_2O$, $CaHAsO_4 \cdot 2H_2O$ and $CaH_4(AsO_4)_2$ are formed during early stage of hydration of OPC. After 10 years, oxyanion-substituted ettringite and $CaHAsO_4 \cdot 2H_2O$ disappeared and two new phases, $NaCaAsO_4 \cdot 7.5H_2O$ and $Ca_5(AsO_4)_3(OH)$, were identified. Du et al. (2014a) investigated the effect of Zn concentration on the immobilization mechanisms of OPC-treated zinc-contaminated kaolin. Results showed that at low Zn levels, no specific crystalline zinc compounds were identified while increasing Zn concentration led to the formation of $CaZn_2(OH)_6 \cdot 2H_2O$. However, these studies are conducted in the laboratory, which may be vastly different from the field conditions. In addition, even if the chemical equilibrium has been reached, in situ S/S-treated soils are subject to ever-changing environment, which may disturb the system significantly. Other possible interactions include the pozzolanic reaction between the cement and clay minerals (see above) and alkali silica reaction (ASR) between the cement and the sand particles (Wang and Gillott, 1991), which may induce swelling and subsequently undermine the structural integrity of the treated soils.

4.3 Environmental stresses

A number of environmental conditions will affect the effectiveness of S/S treatment including acid attack, sulfate attack, wet–dry cycles, freeze–thaw cycles, carbonation, and microbial activities.

4.3.1 Acid attack

Under acidic conditions, the alkalinity of the OPC pore water is reduced, which destabilizes hydration products, leading to their complete dissolution and consequently resulting in increased porosity and permeability. Depending on the immobilization mechanism, the contaminants adsorbed, precipitated, or encapsulated will be leached out from the S/S matrix under the acid attack. The acid could come from the acid rain infiltration, industrial water runoff, underground water and may be produced by microbial activities (see Section 4.3.5).

The resistance of S/S materials to acid is usually evaluated by leaching tests. A review of commonly used leaching tests is provided in (Washington State Department of Ecology, 2003). Batch leaching tests such as Toxicity Characteristic Leaching Procedure (TCLP) (USEPA, 1992) and Synthetic Precipitation Leaching Procedure (SPLP) (USEPA, 1994) use crushed samples with different acid composition/concentration, which is based on the assumption that the physical integrity of the materials will be destroyed by various reasons and the chemical equilibrium was reached at the end of the tests. Some modifications are made on the batch leaching tests in order to better simulate the real disposal scenario. For instance, (Webster and Loehr, 1996) used real seawater as the extraction fluid in a batch leaching test and constantly renewed the seawater to extract the Cr sequentially. Nonetheless, they are still assuming that the S/S materials will be subject to one single disposal scenario throughout the service life, so the batch leaching tests are mainly used to screen/categorize materials (Washington State Department of Ecology, 2003) or quickly assess the effectiveness of in situ S/S treatment against predetermined criteria, while they do not provide any information about the leaching mechanisms or kinetics.

Flow-around/tank leaching tests assess the leaching rate of the contaminants from the S/S matrix over time at a specific liquid/solid (L/S) ratio. The leachate can be adjusted depending on the expected exposure environment. For example, Liu et al. (Liu et al., 2013) used a tank leaching test to study the long-term (up to 123 d) leaching behavior of phenol from PC-stabilized soils using activated carbon as the adsorbent. Three leachants with different pH values and chemical compositions were used to simulate different environmental scenarios. It is found that AC enhanced the immobilization of phenol at an optimal content of 2% while too much AC leads to increased porosity of the S/S products, which potentially increases the leaching of phenol. Du et al. (2014b) prepared simulated acid rains with different pH values in the tank leaching test and concluded that D_e of Pb from an OPC-stabilized soil is dependent on the cement dosage and leachant pH. Both studies confirm that the success of S/S for contaminants depends on not only the physical and chemical properties of the final product but also the leachant type, which shed lights on the effect of environmental stresses on the long-term performance of the in situ S/S.

Flow-through leaching tests can be carried out on either monolithic samples (Poon et al., 2001) or crushed and packed samples (Shackelford and Glade, 1997). They enable the monitoring of the leachate composition versus the volume of leachant passing through the samples and are used to imitate field conditions where both dissolution/precipitation and diffusion processes occur. Monolithic samples are more representative to the in situ S/S-treated soils that have not experienced physical degradation while crushed/packed samples could simulate the infiltration through the cracked zones (Catalan et al., 2002) due to various reasons including poor mixing.

4.3.2 Wet–dry and freeze–thaw cycles

The wet–dry action could be induced by the rainfall infiltration and subsequent evaporation or surface runoff as well as the fluctuation of the groundwater table, which is well known to undermine the cemented soil properties over time (Du et al., 2015). Leaching of Pb and Zn was found to increase with the wet–dry cycles from the OPC or OPC-PFA-stabilized soils (Zha et al., 2013; Li et al., 2014). Fitch and Cheeseman investigated the environmental effect on the PC-PFA-stabilized waste exposed outdoor for 10 years (Fitch and Cheeseman, 2003) and found that the wet–dry cycles may play an important role in the S/S performance due to the precipitation of $Ca(OH)_2$ in the surface region, surface runoff of heavy metals as well as the migration of the heavy metals to further depth into the S/S matrix.

Freeze–thaw cycles are common in cold regions, and OPC develops cracks and eventually disintegrates under the freezing–thawing action, due to the hydraulic pressure exerted by the dilation of water upon freezing and the osmotic pressure from the partial freezing of the pore water containing various ions (Powers, 1959).

Freeze–thaw cycling is found to be more detrimental to the S/S-treated mercury-containing sludge than the wet–dry cycling, leading to the destruction of the monolith and significant increase of mercury leachability (Yang, 1993). Al-Tabbaa and King (1998) investigated the durability of OPC-PFA-bentonite stabilized sand containing three types of contaminants, i.e., NaCl, polyethylene glycol, and glyphosate at two concentrations under wet–dry and freeze–thaw actions. They evaluated the degradation of the S/S products using mass loss and permeability and found that both actions undermined the physical integrity of the stabilized soils, but the degradation rate was highly dependent on the sample curing time, type and dosage of the contaminants.

The resistances of in situ S/S soils to freeze–thaw and wet–dry cycles are usually required, which are evaluated following ASTM D 4842 and ASTM D-4843, respectively. It is generally required that the mass loss after 12 cycles of wetting–drying or freezing–thawing be $<10\%$ (Fleri et al., 2005). Increasing cement dosage, controlling water/cement ratio, and adding air entrainment agents proved to be effective measures to improve the freeze–thaw and wet–dry resistance (Conner). In reality, most case studies reported that the S/S materials exhibited excellent resistance to freezing–thawing and wetting–drying cycles with only $<1\%$ weight loss after the tests (De Percin and Sawyer, 1991; Stegemann et al., 1997). Nevertheless, it should be noted that the indicator of the two tests, i.e., weight loss, cannot adequately assess the leachability of the contaminants, as indicated in (Fitch and Cheeseman, 2003). In addition, the number of cycles may not represent the real field conditions.

It is worth noting that an accelerated rainfall weathering test was designed in Humez and Prost (1999), Humez et al. (1997), Badreddine et al. (2004), which is able to simulate the weathering of S/S materials exposed to alternating periods of saturation and below-saturation moisture contents and quantify the effect of wet–dry cycles on the leaching kinetics of contaminants. The basic concept is to compare the amount of water that passes through the sample in the water extractor to the annual rate of precipitation. In this way, the authors are able to simulate the rainfall weathering (wet–dry cycles) to the S/S materials in 90 days, predicting the long-term behavior in several centuries. The kinetics of release of pollutants are measured, and the evolution of the structure and physicochemical properties of the S/S wastes can be examined via microstructural analyses. The authors found that the bulk of the trace metals was incorporated in the matrix of the C–S–H phases via various analytical techniques, e.g., XRD, FTIR, SEM/EDS, and NMR (Rose et al., 2000; Lequeux et al., 2001; Pomies et al., 2001; Moulin et al., 1999). In addition, rate-limiting mechanisms for leaching of metals were found to be: (1) diffusion at the beginning of the weathering process; and (2) surface reactions in the long-term.

4.3.3 Sulfate attack

Most soils contain some sulfates, such as gypsum in natural soils, ammonium sulfate in agricultural soils, and sulfuric acid due to the biogenic processes. In addition, the groundwater and seawater usually contain a significant amount of potassium, sodium, and magnesium sulfates. OPC is particularly susceptible to sulfates, due to the presence of portlandite and aluminum-bearing phases, which readily react with sulfates to generate expansive gypsum and ettringite, leading to reduced pH of the system and loss of mechanical strength and subsequent cracking of the cement structure. Consequently the permeability increases, which facilitates the penetration of sulfate ions, resulting in accelerated deterioration (Gollop and Taylor, 1992; Santhanam et al., 2003). It should be noted that the evaluation of sulfate attack on concrete materials is mainly based on the accelerated sulfate immersion test, and a high concentration/continuous supply of sulfate is present, which may not be applicable for in situ S/S-treated soils. Nevertheless, there is not much research on the effect of sulfate attack on the effectiveness of S/S-treated soils especially considering the leaching of contaminants.

A number of studies investigated the long-term properties of ex situ PC-stabilized wastes exposed to seawater, which provides implications for the durability of in situ S/S materials in the possible event of seawater intrusion due to groundwater abstraction in coastal regions (Ferguson and Gleeson, 2012). Guo et al. (2001) investigated the behavior of OPC (type II) and Class C PFA-stabilized phosphogypsum (PG) waste subjected to a 1.5-year field submergence test and a 28-day saltwater dynamic leaching study. It was found that all field composites survived with no signs of degradation. The effective diffusion coefficients for Ca are 1-2 orders of magnitude smaller than those estimated for PG: PC type II: lime composites, which exhibited severe deterioration within 2 months of saltwater submergence (Guo et al., 1999). Microstructural analyses confirmed that the high resistance to seawater attack was due to the reaction between the composites and the saltwater, resulting in the precipitation of calcite, which encapsulates and protects the composites from further dissolution.

Hockley and van der Sloot (1991) studied the durability of PC- and lime-stabilized coal combustion wastes after 8-year exposure to a marine environment. They demonstrated that precipitation and dissolution processes moved as a sharp boundary from the surface toward the center of the stabilized block, penetrating about 10–20 mm in the 8 years. The pore refinement attributed to the precipitation (e.g., brucite, calcite) in the matrix pores of the boundary region restricted diffusion further into the stabilized block, which may also restrict degradation of the waste block matrix and the leaching of contaminants. Two implications are obtained from the study: (i) the commonly used tank leaching test cannot adequately assess the diffusion of contaminants due to the pore refinement process, and (ii) there is potential of facilitating the pore refinement process through mix design in the S/S technology, leading to better long-term durability exposed to specific chemical attacks.

4.3.4 Carbonation

Atmospheric carbon dioxide interacts with cement hydration products, which leads to the consumption of portlandite, forming calcium carbonate and the decalcification of the C–S–H gel (Morandeau et al., 2014). The carbonation process alters both the chemical and physical properties of cement, concrete, and S/S wastes. Several studies (Yousuf et al., 1993; Lange et al., 1996a; Lange et al., 1996b, 1997; Sweeney et al., 1998) reported that accelerated carbonation of PC-based reagents (different types of PC with/without pozzolans) increased calcite contents, improved strength, and led to a significant reduction in a wide range of leachable metals extracted compared to air-cured samples. Contradicting results were reported showing increased leachability of SO_4^{2-} and heavy metals such as Zn and Cr after carbonation (Alba et al.,

2001), which can be ascribed to the carbonation of ettringite (to gypsum) and the decrease of pH and polymerization of C–S–H, which corroborates with (Valls and Vâzquez, 2000). In a study to evaluate carbonation on the long-term performance of cementitious waste forms, Sr was found to be more tightly fixed while increased release of Cd, Pb, and Co was observed (Bin Shafique et al., 1998). It is concluded that the carbonation effect on immobilization depends on the specific species of concern.

However, it should be noted that these conclusions are based on accelerated carbonation tests, where pure or high concentrations (>10%) of CO_2 were used to cure the samples. It is well recognized that the chemical composition and the microstructure of cement-based materials carbonated using a high CO_2 concentration were different from those carbonated under atmospheric conditions. The CO_2 concentration that would preserve the nanostructure of C–S–H gel in comparison with natural conditions is around 3% (Castellote et al., 2008, 2009). Therefore, the results from the improperly designed accelerated carbonation tests may not represent what happens under the ambient conditions. In addition, most literature attributed the effect of carbonation to the portlandite consumption and the reduction of ANC and later the decalcification of C–S–H. However, it was proven that both portlandite and C–S–H will be attacked under the carbonation simultaneously (Morandeau et al., 2014; Groves et al., 1990), and hence, any contaminants associated with the adsorption/incorporation by C–S–H will be affected, which has been largely omitted by the researchers. In addition, carbonation-induced shrinkage is rarely studied for S/S materials, which can potentially result in pore coarsening, loss of strength, and microcracking in the long term (Ye et al., 2017).

4.3.5 Biological degradation

Microbial activities produce metabolites (e.g., organic or inorganic acids and CO_2), which are mostly detrimental to cement and concrete materials, and microbially influenced degradation (MID) to PC has been studied extensively in the last two decades regarding the deterioration rate/mechanism (Libert et al., 1993; Bertron, 2014; Knight et al., 2002; Nica et al., 2000; Roberts et al., 2002; Parker, 1945; De Windt and Devillers, 2009; Chehade et al., 2009; Kawai et al., 2000). A sophisticated test called the Build-Mat-Bio-test (BMB) was designed in Magniont et al. (2011), which enables the separation of the deterioration caused by the metabolites alone from those caused by the growing bacterial cells plus their metabolites. Results showed that not only the metabolites attack the PC, leading to dissolution of the hydration products, but also the microorganisms exacerbated the deterioration kinetics and intensity.

Compared to concrete materials, in situ S/S-treated soils are more prone to be exposed to microorganisms, which are prevalent in the virgin/contaminated soils and underground water. Nevertheless, there is very limited research regarding the MID to S/S materials. An early study (Rogers et al., 1996) examined the deterioration of OPC solidified low-level radioactive wastes caused by two sulfur-oxidizing bacteria (*T. thiooxidans* and *T. ferrooxidans*). Severe dissolution of the cement hydration products was observed, which was accompanied by the decrease of physical integrity of the S/S products. Though the leaching of the radionuclide contained was not assessed in this work, it is reasonable to deduce that the contaminant immobilization capacity would be significantly undermined due to the physical and chemical degradation of the solidified wastes. Rogers et al. (2003) designed an experiment to investigate the effect of biofilm on the deterioration of a PC-stabilized synthetic Co-containing waste (21% $CoCl_2$/79% type I OPC) and confirmed that *T. thiooxidans* can form a biofilm on the surface of the stabilized waste, which may increase the degradation rates compared to continued exposure to the acid only. A series of studies were conducted to investigate the leaching behavior of Cr from OPC-stabilized synthetic wastes (Idachaba et al., 2001, 2002, 2003, 2004; Carmalin Sophia et al., 2007), which found that MID increased the leaching of Cr significantly compared to the sterile medium with the same pH value, and the leaching kinetics can still be described adequately by a bulk diffusion model (Idachaba et al., 2003; Carmalin Sophia et al., 2007).

It should be noted that current studies only investigated a few single inorganic contaminants (e.g., Co or Cr) and limited types of microorganisms (e.g., *T. thiooxidans, T. ferrooxidans)* under well-controlled laboratory environment. In addition, accelerated tests (high initial concentrations of microorganisms or continuous replenish of acids in the tank leaching test) were used to assess the MID to OPC and OPC-stabilized wastes, and none examined the real-time MID to in situ S/S-treated soils. Given the extremely complex nature of the wastes and their exposure environment in the in situ S/S treatment, it is expected that various types of microorganisms will be present that may produce organic acids, which are more detrimental to PC than inorganic acids at the same pH value (Bertron and Duchesne, 2013). On the other hands, some microorganisms may interact with the contaminants and modify their fates (Islam et al., 2004; Cheung and Gu, 2007; Gadd, 2007), either increasing or decreasing their leaching potential from the S/S products. In any case, microorganisms play an important role in the in situ S/S technology in the long run and warrant further research.

5. Long-term case studies on the effectiveness of in situ S/S

There are mainly three different ways of assessing the long-term effectiveness of in situ S/S materials: laboratory tests, modeling, and field monitoring. The laboratory tests are usually based on the speciation of the contaminants to deduce their leaching potential (see Section 3) or by acceleration of the degradation process (see Section 4). They either provide qualitative assessment on the durability or generate useful model parameters for the prediction of the long-term performance. Field monitoring and modeling feedback each other and improve the accuracy of the prediction for future projects. There are a large number of accelerated tests available but very limited field data to validate the usually oversimplified models developed. The correlation between the lab leaching test and field data proved unsuccessful in many cases. The main drawbacks of using lab-accelerated test and modeling for assessing the long-term effectiveness of in situ S/S include: (i) lab tests are usually based on the dosing of single or combination of a few contaminants into the reagents. However, real wastes are invariably complex mixtures of compounds with contrasting physical and chemical properties that are difficult to characterize (Hills and Pollard, 1997), which may significantly influence both hydration and pozzolanic reactions of the reagents both in the short term and in the long term; (ii) accelerated tests usually do not consider the evolution of the S/S materials with time; and (iii) the site-specific ever-changing environmental stresses (e.g., carbonation, humidity change, temperature, sulfate and acid attack, biological processes) can never be simulated successfully over the long term.

Although a number of in situ S/S projects have been carried out (Bates and Hills, 2015), there is still a dearth of data on the long-term performance of in situ S/S (see Table 10.1). In general, it can be concluded that in situ S/S is an effective soil remediation technology, and the use of novel binders holds significant promise of enhancing the long-term stability of the treated soils exposed to site conditions (Fig. 10.2).

Meanwhile, some important implications can be obtained from the field data, which reiterate many points discussed earlier:

a. Significant heterogeneity was observed, both for the contamination levels and mixing efficiency and hence the test results in the materials properties. In addition, the sampling process may significantly affect the assessment (PASSiFy, 2010);

b. The effectiveness of in situ S/S also depends on the treatment process such that the single columns were found to performance better than the overlapping columns (Al-Tabbaa and Boes, 2002);

c. Interactions among the binder, soil, contaminants, and CO_2 were detected by microstructural and mineralogical analyses (Wang et al., 2014; PASSiFy, 2010);

d. Cement hydration continued for many years in the contaminated soils and clay minerals interacted with cements, which affected the long-term effectiveness of S/S (Jin et al., 2016; Al-Tabbaa and Evans, 2000).

e. Novel cements could perform much better than OPC in terms of both physical encapsulation and contaminants immobilization (Jin et al., 2016; Wang et al., 2016).

f. Leaching test at a certain time point may not reflect the true effectiveness of the S/S treatment since there is no guarantee that no outbreaks occurred

TABLE 10.1 Long-term performance of in situ S/S applications.

Site	Major contaminants (mg/kg)	S/S reagents	Test time (y)	Main findings	References
South 8th street, Arkansas, USA	Pb (22,000) and sulfuric acid ($pH < 1.0$)	25% limestone, 20% OPC, and 10% PFA	5	(i) Average UCS decreased from 0.5 MPa to 0.15 MPa and permeability increased from 10^{-8} m/s to 10^{-7} m/s due to sampling stress; (iii) SPLP lead leachability exceeds the performance criterion, but further investigation found that the S/S is effective in cleaning up the groundwater contamination; (iv) OPC and PFA are partially hydrated and ettringite and gypsum-filled pores.	PASSiFy (2010), Antemir et al. (2010)
Quarry Dump, Georgia, USA	TPH (1200)	25% OPC	12	(i) UCS increased to 1.3 MPa from 0.4 MPa (design value) and permeability was within one order of magnitude of 10^{-7} m/s (design value); (ii) SPLP for TPH is < 300 mg/L; (iii) moderate carbonation, with carbonation products filling pores.	PASSiFy (2010), Antemir et al. (2010)
Columbus MGP, Georgia, USA	PAHs (2400) TPH (5500)	10% OPC	10 & 12	(i) UCS increased from 1.1 MPa at 28 d to 3.1 MPa at 10 y but decreased to 0.9 MPa at 12 y; (ii) permeability increased from 10 $^{-8}$ m/s at 28 d and 10 y to slightly higher than 10^{-7} m/s at 12 y; (iii) SPLP at 12 y for cyanide was 1.4 μg/L and PAH 372.1 μg/L, similar to the performance at 10 y.	PASSiFy (2010), Antemir et al. (2010)

West Drayton, Middlesex, UK	Pb (2345–2801), Cu (626–1264), Zn (295–1800), Ni (24–232), oil (325–1900)	OPC (2%–8%), PFA (0%–8%), lime (0%–0.4%), bentonite (0%–8%)	1.2 & 2.3	(i) UCS increased up to 2.3 y while TCLP leachate pH decreased; (ii) permeability and compressibility decreased from 2 m to 1.2 y and increased slightly at 2.3 y; (iii) PFA particles existed after 1.2 y and carbonation occurred; (iv) mixes with OPC and bentonite outperformed other mixes and satisfied all the design criteria after 2.3 y.	Al-Tabbaa and Evans (2000)
West Drayton, Middlesex, UK	Same as above	Same as above	5	(i) UCS increased up to 5 y for most mixes except for the one with bentonite and the one with lowest cement content; (ii) permeability was (0.21–3.2) * 10^{-9} m/s at 5 y; (iii) TCLP leachate pH decreased from 0.2 to 5 y, which is associated with increased leachability of Cu, Zn, and Pb by 3–100 times at 5 y; (iv) specimens collected from the overlapping columns were not effective compared to those obtained from the single columns.	Al-Tabbaa and Boes (2002)
West Drayton, Middlesex, UK	Same as above	Same as above	17[a]	(i) Hydration was complete at 17 y and further carbonation occurred compared to 5 y; (ii) the mix with lowest cement content performed worst in metal immobilization; (iii) TCLP leachate pH was higher in overlapping columns than in single columns and the pH values	Wang et al. (2014)

Continued

TABLE 10.1 Long-term performance of in situ S/S applications.—cont'd

Site	Major contaminants (mg/kg)	S/S reagents	Test time (y)	Main findings	References
				decreased from 5 y to 17 y; (iv) leachability of the metals is correlated with leachate pH and British standard (EN, 2002) leached less metals compared to TCLP; (v) sequential extraction showed metals are mostly bound to Fe–Mn phases and speciation depends on binder/metal types.	
Castleford, Yorkshire, UK	Pb (95–175), Zn (150–220), As (130–140), Cr (700–1150), Cu (1075–1600), Ni (1170–2200), SVOCs (≤1400), TPH (≤8000)	24 mixes including OPC, PFA, MgO, GGBS and/or modified bentonite (Al-Tabbaaa et al., 2015)	1.5[b]	(i) For all mixes investigated, UCS increased, leachate pH and leachability of the total organics decreased from 2 m to 1.5 y using the British standard (EN, 2002); (ii) Zn leachability was found to increased slightly at 1.5 y while Ni, Cu, and Pb in general decreased; (iii) Ni and Cu have higher leachability compared to Zn and Pb, but the leachate concentrations of all were below the drinking water standard except for Ni in some mixes; (iv) modified clays improved the strength slightly at 1.5 y.	Wang et al. (2015a), Wang et al. (2015b)
Castleford, Yorkshire, UK	Same as above	Same as above	3	Fig. 10.2 showed the time-dependent performance of various mixes: (i) UCS increased from 2 m to 3 y; (ii) permeability is less than 10^{-9} m/s except	Al-Tabbaaa et al. (2015), Jin et al. (2016), Shand et al. (2019)

				for the MgO-only mix; (iii) leachate pH decreased from 2 m to 3 y; (iv) OPC alone is ineffective in immobilizing organics and leachability increased with time while the MgO-based binders are superior to OPC-based ones in immobilizing both organic and inorganic contaminants; (v) modified clay decreased the strength and permeability slightly but its influence on contaminants immobilization was unclear.	

[a]*specimens were cored from the field at 4.5 y and cured in the laboratory up to 17 y.*
[b]*specimens were cored from the field at 1 m and cured in the laboratory up to 1.5 y.*

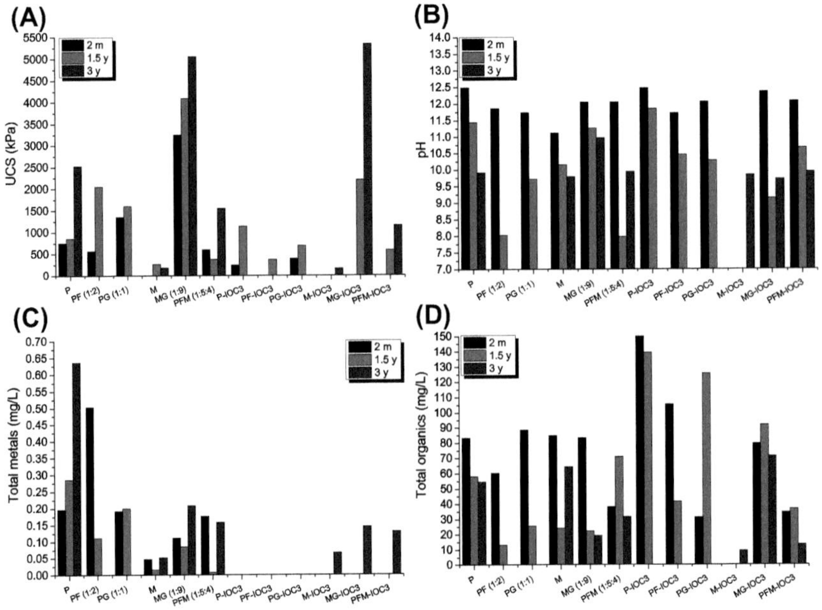

FIGURE 10.2 Time-dependent performance of in situ S/S-treated contaminated soils up to 3 years of (A) unconfined compressive strength; (B) leachate pH values; (C) leachate concentrations of all the metals (Ni, Pb, Cu, Zn, and Cr); (D) leachate concentrations of the total organics. The detailed mix design and test methods can be found in Jin et al. (2016), Wang et al. (2015a). *Credit: Fei Jin.*

before the assessment. More informative monitoring should investigate the time-dependent performance and test the total concentrations/speciation of the contaminants in the specimens, which entails high costs.

6. Conclusions and future prospects

In situ solidification/stabilization (S/S) has many advantages over other soil remediation technologies (Section 2). The current practice is that whenever in situ solidified/stabilized soils pass the quality acceptance criteria (usually strength, permeability, and TCLP leachability), they are rendered satisfactory. However, since the contaminants remain underground, the biggest concern of applying in situ S/S is its long-term effectiveness, particularly under the complex and ever-changing site conditions. This chapter reviewed the main immobilization mechanisms by S/S and the leaching behavior of contaminants from the S/S matrix (Section 3). Most importantly, various factors influencing the long-term effectiveness of in situ S/S and the related assessment methods were discussed in detail to reveal their applicability and limitations (Section 4). With limited field monitoring data, this technology has proven its long-term effectiveness, but it can be seen that the longest assessment time is only 17 years and in situ S/S involves many potential sources of variability in the treated materials (Section 5).

Future studies on improving the long-term effectiveness and our better understanding about the in situ S/S technology may include the following aspects: (i) utilization of novel and sustainable binders and additives. For instance, ferrous sulfate, which is available as an industrial waste, was used as an additive to treat As- and Ba-containing wastes (Fuessle and Taylor, 2000). It improved the metal(loid) retention rate significantly up to 540 days due to the formation of iron arsenic complexes and barium sulfate precipitate; Novel cements such as magnesium phosphate cement was successfully applied to treat highly contaminated soils with better biocompatibility (Du et al., 2014c), and biochar was found to be able to immobilize a wide range of contaminants (Shen et al., 2016); (ii) combination of in situ S/S with other remediation technologies such as bioremediation, oxidation/reduction, etc. (Chrysochoou and Dermatas, 2006; Kogbara et al., 2016); (iii) use of advanced techniques for analysis and model improvement. Microscopic, spectroscopic, and mineralogical analyses together with in-depth quantitative and statistical analyses can show what has been altered, the mode of alteration, alteration pathways, and what are the most significant influencing factors of the in situ S/S materials over time. These results are important not only in directly evaluating the permanence and leaching potential of contaminants under given environmental conditions but also in providing more accurate intrinsic material properties to improve the modeling accuracy; (iv) development of the long-term monitoring protocol. The effectiveness of in situ S/S cannot be evaluated based on one single test, and in order to facilitate further optimization and validation of this technology, a suite of different tests should be employed to provide as much important information as possible; and (v) site investigation and risk assessment linked with post usage of the site. Site investigation provides potential hazards affecting the long-term effectiveness of the treatment, which must be considered prior to the design and application. The assessment on the long-term effectiveness should also consider the post use of the site. For instance, some sites are potentially used for building foundations, and the compressibility should be evaluated apart from strength and permeability. Sites for greening purpose should consider the biocompatibility and its potential exposure pathway to the humans. In any case, a proper risk assessment is important for in situ S/S depending on the post-development plan.

It is believed that with more intensive cross-disciplinary research activities and practical experience on the in situ S/S technology, better long-term effectiveness and higher sustainability can be achieved, leading to stronger confidence and hence wider applications of this soil remediation technology in the future.

References

Akhter, H., Cartledge, F.K., Roy, A., Tittlebaum, M.E., 1997. Solidification/stabilization of arsenic salts: effects of long cure times. Journal of Hazardous Materials 52 (2–3), 247–264.

Al-Tabbaa, A., Boes, N., 2002. Pilot in situ auger mixing treatment of a contaminated site. Part 4. Performance at five years. Proceedings of the ICE - Geotechnical Engineering 155 (3), 187–202.

Al-Tabbaa, A., Evans, C.W., 2000. Pilot in situ auger mixing treatment of a contaminated site. Part 3. Time-related performance. Proceedings of the Institution of Civil Engineers 143 (2), 103–114.

Al-Tabbaa, A., King, S.D., 1998. Time effects of three contaminants on the durability and permeability of a solidified sand. Environmental Technology 19 (4), 401–407.

Al-Tabbaaa, A., Jin, F., O'Connor, D., Abunada, Z., 2015. Deep mixing field trials for land remediation: some 3-year results from the SMiRT project. In: Proceedings of the Deep Mixing 2015 Conference, (DFI).

Alba, N., Vázquez, E., Gassó, S., Baldasano, J.M., 2001. Stabilization/solidification of MSW incineration residues from facilities with different air pollution control systems. Durability of matrices versus carbonation. Waste Management 21 (4), 313–323.

Alver, E., Metin, A.Ü., 2012. Anionic dye removal from aqueous solutions using modified zeolite: adsorption kinetics and isotherm studies. Chemical Engineering Journal 200, 59–67.

American National Standards Institute (ANSI), 1986. Measurement of the Leachability of Solidified Low-Level Radioactive Wastes by a Short-Term Test Procedure (ANSI/ANS-16.1).

Antemir, A., Hills, C.D., Carey, P.J., Gardner, K.H., Bates, E.R., Crumbie, A.K., 2010. Long-term performance of aged waste forms treated by stabilization/solidification. Journal of Hazardous Materials 181 (1–3), 65–73.

Badreddine, R., Humez, A.N., Mingelgrin, U., Benchara, A., Meducin, F., Prost, R., 2004. Retention of trace metals by solidified/stabilized wastes: assessment of long-term metal release. Environmental Science and Technology 38 (5), 1383–1398.

Baker, P.G., Bishop, P.L., 1997. Prediction of metal leaching rates from solidified/stabilized wastes using the shrinking unreacted core leaching procedure. Journal of Hazardous Materials 52.

Bates, E., Hills, C., 2015. Stabilization and Solidification of Contaminated Soil and Waste: A Manual of Practice.

Bertron, A., 2014. Understanding interactions between cementitious materials and microorganisms: a key to sustainable and safe concrete structures in various contexts. Materials and Structures 47 (11), 1787–1806.

Bertron, A., Duchesne, J., 2013. Attack of cementitious materials by organic acids in agricultural and agrofood effluents. In: Performance of Cement-Based Materials in Aggressive Aqueous Environments. Springer, pp. 131–173.

Bin Shafique, M.S., Walton, J.C., Gutierrez, N., Smith, R.W., Tarquin, a J., 1998. Influence of carbonation on leaching of cementitious wasteforms. Journal of Environmental Engineering 124 (5), 463–467.

Bone, B.D., Barnard, L.H., Boardman, D.I., Carey, P.J., Hills, C.D., Jones, H.M., MacLeod, C.L., Tyrer, M., 2004. Review of Scientific Literature on the Use of Stabilisation/Solidification for the Treatment of Contaminated Soil , Solid Waste and Sludges.

Briggs, D., Seah, M.P., 2003. Practical Surface Analysis: by Auger and X-ray Photoelectron Spectroscopy. Wiley.

B. S. EN, 2002. 12457-2, Characterisation of Waste–Leaching–Compliance Test for Leaching of Granular Waste Materials and Sludges. UK.

Carmalin Sophia, A., Swaminathan, K., Sandhya, S., 2007. Microbially-influenced degradation of solidified/stabilized metal waste. Bioresource Technology 98 (13), 2562–2567.

Castellote, M., Andrade, C., Turrillas, X., Campo, J., Cuello, G.J., 2008. Accelerated carbonation of cement pastes in situ monitored by neutron diffraction. Cement and Concrete Research 38 (12), 1365–1373.

Castellote, M., Fernandez, L., Andrade, C., Alonso, C., 2009. Chemical changes and phase analysis of OPC pastes carbonated at different CO_2 concentrations. Materials and Structures 42 (4), 515–525.

Catalan, L.J.J., Merlière, E., Chezick, C., 2002. Study of the physical and chemical mechanisms influencing the long-term environmental stability of natrojarosite waste treated by stabilization/solidification. Journal of Hazardous Materials 94 (1), 63–88.

Cavani, F., Trifiro, F., Vaccari, A., 1991. Hydrotalcite-type anionic clays : preparation, properties and applications. Catalysis Today 11 (2), 173–301.

Celis, R., Carmen Hermosín, M., Cornejo, J., 2000. Heavy metal adsorption by functionalized clays. Environmental Science and Technology 34 (21), 4593–4599.

Chehade, M.H., Lors, C., Damidot, D., 2009. Design of a simplified and accelerated test for reinforced concrete biodeterioration. Proceeding RILEM TC 334.

Chen, Q., Zhang, L., Ke, Y., Hills, C., Kang, Y., 2009. Influence of carbonation on the acid neutralization capacity of cements and cement-solidified/stabilized electroplating sludge. Chemosphere 74 (6), 758–764.

Cheung, K.H., Gu, J.-D., 2007. Mechanism of hexavalent chromium detoxification by microorganisms and bioremediation application potential: a review. International Biodeterioration & Biodegradation 59 (1), 8–15.

Chrysochoou, M., Dermatas, D., 2006. Evaluation of ettringite and hydrocalumite formation for heavy metal immobilization: literature review and experimental study. Journal of Hazardous Materials 136 (1), 20–33.

Cocke, D.L., 1990. The binding chemistry and leaching mechanisms of hazardous substances in cementitious solidification/stabilization systems. Journal of Hazardous Materials 24, 231–253.

Conner, J.R., 1997. Guide to Improving the Effectiveness of Stabilization/Solidification. Portland Cement Association, Skokie, IL.

Conner, J.R., 1990. Chemical Fixation and Solidificaton of Hazardous Wastes. Van Nostrand Reinhold, New York.

Conner, J.R., Hoeffner, S.L., 1998. A critical review of stabilization/solidification technology. Critical Reviews in Environmental Science and Technology 28 (4), 397–462.

Conner, J.R., Hoeffner, S.L., 1998. The History of Stabilization/Solidification Technology. Critical reviews in Environmental Science and Technology 28 (4), 325–396.

Côté, P.L., Bridle, T.R., 1987. Long-term leaching scenarios for cement-based waste forms. Waste Management and Research 5 (1), 55–66.

Côte, P.L., Bridle, T.R., Benedek, A., 1987. An approach for evaluating lon-term leachability from measurement of intrinsic waste properties. Hazardous and Industrial Solid Waste Testing and Disposal. ASTM STP 933 6, 63–78.

De Percin, P.R., Sawyer, S., 1991. Long-term monitoring of the hazcon stabilization process at the Douglassville, Pennsylvania Superfund site. Journal of the Air and Waste Management Association 41 (1), 88–91.

De Windt, L., Badreddine, R., 2007. Modelling of long-term dynamic leaching tests applied to solidified/stabilised waste. Waste Management 27 (11), 1638–1647.

De Windt, L., Devillers, P., 2009. Coupled modeling of cement paste degradation by microbial activity. Concrete in Aggressive Aqueous Environments; Performance and Testing 196–203.

De Windt, L., Badreddine, R., Lagneau, V., 2007. Long-term reactive transport modelling of stabilized/solidified waste: from dynamic leaching tests to disposal scenarios. Journal of Hazardous Materials 139 (3), 529–536.

Du, Y.J., Jiang, N.J., Liu, S.Y., Jin, F., Singh, D.N., Puppala, A.J., 2014. Engineering properties and microstructural characteristics of cement-stabilized zinc-contaminated kaolin. Canadian Geotechnical Journal 51 (4).

Du, Y.J., Wei, M.L., Reddy, K.R., Liu, Z.-P., Jin, F., 2014. Effect of acid rain pH on leaching behavior of cement stabilized lead-contaminated soil. Journal of Hazardous Materials 271, 131–140.

Du, Y.J., Wei, M.L., Reddy, K.R., Jin, F., Wu, H.L., Liu, Z.B., Dec. 2014. New phosphate-based binder for stabilization of soils contaminated with heavy metals: leaching, strength and microstructure characterization. Journal of Environmental Management 146 (0), 179–188.

Du, Y.J., Bo, Y.L., Jin, F., Liu, C.Y., 2016. "Durability of reactive magnesia-activated slag-stabilized low plasticity clay subjected to drying–wetting cycle. European Journal of Environmental and Civil Engineering 22 (2), 215–230.

Ferguson, G., Gleeson, T., 2012. Vulnerability of coastal aquifers to groundwater use and climate change. Nature Climate Change 2 (5), 342.

Fitch, J.R., Cheeseman, C.R., 2003. Characterisation of environmentally exposed cement-based stabilised/solidified industrial waste. Journal of Hazardous Materials 101 (3), 239–255.

Fleri, M.A., Whetstone, G.T., 2007. In situ stabilisation/solidification: project lifecycle. Journal of Hazardous Materials 141 (2), 441–456.

Fleri, M.A., Whetstone, G.T., Bauman, J.P., 2005. "Stabilisation/solidification of manufactured gas plant wastes: Part 1–treatability study. In: Proceedings of the International Conference, Stabilisation/Solidification Treatment and Remediation–Advances in S/S for Waste and Contaminated Land, pp. 205–213.

Fuessle, R.W., Taylor, M.A., 2000. Stabilization of arsenic- and barium-rich glass manufacturing waste. Journal of Environmental Engineering 126 (3), 272–278.

Gadd, G.M., 2007. Transformation and mobilization of metals, metalloids, and radionuclides by microorganisms. In: Biophysico-Chemical Processes of Heavy Metals and Metalloids in Soil Environments. Wiley, Hoboken, pp. 53–96.

Glasser, F.P., 1993. Chemistry of cement-solidified waste forms. Chemistry and Microstructure of Solidified Waste Forms 1–39.

Glasser, F.P., 1997. Fundamental aspects of cement solidification and stabilisation. Journal of Hazardous Materials 52, 151–170.

Gollop, R.S.R., Taylor, H.F.F.W.H.F.W., 1992. Microstructural and microanalytical studies of sulfate attack. I. Ordinary portland cement paste. Cement and Concrete Research 22 (6), 1027–1038.

Gougar, M.L.D., Scheetz, B.E., Roy, D.M., 1996. Ettringite and C_S_H Portland cement phases for waste ion immobilization: a review. Waste Management 16 (4), 295–303.

Groves, G.W., Rodway, D.I., Richardson, I.G., 1990. The carbonation of hardened cement pastes. Advances in Cement Research 3 (11), 117–125.

Guettala, A., Abibsi, A., Houari, H., 2006. Durability study of stabilized earth concrete under both laboratory and climatic conditions exposure. Construction and Building Materials 20 (3), 119–127.

Guo, T., Hawke, A.S., Rusch, K.A., 1999. Determination of calcium diffusion coefficients as an estimator of the long-term dissolution potential of phosphogypsum:Cement:Lime composites. Environmental Science and Technology 33 (18), 3185–3192.

Guo, T., Malone, R.F., Rusch, K.A., 2001. Stabilized phosphogypsum: class C fly ash: portland type II cement composites for potential marine application. Environmental Science and Technology 35 (19), 3967–3973.

Guo, G., Zhou, Q., Ma, L.Q., 2006. Availability and assessment of fixing additives for the in situ remediation of heavy metal contaminated soils: a review. Environmental Monitoring and Assessment 116 (1–3), 513–528.

Hansen, P.F., Pedersen, E.J., 1977. Maturity Computer for Controlled Curing and Hardening of Concrete.

Harbottle, M.J., Al-Tabbaa, A., Evans, C.W., 2007. A comparison of the technical sustainability of in situ stabilisation/solidification with disposal to landfill. Journal of Hazardous Materials 141, 430–440.

Hills, C.D., Pollard, S.J.T., 1997. The influence of interference effects on the mechanical, microstructural and fixation characteristics of cement-solidified hazardous waste forms. Journal of Hazardous Materials 52, 171–191.

Hinsenveld, M., 1993. A Shrinking Core Model as a Fundamental Representation of Leaching Mechanisms in Cement Stabilized Waste. University of Cincinnati.

Hockley, D.E., van der Sloot, H.A., 1991. Long-term processes in a stabilized coal-waste block exposed to seawater. Environmental Science and Technology 25 (8), 1408–1414.

Humez, N., Prost, R., 1999. A new experimental approach to study the longterm behaviour of solidified/stabilized wastes. Chemical Speciation and Bioavailability 11 (1), 1–24.

Humez, N., Humez, A.L., Juste, C., Prost, R., 1997. A new assessment of mobility of elements in sediments and wastes. Chemical Speciation and Bioavailability 9 (2), 57–65.

Idachaba, M.A., Nyavor, K., Egiebor, N.O., Rogers, R.D., 2001. A refinement of the biofilm formation method for waste forms stability evaluation. Journal of Hazardous Materials 84 (1), 95–106.

Idachaba, M.A., Nyavor, K., Egiebor, N.O., 2002. Evaluation of microbial stability of simulated solid and liquid waste forms using a refined biofilm formation method. Journal of Hazardous Materials 90 (3), 279–295.

Idachaba, M.A., Nyavor, K., Egiebor, N.O., 2003. Kinetic analysis of data obtained from studies on microbial degradation of cement waste forms, using shrinking core models. Journal of Hazardous Materials 99 (1), 57–69.

Idachaba, M.A., Nyavor, K., Egiebor, N.O., 2004. The leaching of chromium from cement-based waste form via a predominantly biological mechanism. Advances in Environmental Research 8 (3–4), 483–491.

Islam, F.S., Gault, A.G., Boothman, C., Polya, D.A., Charnock, J.M., Chatterjee, D., Lloyd, J.R., 2004. Role of metal-reducing bacteria in arsenic release from Bengal delta sediments. Nature 430 (6995), 68.

Jin, F., Al-Tabbaa, A., Dec. 2014. Evaluation of novel reactive MgO activated slag binder for the immobilisation of lead and zinc. Chemosphere 117, 285–294.

Jin, F., Wang, F., Al-Tabbaa, A., Feb. 2016. Three-year performance of in-situ solidified/stabilised soil using novel MgO-bearing binders. Chemosphere 144, 681–688.

Kawai, K., Morinaga, T., Tasawa, E.I., 2000. The mechanism of concrete deterioration caused by aerobic microorganisms. In: First International RILEM Workshop on Microbial Impact on Building Materials, pp. 6–7.

Kirk, D.R., 1996. Summary of US EPA research on solidified/stabilized waste form long-term durability. In: Stabilization and Solidification of Hazardous, Radioactive, and Mixed Wastes, vol. 3.

Kjellsen, K.O., Detwiler, R.J., Gjørv, O.E., 1991. Development of microstructures in plain cement pastes hydrated at different temperatures. Cement and Concrete Research 21 (1), 179–189.

Klich, I., Batchelor, B., Wilding, L.P., Drees, L.R., 1999. Mineralogical alterations that affect the durability and metals containment of aged solidified and stabilized wastes. Cement and Concrete Research 29 (9), 1433–1440.

Klich, I., Wilding, L.P., Drees, L.R., Landa, E.R., 1999. Importance of microscopy in durability studies of solidified and stabilized contaminated soils. Soil Science Society of America Journal 63, 1274–1283.

Knight, J., Cheeseman, C., Rogers, R., 2002. Microbial influenced degradation of solidified waste binder. Waste Management 22 (2), 187–193.

Kogbara, R.B., Ayotamuno, J.M., Onuomah, I., Ehio, V., Damka, T.D., 2016. Stabilisation/solidification and bioaugmentation treatment of petroleum drill cuttings. Applied Geochemistry 71, 1–8.

Kosakowski, G., Berner, U., 2013. The evolution of clay rock/cement interfaces in a cementitious repository for low- and intermediate level radioactive waste. Physics and Chemistry of the Earth 64, 65–86.

Landrot, G., Ginder-Vogel, M., Sparks, D.L., 2009. Kinetics of chromium (III) oxidation by manganese (IV) oxides using quick scanning X-ray absorption fine structure spectroscopy (Q-XAFS). Environmental Science and Technology 44 (1), 143–149.

Lange, L.C., Hills, C.D., Poole, A.B., 1996. Preliminary investigation into the effects of carbonation on cement-solidified hazardous wastes. Environmental Science and Technology 30, 25–30.

Lange, L.C., Hills, C.D., Poole, A.B., 1996. The influence of mix parameters and binder choice on the carbonation of cement solidified wastes. Waste Management 16 (8), 749–756.

Lange, L.C., Hills, C.D., Poole, A.B., 1997. Effect of carbonation on properties of blended and non-blended cement solidified waste forms. Journal of Hazardous Materials 52, 193–212.

Lequeux, N., Boch, P., Pomie, M., 2001. Speciation of cadmium in cement Part II. C3 S hydration with Cd 2+ solution. Cement and Concrete Research 31 (1), 571–576.

Li, X.D., Poon, C.S., Sun, H., Lo, I.M.C., Kirk, D.W., Apr. 2001. Heavy metal speciation and leaching behaviors in cement based solidified/stabilized waste materials. Journal of Hazardous Materials A82 (3), 215–230.

Li, J.S., Xue, Q., Wang, P., Li, Z.Z., Liu, L., 2014. Effect of drying-wetting cycles on leaching behavior of cement solidified lead-contaminated soil. Chemosphere 117 (1), 10–13.

Libert, M.F., Sellier, R., Jouquet, G., Trescinski, M., Spor, H., 1993. Effects of microorganisms growth on the long-term stability of cement and bitumen. Scientific Basis for Nuclear Waste Management XVI (August), 267.

Lin, F., Meyer, C., 2009. Hydration kinetics modeling of Portland cement considering the effects of curing temperature and applied pressure. Cement and Concrete Research 39 (4), 255–265.

Liu, J., Nie, X., Zeng, X., Su, Z., 2013. Long-term leaching behavior of phenol in cement/activated-carbon solidified/stabilized hazardous waste. Journal of Environmental Management 115, 265–269.

Lothenbach, B., 2010. Thermodynamic equilibrium calculations in cementitious systems. Materials and Structures 43 (10), 1413–1433.

Lothenbach, B., Matschei, T., Möschner, G., Glasser, F.P., 2008. Thermodynamic modelling of the effect of temperature on the hydration and porosity of Portland cement. Cement and Concrete Research 38 (1), 1–18.

Lowell, S., Shields, J.E., Thomas, M.A., Thommes, M., 2012. Characterization of Porous Solids and Powders: Surface Area, Pore Size and Density, vol. 16. Springer Science & Business Media.

Ludwig, C., Johnson, C.A., Käppeli, M., Ulrich, A., Riediker, S., 2000. Hydrological and geochemical factors controlling the leaching of cemented MSWI air pollution control residues: a lysimeter field study. Journal of Contaminant Hydrology 42 (2–4), 253–272.

Magniont, C., Coutand, M., Bertron, A., Cameleyre, X., Lafforgue, C., Beaufort, S., Escadeillas, G., 2011. A new test method to assess the bacterial deterioration of cementitious materials. Cement and Concrete Research 41 (4), 429–438.

Malviya, R., Chaudhary, R., 2006. Factors affecting hazardous waste solidification/stabilization: a review. Journal of Hazardous Materials 137 (1), 267–276.

Mollah, M.Y., Kesmez, M., Cocke, D.L., 2004. An X-ray diffraction (XRD) and Fourier transform infrared spectroscopic (FT-IR) investigation of the long-term effect on the solidification/stabilization (S/S) of arsenic(V) in Portland cement type-V. The Science of the Total Environment 325 (1–3), 255–262.

Montgomery, D.M., Sollars, C.J., Perry, R., Tarling, S.E., Barnes, P., Henderson, E., 1991. Treatment of organic-contaminated industrial wastes using cement-based stabilization/solidification– ii. Microstructural analysis of the organophilic clay as a pre-solidification adsorbent. Waste Management and Research 9, 113–125.

Morandeau, A., Thiery, M., Dangla, P., 2014. Investigation of the carbonation mechanism of CH and CSH in terms of kinetics, microstructure changes and moisture properties. Cement and Concrete Research 56, 153–170.

Mossop, K.F., Davidson, C.M., 2003. Comparison of original and modified BCR sequential extraction procedures for the fractionation of copper , iron , lead , manganese and zinc in soils and sediments. Analytica Chimica Acta 478, 111–118.

Motsi, T., Rowson, N.A., Simmons, M.J.H., 2009. Adsorption of heavy metals from acid mine drainage by natural zeolite. International Journal of Mineral Processing 92 (1–2), 42–48.

Moulin, I., Stone, W.E.E., Sanz, J., Bottero, J.-Y., Mosnier, F., Haehnel, C., 1999. Lead and zinc retention during hydration of tri-calcium silicate: a study by sorption isotherms and 29 Si nuclear magnetic resonance spectroscopy. Langmuir 15 (8), 2829–2835.

Nica, D., Davis, J.L., Kirby, L., Zuo, G., Roberts, D.J., 2000. Isolation and characterization of microorganisms involved in the biodeterioration of concrete in sewers. International Biodeterioration and Biodegradation 46 (1), 61–68.

Ortego, J.D., Jackson, S., Yu, G.-S., McWhinney, H., Cocke, D., 1989. Solidification of hazardous substances-a TGA and FTIR study of Portland cement containing metal nitrates. Journal of Environmental Science and Health, Part A 24 (6), 589–602.

Panday, K.K., Prasad, G., Singh, V.N., 1985. Copper (II) removal from aqueous solutions by fly ash. Water Research 19 (7), 869–873.

Paria, S., Yuet, P.K., 2006. Solidification-stabilization of organic and inorganic contaminants using portland cement: a literature review. Environmental Reviews 14 (4), 217–255.

Parker, C.D., 1945. The function of thiobacillus concretivorus (Nov. Spec.) in the corrosion of concrete exposed to atmospheres containing hydrogen sulphide. Australian Journal of Experimental Biology and Medical Science 23 (2), 91–98.

PASSiFy, 2010. Performance Assessment of Solidified/stabilised Waste-Forms-An Examination of the Long-Term Stability of Cement-Treated Soil and Waste.

Perera, A.S.R., Al-Tabbaaa, A., Reid, J.M., Stegemann, J.A., 2005. State of practice report UK stabilisation/solidification treatment and remediation, Part IV: testing & performance criteria. In: Proceedings of the International Conference on Stabilization/Solidification Treatment and Remediation, pp. 415–435.

Pollard, S.J.T., Montgomery, D.M., Sollars, C.J., Perry, R., 1991. Organic compounds in the cement-based stabilisation/solidification of hazardous mixed wastes—mechanistic and process considerations. Journal of Hazardous Materials 28, 313–327.

Pomies, M.P., Lequeux, N., Boch, P., 2001. Speciation of cadmium in cement- Part I. Cd2+ uptake by C-S-H. Cement and Concrete Research. 31 (4), 563–569.

Poon, C.S., Chen, Z.Q., Wai, O.W.H., 2001. The effect of flow-through leaching on the diffusivity of heavy metals in stabilized/solidified wastes. Journal of Hazardous Materials 81 (1–2), 179–192.

Powers, T.C., 1958. The Physical Structure and Engineering Properties of Concrete, Volume 90 of Research department bulletin. Portland Cement Association.

Rauret, G., Lopez-Sanchez, J.F., Sahuquillo, A., Rubio, R., Davidson, C., Ure, A., Quevauviller, P., 1999. Improvement of the BCR three step sequential extraction procedure prior to the certification of new sediment and soil reference materials. Journal of Environmental Monitoring 1 (1), 57–61.

Roberts, D.J., Nica, D., Zuo, G., Davis, J.L., 2002. Quantifying microbially induced deterioration of concrete: initial studies. International Biodeterioration and Biodegradation 49 (4), 227–234.

Rogers, R.D., Hamilton, M.A., Veeh, R., McConnell, J., 1996. Microbially influenced degradation of cement solidified low-level radioactive waste forms. In: Stabilization and Solidification of Hazardous, Radioactive and Mixed Wastes: 3rd Volume, ASTM STP 1240, pp. 116–132.

Rogers, R.D., Knight, J.J., Cheeseman, C.R., Wolfram, J.H., Idachaba, M., Nyavor, K., Egiebor, N.O., 2003. Development of test methods for assessing microbial influenced degradation of cement-solidified radioactive and industrial waste. Cement and Concrete Research 33 (12), 2069–2076.

Rose, J., Moulin, I., Hazemann, J.L., Masion, A., Bertsch, P.M., Bottero, J.Y., Mosnier, F., Haehnel, C., 2000. X-ray absorption spectroscopy study of immobilization processes for heavy metals in calcium silicate hydrates: 1. Case of lead. Langmuir 16 (25), 9900–9906.

Roy, S.K., 1999. Durability of concrete—accelerated carbonation and weathering studies. Building and Environment 34 (5), 597–606.

Sánchez, L., Cuevas, J., Ramírez, S., De León, D.R., Fernández, R., Dela Villa, R.V., Leguey, S., 2006. Reaction kinetics of FEBEX bentonite in hyperalkaline conditions resembling the cement–bentonite interface. Applied Clay Science 33 (2), 125–141.

Santhanam, M., Cohen, M.D., Olek, J., 2003. Mechanism of sulfate attack: a fresh look - Part 2. Proposed mechanisms. Cement and Concrete Research 33 (3), 341–346.

Savage, D., Walker, C., Arthur, R., Rochelle, C., Oda, C., Takase, H., 2007. Alteration of bentonite by hyperalkaline fluids: a review of the role of secondary minerals. Physics and Chemistry of the Earth, Parts A/B/C 32 (1–7), 287–297.

Shackelford, C.D., Glade, M.J., 1997. Analytical mass leaching model for contaminated soil and soil stabilized waste. Ground Water 35 (2), 233–242.

Shand, M.A., Al-Tabbaaa, A., Qian, J.S., Mo, L.W., Jin, F., 2019. Magnesia Cements: From Formulation to Application. Elsevier Ltd (Forthcoming).

Shen, Z., Som, A.M., Wang, F., Jin, F., McMillan, O., Al-Tabbaa, A., 2016. Long-term impact of biochar on the immobilisation of nickel (II) and zinc (II) and the revegetation of a contaminated site. The Science of the Total Environment 542, 771–776.

Shi, C., Spence, R., 2004. Designing of cement-based formula for solidification/stabilization of hazardous, radioactive, and mixed wastes. Critical Reviews in Environmental Science and Technology 34 (4), 391–417.

Stegemann, J.A., Caldwell, R.J., Shi, C., 1997. Variability of field solidified waste. Journal of Hazardous Materials 52 (2–3), 335–348.

Sun, D., Zhang, X., Wu, Y., Liu, X., 2010. Adsorption of anionic dyes from aqueous solution on fly ash. Journal of Hazardous Materials 181 (1–3), 335–342.

Sweeney, R.E.H., Hills, C.D., Buenfeld, N.R., 1998. Investigation into the carbonation of stabilised/solidified synthetic waste. Environmental Technology 19 (9), 893–902.

Tessier, A., Campbell, P.G.C., Bisson, M., 1979. Sequential extraction procedure for the speciation of particulate trace metals. Analytical Chemistry 51 (7), 844–851.

Tiruta-Barna, L., Imyim, A., Barna, R., 2004. Long-term prediction of the leaching behavior of pollutants from solidified wastes. Advances in Environmental Research 8 (3–4), 697–711.

USEPA, 1992. Method 1311. Toxicity characteristic leaching procedure. EPA SW-846 Test Methods for Evaluating Solid Waste: Physical/Chemical Methods.

USEPA, 1994. Method 1312: synthetic precipitation leaching procedure. Test Methods for Evaluating Solid Waste: Physical/Chemical Methods-846.

USEPA, 1993. Technology Resource Document-Solidification/Stabilization and its Application to Waste Materials. EPA/530/R-93/012.

Ure, A.M., Quevauviller, P.H., Muntau, H., Griepink, B., 1993. Speciation of heavy metals in soils and sediments. An account of the improvement and harmonization of extraction techniques undertaken under the auspices of the BCR of the Commission of the European Communities. International Journal of Environmental Analytical Chemistry 51 (1–4), 135–151.

USEPA, 2007. Treatment Technologies for Site Cleanup: Annual Status Report.

Valls, S., Vàzquez, E., 2000. Durability study of a sewage sludge-cement-sand system and its environmental impact. In: Waste Management Series, vol. 1. Elsevier, pp. 151–161.

Wang, H., Gillott, J.E., 1991. Mechanism of alkali-silica reaction and the significance of calcium hydroxide. Cement and Concrete Research 21 (4), 647–654.

Wang, F., Wang, H., Al-Tabbaa, A., 2014. Leachability and heavy metal speciation of 17-year old stabilised/solidified contaminated site soils. Journal of Hazardous Materials 278, 144–151.

Wang, F., Wang, H., Al-Tabbaa, A., 2015. Time-dependent performance of soil mix technology stabilized/solidified contaminated site soils. Journal of Hazardous Materials 286, 503–508.

Wang, F., Wang, H., Jin, F., Al-Tabbaa, A., 2015. The performance of blended conventional and novel binders in the in-situ stabilisation/solidification of a contaminated site soil. Journal of Hazardous Materials 285, 46–52.

Wang, F., Jin, F., Shen, Z., Al-Tabbaa, A., 2016. Three-year performance of in-situ mass stabilised contaminated site soils using MgO-bearing binders. Journal of Hazardous Materials 318, 302–307.

Washington State Department of Ecology, 2003. An Assessment of Laboratory Leaching Tests for Predicting the Impacts of Fill Material on Ground Water and Surface Water Quality – A Report to the Legislature. Washington.

Webster, M.T., Loehr, R.C., 1996. Long-term leaching of metals from concrete products. Journal of Environmental Engineering 122 (August), 714–721.

Wei, S., Jiang, Z., Liu, H., Zhou, D., Sanchez-Silva, M., 2013. Microbiologically induced deterioration of concrete - a review. Brazilian Journal of Microbiology 44 (4), 1001–1007.

Yang, G.C.C., 1993. Durability study of a solidified mercury-containing sludge. Journal of Hazardous Materials 34 (2), 217–223.

Ye, H., Radlińska, A., Neves, J., 2017. Drying and carbonation shrinkage of cement paste containing alkalis. Materials and Structures 50 (2).

Yousuf, M., Mollah, A., Hess, T.R., Tsai, Y.N., Cocke, D.L., 1993. An FTIR and XPS investigations of the effects of carbonation on the solidification/stabilization of cement based systems-Portland type V with zinc. Cement and Concrete Research 23 (4), 773–784.

Zha, F., Liu, J., Xu, L., Cui, K., 2013. Effect of cyclic drying and wetting on engineering properties of heavy metal contaminated soils solidified/stabilized with fly ash. Journal of Central South University 20 (7), 1947–1952.

Zhang, M., Yang, C., Zhao, M., Yang, K., Shen, R., Zheng, Y., 2017. Immobilization potential of Cr(VI) in sodium hydroxide activated slag pastes. Journal of Hazardous Materials 321, 281–289.

Chapter 11

Remedial process optimization and sustainability benefits

Jim Leu[1], Deyi Hou[2]

[1]*Parsons Corporation, Walnut Creek, CA, United States;* [2]*School of Environment, Tsinghua University, Beijing, China*

1. Introduction

The remediation of contaminated soil and groundwater is a major environmental challenge today. The U.S. Environmental Protection Agency (USEPA) estimates that the United States contains more than 294,000 contaminated sites, including 77,000 known sites and 217,000 sites likely to be discovered in the future. Cleaning up these contaminated sites will cost more than $200 billion (USEPA, 2004). Other nations face similar challenges. For example, in the United Kingdom, an estimated 50,000 to 200,000 ha representing approximately 100,000 sites may have been contaminated (Pollard et al., 2001). The vast scope of this problem warrants the development of cost-effective solutions in every step of the remediation process.

Value engineering is a systematic method to optimize the cost-effectiveness of goods and services by improving function and/or reducing costs. It was developed during World War II by General Electric Company to reduce cost and improve product, and it has become a standard project management technique. Value engineering is a typical component of the design phase of large environmental remediation projects. Increasingly, it also is applied to smaller remediation projects and to the implementation or operational phases of environmental remediation projects.

The concept of remedial process optimization (RPO) was developed in the 1990s, and the term refers to the application of value engineering to the postdesign components of soil and groundwater cleanup projects. The Interstate Technology and Regulatory Council (ITRC) defines RPO as the "systematic reevaluation of initial objectives and the continuous improvement and optimization of remediation technologies and techniques," which represents an improvement from "a basic 'study, design, build' linear paradigm" (ITRC, 2007). RPO has been promoted within several military, government, and

Sustainable Remediation of Contaminated Soil and Groundwater. https://doi.org/10.1016/B978-0-12-817982-6.00011-2

industry groups, including the Air Force Center for Engineering and the Environment (AFCEE) and ITRC. It has become widely used in both government and private sector projects.

With the rapid development of new technologies and the aging of existing remedial systems, however, RPO is drawing more attention. The implementation of RPO can, in many cases, provide cost-effective improvement in ongoing remediation performance.

RPO is widely applied in remediation projects by industries and the US government. It adapts value engineering to the complexity of environmental remediation projects. While value engineering is a typical component during the design phase, RPO has typically been applied to the implementation phase of environmental remediation projects. RPO is defined as a systematic evaluation of initial objectives followed by the continuous improvement and optimization of remediation processes.

Remedy selection and RPO usually focus on cost, risk reduction, compliance with existing laws and regulations, and other metrics. By including sustainability in an environmental restoration program, several new metrics may become part of the remediation process, such as evaluating carbon emissions, energy consumption, air/water impacts, and resource conservation for land and/or groundwater.

This chapter describes RPO and green and sustainable remediation (GSR), their approaches, tools, and applications in soil and groundwater remedies. Case studies are presented to incorporate GSR into RPO evaluations and summarize GSR benefits from RPO recommendations.

2. Remedial process optimization

RPO is a systematic process of evaluating the performance and effectiveness of existing site remediation systems and proposing cost-effective solutions toward site closeout (AFCEE, 2001). Since its development in the 1990s, RPO has become widely used in both government and private sector projects. It is different from traditional optimization technique in that it continuously evaluates the objectives and approaches focusing on the analyses and evaluation. RPO is becoming more attractive and even necessary as treatment systems become aged and the desire to close sites accelerates. A USEPA study of 20 Superfund remedial systems in 2002 indicated that the annual operation and maintenance (O&M) cost could be reduced by over 30% by implementing RPO (USEPA, 2002).

RPO was not fully implemented in remedial systems in the past because of the following:

- Optimization during implementation phase has drawn less attention than the optimization of initial remedial design, which tends to save significant capital cost.

- Continuous optimization involves uncertainty because of unpredictable changes of site conditions, remedial objectives, and regulatory thresholds.
- Insufficient knowledge of RPO leads to technical difficulties and unnecessarily inflates RPO costs.

With continuous and rapid development of new technologies and aging remedial systems, RPO is drawing more attention than before. The implementation of RPO proves a cost-effective method to improve ongoing remediation performance. The benefits of RPO are summarized as follows:

- Ensure protectiveness of human health and the environment.
- Reevaluate cleanup goals in response to changing regulations, site conditions, and technologies.
- Track remedial progress.
- Reduce O&M and monitoring costs.
- Accelerate site transfer or closure.

2.1 RPO phased approach

The RPO evaluation could be conducted in three phases as follows (AFCEE, 2001):

Phase I is an annual review of system performance, including following main tasks:

- Check cleanup goals.
- Evaluate mass removal rates.
- Check compliance and evaluate contaminants containment.
- Conduct general cost analysis.

Phase I could be easily conducted with on-site documents, including but not limited to investigation, studies, design, remedial action completion documents, O&M manual, and O&M data.

When RPO opportunities are identified through Phase I document review due to changing site conditions, remedial objectives, regulatory thresholds, or innovative technologies, an intensive **Phase II** will be invoked to include the following tasks:

- Review and reevaluate remediation goals based on current conditions.
- Review and evaluate remedial design based on current conditions.
- Evaluate system operation performance.
- Evaluate technology effectiveness.
- Provide recommendations.

Phase II usually requires a team with diverse areas of expertise, including engineers, scientists, geologists, and planners, to review on-site documents and conduct a site visit. The Remedial System Evaluation (RSE) Checklist

(USACE, 2010) is a good resource to conduct Phase II evaluation of each component of the remediation system.

After completing Phase II evaluation, the team will discuss and provide RPO recommendations with a life cycle cost–benefit analysis. The recommendations will be implemented in **Phase III** after they are approved and accepted by the property owners and regulatory agencies. Phase III usually consists of the following tasks:

- Implement Phase II recommendations.
- Continue existing systems if goals will be achieved.
- Change systems.
- Change remedial action objective approach, such as from source containment to remediation, land use controls, etc.

2.2 Long-term monitoring optimization

In addition to remedial process, long-term monitoring usually has another significant cost for ongoing remediation systems, such as groundwater pump and treat (P&T), soil vapor extraction (SVE), or air sparging (AS). Long-term monitoring optimization (LTMO) evaluates long-term monitoring programs, including number and locations of monitoring points, monitoring frequency, monitoring parameters, and sampling and analysis procedures; it provides an optimal monitoring program, which maximizes the monitoring results but minimizes the monitoring cost (USEPA, 2005). Commonly used LTMO tools are geostatistical temporal/spatial (GTS) optimization algorithm, monitoring and remediation optimization system (MAROS), and Parsons' three-tiered LTMO (3TMO). GTS is a quantitative/statistical calculative approach, 3TMO is a qualitative/judgmental heuristics approach, and MAROS is a balanced approach of the two.

Groundwater quality monitoring is an essential component of all systems used to remediate contaminated groundwater. It is also a major cost item in some long-term remediation approaches, such as monitored natural attenuation (MNA) or barrier-controlled MNA. Therefore, a systematic groundwater monitoring optimization approach to project self-monitoring programs often is beneficial.

A USEPA study analyzed the temporal variation of groundwater quality over a 48-week period (Barcelona et al., 1989). The study indicated that too-frequent sampling could result in loss of informative value of the data. Researchers at the Lawrence Livermore National Laboratory developed a framework based on descriptive statistics to reduce sampling frequency of groundwater monitoring wells (Johnson et al., 1995). Their method requires a less extensive statistics background to interpret, compared with traditional inferential statistics.

Despite the importance of groundwater monitoring in the remediation process, it is not common to see quantitative optimizations to the design of groundwater monitoring. A quantitative approach would require (1) complicated statistical analysis with knowledge and experience in statistics and (2) a strong background in environmental chemistry and geohydrology. To optimize groundwater monitoring, one must understand the underlying mechanisms that affect the fate and transport of pollutants.

Two groups of predictive models are available for RPO analyses: deterministic models and simple statistical models. Deterministic models involve complex mechanistic formulations based on theoretically or experimentally derived process functions. Simple statistical models are primarily based on correlative relationships between pollutant levels and environmental variables. These models are more effective for contaminant plumes migrating in highly heterogeneous soil and/or groundwater and involving many types of transportation pathways and degradation mechanisms.

Groundwater monitoring optimization involves refining the number of wells, sampling frequency, and analytes. Many studies have been conducted on the optimization of groundwater monitoring network design (Hudak and Loaiciga, 1993; Mahar and Datta, 1997; Yeh et al., 2006; Kim and Lee, 2007; Chadalavada1 and Datta, 2008). However, these quantitative approaches require intensive, sophisticated modeling and computation and, therefore, are not very practical or cost-effective on relatively small projects.

Most groundwater monitoring network design is performed by qualitative approaches. Unfortunately, these can be subjective and ineffective. However, the following tools can be used during optimization of existing groundwater monitoring networks that are computationally simple and can be implemented by a nonstatistician.

2.2.1 Descriptive and statistic tools

Five commonly used descriptive and statistic tools are discussed as follows.

- Descriptive trend plot. Water quality data are plotted along a time axis to determine the temporal variation of contaminant concentrations in groundwater. Advantages include ease in detecting trends and the ability to convey information to others, such as regulators.
- Trend factor analysis. This descriptive statistic is an average yearly trend factor as defined by the least-squares slope of the regression of time against concentration or log-transformed concentration. It can be calculated for selected analytes at selected monitoring wells. Trend factor analysis provides a basis for quantitative comparison. For example, if the trend factor was based on log-transformed concentrations and yearly periods, a trend factor of 0.01 means an average of 1% increase in contaminant concentration over a period of 1 year, while a trend factor of −0.01 means an average of 1% decrease in contaminant concentration over a period of 1 year.

- Autocorrelation analysis. Groundwater monitoring data at each well are time series. In time series, there usually are repeating patterns, such as the presence of seasonal signals. The repeating patterns found in groundwater monitoring data could have been caused by variation of water levels, precipitation, and other environmental factors. This repeating pattern often causes short-term fluctuations and, consequently, requires more groundwater monitoring to capture the full pattern. However, as mentioned earlier, too-frequent sampling can degrade rather than enhance the information value of data (Barcelona et al., 1989). The ratio of effective independent sample size, n_{ef}, and actual sample size, n, was calculated for time series that would have been obtained by various sampling frequencies. According to Barcelona et al., sampling at interval lengths such that the ratio of effective independent sample size and actual sample size approaches 1.0 is recommended. The application of autocorrelation analysis can significantly reduce the required groundwater monitoring frequency, thus optimizing remediation program.
- Mann-Kendall test. The Mann-Kendall test is a nonparametric statistical test that has been popularly used to assess trend significance in time series. The test compares the relative magnitude of monitoring data instead of the data values themselves. Unlike some commonly used inferential statistics that requires normal distribution of the target variable, the Mann-Kendall test does not require that the data conform to any particular distribution. Use of the Mann-Kendall trend test can detect natural degradation trends and contaminants migration in a timely manner, thus optimizing decision-making during remedial processes.
- Principal component analysis (PCA). PCA is a multivariate analysis approach commonly used to determine the correlation among groundwater monitoring data, specifically the temporal and spatial pattern of all variables. PCA is a mathematical method in which potentially correlated variables are transformed into a smaller number of uncorrelated variables called principal components. These principal components often contain the "most important" aspect of the data matrix. The temporal and spatial pattern given by the PCA can provide meaningful insight into the underlying mechanism of constituent of concern (COC) transportation and degradation and, consequently, establish a basis for groundwater monitoring optimization.

2.2.2 Deterministic tools

Two commonly used deterministic tools are described and discussed as follows.

- Qualitative analysis of hydraulic capture and plume migration. Qualitative and quantitative analyses are performed to evaluate the hydraulic capture of groundwater extraction treatment system (GWETS) and contaminant plume stability. These analyses are based on the site conceptual model and a comprehensive evaluation of data trends.

- Groundwater flow and contaminant transport modeling. Results from groundwater flow and contaminant transport modeling are used to understand the site condition and optimize the remedial system. These deterministic models can provide simulation for various scenarios, thus providing a basis for decision-making in selecting and optimizing remedies. In addition, unlike groundwater monitoring, which only provides discrete results, deterministic models provide a basis for evaluating the full extent of contaminant distribution and migration. Combining real monitoring data with deterministic modeling can provide a holistic approach to determining the most effective remedial process optimization.

For sampling reduction efforts, qualitative analyses such as descriptive trend plots are among the most widely used approaches because they are easy to use and interpret. The quantitative tools are less commonly used because they require more expertise and up-front labor cost. The Mann-Kendall test is among the most often used quantitative approaches because it does not require much statistical knowledge to interpret the results.

2.3 RPO applications

In addition to optimizing remedial process and monitoring program, RPO can be integrated into different approaches to help optimize the remedy system. A three-tier approach was developed using RPO to optimize the existing remedial system and applied to a P&T system at a remediation site (Hou and Leu, 2009). A P&T system usually becomes expensive for removing contaminant mass after operation for an extended time because the cost of P&T stays high while concentration of contaminants in extracted groundwater decreases insignificantly. However, the P&T system may have to continue until contaminants in groundwater reach the cleanup goals. Therefore, the optimization of the existing remedial system and an exit strategy will be necessary and beneficial to all stakeholders. Hou and Leu applied RPO to reduce pumping wells and monitoring frequency and develop an exit strategy. Based on the tiered analysis, RPO is expected to save 68% of groundwater monitoring cost and 30% of O&M cost.

RPO can be applied to compare several similar remedial sites to assist in deciding systematic remedial process selection and optimization for different types of sites (Hou et al., 2009). In their studies, a systematic approach to RPO was applied at various sites, including groundwater extraction and treatment systems, SVE systems, hydraulic barrier systems, and natural attenuation with long-term groundwater monitoring. Various RPO recommendations considered for implementation include treatment system curtailment, treatment time reduction, monitoring program reductions, pumping optimizations, and treatment system performance optimizations. Potential RPO recommendations were identified and evaluated by comparing groundwater monitoring optimization. Treatment systems were recommended to evaluate routinely over a consistent period. Often, monitoring programs and treatment systems were evaluated for RPO recommendations yearly or biyearly. Once RPO recommendations are identified, they must be evaluated against the suggested criteria: feasibility for implementation, potential for cost savings, remedial process improvements, and regulatory approval.

3. Green and sustainable remediation

Green and sustainable remediation (GSR) is defined to protect human health and the environment while maximizing the environmental, social, and economic benefits throughout the project life cycle. It considers the environmental impacts and benefits during investigation and remediation implementation activities. Incorporating GSR into an environmental restoration project requires evaluating and selecting a remedy that minimizes the environmental impacts yet provides the maximum environmental benefits through the entire life of the remedial project.

GSR resources can be obtained from the U.S. Sustainable Remediation Forum (SuRF) (SuRF, 2009), ITRC, USEPA, American Society for Testing and Materials (ASTM), and others. Overall, the goals of GSR are as follows:

- Achieve remedial action goals.
- Support use and reuse of remediated lands.
- Increase operation efficiencies.
- Reduce total pollutant and waste.
- Minimize degradation and enhance ecology of the site.
- Reduce air emissions and greenhouse gas (GHG) production.
- Minimize impacts to water quality and water cycles.
- Conserve natural resources.
- Achieve long-term financial return from investments.

3.1 GSR metrics and tools

Green and sustainability metrics can be used to help determine relative benefits versus negative impacts during various phases of remediation, including remedy selection and RPO. Metrics have been developed to consider as part of GSR, as follows (USEPA, 2008a):

- Energy consumption
- GHG emissions
- Criteria pollutant emissions

- Water impacts
- Ecological impacts
- Resource consumption

After the GSR metrics are identified, GSR tools are used to evaluate the GSR metrics. Best management practices (BMPs), qualitative tools, and advanced quantitative tools were developed by the government and industries to assist GSR evaluation in estimating GSR metrics, identifying environmental footprint, and reducing the most significant footprint.

Most commonly used BMPs can be found in USEPA's Greener Cleanup (EPA, 2008b), ASTM standards, the SuRF framework, and ITRC guidance. The advantages of the BMPs are that they are simple, cost-effective, and easy to implement. The disadvantage is that they do not evaluate trade-offs.

Qualitative tools are developed to identify and implement commonsense GSR-related BMPs. The Green Remediation Evaluation Matrix, for example, is a simple Excel tool developed by the California Department of Toxic Substances Control. The advantages of the qualitative tools are that they can evaluate multiple metrics and make simple calculations. The disadvantage is that they require a scoring method.

Advanced quantitative tools use a quantitative comparison across the remedial options and remedy components to measure the footprint reductions gained from the BMPs selected and implemented. Advanced quantitative tools developed by the government include USEPA's Spreadsheets for Environmental Analysis (SEFA), the Air Force's Sustainable Remediation Tool (SRT) (AFCEE, 2009), and SiteWise from the Navy and the Army. Industry-developed tools, such as SimaPro life-cycle assessment, are also available. The advantages of the advanced quantitative tools are quantifying multiple metrics and tracking impacts from cradle to grave. The disadvantages are that they are costly and time-consuming and require scoring methods and more data.

3.2 GSR application in RPO

To reduce environmental, economic, and social impacts, it is better to consider GSR approaches in the early remediation phases, such as remedial investigation, feasibility study, and remedial design phases. However, more site-specific operational data are available in the RPO phase than the early remedial phases when conducting a GSR analysis, so a GSR evaluation using the actual operational data in the RPO phase is more precise than in the remedy evaluation and selection phase. In addition, RPO provides an excellent opportunity to evaluate and improve the sustainability of the existing remedial system and to consider more sustainable remedial systems if the remedial system needs to be replaced through RPO evaluation.

The application of GSR elements in the three RPO phases focuses on minimizing the impacts associated with the current and future remedies required to achieve site closure. The GSR approaches can be incorporated in three RPO phases, as shown in Table 11.1.

4. Case studies

GSR can be applied in different remediation phases, including investigation, remedy evaluation and selection, remedy design, remedy construction, operation, maintenance and monitoring, remedy optimization, and closeout. GSR has been considered in remedy selection and RPO recommendations for several years in the US case studies are presented next.

4.1 Petroleum hydrocarbons site, California, USA

The site is a flat hilltop surrounded by agricultural field. Site impacts to the soil include benzene, toluene, ethylbenzene, and xylene (BTEX) and total petroleum hydrocarbons (TPH) in gasoline (TPHg) and diesel (TPHd). Geological investigation indicates that a sandy zone (20–40 ft bgs) is sandwiched by two clay layers (Dominguez et al., 2012).

Bioventing is a commonly used, proven technology for TPH remediation in vadose-zone conditions, where biodegradation is more effective under aerobic conditions than in anaerobic conditions. Bioventing uses oxygen in supplied

TABLE 11.1 GSR approaches in RPO phases.

PRO phases	GSR approaches
Phase I—review system performance	• Review energy use trend • Review GHG emission trend • Review waste generation trend • Review air emission trend • Review water use trend
Phase II—evaluate cleanup goals, effectiveness, and efficiency of the remediation and monitoring systems	• Reduce energy use • Reduce GHG emission • Reduce waste generation • Reduce air emission • Reduce all resource consumption
Phase III—implement RPO recommendations	• Implement BMPs • Communicate with stakeholders about GSR benefit through RPO • Use cost saving from RPO to enhance GSR solutions

air to stimulate aerobic bacteria, increasing the effectiveness of natural biodegradation processes. The technology can be applied either with conventional sources of energy or with natural and sustainable (passive) energy sources such as wind or solar power.

An environmentally sustainable wind-powered bioventing device was designed, installed, and operated to address the TPH impacts in shallow soils at the site. The treatment system consisted of 12-inch-diameter rotating funnels mounted on existing 2-inch-diameter shallow groundwater monitoring wells designed to turn into the wind to facilitate passive air injection into the vadose zone in order to stimulate aerobic biodegradation of COCs in soil (Fig. 11.1). The wind-powered bioventing uses natural and sustainable wind power to directly inject air into the land without the use of a conventional energy source and blower. It provides not only cost savings on capital equipment and O&M of the blower system but also GSR technology by using a natural power source and reducing the environmental carbon footprint.

After 3 months of system operation, oxygen delivery increased to 20% from 2.5% and hydrocarbon concentrations were reduced by 80%–90% with minimal rebound. Measurements of oxygen and carbon dioxide concentrations indicated that the radius of influence is beyond 35 feet. The benefits of the bioventing system included an estimated reduction in electricity consumption by about 20,000 kW h per year and elimination of air pollutants and GHG emissions equivalent to 12 metric tons of carbon dioxide per year. This technology is a sustainable approach to environmental remediation: it has no power requirements, it requires minimal O&M, and it creates no on-site fugitive emissions.

4.2 Four groundwater pump and treat sites, California, USA

Sites A, B, C, and D have groundwater extraction and treatment systems and/or SVE and treatment systems. These treatment systems all have operated for more than 5 years. This comparative analysis discusses how remedial costs

FIGURE 11.1 Wind-powered bioventing system.

vary among sites and how cost saving could be achieved through RPO (Hou et al., 2009).

Table 11.2 presents the cost of components for the current remedial treatment systems at these four sites, as well as the cost and cost saving associated with RPO at each site. All costs are annually based and are expressed in percent of the total annual O&M costs. Costs were derived from actual project expense and project budgets, as follows:

- The laboratory cost is the direct subcontractor cost associated with chemical analyses.
- The discharge cost is the fee paid to a local government or municipal utility district for discharging treated wastewater.
- Subcontractor cost includes activities such as carbon change-out for granulated activated carbon (GAC) systems, mechanical and electric repair, and waste disposal cost.
- Groundwater monitoring and O&M inspection labor include field and office hours spent on the O&M by the primary contractor.
- Groundwater monitoring cost is the sum of all costs associated with groundwater monitoring, including laboratory cost, labor cost, and other direct cost.
- The RPO cost/cost saving includes the cost and benefit associated with RPO evaluation and implementation, except for interim remedial action and treatment system curtailment.

TABLE 11.2 Cost comparison of remedial process optimization at four sites.

Sites	Site A (%)	Site B (%)	Site C (%)	Site D (%)
Laboratory [a]	13	19	3	9
Discharge [a]	11	29	46	58
Subcontractor [b]	18	11	17	5
Groundwater monitoring and O&M inspection labor [a]	51	27	24	10
Current RPO cost [c]	14	2	9	3
RPO saving [c,d]	58	46	13	23

[a] *Annual cost.*
[b] *Annual subcontractor cost includes activities such as carbon change-out, mechanical and electric repair, and waste disposal.*
[c] *Does not include long-lead or unquantifiable cost and cost savings.*
[d] *Annual cost saving for 5 years is converted to present value by assuming a discount rate of 7% (USEPA, 1988).*

At Site A, labor costs associated with groundwater sampling, weekly O&M inspections, and reporting account for 51% of the total O&M cost. This indicates that there is substantial cost-saving potential associated with groundwater monitoring optimization and strategic O&M planning. Other major cost items include laboratory service, discharge fees, and subcontractor costs, but, except for laboratory service, these could not be reduced under the current site conditions. The total cost saving of the proposed RPO over 5 years could reduce 58% of the annual O&M cost.

At Site B, labor costs associated with groundwater sampling, weekly O&M inspection, and reporting account for 27% of the total O&M costs. In comparison, the discharge fee is quite high, accounting for 29% of the total O&M cost. Groundwater monitoring remains the largest portion of the total O&M cost at Site B The RPO based on groundwater sampling reduction led to a 46% cost saving.

At both Site C and Site D, water discharge fees account for the largest portion of total O&M costs, at 46% and 58%, respectively. This cost can be significantly reduced by groundwater extraction network optimization. Groundwater monitoring accounts for only 24% and 10% of the total O&M costs at Site C and Site D, respectively. Because the RPO cost saving do not include cost savings associated with treatment system curtailment or extraction network optimization, it accounts for only 13% and 23% of the total O&M cost at these sites. If treatment system curtailment and/or optimization is successfully implemented, the cost saving is expected to significantly increase.

4.3 Petroleum hydrocarbons site, California, USA

The site operated as a natural gas processing station. Environmental characterization and remediation activities conducted at the site identified site COCs as unrefined TPH and BTEX in both soil and groundwater (Leu et al., 2016; Leu and Morales, 2016).

Several historical remedial activities on site include an ongoing GWETS and 5 years of AS/SVE operation. The GWETS has five extraction wells for hydraulic containment and mass removal of contaminants in groundwater. The AS/SVE consisted of five AS wells and 12 SVE wells. Both systems reached practical limits of COCs mass removal.

RPO was conducted to (1) evaluate the historical and current effectiveness of site remedial activities and (2) recommend an approach to optimize the remedial activities.

4.3.1 RPO evaluation

RPO was conducted to evaluate hydraulic capture, natural attenuation, and mass removal. The RPO evaluation reached the following conclusions:

- To ensure proper hydraulic containment, the average groundwater pumping rate should be greater than 3.2 gallons per minute.

- TPHd is known to undergo natural attenuation under ideal conditions, but the process requires more time for complete degradation of this COC.
- Neither the groundwater extraction system nor the AS/SVE system can be expected to make further significant progress toward contaminant reduction by mass recovery in their current configurations, since they have achieved 98.9% and 99.7%, respectively, of the theoretical recoverable mass.

4.3.2 RPO recommendation

Various options were considered to facilitate more rapid attainment of cleanup criteria than would occur naturally. The RPO evaluation concluded that the existing systems had met their limits and recommended in situ chemical oxidation to accelerate site closure. Ozone sparging was selected to treat the on-site downgradient plume, and persulfate injection was selected to treat the off-site upgradient plume beneath a residence due to concerns of ozone off-gassing.

4.3.3 RPO implementation

Bench-scale treatability studies were conducted to estimate persulfate and ozone dosages and evaluate COCs removal and impacts of by-products such as chromium VI and bromate. An ozone sparging pilot test was conducted to determine operating parameters and radius of influence for the system design and evaluate the by-products generation and attenuation. A 60-minute injection cycle and 20-ft radius of influence were obtained by the pilot study and were recommended for the full-scale system. Chromium VI and bromate concentrations increased to 44 and 110 μg per liter, respectively, after the ozone injection, but both concentrations decreased to background concentrations 3 months after the ozone injection.

A full-scale system was implemented to include six ozone injection wells supplied with a 27 pounds per day ozone generator and 12 direct push boreholes each injected with 1400 gallons of 10% sodium persulfate. The interaction between persulfate and ozone created a transition zone of low pH with high dissolved oxygen (DO) and oxygen-reduction potential (ORP), which could generate more radicals to expedite COCs removal. After application of 50% design dosages, the gasoline-range organics (GRO) was 95% removed, as shown in Figs. 11.2 and 11.3. Another year of rebound monitoring indicated the COCs decreasing trend, which was confirmed by BIOSCREEN modeling.

4.4 Chlorinated solvent site, California, USA

The site is a large former manufacturing facility in the San Francisco Bay Area. Historical industrial operations from the 1950s to the 1990s resulted in impacts to soil and groundwater across the site. The case study focuses on one

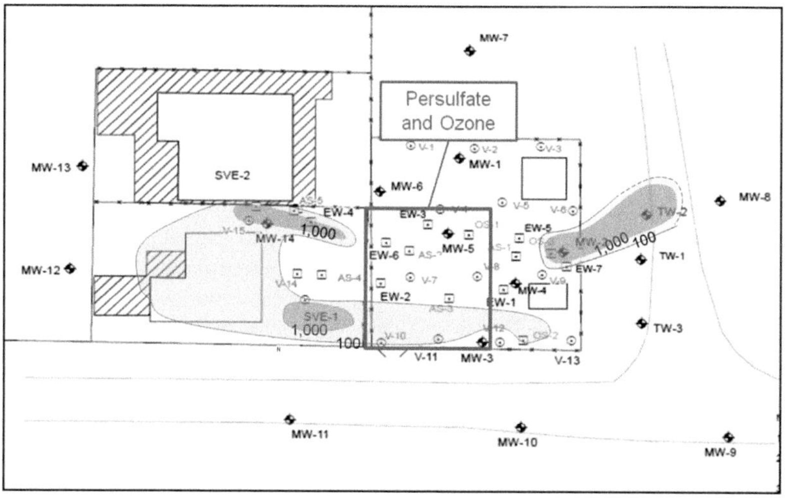

FIGURE 11.2 GRO concentration contour before RPO implementation.

of three major groundwater plumes on site, with primary constituents consisting of tetrachloroethylene (PCE) and its breakdown products. The impacted groundwater is under tidal influence. The remediation objective is to protect human health and the environment. The primary short-term goal is to reduce contaminant mass in groundwater and soil gas, and the long-term goal is to achieve drinking water quality standards. The life cycle was assessed to determine the sustainability of various remedial alternatives. Based on the

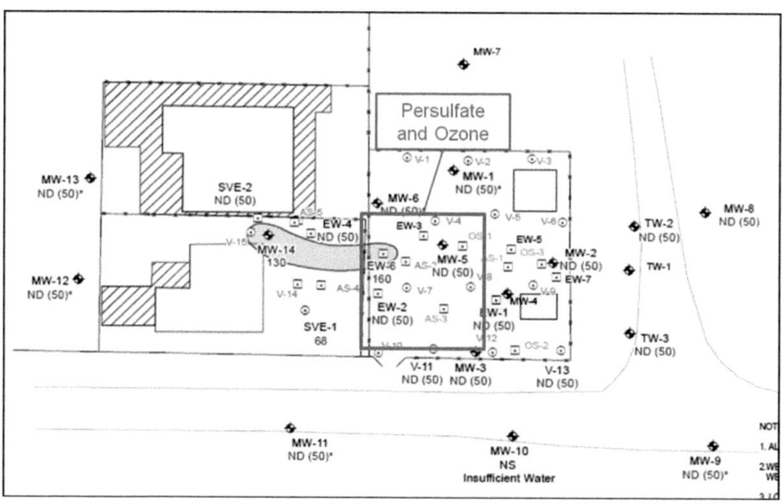

FIGURE 11.3 GRO concentration contour after RPO implementation.

evaluation of sustainability among other balance factors, enhanced in situ bioremediation (EISB), phytoremediation, and MNA were selected to treat source areas, dissolved plume, and plume fringes, respectively (Schulman et al., 2016).

Bench studies and pilot studies were conducted for EISB and phytoremediation, and the results were used to design the full-scale remediation systems. These systems were constructed and operated for more than 3 years. In the source areas, emulsified vegetable oil and bioaugmentation were injected into 15 wells to enhance in situ biodegradation of the chlorinated solvent compounds. In the dissolved plume, a phytoremediation including approximately 550 eucalyptus and willow trees with two solar-powered irrigation stations was applied in a barrier configuration to remove the contaminants, as shown in Figs. 11.4 and 11.5. Eight MNA wells were selected to manage the low-concentration plume fringes.

The remediation systems were implemented, and quarterly/annual monitoring has been conducted to monitor and document remedy performance and compliance since inception. Field measurements of neutral pH, negative ORP, and low DO (<1 mg per liter) indicate that anaerobic conditions conducive to biodegradation of PCE have been induced in the treatment zone. The concentrations of PCE and its daughter products have decreased significantly to nondetection levels in the source area treatment zone since the EISB implementation, indicating that the EISB system is successfully destroying contaminant mass in situ. Order of magnitude increases in microbes and functional gene populations support the conclusion that active biodegradation is occurring in the EISB treatment zone. Phytoremediation plot trees are growing successfully with low mortality rates (<5%) despite high soil salt concentrations.

FIGURE 11.4 Solar-powered irrigation system.

FIGURE 11.5 Phytoremediation.

4.5 Petroleum hydrocarbons site, California, USA

The natural gas processing site is impacted with TPH and BTEX in both soil and groundwater. Maximum impact was detected in the smear zone. An SVE and an AS system were implemented to remove TPH in soil and groundwater, respectively (Fig. 11.6). After gasoline-range organics (GRO) and light-end diesel-range organics (DRO) were removed, the AS/SVE system reached its technology limit using volatilization mechanism. To expedite the removal of

FIGURE 11.6 Soil vapor extraction and air sparging systems.

the remaining heavy-end DRO, the AS/SVE system was optimized to operate as a biosparging/bioventing system focusing on biodegradation mechanisms. Following system optimization, the site is nearing no-further-action status (Leu et al., 2015).

An SVE system was implemented to remove COCs in the vadose zone when the groundwater elevation declined, resulting in an exposed smear zone. The SVE system consisted of six SVE wells and three 2000-pound GAC vessels for vapor treatment. Two years later, an AS system was implemented to remove COCs in groundwater. The AS system consisted of two AS wells and four groundwater monitoring wells. Field observation indicated that the cumulative COC removed by SVE reached an asymptotic level. An early breakthrough of the GAC vessels was diagnosed as low-carbon-affinity polar compounds generated from DRO degradation. GRO was not detected in groundwater monitoring wells. These observations indicate that biodegradation mechanisms gradually dominated the COC removal. The AS/SVE system was optimized to a more sustainable biosparging/bioventing system by optimizing AS flow rate to sustain air required for biosparging in the saturated zone and shutting down low-concentration SVE wells to allow air migration through the vadose zone for bioventing.

Optimization to biosparging/bioventing system from existing AS/SVE system reduces a significant SVE O&M cost, lowers energy requirements, and is more cost-effective in removing heavy parts of DRO through biodegradation mechanisms. After operating the SVE system for 2.5 years, the average VOC concentrations were reduced by approximately 75% in the vadose zone. After 9 months of AS operation, the groundwater DRO concentrations decreased by approximately 70%.

4.6 Summary of case studies

The RPO recommendations and GSR benefits of the aforementioned case studies are summarized in Table 11.3. RPO focuses on minimizing life-cycle cost to achieve cleanup goals, and GSR focuses on maximizing environmental, social, and economic benefits from cleanup action over system lifetime. GSR benefits are derived from RPO recommendations through continuous RPO evaluations over the project life cycle.

5. Conclusions

RPO applies the principles of value engineering to the implementation components of soil and groundwater remediation projects. However, continuous improvement is a key feature of RPO, and conditions at remediation sites change over time, so the application of RPO only once or twice in the life of a 30-year project will likely miss opportunities to reduce costs or hasten site closure. More complex RPO requires more detailed up-front risk–benefit

TABLE 11.3 RPO results in GSR practices.

RPO recommendation	GSR benefit
Use passive wind-driven bioventing.	Lower energy consumption, lower waste generation, lower GHG emission.
Reduce long-term monitoring well number and sampling frequency.	Reduce energy consumption, reduce waste generation.
Turn off GWETS and AS/SVE when technology limits are reached.	Reduce GHG emission, reduce energy consumption.
Use solar energy for remote irrigation.	Use renewable energy source, lower GHG emission.
Optimize from AS/SVE to biosparging/bioventing.	Reduce energy consumption, reduce air emissions.

analyses to determine if implementation of the RPO is feasible. And because regulatory approval is an important component of the RPO process, criteria for evaluating RPO opportunities must include the feasibility of obtaining such approval balanced with the potential cost incurred.

RPO can indicate the need for alternative methods of achieving site cleanup goals. If RPO suggested that GWETS may not be the most effective solution in removing the COC mass, an interim remedial action was advocated for source removal. The need to augment the existing system would not have been detected without the RPO investigation. Alternative cleanup measures, such as the interim remedial action, can have substantial up-front capital costs but can reduce costs over the life of the project by shortening the time to achieve the site cleanup goals. However, there is no guarantee that implementation of the measure would be successful, just as there is no guarantee that the original method would meet cleanup goals in the anticipated timeframe.

Groundwater monitoring presents multiple opportunities for optimization, including removing wells from the sampling program, reducing sampling frequency, and reducing the analysis suite to include only those compounds that continue to be detected in that part of the site at concentrations above maximum contaminant levels. Based on the case studies, the costs for conducting an RPO analysis for groundwater monitoring range between 2% and 14% of annual groundwater monitoring costs, and once implemented, the RPO can result in considerable cost savings.

The cost savings from implementing the measures recommended by the RPO analysis vary with site-specific conditions rather than project size. However, the case studies indicate that the cost savings exceeded their corresponding RPO implementation costs several-fold in the first year after implementation. O&M costs include multiple components, such as laboratory costs, discharge fees, and labor for sampling and maintenance. These vary among sites as a percentage of total costs. Components that account for the largest segments of the total O&M cost often represent the largest opportunities for cost savings through RPO.

RPO requirements and programs are highly variable and site-specific. The optimization addresses specific activities programmatically. Optimization requires baseline determination, alternatives evaluation, and technical/cost analyses. RPO cost savings generally exceed 20% of life-cycle cost. A life cycle cost−benefit analysis needs to be conducted for the RPO recommendations before implementation.

GSR should be considered during the site investigation, remedy evaluation and selection, remedy implementation, O&M, and monitoring activities. RPO provides an opportunity to incorporate GSR to the existing remedial system. GSR approaches can be incorporated into three phases of RPO. RPO evaluations should consider green/sustainability aspects when developing RPO recommendations to maximize environmental, economic, and social benefits, to minimize life−cycle cost, and to expedite site closure. Case studies indicate that RPO recommendations can result in GSR benefits. RPO is an effective way of reducing the life-cycle environmental footprint of remedial operations, thus making remediation greener and more sustainable.

RPO focuses on minimizing life-cycle cost of environmental remediation, which usually does not consider the cost associated with social and economic benefits from GSR. A systematically social and economic cost estimate from GSR would help expand RPO benefits and improve life-cycle cost of RPO recommendation.

References

Air Force Center for Engineering and the Environment (AFCEE), 2001. Remedial Process Optimization (RPO) Handbook.

Air Force Center for Engineering and the Environment, 2009. Sustainable Remediation Tool. http://www.afcee.af.mil/resources/technologytransfer/programsandinitiatives/sustainableremediation/srt/index.asp.

Barcelona, J.M., Wehrmann, H.A., Schock, R.M., Sievers, E.M., Karny, R.J., 1989. Sampling Frequency for Groundwater Quality Monitoring. EPA/600/S4-89/032.

Chadalavadal, S., Datta, B., 2008. Dynamic optimal monitoring network design for transient transport of pollutants in groundwater aquifers. Water Resources Management 22 (6).

Dominguez, R., Leu, J., Bettahar, M., 2012. Sustainable wind-driven bioventing at a petroleum hydrocarbon-impacted site. Remediation: The Journal of Environmental Cleanup Costs, Technologies, & Techniques 22 (3).

Hou, D., Keith, R., Leu, J., et al., 2009. A comparative analysis of remedial process optimization. In: 19th Annual AEHS Meeting and West Coast Conference on Soils, Sediments, and Water, March.

Hou, D., Leu, J., 2009. Optimizing the remedial process at a petroleum hydrocarbon contaminated site using a three-tier approach. Journal of Environmental Engineering 135 (11), 1171. ASCE.

Hudak, P.F., Loaiciga, H.A., 1993. An optimization method for monitoring network design in multilayered groundwater flow systems. Water Resources Research 29 (8), 2835–2845.

Interstate Technology & Regulatory Council (ITRC), 2007. Improving Environmental Site Remediation Through Performance-Based Environmental Management. Interstate Technology & Regulatory Council, Remediation Process Optimization Team, Technical and Regulatory Guidance, RPO-7. Washington, D.C. http://www.itrcweb.org/Documents/RPO-7.pdf.

Johnson, M.V., Tuckfield, R., Ridley, N.M., Anderson, R.A., 1995. Reducing the sampling frequency of groundwater monitoring wells. Environmental Science & Technology 30 (1), 355–358.

Kim, K.-H., Lee, K.-K., 2007. Optimization of groundwater-monitoring networks for identification of the distribution of a contaminant plume. Stochastic Environmental Research and Risk Assessment 21 (6), 785.

Leu, J., Lin, J., Morales, M., Ferris, S., 2015. Transition to sustainable biosparging/bioventing from air sparging/SVE at a petroleum hydrocarbon-impacted site. In: 3rd International Symposium on Bioremediation and Sustainable Environmental Technologies, Miami, Florida. May.

Leu, J., O'Connell, S., Bettahar, M., 2016. Remedial process optimization and ozone sparging for petroleum hydrocarbon-impacted groundwater. Remediation: The Journal of Environmental Cleanup Costs, Technologies, & Techniques 26 (3).

Leu, J., Morales, M., 2016. Upgradient persulfate injection and downgradient ozone sparging to accelerate closure of a petroleum hydrocarbons-impacted site. In: 23rd International Petroleum Environmental Conference, New Orleans, Louisiana. November.

Mahar, P.S., Datta, B., 1997. Optimal monitoring network and ground-water–pollution source identification. Journal of Water Resources Planning and Management 123 (4), 199.

Pollard, S.J.T., Lythgo, M., Duarte-Davidson, R., 2001. The Extent of Contaminated Land: Problems and the Scientific Response. The Royal Society of Chemistry. http://www.rsc.org/ebooks/archive/free/BK9780854042753/BK9780854042753-00001.pdf.

Schulman, M., Griffiths, D., McGlochlin, L., Brown, D., Leu, J., Hou, D., Garon, K., 2016. Using in-situ bioremediation and phytoremediation to treat large chlorinated solvent plumes: a case study of implementing green and sustainable remediation. In: Tenth International Conference on Remediation of Chlorinated and Recalcitrant Compounds. Palm Springs, California. May.

U.S. Army Corp of Engineers (USACE), 2010. Remediation System Evaluation (RSE) Checklists. http://www.environmental.usace.army.mil/rse_checklist.htm.

U.S. Environmental Protection Agency (USEPA), 1988. Guidance for Conducting Remedial Investigations and Feasibility Studies Under CERCLA, Interim Final. EPA/540/G-89/004.

USEPA, 2002. Pilot Project to Optimize Superfund-Financed Pump and Treat Systems: Summary Report and Lessons Learned. EPA 542-R-02-008a. November.

USEPA, 2004. Cleaning Up the Nation's Waste Sites: Markets and Technology Trends, 2004 Edition. EPA 542-R-04-015.

USEPA, 2005. Roadmap to Long-Term Monitoring Optimization. EPA 542-R-05-003. May.

USEPA, 2008a. Green Remediation: Incorporating Sustainable Environmental Practices into Remediation of Contaminated Sites. EPA 542-R-08-002.

USEPA, 2008b. Green Remediation: Best Management Practices for Excavation and Surface Restoration. EPA 542-F-08-012. December.

U.S. Sustainable Remediation Forum (SuRF), 2009. Sustainable remediation white paper: integrating sustainable principles, practices, and metrics into remediation projects. Remediation 19 (3), 5.

Yeh, M.-S., Lin, Y.-P., Chang, L.-C., 2006. Designing an optimal multivariate geostatistical groundwater quality monitoring network using factorial kriging and genetic algorithms. Geo 50 (1), 101.

Chapter 12

Landscape architecture and sustainable remediation

Xiaodi Zheng, Niall G. Kirkwood
School of Architecture, Tsinghua University, Beijing, China; Department of Landscape Architecture, Harvard Graduate School of Design, Harvard University, Cambridge, United States

1. Overview

In recent years, under the background of China's ecological civilization construction, the remediation and regeneration of brownfields have gained unprecedented attention and advancement. "Brownfield" is an important concept put forward by Western scholars in the 20th century. Different countries worldwide have not reached an agreement on the definition of "brownfields," while most emphasize on land redevelopment and potential contamination. In China, brownfields mean abandoned, idle, production-limited industrial or commercial sites, whose expansion or redevelopment becomes complicated due to the risks of existing or potential environmental contamination (China National Committee for Terms in Sciences and Technologies, 2014).

In 2005–13, a national wide soil pollution survey was conducted. According to the bulletin of the national soil pollution survey in 2014, the overall soil pollution status in China is severe, especially in industrial brownfield. The national total point exceeding the standard rate is 16.1%. Most pollutants are inorganic pollutants, mostly heavy metal, accounting for more than 80%. And the highest amount of pollutants is cadmium. Among the 5846 heavily polluted enterprises and the surrounding soil sites, the overstandard points accounted for 36.3%. Since then, China has successively promulgated a series of guidelines and policy documents for the treatment of contaminated sites, which have played an important guiding role in relevant practices, such as <the Technical Guidelines for Risk Assessment of Contaminated Sites> and <the Technical Guidelines for Soil Remediation of Contaminated Sites> and so on. In particular, the introduction of <the 2016 Soil Pollution Prevention Action Plan> and <the Measures for Soil Environmental Management of Polluted Lands (Trial)> has greatly promoted the practice of brownfield regeneration projects. The soil law as adopted at the fifth session of

Sustainable Remediation of Contaminated Soil and Groundwater. https://doi.org/10.1016/B978-0-12-817982-6.00012-4

the standing committee of the 13th National People's Congress of the People's Republic of China on August 31, 2018, was issued with effect from January 1, 2019.

Research and practice on brownfield in China basically started in the early 21st century. Compared with developed countries, China's brownfield remediation and regeneration started late, but it has rapidly developed into the frontier and important topic of practice and research in landscape architecture and environmental engineering. There are three issue with brownfields as the theme in "Chinese Landscape Architecture," namely "Brownfields Restoration" in February 2013 and "Brownfields Regeneration" in April 2015, which were both organized by the first author. Three other issues also focused on similar areas, including "Waste Land Renovation and Reconstruction" in February 2008, "Ecological Restoration" in August 2011, and "Mining Heritage" in July 2012 (Zheng, 2017a,b) and "Brownfield Restoration and Design" in February 2019. According to the three key professional journals of landscape architecture in China, referring to "Chinese Landscape Architecture," "Landscape Architecture," and "Landscape Architecture Frontiers,"[1] the number of articles focused on brownfields regeneration shows an increasing trend, and the research mainly focused on individual brownfields sites. Among the articles of brownfields case study, 77% are industrial brownfields, including vacant lands of industrial and infrastructure facilities and mining wastelands. Moreover, among important professional awards given by CHSLA and ASLA, many were awarded to brownfields projects. Through analyzing the ASLA awards for the past 5 years, it can be found that almost half awards-winning projects are brownfields regeneration efforts (Zheng, 2017a,b).

2. Brownfields regeneration and landscape architecture

2.1 Why and how landscape architecture plays an important role?

Compared to ordinary real estate development, open space is more favored as the end use in brownfields regeneration because it exhibits four distinctive advantages (see Fig. 12.1):

① Open space adapts better to the special site conditions of brownfields. Due to complex site conditions caused by soil and groundwater contamination

---

1. <Chinese Landscape Architecture> was founded in 1985 and hosted by CHSLA, and it is a "Chinese Core Journal" and "Chinese Core Journal of Science and Technology"; <Landscape Architecture> is organized by the Ministry of Education and hosted by BJFU, and it was founded in 1985; <Landscape Architecture Frontiers> is organized by the Ministry of Education, hosted by Higher Education Press and Peking University, and prepared by the School of Architecture and Landscape Design of Peking University, and it(bimonthly) was founded in February 2013.

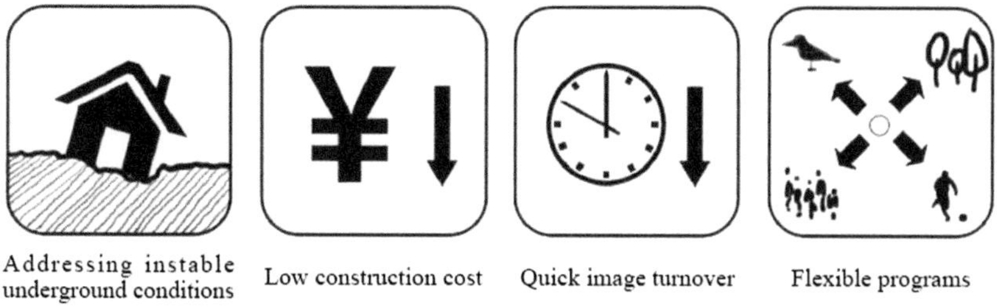

FIGURE 12.1 Four advantages of landscape as end use for brownfields regeneration.

and potentially unstable underground situation, often times buildings are difficult, or impossible in some cases, to be constructed in brownfields. In many cases, open space may become the only choice for end use.

② The construction cost of landscape projects is relatively low.

③ Landscape projects have the advantages of quick returns and short cycles, and thus are favored by the local governments or related stakeholders to present the positive image of brownfields regeneration.

④ Open spaces are more flexible to adapt to various programs and the long-term remediation process. As remediation work progresses, the bearing capacity and risk level of the site would be changing, which in turn would allow for different transitional and/or permanent programs. Open spaces can more easily adapt to this dynamic process (Zheng, 2017a,b).

It is not a coincident that landscape architecture becomes one of the leading disciplines to focus on brownfields regeneration. The fact that landscape architecture is able to address the complicated issues involved in the process of brownfields regeneration from all three aspects of ecological process, social humanities, and physical environment explains this phenomenon (see Fig. 12.2):

① Brownfields regeneration involves complicated and dynamic ecological processes, so it is necessary to take into full consideration of various ecological elements, such as water, vegetation, animals, and habitats.

② Brownfields regeneration is inseparable from social and cultural activities. It shall strive for environmental justice through public participation process.

③ Brownfields regeneration will eventually be implemented through transforming and shaping the physical environment, so as to regain vitality and increase value.

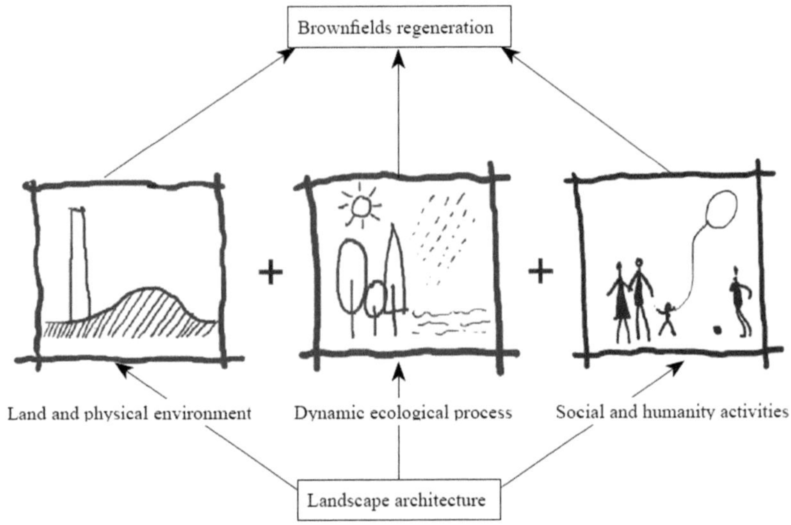

FIGURE 12.2 Key aspects of brownfields regeneration tackled by the discipline of landscape architecture.

Landscape architecture is one of the few disciplines to practice and research by synthesizing all three aforementioned aspects, which is precisely the reason why landscape architects often play a leading role in brownfields regeneration projects (Zheng, 2017a,b).

Brownfields regeneration is a complicated and comprehensive subject, which is impossible to be tackled by any one single discipline, but involves at least the following fields: law, politics, economy, ecology, environmental engineering, landscape architecture, and social humanities. Among them, landscape architecture plays an important role as a "bridge" and coordinator to connect principles and technologies of science and engineering disciplines, such as environmental science, ecology, hydrology, and geotechnical engineering, with theoretical knowledge of liberal arts disciplines, such as public policy, economic management, and sociology. Eventually, inputs from all disciplines shall be synthesized through landscape planning and design process and reflected in the physical space formed (see Fig. 12.3) (Zheng, 2017a,b).

2.2 Landscape practice in brownfield regeneration

The practice related to the remediation and regeneration of brownfields worldwide has made remarkable achievements in the past 10 years. The practice is mainly based on three types of sites: industrial and infrastructure abandoned land, mining wasteland, and landfills.

Industrial and infrastructure abandoned land refers to abandoned, idle industrial and infrastructure land, including industrial and mining enterprises'

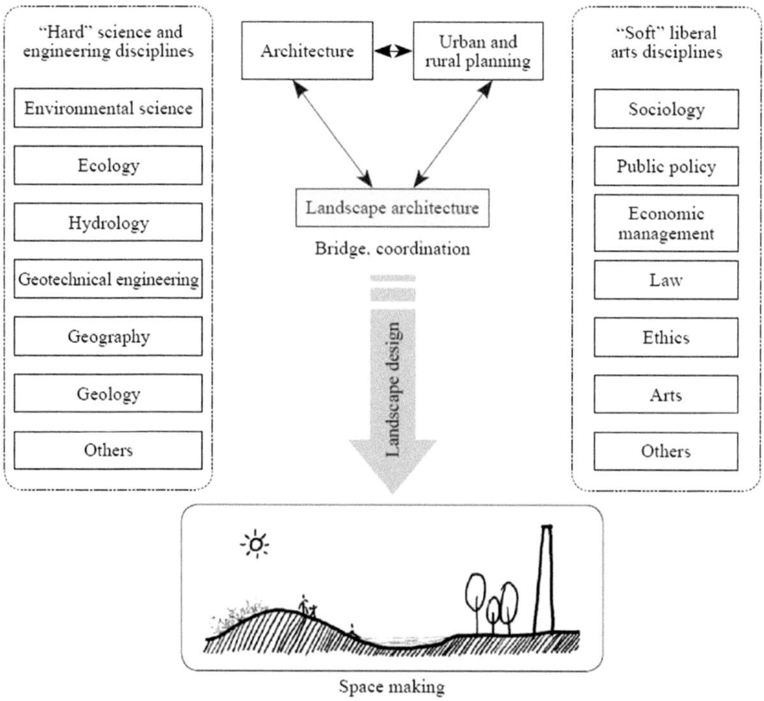

FIGURE 12.3 The role of landscape architecture in brownfields regeneration.

production workshops, warehouses and their ancillary facilities, railways, gas stations, terminals, airports, and other land. Such brownfields have the characteristics of flat terrain and little change in elevation on the site features. According to the target use of such brownfield reuse, it can be divided into cultural and creative industrial parks, industrial heritage parks, and urban comprehensive development zones (see Fig. 12.4).

Mining wasteland refers to "land that cannot be reused due to mining activities destroyed and occupied, before rectified, including bare mining rock, waste soil (stone, slag) heap, coal gangue heap, tailings pond, abandoned plant and so on" (Wang, Zhang, 2015), the mining subsidence area, and various wasteland sites with potential dangers around the mine are also included. According to the degree of regeneration of mining wasteland and the main characteristics of development, the practice of mining wasteland could be divided into three types: land reclamation, mining parks, and urban parks (see Fig. 12.5).

As the city continues to expand, landfills become an indispensable infrastructure for the city. However, as many old landfills have been designed to be replenished or abandoned due to the construction of new waste treatment facilities, some landfills have gradually entered the stage of closure. Some of the sites, with the cooperation of environmental engineering and landscape majors, gradually evolved into a popular space for the public. According to the functional orientation of the site after renovation, the brownfield regeneration of landfills in China is mainly divided into the following four categories: Expo Parks, sports and recreational parks, country parks, and ordinary urban parks (see Fig. 12.6).

3. Brown earth-work

3.1 Concept

Despite the complication and challenges of contamination and remediation, there are many successful landscape projects in brownfields regeneration. Although these sites are all brownfields, the landscape strategies for their regeneration are either identical or dramatically different. Thus a question is raised: could there be a common core element among various brownfields, which may decisively influence the choice of landscape regeneration strategies and the final spatial morphology of the site? The answer is YES, and this core element is identified as "Brown Earth-Work" by the author.

Brown Earth-Work refers to the contaminated (or potentially contaminated) soil and other soil-like substances in brownfields, including

FIGURE 12.4 (1) cultural and creative industrial parks—Guangzhou Red Brick Factory, Guangzhou, China,[2] (2) cultural and creative industrial parks—Beijing 798 Art District, Beijing, China,[3] (3) industrial heritage parks—Zhongjiang Qijiang Park, (4) industrial heritage parks——Nord Duisburg Landscape Park, Duisburg, Germany, (5) industrial heritage parks—Xiamen Railway Cultural Park, Xiamen, China,[4] (6) urban comprehensive development zones——Capital Steel Park, Beijing, China. *(3) Photograph by Zheng Xiaodi, (4) Photograph by Fu Quanchuan, (6) Photograph by Fu Quanchuan.*

2. 百度图片
3. 百度图片
4. 百度图片

FIGURE 12.4 cont'd

FIGURE 12.4 cont'd

contaminated soil, slags, tailings, wastes, and incineration ashes, etc., whose contamination shall be investigated, assessed, and remediated in the process of brownfields regeneration.

The most essential feature that distinguishes brownfields from general sites is the "contamination" existing or potentially existing on site. In the fields of environmental engineering and chemistry, "contamination" is often presented in forms of complex chemical formulas or heavy metal elements, which indicates the abstract existence of pollutants, without recognizing the material existence of pollutants in physical environment. In reality, pollutants or contaminants are attached to material carriers in the physical environment, which may be gradually reduced along with the remediation process; however, is difficult to be completely eliminated. For example, even if domestic waste is treated by incineration, there are still ashes left to be disposed of; even if contaminated soil is remediated by physical or chemical methods, there would still be residual materials; and the body mass of contaminated soil treated with phytoremediation would not be reduced at all.

This material carrier of "contamination" is Brown Earth-Work. Identifying and coining this new term is not to further complicate the already complex brownfields regeneration process, but to provide the spatial interpretation of site contamination, to clarify and spatialize the core element in current practice of brownfields regeneration, and to unify both the abstract and the material existence of contamination in three-dimensional physical environment (see Fig. 12.7) (Zheng, 2017a,b).

3.2 Brown earth-work in landscape architecture and environmental engineering

Among the disciplines involved in brownfields regeneration, landscape architecture and environmental engineering are the two that have decisive influences over the physical morphology of brownfield sites in practice. The former is to create a sustainable ecosystem and an attractive space, while the latter is to clean up pollutants through remediation. In one single project, these two disciplines have impacts on the same site and influence over each other despite their different objectives (Zheng, 2017a,b).

The landscape architecture system consists of three basic natural elements: topography, water, and vegetation. The construction of the landscape

FIGURE 12.5 (1) land reclamation—Land Reclamation of Shuozhou Mining Area, Shuozhou, China, (2) land reclamation—Nine Miles Run, Pittsburgh, USA, (3) mining parks——Huangshi National Mining Park, Huangshi, China, (4) mining parks—Xuyi Xiangshan National Mining Park,[5] (5) urban parks——Tangshan South Lake Park, Tangshan, China, (6) urban parks—Lake Park Lünen, Lünen, Germany. *(1) Photograph by Niall Kirkwood, (2) Photograph by Fu Quanchuan, (3) Photograph by Fu Quanchuan, (5) Photograph by Zheng Xiaodi, (6) Photograph by Fu Quanchuan.*

5. 百度图片

FIGURE 12.5 cont'd

FIGURE 12.5 cont'd

architecture system in brownfield regeneration is closely related to the "Brown Earth-Work" (see Fig. 12.8):

① "Brown Earth-Work" and topography: regardless of the method used to control brown earthwork, the amount of earthwork will significantly affect the shaping of the site's spatial topography; and the requirement to absorb brown earthwork in the site often becomes an opportunity to shape the terrain.

② "Brown Earth-Work" and water: when the rainwater flows through the brown earth, it will carry the pollutants, leading to pollution of the downstream water system and the groundwater body; therefore, the design of the water system in the brownfield needs to cut off the contact between the rainwater and the brown earth, combined with the sustainable rainwater collection and purification and recycling system.

③ "Brown Earth-Work" and vegetation: the brown earth in the repair process still contains certain harmful substances, and only certain species can grow on it; and even if the soil is covered by the isolated brown earth, the thickness of the soil layer is limited, as a result, the species planted should be specially selected according to the specific conditions.

According to the specific pollutants attached to the "Brown Earth-Work," it could reflect to the corresponding alternative remediation technology in environmental engineering (see Fig. 12.9). Due to the complexity of the original industrial activities and site use on the brownfield, the pollutants in the site are often presented as multiple types of mixed pollution, which requires a variety of different repair techniques for pollution control.

3.3 Contaminated characteristics and spatial characteristics

The composition of brown earth in industrial and infrastructure brownfields is dominated by contaminated soil and may also be accompanied by industrial waste, construction waste, and other materials. According to different presite uses and sources of pollution, the brown earth in the brownfields has different pollution components. In comparison, the industrial brownfields are relatively heavily polluted, and the infrastructure brownfields are relatively less polluted. In terms of spatial characteristics, the sites where industrial and infrastructure are idle are generally flat and there is no dramatic height difference (see Chart 1).

The brown earth in the wasteland of the mining industry is dominated by tailings, waste rock, and slag, and will also contain contaminated soil.

FIGURE 12.6 (1) Expo Parks—Xi'an World Horticultural Exposition 2011[6] (2) sports and recreational parks—Yingshan County Sports Park, Nanchong, China,[7] (3) country parks——Tianziling Ecological Park, Hangzhou, China, (4) ordinary urban parks——Fresh Kills Park, New York, USA. *(3) Photograph by Zheng Xiaodi, (4) Photograph by Zheng Xiaodi.*

6. 百度图片
7. 百度图片

FIGURE 12.6 cont'd

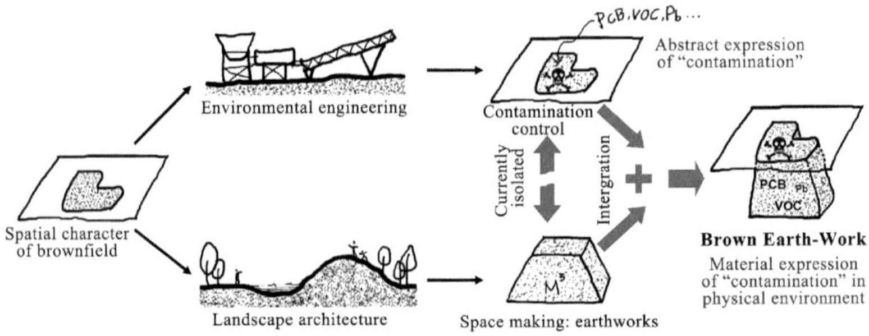

FIGURE 12.7 The concept of "Brown Earth-Work."

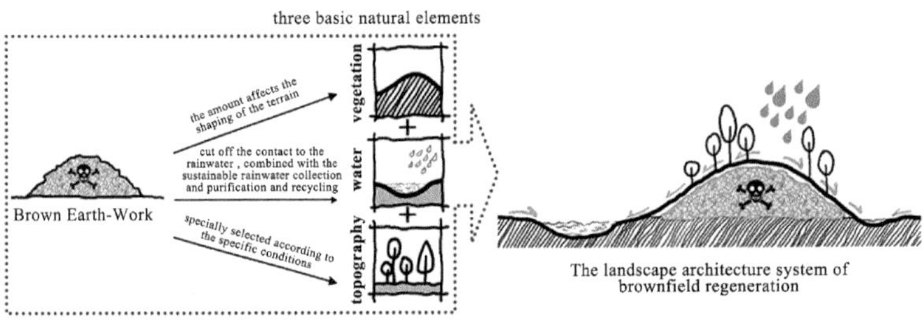

FIGURE 12.8 "Brown Earth-Work" and the construction of the landscape architecture system.

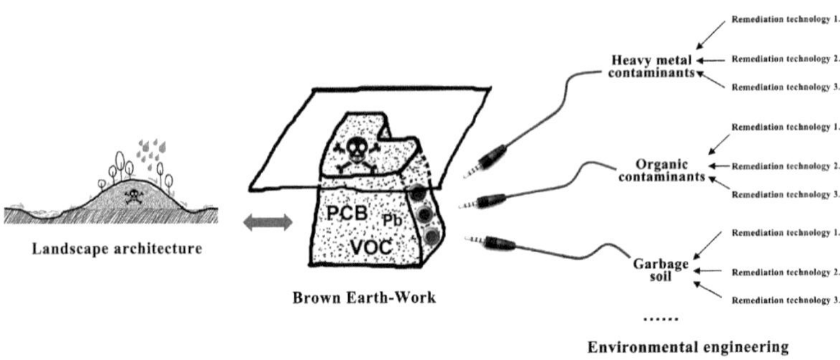

FIGURE 12.9 "Brown Earth-Work" and environmental engineering.

Abandoned mining pits are often used as dumping grounds, so brown earth-work may also contain domestic waste, construction waste, and industrial waste. Among the various types of mining wasteland, the pollution of the earth and gravel mining site is relatively light, while the coal mines and metal mines wasteland contain many heavy metal pollution or acidic, alkaline, and toxic pollutants. In terms of spatial characteristics, huge mining pits are the most unique spatial elements in the wasteland of the mining industry. The terrain changes drastically, which is both a challenge and a unique opportunity for regeneration. The tailings pond and dumping site are mostly mountain-shaped or terraced. Underground mining activities have caused mining subsidence areas (see Chart 2).

The brown earthwork of a landfill is rubbish, that is, solid waste. From the perspective of geotechnical engineering, it can also be called garbage soil. Garbage soil refers to a new special soil formed by mixing municipal solid waste and its covered soil dumped in a landfill; and it is a granular structure containing degradable components and fibrous structure reinforcement (Yan, 2004). Different solid wastes have different pollution characteristics. Medical waste, hazardous waste, and toxic and hazardous industrial waste need to be transported and stored separately and treated specially. It is not uncommon for these special treatment sites to be used for new purposes. The landfill sites for domestic waste and construction waste are the main objects we discuss (see Chart 3).

4. Sustainable remediation combined with design

4.1 Landscape architecture strategies

Based on the risk assessment, in order to avoid the harm of brown earth to human health, the simplest and straightforward method is to isolate the brown earth to prevent the public from contacting and using. When the activities of humans in the brownfield regeneration overlap with the scope of the brown earthwork, two specific means can be adopted to achieve the purpose of isolation: one is to overhead the space of human activities to isolate brown earthwork form people, referred to as "overhead"; the other is to cover the brown earth with insulation material, referred to as "covering." When brown earthwork is heavily polluted, it cannot be effectively controlled by isolation, or there is not enough space in the site to isolate, or "isolation" is not the best choice from the perspective of economy and project cycle, the brown earth can be "remediated." Based on whether the brown earthwork is moved during the remediation process, the remediation strategy can be divided into two types: "in situ" and "ex situ." Isolation and remediation strategies work in industrial and infrastructure brownfields regeneration, while isolation, remediation, backfilling, and comprehensive utilization strategies work in mining brownfields regeneration, and at the same time, isolation and reduction strategies work in landfill brownfields regeneration. Furthermore, combined with landscape architecture system construction, different types of brownfield have different regeneration strategies (see Fig. 12.10).

There are so many brownfield regeneration practices that could be interpreted by landscape strategies based on "Brown Earth-Work"; some present similar spatial characteristics, such as the rolling hills of the Kite Mountain in Seattle Gasworks Park in the United States, the conical landmark mountains in Sydney Olympic Games Millennium Park in Australia, and the 10 m-high

CHART 1

	Industrial brownfields	Infrastructure brownfields
Degree of pollution "Brown Earth-work"	Heavy Mainly consists of contaminated soil, and maybe industrial waste, etc.  	Light Mainly consists of contaminated soil 
Spatial characteristics	Huge industrial buildings with flat terrain  	Flat terrain 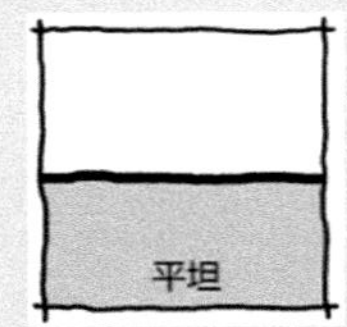

CHART 2

	Coal mining and metal mining wasteland	Gravel mining wasteland
Degree of pollution "Brown Earth-work"	Heavy Mainly consists of tailings and waste rock and maybe some construction waste, industrial waste, or domestic waste; and the soil is heavily polluted by heavy metals 	Light Waste rock, dumped garbage, and waste 
Spatial characteristics	Concave or depressed sites; tailings ponds and dumping sites are mostly mountain-shape or platform-shape 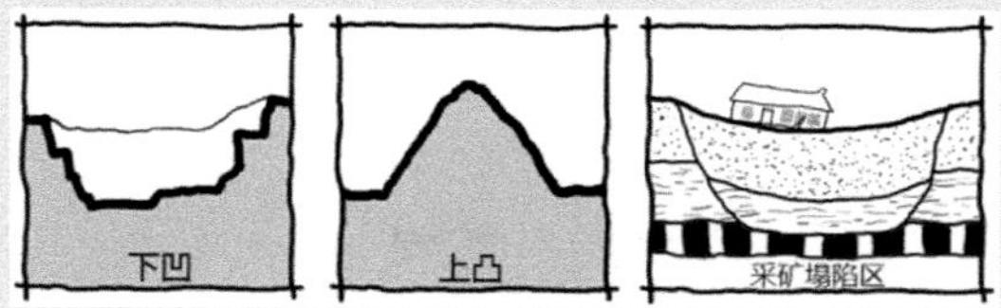	Concave 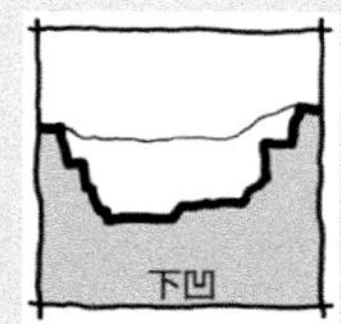

CHART 3

<table>
<tr><td></td><td>Domestic waste</td><td colspan="2">Informal landfill</td></tr>
<tr><td rowspan="2">Degree of pollution
"Brown Earth-work"</td><td rowspan="2">Heavy(with control)
Domestic waste (garbage soil)

</td><td>Heavy</td><td>Light</td></tr>
<tr><td>Mainly domestic waste

</td><td>Construction waste

</td></tr>
<tr><td>Spatial characteristics</td><td>Mountain-shaped or concave with leachate and biogas drainage treatment facility
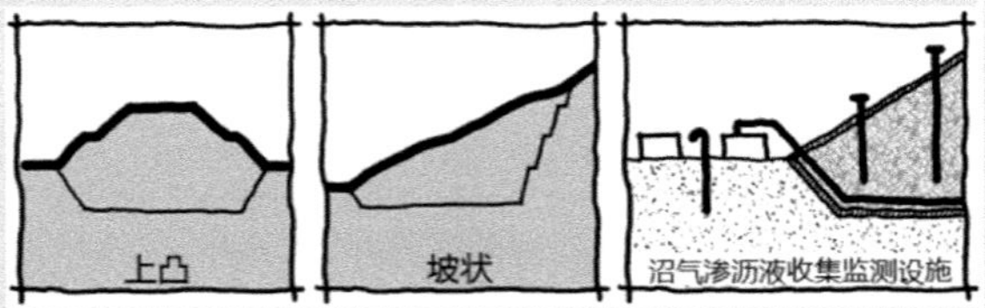
</td><td colspan="2">Mountain-shaped or concave
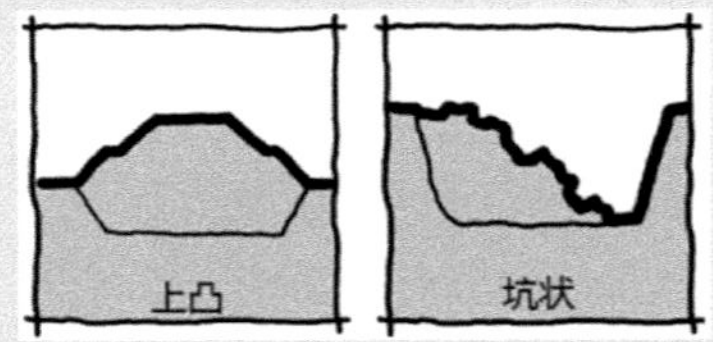
</td></tr>
</table>

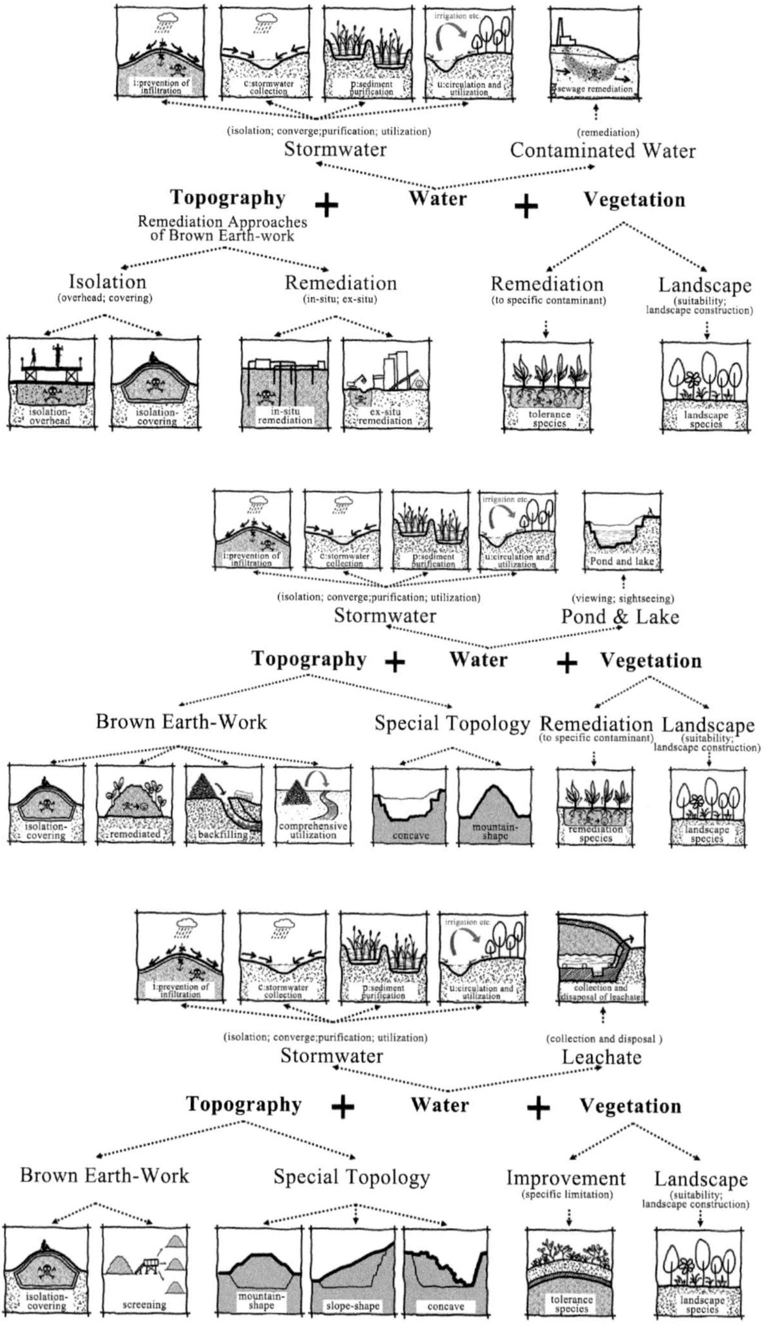

FIGURE 12.10 industrial and infrastructure brownfield regeneration and landscape architecture system construction, 2 mining brownfield regeneration and landscape architecture system construction, (landfill brownfield regeneration and landscape architecture system construction.

circular hill in Minneapolis Gold Medal Park in the United States; some projects appear very differently, such as the huge elevated hardwood platform in Civita Park of Sao Paulo in Brazil and the inaccessible bunker gardens in Duisburg-Nord Park in Germany (Zheng, 2017a,b).

The Seattle Gasworks Park is located on the north shore of Lake Union in Seattle, USA. It covers an area of 9 ha. It was used as a gas production site for the former Seattle Gas Works and equipment storage space for the Gulf of Telecommunications in the 20th century. In 1970, the design and regeneration began. When the gas plant was operated, a large amount of pollutants was discharged into the environment. After the transformation into a park, the environmental survey and pollution control process was tortuous and long. It is still in progress, mainly adopting the treatment strategy of isolation and in situ remediation. In order to control the pollution still existing in the underground 18m, the project finally used biodegradation and deep cultivation methods to repair the contaminated soil (Sun and Liu, 2004), and covered with new soil at the surface. After the demolition of industrial buildings, the concrete construction waste left over from the site was piled up into a mountain about 14m high, becoming Kite Hill overlooking the Seattle skyline. Because the thickness of the overlying soil is limited after the brown earth is covered, the vegetation in the park is dominated by lawn (see Fig. 12.11).

After opening to the public at the first time in 1976, the Seattle Gasworks Park experienced two closures and regovernance. Its operations also fully reflected the brownfield management approach based on risk assessment. The

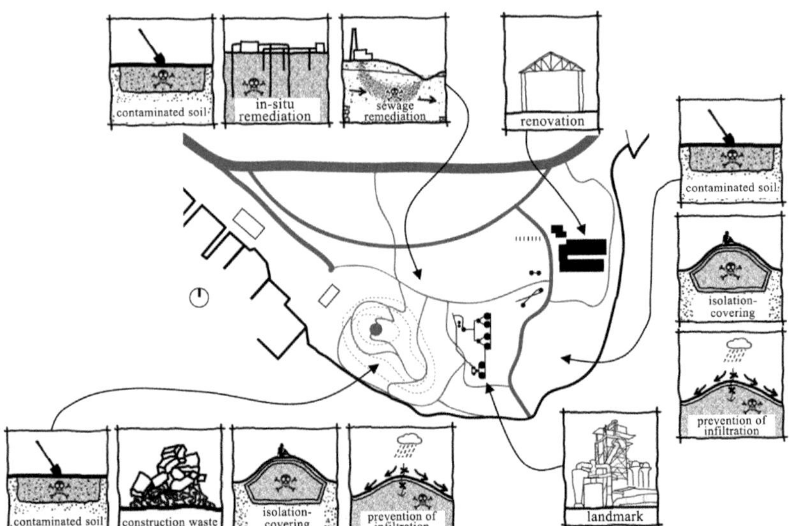

FIGURE 12.11 Seattle Gasworks Park and "Brown Earth-Work" analysis.

FIGURE 12.12 Seattle gasworks park, Seattle, USA. *Photograph by Zhao Wei.*

premise that the park is open to the public is not that the pollution in the site has been completely eliminated, but to clarify how long and how often the use intensity will cause fatal harm to human health. The park is reopened to the public when it is proven that the park is not contaminated heavily enough to be harmful to human health. However, certain function is conditional, requiring the public not to have direct access to the soil and water bodies in the site. The design of the park makes full use of the existing resources of the site. The original industrial equipment has been preserved. The sturdy industrial aesthetics and the undulating natural landscape of the waterfront are both a commemoration of the industrial age and a good public open space (see Fig. 12.12).

4.2 Other sustainable remediation strategies

Despite typical landscape strategies based on "Brown Earth-Work," there still exist some other sustainable remediation strategies in brownfield regeneration, for example, industrial nature in Natur-Park Südgelände, staged transformation in Fresh Kills Park, phytoremediation in PHYTO GARDEN of the International Horticultural Expo in 2016 in Tangshan.

The theme of PHYTO GARDEN denotes the need to bring life, beauty, and people to distressed and polluted land through landscape design. Plants are not only ornamental but can also be used for remediation of pollutants and crop production, such as biomass energy plants, papermaking plants, and timber production plants. These plants not only create economic value but contribute to the development of the city and the health of the region at the same time. The design used a number of the existing site elements including concrete pads and ground surfaces. The intention was to demonstrate that former sites,

FIGURE 12.13 Phyto Project, Expo 2016, Tangshan, Hebei Province, China. *Photograph by Niall Kirkwood (2016).*

however benign could contain unhealthy and poisonous soils and groundwater but that we can remain optimistic about their recovery. A series of planting typologies were used with a range of plant species. The typologies will include aspects of the following: planted stabilization mats, groundwater migration tree stands, interception hedgerows, degradation bisques, degradation hedge and living fence, and extraction plots (see Fig. 12.13).

There is a great need to repair the soils, groundwater, and subsurface conditions of these places. Some hyperaccumulator and remediation plants are featured in the garden that have the ability of extraction, sequestration, and destruction of common industrial pollutants in soil and groundwater. These plants not only provide beautiful visual and aesthetic enjoyment, but also can improve the polluted urban land, acting as a boundary between the polluted area and the nonpolluting area.

The PHYTO GARDEN portrays the power and ability of plants to extract, sequester, and destroy industrial pollution commonly found in soils and groundwater. Visitors to the Garden Project can walk through a series of rooms that focus on the major toxins and see the range and beauty of horticultural species who have "special properties" that allow them to transform contaminated urban land and provide a more sustainable choice for land planning and neighborhoods. They can also be used to help buffer the noncontaminated site, where the risk of pollution could be possible, create fuel as biomass renewable energy, and participate in ensuring food crop safety. The author showed how plants can be more than decoration and that they can assist cities, towns, and districts in a productive way to make the environment safe for all citizens.

5. Sustainable development and prospects for the future

While in China, not only effective collaboration between professionals of landscape architecture and environmental engineering is still limited in practice, but also the consensus on the necessity of such collaboration has not yet been established. A well-known expert in soil remediation that the author interviewed even stated frankly that "there is no need for cross-disciplinary collaboration between environmental engineers and designers, since there are just two scenarios in practice to deal with this issue: one is that designers are to design the site after it has been remediated by environmental engineers, the other is that environmental engineers are to remediate the site in accordance with the plan made by designers." This kind of view is not uncommon among environmental engineering professionals and quite some interviewees even doubt about why landscape architects shall acquire any environmental engineering knowledge.

In the field of landscape architecture, the situation is not optimistic either. The research and practice of brownfields regeneration in China are mostly carried out in the past 10 years. Although there gradually is a consensus on the necessity and significance of brownfields regeneration within the field of landscape architecture, most landscape architects have little or no knowledge of the seriousness and complexity of contamination, the dynamic and long-term nature of site remediation, and the spatial impacts of specific remediation technologies on the physical site. Many projects are only to transform brownfield sites from "brown" to "green" in appearance and take it as the only target and standard to evaluate whether the transformation is successful or not. Landscape architects who lack related environmental engineering knowledge are unable to engage in a dialogue with environmental engineers working on the same project, let alone to achieve an effective and close collaboration.

The number of brownfields regeneration projects increased drastically in China in recent years. More and more landscape architects are involved in this field. Thus, it is important and urgent to clarify the critical relationship between site remediation and landscape design during the process. Only by clearly understanding the subject can we develop an efficient system to further guide practice (Zheng, 2017a,b).

In the 2016 Brownfields Conference held in Tsinghua, almost 300 scholars, experts, industry leaders, government officials, and students gathered together, reaching a consensus that brownfields regeneration is a complex issue, which calls for participation of different disciplines and different government sectors, and calls for visions from larger perspective beyond a single site. In 2018, the second International Conference on Brownfields Regeneration and Ecological Restoration was held in BEIJING. The conference works closely with the China Sustainable Environmental Remediation Conference with the support from China Association of Environmental Protection Industry, aiming to further strengthen the communication and collaboration between landscape architecture and environmental engineering in the process of brownfield regeneration so as to better respond to the challenges of sustainable development and ecological civilization construction. By this way, we are trying to establish a platform for landscape architects and environmental engineers to communicate with each other, and to develop an efficient system to guide practice in a more sustainable way together.

References

China National Committee for Terms in Sciences and Technologies, 2014. Chinese Terms in Architecture [Z].

Sun, X., Liu, X., 2004. Constructing back-to-nature spiritual home-contemporary American great landscape architect richard haag. Journal of Chinese Landscape Architecture 20 (3), 11–15.

Wang, J.P., Zhang, Y.C., 2015. From wasteland to the wonderful land: the ecological restoration and landscape reconstruction of chalk mine land in suburb. Journal of Chinese Landscape Architecture 31 (04), 20–24.

Yan, S., 2004. Research on the Engineering Properties of Municipal Solid Waste [D]. Institute of Rock& Soil Mechanics, The Chinese Academy of Sciences.

Zheng, X.D., 2017a. "Brown earth-work": a new landscape paradigm and research approach for brownfields regeneration. Journal of China City Planning Review 26, 31–39.

Zheng, X.D., 2017b. Upsurge, misconception, and prospect: brownfields regeneration in landscape architecture in China. Journal of Chinese Landscape Architecture 33 (05), 10–14.

Chapter 13

Phytoremediation value chains and modeling

Thomas Upcraft, Miao Guo
Department of Chemical Engineering, Imperial College London, London, United Kingdom

1. Introduction

In this chapter, a new concept is presented: phytoremediation value chains. Value chains are composed of a full set of functional value-adding activities, incorporating resources, processes, and technologies that produce and distribute commodities. Phytoremediation value chains represent a complex system, where plant species interlink with environmental variables, including soil metal contamination. Biorefining is integrated in the latter stages of the value chain to further convert plant species into value-added products. This chapter will discuss the complexity and cross-disciplinarily of this topic, while highlighting the emerging research gaps, which can be bridged by systems modeling approaches at the science—engineering interface. Phytoremediation can be applied for soil and groundwater remediation. Both applications offer potential to be integrated with biorefining under value chain concept. This chapter has a particular focus on soil heavy metal remediation.

In contrast to phytoremediation, biorefinery value chains have been widely studied. Biorefinery value chains concern multisite operational decisions on biorefinery and distribution, including upstream biomass cultivation and collection, pretreatment and conversion, downstream distribution, and storage. Research interest on biorefinery value chain design and optimization has grown significantly in recent years owing to increased attention that bioproducts have received, particularly biofuels, as society seeks to reduce its reliance on fossil fuels. Such a shift in energy sourcing brings with it its own challenges, one of which is the inherent uncertainty present at each stage of the value chain. Factors such as feedstock yield caused by seasonal weather conditions and fluctuating market prices for the array of derivable products introduce a large degree of uncertainty, which poses unique challenges to long-term planning decisions (Kazemzadeh and Hu, 2013).

Sustainable Remediation of Contaminated Soil and Groundwater. https://doi.org/10.1016/B978-0-12-817982-6.00013-6

Phytoremediation is a promising green soil remediation technique that utilizes certain plant processes (phytoextraction, phytostabilization, phytovolatilization, phytodegradation, rhizofiltration) to remove contaminants from soil or reduce their bioavailability. Although more cost-effective than many widely used remediation techniques, a significant hurdle to its widespread utilization is the time required for sufficient decontamination. Additional soil amendments to aid phytoremediation are an alternative; however, timescales may still remain in the order of decades for complete remediation. Two specific limitations determine the strategy's success: remediation time and cost. One could hypothesize that a technological advancement would reasonably reduce the timescale, especially considering that research interest in this area is growing; however, such an advancement is never guaranteed. Therefore, if the time required for phytoremediation is assumed not to improve significantly in the coming decades, another angle for advancement must be taken. Such a direction becomes obvious when two contemporary and increasingly active research areas are bridged; namely, phytoremediation and downstream utilization as part of a fully realized biorefinery where phytoremediating-terrestrial plants can be converted into value-added bioproducts.

Despite empirical effort to combine phytoremediation and the biorefinery concept (Sotenko et al., 2017; Asad et al., 2017), and cropping system planning for hyperaccumulator species and metal removal (Tang et al., 2012), the value chain (VC) of phytoremediation remains an open challenge. Notably, phytoremediation value chains are complicated by their interconnection with soil metal contamination and other environmental variables as well as multi-echelon complexity with new nodes and VC structures evolved from the phytoremediation transition. A number of reviews have been published on value chain optimization modeling in general (Shah et al., 2004; Papageorgiou, 2009; Garcia and You, 2015; Yue et al., 2014; Sharma et al., 2013). Instead of a broad coverage of modeling studies, this chapter will discuss the multiechelon and multiscale phytoremediation value chain characteristics and design challenge and proposes future modeling research opportunities.

This chapter is presented in three sections. An overview of phytoremediation processes and mechanisms addresses the challenge and state-of-the-art empirical research in soil metal remediation by terrestrial plant species. A critical review of the phytoremediation modeling research highlights emerging research gaps, which is followed by the identified future research directions in the field of phytoremediation value chain and system modeling.

2. Overview of soil contamination and phytoremediation

2.1 The scope of soil contamination

Soil is a dynamic system that serves a wide range of functions that both support and are essential to human activities and ecosystems. These functions, set out in the European Union (EU) Thematic Strategy on Soils (EC, 2006) (summarized in Table 13.1), range from acting as a utility, such as a raw material source, or providing a service, such as a carbon sink, to playing host to a vast range of organisms (soil biota). The biota, along with other natural capital resources, provides both ecosystem services that can be classified broadly into the following four categories; supporting, provisioning, regulating, and cultural ecosystem services. Ecosystem services include, but are not limited to, carbon and nutrient cycling, water storage, and food supply (Jeffery et al., 2010), all of which are underpinned by soil systems. In essence, the importance of soil, considered a nonrenewable resource (Jeffery et al., 2010; EU, 2004), which can take millennia to form and be degraded in a relative instant, should not be undermined. Therefore, it is wise to afford soil the same level of consideration and protection as other, vital, life-supporting resources, such as water, as the direct impacts of the consequences to local biodiversity, human health and food security are likely to be considerable (EU, 2004).

Soil degradation raises concerns due to the relatively short time span in which fertile soil can become arid. The eight key threats to soil quality (in the EU) are defined in Table 13.1. These processes are not exclusive, and each can be heavily influenced by the occurrence of another. For example, contamination of soil with pollutants would introduce toxicity to organisms, thereby impacting the ecosystem. This in turn would induce a reduction in soil biodiversity and organic matter content, consequentially resulting in a nutrient

TABLE 13.1 Soil functions and processes attributed to soil degradation as defined as part of the EU Thematic Strategy on Soils (EC, 2006; EU, 2004).

Soil functions	Soil degradation processes
Biomass production, including in agriculture and forestry	Erosion by water or wind
Storing, filtering and transforming nutrients, substances, and water	Organic matter decline
Biodiversity pool, such as habitats, species, and genes	Salination
Physical and cultural environment for humans and human activities	Compaction
Source of raw materials	Contamination
Acting as a carbon pool	Sealing
Archive of geological and archaeological heritage	Loss of soil biodiversity
	Landslides and flooding

imbalance and soil erosion (Pérez and Eugenio, 2018). It is from this list that contamination of soil will be the focus of the rest of this chapter.

2.1.1 Extent of contamination

Contamination of soil, as a consequence of many modern-day industrial and domestic practices, has led to concerns regarding potential health and environmental impacts (Pérez and Eugenio, 2018). Soil contaminants are classified as organic or inorganic pollutants. Examples of organic pollutants are fuel oil, plastic products, and polycyclic aromatic hydrocarbons (PAHs). Improper waste disposal is a large source of organic pollutants (Valentín et al., 2013). Inorganic pollutants cover a range of compounds and elements, such as salts and nutrients; however, the type of greatest concern are toxic heavy metals. Toxic metal contaminants are often divided into two groups: radionucleotides and heavy metals (Salt et al., 1995). Worldwide, at least 10 million sites have been identified as polluted, with over half citing heavy metals as one of the contaminants. This alone results in an estimated annual global economic burden of $10 billion (He et al., 2015). This chapter focuses on soil heavy metal remediation.

2.1.1.1 Europe

Analysis of the 2012 Land Use/Land Cover Area Frame Topsoil Survey (LUCAS), in which approximately 20,000 soil samples were analyzed across 27 EU member states (Tóth et al., 2013; Hiederer, 2018), showed that 6.24% (137,000 km^2) of agricultural land is in need of local assessment and subsequent remediation due to its potential health or ecological risks (Toth et al., 2016). A key observation made by the European Environment Agency, based on a data collection request from 2011, is that, on average, Europe is host to 5.7 confirmed contaminated sites per 10,000 people, with a further 4.2 potentially contaminated sites for every 1000 inhabitants. This results in an estimated 2.5 million potentially contaminated sites with 340,000 of these sites presumed contaminated (extent of soil contaminant, 2014). In total, heavy metals account for 35% of total soil contamination, although between individual nations, this can fluctuate significantly, such as in the Former Yugoslav Republic of Macedonia (FYROM), where 89% of soil contamination is a result of heavy metal contamination. This is in stark contrast to Belgium, where mineral oils (50%) are the greatest contaminant, with heavy metals accounting for only 9%. These figures reflect the large variation in anthropogenic activities between countries, resulting in differing pollutant composition profiles. The contaminating activities that are common to most European nations are waste treatment and disposal and industrial and commercial activities (extent of soil contaminant, 2014).

2.1.1.2 China

Having undergone rapid industrialization and economic development in recent decades, China now faces significant challenges with regard to topsoil contamination from heavy metals (Zhao et al., 2015). As the availability of arable land also reduces, the continued use of contaminated farmlands for agriculture is of great concern, as crops with excessive levels of contaminants are finding their way into the food chain (Xiao et al., 2017). Estimates of arable land contamination vary. Zhang et al. (2015) collated 465 articles investigating heavy metal pollution rates in farmland soil and found that 10.18% of arable land was contaminated, with heavy metal pollution affecting 13.86% of grain production. In 2014, the Chinese government published the results of the first nationwide soil pollution survey. The report told the same story as many research articles that preceded. Having surveyed more than two-thirds of the nation's land area, 16.1% of all samples (inclusive of farmland, forestland, grassland, and unused land) exceeded the Class II standard, set as the benchmark signifying contamination (inorganic and organic) (Zhao et al., 2015). Worryingly, of all arable land surveyed, 19.4% was determined to be contaminated (The Ministry of Environmental Protection, 2014; China Water Risk, 2014). In the same report, inorganic pollutants were deemed overwhelmingly as the most significant, as inorganic contamination was found in 82.8% of the polluted sampling (China Water Risk.P & M, 2014; Li et al., 2019). Cadmium (Cd) was found to have the highest pollution rate, ranging from 7% to 7.75% (Zhang et al., 2015; The Ministry of Environmental Protection, 2014; Li et al., 2019), which is particularly worrying due also to its high bioaccessibility (Gu et al., 2016). In total, 84.85% of farmland heavy metal soil pollution is accounted for by the elements Cd, mercury (Hg), copper (Cu), and nickel (Ni) (Zhang et al., 2015). Based on the same national soil survey, Chen et al. (2015) combined pollution indicators and an exposure risk model to quantify the risk to human health. From this, the authors highlighted the south/south-western regions of China as priority contamination control areas, where concentrated mining activities lead to high metal concentration in soil (Cd; 3.8%, As; 4.0%, and Ni; 6.1%) and high exposure risks.

The problem of soil contamination also extends into urban areas. A study found that although within human-safe levels, soils from urban parks in Guangzhou, the capital city of the Guangdong Province, showed elevated levels of a series of heavy metals (Cd, Lead (Pb), Ni, Cu, Zinc (Zn) and Manganese (Mn)) compared to background readings in the surrounding province (Gu et al., 2016); this suggests anthropogenic enrichment as a cause. A number of parks fell into the "slightly and moderately polluted classifications" (China; GB15618-1995 Class II and Class III). This, however, should still heed alarm, due to the direct risk of exposure to the heavy metals through "ingestion, inhalation and dermal contact" compared to indirect exposure from contaminated agriculture land, which propagates through the food chain (Gu et al., 2016). In the north-eastern Chinese industrial city of Anshan, in Liaoning province, a similar array of heavy metal contaminants were found to be above background values (Qing et al., 2015).

2.2 Contamination indicators

Determining soil contamination includes compositional analysis as well as quantitative comparison with a maximum allowable concentration threshold value. Common to the most critical research articles is the careful consideration of the local geochemical composition and appropriate selection of a pollution indicator. Often, to ensure that a high reading of a contaminant is not due to the natural composition of the soil in that region, these values are normalized by background concentrations of the compound of interest from regions that are geochemically representative of the area under investigation. Furthermore, values obtained from the indicators should be sceptically evaluated against a reference system with consideration of region-specific factors (such as previous land use or soil type) that may affect the robustness of the determined degree of contamination. Finally, comparison of pollution levels with different countries should be carried out carefully, as it is often the case that governments define threshold values differently. It would, therefore, be a mistake to superficially compare conclusions on degree of soil contamination from two different surveys, without understanding, and being critical of, the methodology behind each conclusion (Zhao et al., 2015).

2.2.1 Geoaccumulation index

Two commonly used indicators for assessing soil contamination are the geoaccumulation index (I_{geo}) (Chen et al., 2015; Qing et al., 2015; Abrahim and Parker, 2008; Zhang et al., 2007) and the enrichment factor (*EF*) (Gu et al., 2012, 2016; Chen et al., 2015; Abrahim and Parker, 2008; Zhang et al., 2007). The geoaccumulation index, defined by Eq. (13.1), aims to assess the pollution level of heavy metals in soil.

$$I_{geo} = \log_2 \left[\frac{C_n}{1.5B_n} \right] \tag{13.1}$$

C_n is defined as the concentration of metal ion in the soil and B_n is the background concentration of the same ion. The factor 1.5 is to reduce effects of variations in background values (Müller, 1969). Once determined, the degree of pollution is then classified by comparing the I_{geo} of a contaminant to the criteria described in Table 13.2.

2.2.2 Enrichment Factor

The Enrichment Factor is a method to determine the origins of heavy metal contaminants, whether from natural or anthropogenic sources (human

TABLE 13.2 Degree of contamination according to value of Geo-accumulation Index (Zhang et al., 2015).

Geoaccumulation index	Classification
$I_{geo} \leq 0$	Unpolluted
$0 < I_{geo} \leq 1$	Unpolluted to moderately polluted
$1 < I_{geo} \leq 2$	Moderately polluted
$2 < I_{geo} \leq 3$	Moderately to heavily polluted
$3 < I_{geo} \leq 4$	Heavily polluted
$4 < I_{geo} \leq 5$	Heavily to extremely polluted
$I_{geo} > 5$	Extremely polluted

interference). The *EF* can be determined by Eq. (13.2) (Abrahim and Parker, 2008):

$$EF = \frac{\frac{M_{sample}}{N_{sample}}}{\frac{M_{baseline}}{N_{baseline}}} \tag{13.2}$$

where M_{sample} and N_{sample} represent the concentrations of contaminant and normalizer elements in the sample region, respectively, and $M_{baseline}$ and $N_{baseline}$ denote the concentrations of contaminant and normalizer elements in the chosen background regions.

As mentioned previously, normalization is important as it enables natural variation of elemental abundance due to grain size and mineralogical composition of sediments to be accounted for. An ideal candidate element for normalization is one that is present in soil at high concentration, thus is less impacted by anthropogenic contamination, and shares a relationship with the trace metal of interest. Fe is commonly used for normalization due to its natural abundance, and any enrichment from relatively small anthropogenic sources is minimal (Abrahim and Parker, 2008; Zhang et al., 2007).

Upon calculating the *EF*, the value is then assessed among the criteria in Table 13.3. The origin of contamination can be determined, as to whether primarily from natural or anthropogenic sources. Furthermore, the degree of contamination can also be quantified.

TABLE 13.3 Degree of soil contaminant enrichment according to value of Enrichment Factor (Chen et al., 2015).

Enrichment factor	Origin of contamination
$EF < 0.5$	Crustal material
$0.5 \leq EF \leq 1.5$	Either entirely from crustal materials or natural weathering processes
$EF > 1.5$	Likely anthropogenic source
	Degree of contamination
$EF < 2$	Minimal enrichment
$2 \leq EF \leq 5$	Moderate enrichment
$5 \leq EF \leq 20$	Significant enrichment
$20 \leq EF \leq 40$	Very high enrichment
$EF > 40$	Extremely high enrichment

2.3 Land remediation techniques

Remediation techniques vary with their effectiveness and impact on the area considered. Several remediation techniques have been developed and referenced in literature (Table 13.4), with some used extensively, while others remain at small-scale development stage. Remediation strategies are classified depending on whether excavation is required prior to treatment (*ex situ*), or remediation can be achieved with minimum soil disturbance (*in situ*) (Wuana and Okieimen, 2011). Furthermore, based upon the nature of the remediation process, they can be categorized as physical, chemical, electrical, thermal, or biological (Liu et al., 2018).

The common heavy metal remediation processes are detailed in Table 13.4. Each strategy has a particular set of strengths and weaknesses. Until 1984, land filling was the most common in the United States (FRTR, 2012a); however, with estimates ranging between \$270 and \$460 per ton, it is costly. Therefore, it remains reserved for soils that are heavily contaminated and small in area (Khalid et al., 2017). As concerns over land availability grow, and with an increase in the public conscious for the shift toward a sustainable economy, it is vital that remediation techniques return the soil as near to its original decontaminated state as possible. It is evident that current-day practices, such as land filling, only leave this issue for future generations to consider and are clearly not sustainable business practices.

Ex situ techniques, such as soil washing, offer the ability to drastically reduce contaminant concentrations and can return soil to below-threshold

TABLE 13.4 Comparison of soil remediation methods. Adapted from (Liu et al., 2018; Khalid et al., 2017).

	Remediation method		Description	Advantages	Disadvantages	Relative cost (FRTR, 2012b)	Reference
In situ	Physical	Surface capping	Covering the exposed contaminated soil with a water impermeable layer, thereby eliminating risk of exposure to contaminated soil.	Easy and fast to implement, low cost. Land can be built upon for other uses.	Not explicitly remediation, soil remains contaminated underneath. Loss of land use for crops. Limited to small areas.	Low	Lithos (2017)
		Encapsulation	Similar to surface capping. Physical barriers are used to contain the soil in the contaminated region. Vertical barriers placed within the soil stop horizontal dispersion of contaminants to the surrounding soil.	Easy to implement.	High cost, loss of land use for crops. Limited to small areas.	Low	Liu et al (2018)

Continued

TABLE 13.4 Comparison of soil remediation methods. Adapted from (Liu et al., 2018; Khalid et al., 2017).—cont'd

Remediation method			Description	Advantages	Disadvantages	Relative cost (FRTR, 2012b)	Reference
	Chemical	Soil flushing	Use of an extraction fluid that flows through the soil and extracts contaminants. To mobilize contaminants, chelating agents or acidic solutions are typically used to manipulate the solubility of the metal ions.	Easy to implement, low cost. Land use function remains.	Risk of groundwater contamination.	Medium	Wasay et al (2001)
		Immobilization (Stabilization/ Solidification)	Two techniques to achieve immobilization; stabilization and solidification. Stabilization introduces chemicals to the soil	Easy and fast to implement. Low cost. Land use function remains. Easy and fast to implement. High effectiveness.	Only temporary mitigation of problem. Contaminants still in soil. Reagents specific to metal. High cost. Loss	Low Low	Bone et al. (2005)

			to reduce the mobility and bioavailability of the heavy metals due to physiochemical interactions between components, such as precipitation of metals or complexation with soil amendments. Solidification introduces a binding agent (commonly cement or asphalt) to immobilize the soil as a solid block.		of land function and soil as resource. Further treatment to regain land use more expensive.		
	Thermal	Electrokinetic	Application of electrodes to establish an electric field by supplying a DC voltage across the soil. Separation occurs due to electrophoresis, electromigration, diffusion, and electro-osmosis.	Land-use function remains. Minimal soil disturbance. Simple.	Works over long time. Low efficiency. Best for low permeability soils and low pH	High	Virkutyte et al. (2002), Yan et al. (2018), Lopez-Vizcaino et al. (2016)

Continued

TABLE 13.4 Comparison of soil remediation methods. Adapted from (Liu et al., 2018; Khalid et al., 2017).—cont'd

Remediation method			Description	Advantages	Disadvantages	Relative cost (FRTR, 2012b)	Reference
			Heavy metals migrate toward the cathode from where they can be removed.				
		Vitrification	Application of high temperature treatment to the soil in order to initiate vitrification (formation of glassy material). Heavy metals are trapped within the vitrified soil and organic contaminants are destroyed. Electrodes are commonly used for increasing temperature *in situ.*	Highly effective	Loss of land function and soil as resource. High cost. Limited to small areas.	Medium	Dellisanti et al (2009)

	Biological	Microbial remediation	The use of microorganisms to detoxify soil. More commonly used for organic contaminants and typically used in cooperation with another technique. Microorganisms reduce the mobility of heavy metals by processes leading to biosorption, precipitation, and can undertake volatilization.	Low cost, easy to implement. Very little soil disturbance.	Effective when used in combination with other techniques. Alone, it mostly reduces bioavailability, meaning contaminants remain in soil.	Low	Ojuederie and Babalola (2017), Dong et al. (2013)
		Phytoremediation	Use of plants grown on contaminated land to remove heavy metals from soil (phytoextraction, phytovolatilization) or to reduce their mobility and bioavailability (phytostabilization). Specific processes	Cost-effective. Green alternative. Good for large areas.	Long time required to remediate. Metal-specific. Limited to shallow contamination.	Low	Salt et al (1995)

Continued

TABLE 13.4 Comparison of soil remediation methods. Adapted from (Liu et al., 2018; Khalid et al., 2017).—cont'd

Remediation method			Description	Advantages	Disadvantages	Relative cost (FRTR, 2012b)	Reference
			are discussed further in this chapter.				
Ex situ	Physical	Soil replacement (+landfilling)	Removal of contaminated soil and (full or partial) replacement with noncontaminated soil, thereby diluting concentration of heavy metals in the soil. The excavated soil is disposed of, or treated with other *ex situ* methods. The contaminated soil is placed within a secure landfill that prevents contamination of surrounding soil and groundwater.	Results immediate, highly effective.	High cost, additional land required, land filling means soil still contaminated.	Low–high	Liu et al (2018)

			Disposal, otherwise known as "dig and dump," is a very common technique.				
	Chemical	Soil washing	Extractants are used to remove contaminants from excavated soil. Soil is screened after crushing to remove coarse materials (typically less contaminated). Fine material is then mixed well with washing solution. Ideal washing solutions increase solubility and mobility of contaminants with little effect on remaining soil constituents.	Quick results. Metals can be completely removed.	Can be metal specific. Extractants need careful selection to minimize environmental impacts. Extreme soil disturbance.	Medium	Mann (1999)
		Immobilization/ Solidification	Similar concept to *in situ* chemical immobilization. In stabilization, the binding agents act to	Easy and fast to implement. Low cost. Easy and fast to implement. High effectiveness.	Only temporary mitigation of problem. Contaminants still in soil.	Low High	CL:AIRE (2005)

Continued

TABLE 13.4 Comparison of soil remediation methods. Adapted from (Liu et al., 2018; Khalid et al., 2017).—cont'd

Remediation method			Description	Advantages	Disadvantages	Relative cost (FRTR, 2012b)	Reference
			immobilize only the contaminants Solidification uses a binding agent to encapsulate the soil to form a solid block.		Reagents specific to metal. High cost. Loss of land function and soil as resource. Further treatment to regain land use more expensive.		
	Thermal	Vitrification	Application of high temperature treatment to the soil in order to initiate vitrification (formation of glassy material). Heavy metals are trapped within the vitrified soil and organic contaminants are destroyed.	Highly effective	Loss of soil as resource. High cost (including excavation). Limited to small volumes.	Medium	Meegoda et al (2000)

regulatory standards, allowing its environmental functions to be retained. However, this process is metal-specific and the design of a single washing solution applicable to a range of metal contaminant compositions is difficult. Care is also required in selection of suitable washing solutions, to ensure minimal environmental impact if traces remain (Khalid et al., 2017). *Ex situ* techniques, however, still require excavation, in which costs can rise sharply as remediation area increases. *In situ* techniques mitigate this problem, providing the benefit of minimal soil disturbance. Additionally, a degree of land-use function may be retained during remediation.

Of the *in situ* strategies that are available for heavy metal soil remediation, some are effective at quickly mitigating the risk of exposure (surface capping, encapsulation, solidification, and vitrification), but at the expense of loss of soil environmental function, signifying extensive future treatment if complete environmental function is to be returned. Similarly, a technique such as stabilization, although not significantly altering soil composition, is only a temporary remedy as contaminants remain immobilized in the soil. Electrokinetic remediation is a potentially powerful technique, with minimal soil amendments required; however, development is still at an early stage and is best suited for low-permeability soils over a small area as a long-term strategy. In consideration of the aforementioned, a niche exists for a sustainable remediation solution for lowly contaminated soils that can be applied across large areas, requires low maintenance, is cost-effective, and retains the environmental functions of the soil. This is the gap that phytoremediation aims to fill.

2.4 Phytoremediation as a green alternative for soil remediation

Phytoremediation is a green strategy that is cost-effective, can be applied over large areas, and is highly acceptable among the general public. Phytoremediation is the umbrella term covering a range of more specific processes; Phytodegradation, Phytovolatilization, Phytoextraction, Phytostabilization, and Rhizofiltration (Salt et al., 1998), as described in Fig. 13.1.

Plants naturally possess an ability to uptake heavy metals that have a known biological function and are essential micronutrients (e.g., Iron (Fe), Mn, Zn, Cu, Magnesium (Mg), Molybdenum (Mo), Ni (Sarwar et al., 2017)), alongside those that have unknown purpose (e.g., Cd, Chromium (Cr), Pb, Cobalt (Co), Silver (Ag), Selenium (Se), Hg, Arsenic (As) (Sarwar et al., 2017)). However, at elevated levels, the heavy metals can become toxic to the plant (Sarwar et al., 2017). Plants capable of tolerating high levels of heavy metals are known as *Metallophytes.* This group can be classified further; those found endemic to metal-enriched (metalliferous) soil regions are known as *Obligate Metallophytes*, whereas *Facultative* metallophytes are not exclusive to metalliferous soils and can exist on both metal-enriched and nonenriched soils (Pollard et al., 2014). Furthermore, metallophytes can also be classified

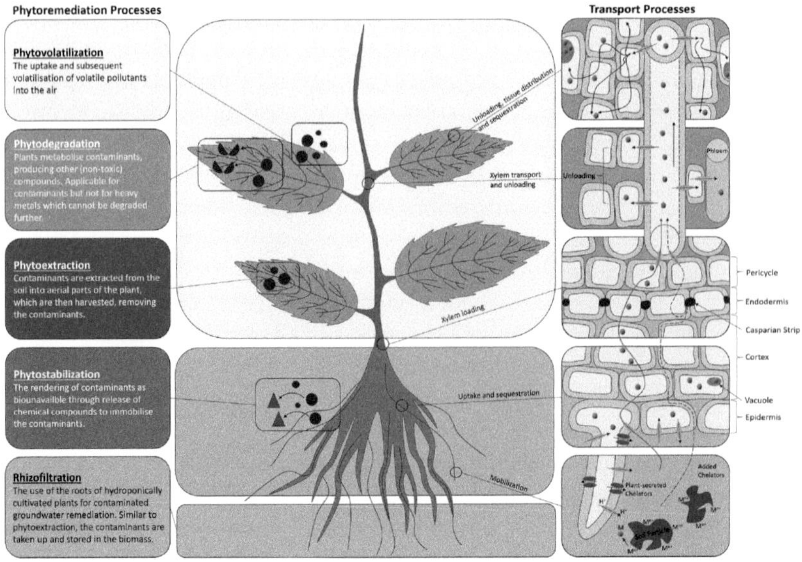

FIGURE 13.1 (left) Schematic of the types of phytoremediation processes exhibited by plants for soil remediation (right: adapted from Clemens et al. (2002)). Demonstrates the processes involved in metal accumulation in plants. **Mobilization**: secretion of natural chelators from plant roots and acidification of the rhizosphere increases soil mobility of metal ions by formation of hydrated metal ions and metal complex ions. **Uptake and sequestration**: Ions are taken up into the roots and transported along through the symplastic pathway (solid-green line) or apoplastic pathway (red-dashed line). Excess metals are sequestered to vacuoles. **Xylem loading**: Ions are transported into the xylem through membrane-bound transport proteins. Ions traveling initially through the apoplastic pathway are blocked by the Casparian strip and sequentially travel via the symplast to the Xylem. **Xylem transport and unloading**: Ions are unloaded throughout the length of the xylem. Xylem—phloem transport occurs, mediated by transfer cells. **Unloading, tissue distribution, and sequestration**: The metal ions enter the symplast of the leaves and are distributed. Excess essential metals and nonessential metals are sequestered in vacuoles.

according to their physiological metal tolerance mechanisms, as *Accumulators* or *Excluders* (Baker, 1981; Bothe, 2011).

Excluders deal with metal toxicity through minimizing uptake of heavy metals and transport into aerial parts of the plant, signifying a low translocation factor. *Accumulators,* on the other hand, uptake and translocate heavy metals into aerial regions. *Hyperaccumulators* can accumulate 10—500 times greater amounts of contaminants than those accumulators considered "average," although there is no fixed definition and the accumulation limit required to be classified as such varies for each heavy metal species. The most basic definitions set the need for a hyperaccumulator to uptake at least an order of magnitude greater than average values, but defined minimum values are usually much larger than this (Chaney et al., 1997; Baker and Brooks, 1989), with some researchers arguing for the definition to be defined as achieving an uptake of 1 wt% for some metal species (>10,000 mg/kg) (Baker and Brooks, 1989). The defined threshold for classification as a hyperaccumulator varies with metal species, which may depend on their abundance within the soil (see Table 13.5). Over 400 metal hyperaccummulator plant species (e.g., oats, barley, wheat, sunflower, soybean, rye) have been identified (Wang et al., 2018). They are capable of accumulating heavy metals including, As, Co, Cu, Mn, Ni, Pb, Se, and Zn, to between 0.1% and 4% of their root and shoot dry weight (Cappa and Pilon-Smits, 2014).

Accumulation capacity can be associated with plant genotype; however, plants can evolve as a new phenotype (hyperaccumulator (Pollard et al., 2014)) in response to limiting growth conditions. The exact evolutionary theory as to why hyperaccumulation traits benefit plants is still unknown. A well-cited paper in the 90s summarized the trending hypotheses at the time as follows (Boyd and Martens, 1992; Boyd, 2004):

TABLE 13.5 Abundances of Selected Heavy Metals in Plants and the Defined Accumulation Value for Classification as a Hyperaccumulator of that Particular Element.

Species	Nonhyperaccumulator levels (mg/kg dry biomass) background levels (Dunn and Milkereit, 2007; Markert, 1994)	Hyperaccumulator level (mg/kg dry biomass)
Mn	200	>10,000 (van der Ent et al., 2012)
Zn	50	>10,000 (Baker and Brooks, 1989), 3000 (van der Ent et al., 2012)
Cu	10	>1000, 300 (van der Ent et al., 2012)
Ni	1.5	>1000 (Chaney et al., 1997; Baker and Brooks, 1989)
Cd	0.05	>100 (van der Ent et al., 2012)
Cr	1.5	>1000 (Baker and Brooks, 1989), 300 (van der Ent et al., 2012)
Pb	1	>1000 (Baker and Brooks, 1989)
Co	0.2	>1000 (Baker and Brooks, 1989), 300 (van der Ent et al., 2012)
Se	0.02	>100 (van der Ent et al., 2012)
As	0.1	>1000 (van der Ent et al., 2012)

1. Accumulation in tissues (tolerance) and sequestration of metal contaminants into aerial tissues such as leaves (disposal) as a metal tolerance mechanism.
2. Allelopathic interference of surrounding soil in order to prevent establishment of competing species.
3. Drought resistance, through the ability of the accumulated metals to assist the plant to withstand drought.
4. An unintended consequence of other transport processes selective for other macroelements, such as essential nutrients
5. As a defense mechanism against pathogens and herbivores.

In the 20 years since the aforementioned hypotheses were published, the ideas have been challenged. With little evidence to support the first four hypotheses, the final has, on the other hand, acquired some supportive findings. Further, difficulties arise when explaining the hyperaccumulation nature of facultative hyperaccumulators, in which hyperaccumulation is an apparent species-wide trait. Although interesting, further discussion on the evolutionary reasons for the appearance of hyperaccumulators is beyond the scope of this Chapter. A review by Pollard et al. (2014) that explores these ideas applied to facultative hyperaccumulators and proposed new hypotheses is provided should it interest the reader.

2.4.1 Phytoextraction uptake mechanisms

The rate of accumulation in the aerial parts of plants is influenced and determined by the following subsequent processes: (a) Mobilization, (b) uptake and sequestration, (c) xylem transport, (d) unloading and tissue distribution, and (e) trafficking and sequestration (Fig. 13.1) (Clemens et al., 2002).

2.4.1.1 Mobilization

The aim of mobilization is to increase the solubility and thus, the bioavailability of metals in the soil. Many metals may be rendered biounavailable due to being immobilized as precipitates or bound to soil constituents. Although still not fully understood, it is thought that plants possess the natural ability to secrete phytosiderophores that behave as chelating agents and solubilize the metals within the rhizosphere (Salt et al., 1995). Specific metal reductases bound to the plasma membrane of roots provide plants with the ability to reduce nonbioavailable metals. The roots can also lower the pH of the local soil through acidification, which increases the solubility of metal ions in solution (Mahmood, 2010).

2.4.1.2 Uptake and sequestration

Most metal ions are understood to be actively transported into plant cells through the symplastic pathway by both specific and generic metal ion carriers. However, competition through these carriers occurs between essential and nonessential metals, which is thought to explain how nonessential metal ions are capable of transport against a concentration gradient. While in the apoplastic pathway, metal ions and metal–chelate complexes migrate into the root via intercellular spaces (Mahmood, 2010), once within the plant, metal ions are either stored in the roots, or transported toward the shoot and aerial parts of the plant.

2.4.1.3 Xylem loading

The translocation of ions from the roots to the shoot typically occurs through the xylem. The presence of the Casparian strip blocks ions from traveling further through the apoplastic pathway, therefore ions consequentially follow the symplastic (Zhu et al., 2006). The transport from root cells into the xylem establishes a rate-limiting step due to the restricted control by membrane transport proteins (Clemens et al., 2002; Mahmood, 2010).

2.4.1.4 Xylem transport and unloading

Xylem cell walls possess a high ion-exchange capacity so transport of metal cations can be severely retarded (Salt et al., 1995). Therefore, transport occurs with metal ions complexed with a range of small organic molecule ligands; evidence of organic acids, such as citrate, has been extensively reported (Kutrowska and Szelag, 2014; Alvarez-Fernandez et al., 2014; Haydon and Cobbett, 2007), although complexes with other molecules exist. Ions are unloaded throughout the length of the xylem, and xylem–phloem transport of ions occurs via transfer cells (Sattelmacher, 2001), the significance of this transfer and subsequent role that phloem transport plays in heavy metal accumulation is an area of interest (Fujimaki et al., 2010; Kobayashi et al., 2013).

2.4.1.5 Unloading, tissue distribution, and sequestration

After uptake into the symplast of leaves via transporters, the metal ions are distributed throughout the leaf via both the symplastic and apoplastic pathway. From here, essential heavy metals in excess and nonessential metals are sequestered into cell vacuoles (Clemens et al., 2002). A useful characterization to quantify the degree in which this transport chain occurs is the translocation factor. The Translocation Factor (TF) is the ratio between the concentration of contaminant stored in the shoot and that stored in the root; therefore, a high TF, indicative of accumulation (greater than 1), is desirable in a phytoremediation strategy (Alkorta et al., 2004).

2.4.2 Phytoremediation strategies

Two main strategies of phytoextraction have been established as the field of phytoremediation has grown: induced phytoextraction and continuous phytoextraction (Salt et al., 1998). Continuous phytoextraction utilizes hyperaccumulators to absorb heavy metals as the plant grows; however, their low biomass yields and growth rates present difficulties for the success of their applications (Mahmood, 2010; Alkorta et al., 2004; Bosiacki et al., 2014). In contrast, in induced phytoextraction, plants capable of high biomass yields are used, in combination with additional assistance, such as chelating agents or microorganisms to enhance the accumulating ability of the plant (Bosiacki et al., 2014). The chelating agents act to increase the solubility in water of the contaminating heavy metals, therefore increasing their bioavailability for uptake by plants. The accumulation efficiency of a plant for a certain metal is directly related to the affinity of the chelate for the metal of interest (Salt et al., 1998). Some chelating agents are biodegradable, while others can cause further pollution, leach to ground waters, or interrupt the ecosystem of the soil (Bosiacki et al., 2014). Therefore, critical evaluation of the phytotoxicity of the chelating agent is essential.

As phytoremediation begins to prove itself in the field as a green and cost-conservative technology for land remediation, several key problems require particular attention. The present spectrum of potential phyto-candidates would certainly allow phytoremediation strategies to be implemented; however, potential deal-breaking compromises are required. The rest of this chapter gives an overview of modeling methods applied in phytoremediation value chains and proposes a modeling research frontier with promise of increasing the attractiveness of phytoremediation for soil remediation.

3. Phytoremediation modeling

This section reflects the state-of-the-art systems modeling approaches and their applications in the field of phytoremediation value chains, including process-based simulation, mathematical optimization, evaluation, spatial data analyses, and machine learning techniques.

3.1 Mathematical optimization

A mathematical model is a structure that is used to represent the characteristics of an object, process, or system. The structure of the model is built through a set of mathematical relationships. The mathematical relationships, for example, equations and inequalities, are representative of physical laws and constraints of the real-world system being described. Models exhibit desirable features in that they:

1. Can demonstrate to users the inner workings of the real system, providing a more comprehensive understanding. It is often the case that relationships are revealed that are not apparent when solely looking at the real system.

2. Can be used for experimentation, saving valuable time, resources and avoiding undesirable experimental situations, such as those experiments that may bring unwanted risk of failure, or raise ethical questions. Particularly in application to phytoremediation, in which field trials are lengthy, and experiments seasonal, mathematical models provide are an attractive alternative.
3. Can be used for optimization (mathematical programming) purposes, in which knowledge of the direction that system conditions should be changed to improve upon a certain aspect of the system can be achieved. For example, rather than process operatives changing the temperature of an on-line reactor in order to achieve a maximum conversion, raising potential safety concerns, a model of the reactor can be used to determine the optimum temperature.

For a model to provide useful information, it must have been developed in a way to ensure that it is truly representative of the real-world system. This involves selecting appropriate mathematical relations, which include the variables of interest. Often, experiments are designed and undertaken to gather and define parameters for the model and confirm its applicability; this is the essence of parameter estimation.

3.1.1 Single Objective Optimization and Multiple Objective Optimization

Single Objective Optimization is an effective approach to achieve a "best" solution, where a single objective is maximized or minimized. In comparison, Multiple Objective Optimization can derive a set of nondominated optimal solutions that provide understanding of the trade-offs between conflicting objectives. This set of optimal solutions are known as Pareto-optimal solutions, among which, the selection of one solution in the set over another would result in the improvement of one objective at the expense of a reduction of a conflicting objective (such as an economic vs. sustainability trade-off). To visualize this set, when mapped graphically, the boundary that the Pareto-optimal set forms is known as the Pareto-optimal front. Multiobjective optimization problems can be formulated into an SOO problem by introducing weighting factors that transform conflicting objectives into a weighted single objective. MOO has received increasing research attention in recent decades due to rising concerns about the environmental sustainability of products and processes. MOO provides a tool in which economic decisions are influenced by environmental performance indicators. A quantitative evaluation tool for such an integration of performance indicators is life cycle sustainability assessment which will be discussed in more detail later.

A variety of methods exist for selecting MOO optimal solutions, which can be broadly classified into the following: No preference methods are those that are not affected by the decision-maker (decision-maker (DM) rejects or

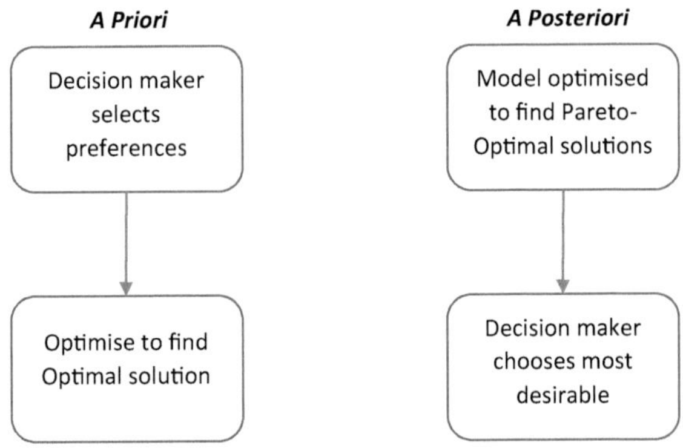

FIGURE 13.2 Visual comparison between *A Priori* and *A Posteriori* methods.

accepts the presented solution) (Miettinen, 1998a). *A Priori* methods require DMs to articulate their preferences, in the form of weightings before optimization commences (Miettinen, 1998b). However, due to limited understanding of the system, the DM may not know what can realistically be achieved or understand the feasibility of their preferences. *A Posteriori* methods generate a set of pareto-optimal solutions from which the DM selects the preferred solution (Miettinen, 1998c). *A Posteriori* methods generate and allow understanding of the whole pareto front; however, as preferences are not highlighted beforehand, more solutions are evaluated, therefore, optimization can be resource intensive. Furthermore, the DM then has a larger amount of solutions to choose from (Fig. 13.2). In the case of many conflicting objectives, interactive optimization approaches can be applied. Interactive methods represent a "human in the loop approach," dynamically involving the decision-makers by alternating between optimization and preference allocation, iteratively and progressively refining the solution search toward the region of interest (Miettinen, 1998d). Despite the interactive methods' ability to provide insights to DMs about the optimization topology, computational complexity hinders its wider application in value chain optimizations.

3.2 Metal biogeochemical cycles and modeling

Phytoremediation processes are complexities that involve the interaction of plants with climate and soil underpinned by photosynthesis and biogeochemical cycles. Such complexities can be simulated by a process-based modeling approach, which embeds C3/C4 photosynthesis models, plant growth, and soil biogeochemical process simulation suites. Here C3 (Calvin–Benson–Bassham Cycle) and C4 cycles (Hatch–Slack cycle) represent different photosynthetic pathways. The photosynthesis models capture leaf biochemistry, canopy development, and ecosystem exchange in response to environmental variables (von Caemmerer and Farquhar, 1981; Long, 1991). Plant growth is closely related to wider environmental drivers, e.g., soil biogeochemical processes and hydraulic dynamics; this simulation is achieved by linking terrestrial crop physiological processes and soil biogeochemistry (Zhang et al., 2002). Thus, process-based modeling approaches convert primary drivers (e.g., climate, soil properties, anthropogenic activity) to soil environmental factors and simulates element (e.g., C/N) transformations mediated by the soil microbes and further enables computation of their transport and transformations in plant–soil–climate interaction. There are over 30 models (Brilli et al., 2017; Peltoniemi et al., 2007) that evaluate the carbon and nitrogen cycles and water flow that control the flux of nutrients in soil. These range from models that consider the decomposition of organic carbon in soils like the RothC (Coleman et al, 1997; Jenkinson and Coleman, 2008) and the carbon cohort models (developed by Bosatta and Ågren) (Bosatta and Ågren, 1996; Bosatta and Berendse, 1984; BOSATTA and ÅGREN, 1995; Bosatta and Ågren, 1994; Bosatta and Ågren, 1985; Bosatta and Ågren, 1999; Bosatta and Agren, 1991), models that only consider the nitrogen cycle of nitrification/denitrification and plant uptake like Sundial (Bradbury et al., 1993) and hydrology models like SWAT (Devia et al., 2015), to models that include all processes such as DayCent (Parton et al., 1998), ECOSSE (Smith et al., 2010a, 2010b), and DNDC (Li et al., 1992, 1994, 2006; Li, 2000). In comparison with widely studied C and N cycling, knowledge gaps emerge in in-depth understanding and modeling of metal uptake kinetics and the transport and transformation of metallic elements in the biogeochemical cycling (Wang et al., 2018; McGrath and Zhao, 2003). Despite the empirical advances on phytoremediation mechanisms, including metal uptake kinetics and metal partitioning (e.g., arsenic, cobalt, chromium) in plant roots and shoots (Wang et al., 2002; Guterres et al., 2019; Lotfy and Mostafa, 2014; Mao et al., 2016; Barzanti et al, 2011), limited research has been published on the kinetic modeling of phytoextraction (Guala et al., 2011; Tunali Akar et al., 2016; Seuntjens et al., 2004; Lugli and Mahler, 2016). These modeling studies have a particular focus on a given metal contamination and plant species (Yu and Gu, 2007), except for the research by Lugli and Mahler (2016) and Verma et al. (2006) Notably, Lugli and Mahler (2016) presented a numerical plant-soil model based on the kinetics for hydraulic phenomena, metal transport, and uptake; however, plant growth and its effects on transpiration were not considered, which means that the climate–plant interaction was not captured and environmental variables such as temperature, solar radiation, and precipitation were eliminated from the model. As highlighted, a process-based modeling approach can address the interlinkage between terrestrial plants and environmental factors in a species-specific manner. Previous research on land remediation focused on the C3 and C4 higher plant species as shown in Table 13.6 (e.g., C4 grasses, C3 food and nonfood crops). Thereby, process-based biogeochemical modeling represents a clear unexplored area with

TABLE 13.6 Summary of RSM studies applied to soil heavy metal contamination by phytoremediation. The response variables and the independent variables have been listed, alongside the RSM design of choice and number of experiments conducted.

Species	RSM design	No. of experiments	Response variable	Independent variables	Ref
Ludwigia octovalvis	BBD	17	Arsenic removal	Contaminant soil concentration, sampling day, aeration rate	Titah et al (2018b)
Chrysopogon zizanioides L.	CCD	20	BOD and COD removal	Palm oil mill secondary effluent (POMSE) concentration, vetiver plant density, time	Darajeh et al (2016)
L. gibba	CCD	11	Growth parameters, photosynthetic pigment and heavy metal removal	Initial Cd and Ni concentration	Demim et al (2013a)
Spinacia oleracea L.	CCD	20	Pb removal efficiency	Pb concentration, sampling day, number of spinach plants present	Farraji et al (2014)
L. gibba	CCD	12	Heavy metal removal of Cd, Cr, Cu, Zn, and Ni	Concentration of: Cd, Cr, Cu, Zn, and Ni, incubation period, frond number	Demim et al (2014)
Pteris vittata L.	CCD	15	Plant growth, As removal	pH, As concentration, and P concentration	Tu and Ma (2003)
L. gibba	CCD	11	Growth parameters, photosynthetic pigment	Concentration of Cd, Cr, Cu, Zn, and Ni	Demim et al (2013b)
Typha domingensis	CCD	13	Pb removal, Ni removal, Cd removal	Number of plants transplanted, sample time (hours)	Mojiri et al (2013)
Napier grass	BBD	15	Pb removal	Pb concentration, NaCl concentration, time	

					Hongsawat et al (2018)
Melastoma malabathricum L.	BBD	17	Pb removal	Pb concentration in sand, exposure time (days), aeration rate	Selamat et al (2017)

tremendous opportunity to advance the understanding of phytoremediation effects of different plant species and genotypes.

3.3 Response Surface Methodology and optimization application in phytoremediation

There are a range of factors that determine and affect the extent and success of phytoremediation. The combination of the numerous variables with many plants that are potential candidates for a phytoremediation strategy lends itself for optimization. A standard variable-by-variable approach can be adopted, but this is time-consuming, costly, and fails to identify any possible interactions between variables. In preference, statistical and mathematical experimental design approaches can be considered. Many exist, with Response Surface Methodology (RSM) being one of the most commonly used.

RSM is a useful technique in experimental design to maximize the information that can be deduced from an experiment in the least number of experimental runs, saving empirical effort. It aims to extract independent and interactive effects of several independent variables (IVs) on one or more dependent variables (DVs). Not only does this improve the efficiency of arriving at conclusions, but it is also a tool that can be used to determine the optimal operating parameters of a process or system. As summarized in Table 13.6, RSM has been widely adopted in the experimental design of phytoremediation research, where the most commonly applied multilevel design approaches include Box−Behnken design (BBD) introduced by Box and Behnken (1960) and Composite Central Design (CCD) developed by Box and Wilson (1951). Additionally, previous research combined RSM and heuristic optimization, e.g., artificial neutral network (ANN), to project and maximize heavy metal removal. Titah et al. (2018a), investigated the arsenic removal by *Ludwigia octovalvis* using RSM-guided experimental design in comparison with ANN analyses, in which ANN delivered improved projection power.

3.4 Application of machine learning techniques in phytoremediation

Machine learning techniques (e.g., principal component analysis, support-vector machine, ANN) have been adopted in the field of phytoremediation to identify the data structure and correlation from complexity and generate predictive models for phytoremediation including plant adsorption, transport, accumulation, and translocation of contaminations in soil and water. Heuristic algorithms, such as Genetic Algorithm (GA) and ANN, were applied in several studies to predict the metal uptake and translocation (Bagheri et al., 2019), model metal toxicity, and plant−microbe interaction (Titah et al., 2018a; Hattab and Motelica-Heino, 2014; Zloch et al., 2017) and to optimize the biogeochemical process for metal removal (Jiang et al., 2016).

The most applied machine learning technique in phytoremediation research is principal component analysis (PCA). PCA is a multivariate procedure that can be applied to reduce dimensionality, e.g., analytical dataset, and identify the variables (i.e., principal components) that best represent the underlying data variation and contribute most to the phytoremediation property differentiation. PCA defines principal components. The dataset samples can be plotted in a new vector space defined by principal components; the most important variables influencing the metal extraction/stabilization properties are descriptors, which can be further processed by heuristic algorithms to define their correlations and project metal transport, removal, and translocation. PCA has been applied in over 20 previous empirical studies, in particular, on interactions of plant microbes and their impacts on metal uptake. Sun et al. (2018) combined experiments and PCA to investigate the correlations of switchgrass quality and composition (lignin, cellulose, and hemicellulose) with driving factors (e.g., cadmium level, arbuscular mycorrhizal fungi inoculation). Garcia-Gonzalo et al. (2017) studied bacterial-assisted phytoremediation by *Silene vulgaris* where PCA was applied to analyze the variation in genotypes in response to chromium contamination. Ali et al. (2017) investigated the sorghum phytoremediation assisted by *Streptomyces pactum* and used PCA to understand the relationships between factors like metal uptake that assist with translocation of metals (e.g., Zn, Pb, Cd, Cu) and their impacts on sorghum growth. Krgovic et al. (2015) used PCA in conjunction with partial least squares discriminant analysis to correlate the metal bioavailability (e.g., Pb, b, Cd, As, Cr, Cu, Co, Ni, Zn, Ba, Fe, Al, and Ag) and their partitioning in different part of *Erigeron canadensis* L. Research has also been carried out on cover crops' interaction with soil bacteria (structure, activity) and their capacity for Cu phytoextraction and PCA effectively identified key drivers in Cu removal. Stojanovic et al. (2016) developed experimental research and used PCA to evaluate the uranium uptake effects by different varieties of corn, sunflower, and soybean.

PCA along with other machine learning techniques offers a powerful approach to data analysis, which can be integrated with other modeling tools to project and optimize phytoremediation systems. This will be further discussed in following sections.

3.5 Phytoremediation modeling in geographical information systems

The use of Geographic Information System (GIS) is an efficient approach to analyze complex spatial phenomena and has been used to model various applications for terrestrial plants. GIS allows for synchronization of spatial and descriptive data in a combination of several layers, with each layer representing data on different features of the spatial region, such as soil data, land use, etc. Despite the capability of spatial data acquisition, storage, processing, and analysis, GIS alone does not allow for incorporating the decision-maker's preferences and heuristics into the problem-solving process. Thus a range of GIS-aided methodologies, e.g., multicriteria decision analysis, fuzzy multicriteria decision making, and mixed integer programming models have been developed and applied to spatial decision support. Such GIS-based models have been developed to optimize and evaluate biomass potential for phytoremediation of contaminated land and ground water (Voets et al., 2013; Schreurs et al., 2011; Chaudhary et al., 2019). Wong et al. (2006) presented a GIS map of urban contamination of trace metals for Hong Kong areas. Schreurs et al. (2011) showed GIS-based scenario analyses of potential energy crops for extraction of Cd, Zn, Pb, and As in a Belgian region, where the environmental variables (soil quality and climate) have been factored in to select the candidate energy crops. The same research team later presented a GIS evaluation to screen the optimum locations for willow phytoremediation (Voets et al., 2013). Heckenroth et al. (2016) presented a GIS-based analysis, which, combined with multivariate analyses to identify the hyperaccumulator species, demonstrated the variation in phytoremediation effects of native plant species. Chaudhary et al. (2019) coupled GIS with PCA to map out microbial-assisted phytoremediation. Considering the phytoremediation system features (e.g., linked with geographically varied environmental drivers), GIS coupled with remote sensing undoubtedly offers an effective way to acquire, process, and analyze spatial data. Thus, one of the promising research directions lies in GIS-aided model integration, covering the multigeographical level decision-making (cultivation combined with regional network) and linking the spatially explicit ecosystems with the biorefinery conversion technosphere.

3.6 Value chain optimization

Six phytoremediation value chain echelons can be generalized—phytoremediation and metal removal, biomass cultivation and supply, refinery for bioproducts, storage, distribution, and demand. Six echelons have been selected based on the key functional value-adding activities involved in the supply chains. Each echelon concerns different decision spaces across temporal and spatial scales. Multiple spatial and temporal scales need to be coordinated in value chain decision-making, ranging from decisions across different time steps (day, month, year) to decisions on site location and network at each echelon. At the phytoremediation stage, long-term planning is reflected by decisions on plant species screening (hyperaccumulators, low-accumulating plants), phytoremediation strategy (e.g., chelating agents, microbe-assisted phytoremediation), land use patterns, cropping systems (e.g., multiple cropping), trace metal translocation and transport, whereas the short-term operational decisions concern field operations, concentration and applications of chelating agents, etc.

Phytoremediation value chains underpinned by plant−environment interaction and biogeochemical cycles (carbon nitrogen and metals) are not only regulated by environmental variables and plant biophysical traits but also constrained by environmental boundaries and soil constraints (e.g., land availability, soil contamination level). Thus, the phytoremediation design and strategy may vary significantly at temporal and spatial scales due to the variability in environmental factors (e.g., climate, soil quality); phytoremediation value chains also need to consider the economic feasibility and environmental sustainability simultaneously. Mathematical optimization of phytoremediation value chains to address such complexity represents a research frontier, which has not yet been explored. Considering the phytoremediation value chain complexity and decision spaces with multiple design criteria, multiobjective optimization based on mathematical programming is worth exploring.

4. The proposed modeling framework for the future research frontier

Several potential research areas have been identified to unlock the complexity of phytoremediation value chains. We address this in the following sections, which are underpinned by cross-disciplinary approaches to bring phytoremediation technology innovations into overarching modeling tools.

4.1 Spatial−temporal phytoremediation system modeling

To project the phytoremediation potential at spatial−temporal scales, an interesting yet unexplored direction is to couple metal biogeochemical simulation with statistical and mathematical programming tools. As highlighted previously, knowledge gaps emerge in in-depth understanding and modeling of metal uptake kinetics and the transport and transformation of metallic elements in the biogeochemical cycling. Thereby, an emerging research area is to further develop process-based biogeochemical models for phytoremediation and metal cycling in the biosphere and the environment. This can be achieved by building on modeling and empirical advances, e.g., analytical data on metal extraction/stabilization, crop physiological process, and soil biogeochemistry modeling. Biogeochemical simulations address the interaction of plant with soil and climate; a validated biogeochemical simulator (for given sites) can be scaled up to regional and national levels and used to project future phytoremediation in response to environmental change based on GIS soil data and climate projection (e.g., EU WATCH, Carbon Dioxide Information Analysis Center).

However, the biogeochemical model development and validation depend on phytoremediation experiments, which are often slow and can be labor-intensive and cost-ineffective (Sotenko et al., 2017). An alternative approach is to apply machine learning and mathematical modeling techniques, which highlights the multidimensional linear/nonlinear characteristics of metal extraction/stabilization based on analytical data patterns, in order to develop predictive capability. Such an approach can define the correlations between phytoextraction/photostability effects with driving components, e.g., plant species and traits, the metal bioavailability in soil (McGrath and Zhao, 2003), and metal partitioning and accumulation rate (Lotfy and Mostafa, 2014; Cui et al., 2007; Prasad, 2003). By softlinking the projection with plant growth models, metal extraction/stabilization in response to climate and soil variation can be estimated.

4.2 Phytoremediation-biorefinery value chain design

Despite a small number of empirical studies on the phytoremediation-biorefinery concept, the refinery process design and value chain modeling remain unexplored. Thus, a challenging research direction lies in the simulation and optimization of refinery processes converting hyperaccumulator biomass into value-added bioproducts (e.g., bioenergy, platform chemicals, biofuels). To advance the understanding of technical feasibility and sustainable design solutions of such phytoremediation-biorefinery value chains, we propose to adopt a simulation—evaluation—optimization approach.

In general, biorefinery process simulation is a model-based representation of chemical, physical, biological, and other technical processes and unit operations in software. It is used for the design, development, analysis, and optimization of technology processes. The advantages of process simulation are to (a) reduce design time by enabling various refinery process configuration tests and (b) help improve scaling-up performances by answering "what if" questions, determining optimal process conditions within given constraints, and assisting in locating performance-limiting steps. Two types of models could be applied for simulation: kinetic and thermodynamic equilibrium models. The former is used to predict the progress and the product compositions along the reaction, whereas the latter, also denoted as zero-dimensional, can be used to project the theoretical efficiency and achievable yields of desired end-reaction products based on assumption that reacting systems reach steady state with minimized Gibbs Free Energy (maximized entropy) (Ahmad et al., 2016; Ahmed et al., 2015; Arora et al., 2017). A variety of commercial process simulators, e.g., Aspen Plus, Aspen Hysys, SuperPro Designer provide platforms to achieve biorefinery conceptual design, where physical relationships, thermodynamic equilibrium, and rate equations can be built to enable the process flowsheeting, refinery-wide process behavior projection.

A quantitative evaluation approach referred to as life cycle sustainability assessment (LCSA) can be applied to evaluate environmental, social, and economic impacts of phytoremediation value chains and support multicriteria decision toward sustainability (UNEP/SETAC Life Cycle Initiative, 2011).

Initiated from life-cycle assessment (ISO, 2006), the life-cycle thinking approach has been extended since 2002 to form an LCSA methodology framework, which consists of three pillars: environmental life-cycle assessment (LCA), life-cycle costing (LCC), and social-LCA (SLCA) (UNEP/SETAC Life Cycle Initiative, 2011). As generalized in Eq. (13.3) (Lin et al., 2016; Guo, 2012), LCSA accounts for all input−output flows occurring at each phytoremediation and refinery process and upstream/downstream stages throughout the "cradle-to-grave" life cycle.

$$EI_{kpi} = \sum_r \sum_s EIf^{in}_{r,kpi} X^{in}_{r,s} F^{in}_{r,s} + \sum_c \sum_s EIf^{out}_{c,kpi} X^{out}_{c,s} F^{out}_{c,s} \tag{13.3}$$

where the variable EI_{kpi} denotes the total sustainability impacts of a given phytoremediation-biorefinery system (per functional unit) expressed as key performance indicator kpi (e.g., greenhouse gas and cost). EI_{kpi} is determined by the characterization impact factors for input resource r $\left(EIf^{in}_{r,kpi}\right)$ or emitted compound c $\left(EIf^{out}_{c,kpi}\right)$ and the input or output flows $\left(F^{in}_{r,s}\ or\ F^{out}_{c,s}\right)$ and concentrations $\left(X^{in}_{r,s}\ or\ X^{out}_{c,s}\right)$ at phytoremediation-biorefinery process s and/or associated life cycle stage s.

By feeding the lab-derived data and conceptual design into the simulators, the proposed approach can generate refinery process flowsheets, including mass and energy flow rates, elemental or chemical flows. By further feeding the flowsheets into LCSA, simulation evaluation can highlight the phytoremediation merit of plant species and refinery design options for a sustainable phytoremediation system and identify the performance-limiting "hotspots," which require further research attention to achieve the overall value chain sustainability.

To address the value chain design challenge, the aforementioned simulation−evaluation approach can be incorporated into mathematical optimization (generated as Eq. 13.4) using surrogate-based modeling techniques (Boukouvala and Floudas, 2017; Caballero and Grossmann, 2008). The optimization algorithms need to: (a) reflect decision spaces across spatial and temporal scales and six phytoremediation-biorefinery SC echelons, and (b) incorporate resource vectors (plant species, metal species, bioproducts), metal extraction and transport, refinery processes, land and soil, environmental and economic decision criteria, to optimize the phytoremediation-biorefinery value chain.

$$\begin{aligned} &Min\ (f_1(x,\ y)\ldots\ldots f_k(x,\ y)) \\ &s.t.\ \ h(x,y) = 0 \\ &g(x,y) \leq 0 \\ &x \in \mathbb{R}^n \\ &y \in \{0,1\}^m \end{aligned} \tag{13.4}$$

where $f_1(x,\ y)\ldots f_k(x,\ y)$ denotes the functions consisting of conflicting multiobjectives (e.g., economic targets and environmental sustainability). The continuous variables x are normally nonnegative, presenting the operational design and strategic planning decisions across value chains, such as metal transport and transformation, feedstock composition, flowrates, and reaction temperature at refinery; discrete variables y define the selection of plant species, refinery technologies and location, value chain network and their interconnections over temporal–spatial scales. Both continuous and discrete variables follow the equality constraints $h(x, y)$ (e.g., energy and mass balance) and inequality constraints $g(x, y)$ design specifications (e.g., environmental boundaries and soil constraints) and logical constraints.

5. Concluding remarks

Conservation of soil as a resource is crucial to guaranteeing food security and preserving ecosystems, among other crucial functions that it serves. The extent of heavy metal soil contamination worldwide is evident, and the anthropogenic impact due, in part, to unchecked industrial activities is clear. Phytoremediation lends itself as a green alternative to traditional soil remediation activities, with particular usefulness when applied to lowly contaminated soils covering a large area, due to its low cost. However, if contaminated land is valuable for use other than its environmental or cropping functions, then "quick-fix" remediation methods may be used, losing a nonrenewable resource for good. In order to improve the attractiveness and productivity of phytoremediation strategies, a variety of modeling tools can be applied to inform the decision-making on phytoremediation. A sample of these models, their methodology, and future research directions have been discussed in this chapter. Taking inspiration from biorefinery value chains, the establishment of phytoremediation value chains offers a potential avenue to add monetary value to phytoremediation strategies through biomass-derived value-added products. Phytoremediation value chains consider six echelons: phytoremediation and metal removal, biomass cultivation and supply, refinery for bioproducts, storage, distribution, and demand. Research in this domain is minimal, and with many interacting areas of the value chain, there is significant scope to fill such research gaps. We proposed a process systems engineering (PSE) approach to assist decision-making throughout value chain design. This includes implementation and integration of modeling approaches at each echelon of the value chain, from improving phytoremediation and metal removal through process-based biogeochemical models, to the use of geographic information systems (GIS) to project future phytoremediation and supply chain patterns and chemical process simulation to model refinery performance. With such an approach, key performance indicators, such as heavy metal removal, economic and environmental indicators can be optimized through MOO, thereby assisting decision-makers in the implementation of phytoremediation strategies. Beyond soil remediation, another important application of phytoremediation is groundwater decontamination. The phytoremediation value chain modeling of groundwater remediation, e.g., chlorinated solvent removal, represents one of the future research directions worth exploring. In particular, from a whole systems perspective, the interdependency of natural capital resources, including water and soil, with environmental systems is studied. Thus a sustainable water-soil-phytoremediation value chain, if fully realized, will enable step change in remediation and resource management strategies, contributing to sector transition toward economically viable and environmentally beneficial solutions.

References

Abrahim, G.M., Parker, R.J., 2008. Assessment of heavy metal enrichment factors and the degree of contamination in marine sediments from Tamaki Estuary, Auckland, New Zealand. Environmental Monitoring and Assessment 136 (1–3), 227–238.

Ahmad, A.A., et al., 2016. Assessing the gasification performance of biomass: a review on biomass gasification process conditions, optimization and economic evaluation. Renewable and Sustainable Energy Reviews 53, 1333–1347.

Ahmed, A.M.A., et al., 2015. Review of kinetic and equilibrium concepts for biomass tar modeling by using Aspen Plus. Renewable and Sustainable Energy Reviews 52, 1623–1644.

Ali, A., et al., 2017. Streptomyces pactum assisted phytoremediation in Zn/Pb smelter contaminated soil of Feng County and its impact on enzymatic activities. Scientific Reports 7, 13.

Alkorta, I., et al., 2004. Chelate-enhanced phytoremediation of soils polluted with heavy metals. Reviews in Environmental Science and Biotechnology 3 (1), 55–70.

Alvarez-Fernandez, A., et al., 2014. Metal species involved in long distance metal transport in plants. Frontiers of Plant Science 5, 105.

Arora, P., et al., 2017. Compartment model for a dual fluidized bed biomass gasifier. Chemical Engineering Research and Design 117, 274–286.

Asad, M., et al., 2017. Pretreatment of trace element-enriched biomasses grown on phytomanaged soils for bioethanol production. Industrial Crops and Products 107, 63–72.

Bagheri, M., et al., 2019. A deeper look at plant uptake of environmental contaminants using intelligent approaches. The Science of the Total Environment 651, 561–569.

Baker, A.J.M., 1981. Accumulators and excluders-strategies in the responses of plants to heavy metals. Journal of Plant Nutrition 3 (1–4), 643–654.

Baker, A.J.M., Brooks, R.R., 1989. Terrestrial higher plants which hyperaccumulate metallic elements - a review of their distribution, ecology and phytochemistry. Biorecovery 1, 81–126.

Barzanti, R., et al., 2011. Cadmium phytoextraction potential of different Alyssum species. Journal of Hazardous Materials 196, 66–72.

Bone, B.D., et al., 2005. "UK guidance on stabilisation/solidification for the treatment of contaminated soil.", pp. 89–95.

Bosatta, E., Ågren, G.I., 1985. Theoretical analysis of decomposition of heterogeneous substrates. Soil Biology and Biochemistry 17 (5), 601–610.

Bosatta, E., Agren, G.I., 1991. Dynamics of carbon and nitrogen in the organic matter of the soil: a generic theory. The American Naturalist 138 (1), 227–245.

Bosatta, E., Ågren, G.I., 1994. Theoretical analysis of microbial biomass dynamics in soils. Soil Biology and Biochemistry 26 (1), 143–148.

Bosatta, E., Ågren, G.I., 1995. Theoretical analyses of interactions between inorganic nitrogen and soil organic matter. European Journal of Soil Science 46 (1), 109–114.

Bosatta, E., Ågren, G.I., 1996. Theoretical analyses of carbon and nutrient dynamics in soil profiles. Soil Biology and Biochemistry 28 (10), 1523–1531.

Bosatta, E., Ågren, G.I., 1999. Soil organic matter quality interpreted thermodynamically. Soil Biology and Biochemistry 31 (13), 1889–1891.

Bosatta, E., Berendse, F., 1984. Energy or nutrient regulation of decomposition: implications for the mineralization-immobilization response to perturbations. Soil Biology and Biochemistry 16 (1), 63–67.

Bosiacki, M., Kleiber, T., Markiewicz, B., 2014. Continuous and Induced Phytoextraction—Plant-Based Methods to Remove Heavy Metals from Contaminated Soil.

Bothe, H., 2011. Plants in Heavy Metal Soils, vol. 30, pp. 35–57.

Boukouvala, F., Floudas, C.A., 2017. ARGONAUT: AlgoRithms for global optimization of coNstrAined grey-box compUTational problems. Optimization Letters 11 (5), 895–913.

Box, G.E.P., Behnken, D.W., 1960. Simplex-sum designs: a class of second order rotatable designs derivable from those of first order. The Annals of Mathematical Statistics 31 (4), 838–864.

Box, G.E.P., Wilson, K.B., 1951. On the experimental attainment of optimum conditions. Journal of the Royal Statistical Society: Series B 13 (1), 1–45.

Boyd, R.S., 2004. Ecology of metal hyperaccumulation. New Phytologist 162 (3), 563–567.

Boyd, R., Martens, S., 1992. The raison d'etre for metal hyperaccumulation by plants. In: The Vegetation of Ultramafic (Serpentine) Soils. University of California, Davis: Andover, UK.

Bradbury, N.J., et al., 1993. Modelling the fate of nitrogen in crop and soil in the years following application of 15N-labelled fertilizer to winter wheat. The Journal of Agricultural Science 121 (3), 363–379.

Brilli, L., et al., 2017. Review and analysis of strengths and weaknesses of agro-ecosystem models for simulating C and N fluxes. Science of the Total Environment 598, 445–470.

Caballero, J.A., Grossmann, I.E., 2008. An algorithm for the use of surrogate models in modular flowsheet optimization. AIChE Journal 54 (10), 2633–2650.

Cappa, J.J., Pilon-Smits, E.A.H., 2014. Evolutionary aspects of elemental hyperaccumulation. Planta 239 (2), 267–275.

Chaney, R.L., et al., 1997. Phytoremediation of soil metals. Current Opinion in Biotechnology 8 (3), 279–284.

Chaudhary, K., Saraswat, P.K., Khan, S., 2019. Improvement in fluoride remediation technology using GIS based mapping of fluoride contaminated groundwater and microbe assisted phytoremediation. Ecotoxicology and Environmental Safety 168, 164–176.

Chen, H., et al., 2015. Contamination features and health risk of soil heavy metals in China. The Science of the Total Environment 512–513, 143–153.

China Water Risk. *MEP & MLR: First Nationwide Soil Pollution Survey* Published. 2014 25/04/2014 14/02/2019]; Available from: http://www.chinawaterrisk.org/notices/mep-mlr-first-nationwide-soil-pollution-survey/.

Clemens, S., Palmgren, M.G., Krämer, U., 2002. A long way ahead: understanding and engineering plant metal accumulation. Trends in Plant Science 7 (7), 309–315.

Coleman, K., et al., 1997. Simulating trends in soil organic carbon in long-term experiments using RothC-26.3. Geoderma 81 (1–2), 29–44.

Cui, S., Zhou, Q., Chao, L., 2007. Potential hyperaccumulation of Pb, Zn, Cu and Cd in endurant plants distributed in an old smeltery, northeast China. Environmental Geology 51 (6), 1043–1048.

Darajeh, N., et al., 2016. Modeling BOD and COD removal from Palm Oil Mill Secondary Effluent in floating wetland by *Chrysopogon zizanioides* (L.) using response surface methodology. Journal of Environmental Management 181, 343–352.

Dellisanti, F., Rossi, P.L., Valdrè, G., 2009. In-field remediation of tons of heavy metal-rich waste by Joule heating vitrification. International Journal of Mineral Processing 93 (3–4), 239–245.

Demim, S., et al., 2013. Cadmium and nickel: assessment of the physiological effects and heavy metal removal using a response surface approach by *L. gibba*. Ecological Engineering 61, 426–435.

Demim, S., et al., 2013. CCD study on the ecophysiological effects of heavy metals on Lemna gibba. Ecological Engineering 57, 302–313.

Demim, S., et al., 2014. Study of heavy metal removal from heavy metal mixture using the CCD method. Journal of Industrial and Engineering Chemistry 20 (2), 512–520.

Devia, G.K., Ganasri, B.P., Dwarakish, G.S., 2015. A review on hydrological models. Aquatic Procedia 4, 1001–1007.

Dong, G., et al., 2013. Formation of soluble Cr(III) end-products and nanoparticles during Cr(VI) reduction by Bacillus cereus strain XMCr-6. Biochemical Engineering Journal 70, 166–172.

Dunn, C.E., 2007. New perspectives on biogeochemical exploration. In: Milkereit, B. (Ed.), Advances in Prospect-scale Geochemical Methods, pp. 249–261.

EC, 2006. Communication from the Commission to the Council, the European Parliament, the European Economic and Social Committee and the Committee of the Regions - tThematic Strategy for Soil Protection [SEC(2006)620] [SEC(2006)1165].

EU, *Proposal for a Directive of the European Parliament and of the Council Establishing a Framework for the Protection of Soil and Amending Directive 2004/35/EC* 2006.

2014 Extent of Soil Contamination.

Farraji, H., et al., 2014. Optimization of phytoremediation of lead-contaminated soil by Spinach (Spinacia oleracea L). International Journal of Scientific Research in Knowledge 2 (10), 480–486.

FRTR. *Remediation Technologies Screening Matrix and Reference Guid*, Version 4.0. 2012 [cited 2018 19/02]; Available from: https://frtr.gov/matrix2/top_page.html.

FRTR, 2012. Remediation Technologies Screening Matrix and Reference Guide, Version 4.0: Table 3-2: Treatment Technologies Screening Matrix. Federal Remediation Technologies Roundtable, Washington, D.C.

Fujimaki, S., et al., 2010. Tracing cadmium from culture to spikelet: noninvasive imaging and quantitative characterization of absorption, transport, and accumulation of cadmium in an intact rice plant. Plant Physiology 152 (4), 1796–1806.

Garcia, D.J., You, F., 2015. Supply chain design and optimization: challenges and opportunities. Computers and Chemical Engineering 81, 153–170.

Garcia-Gonzalo, P., et al., 2017. Different genotypes of Silene vulgaris (Moench) Garcke grown on chromium-contaminated soils influence root organic acid composition and rhizosphere bacterial communities. Environmental Science and Pollution Research 24 (33), 25713–25724.

Gu, Y.G., et al., 2012. Multivariate statistical and GIS-based approach to identify source of anthropogenic impacts on metallic elements in sediments from the mid Guangdong coasts, China. Environmental Pollution 163, 248–255.

Gu, Y.-G., Gao, Y.-P., Lin, Q., 2016. Contamination, bioaccessibility and human health risk of heavy metals in exposed-lawn soils from 28 urban parks in southern China's largest city, Guangzhou. Applied Geochemistry 67, 52–58.

Guala, S.D., Vega, F.A., Covelo, E.F., 2011. Development of a model to select plants with optimum metal phytoextraction potential. Environmental Science and Pollution Research 18 (6), 997–1003.

Guo, M., 2012. Life Cycle Assessment (LCA) of Light-weight Eco-composites. Springer.

Guterres, J., et al., 2019. A new conceptual framework for plant responses to soil metals based on metal transporter kinetic parameters. Journal of Hazardous Materials 364, 449–467.

Hattab, N., Motelica-Heino, M., 2014. Application of an inverse neural network model for the identification of optimal amendment to reduce copper toxicity in phytoremediated contaminated soils. Journal of Geochemical Exploration 136, 14–23.

Haydon, M.J., Cobbett, C.S., 2007. Transporters of ligands for essential metal ions in plants. New Phytologist 174 (3), 499–506.

He, Z., et al., 2015. Heavy metal contamination of soils: sources, indicators, and assessment. Journal of Environmental Indicators 9, 17–18.

Heckenroth, A., et al., 2016. Selection of native plants with phytoremediation potential for highly contaminated Mediterranean soil restoration: tools for a non-destructive and integrative approach. Journal of Environmental Management 183, 850–863.

Hiederer, R., 2018. Data Evaluation of LUCAS Soil Component Laboratory 2009 to 2015 Data on Soil Organic Carbon. Publications Office of the European Union, Luxembourg, p. 63.

Hongsawat, P., et al., 2018. Optimization of lead removal via napier grass in synthetic brackish water using response surface model. IOP Conference Series: Earth and Environmental Science 120, 012020.

ISO, 2006. In: ISO (Ed.), Environmental Management—Life Cycle Assessment—Principles and Framework. British Standards Institution.

Jeffery, S., et al., 2010. European Atlas of Soil Biodiversity. Publications Office of the European Union.

Jenkinson, D.S., Coleman, K., 2008. The turnover of organic carbon in subsoils. Part 2. Modelling carbon turnover. European Journal of Soil Science 59 (2), 400–413.

Jiang, D., et al., 2016. Model development and surface analysis of a bio-chemical process. Chemometrics and Intelligent Laboratory Systems 157, 133–139.

Kazemzadeh, N., Hu, G., 2013. Optimization models for biorefinery supply chain network design under uncertainty. Journal of Renewable and Sustainable Energy 5 (5), 053125.

Khalid, S., et al., 2017. A comparison of technologies for remediation of heavy metal contaminated soils. Journal of Geochemical Exploration 182, 247–268.

Kobayashi, N.I., et al., 2013. Characterization of rapid intervascular transport of cadmium in rice stem by radioisotope imaging. Journal of Experimental Botany 64 (2), 507–517.

Krgovic, R., et al., 2015. Phytoextraction of metals by *Erigeron canadensis* L. from fly ash landfill of power plant "Kolubara". Environmental Science and Pollution Research 22 (14), 10506–10515.

Kutrowska, A., Szelag, M., 2014. Low-molecular weight organic acids and peptides involved in the long-distance transport of trace metals. Acta Physiologiae Plantarum 36 (8), 1957–1968.

Li, C.S., 2000. Modeling trace gas emissions from agricultural ecosystems. Nutrient Cycling in Agroecosystems 58 (1), 259–276.

Li, C., Frolking, S., Frolking, T.A., 1992. A model of nitrous oxide evolution from soil driven by rainfall events: 1. Model structure and sensitivity. Journal of Geophysical Research: Atmospheres 97 (D9), 9759–9776.

Li, C., Frolking, S., Harriss, R., 1994. Modeling carbon biogeochemistry in agricultural soils. Global Biogeochemical Cycles 8 (3), 237–254.

Li, C., et al., 2006. Modeling nitrate leaching with a biogeochemical model modified based on observations in a row-crop field in Iowa. Ecological Modelling 196 (1), 116–130.

Li, T., et al., 2019. Soil pollution management in China: a brief introduction. Sustainability 11 (3), 556.

Lin, Y., et al., 2016. Economic and environmental evaluation of nitrogen removal and recovery methods from wastewater. Bioresource Technology 215, 227–238.

Lithos, 2017. Remediation Strategy: Land at Midland Road. Roysten for Kier Living Northern.

Liu, L., et al., 2018. Remediation techniques for heavy metal-contaminated soils: principles and applicability. The Science of the Total Environment 633, 206–219.

Long, S.P., 1991. Modification of the response of photosynthetic productivity to rising temperature by atmospheric CO_2 concentrations: has its importance been underestimated? Plant, Cell and Environment 14 (8), 729–739.

Lopez-Vizcaino, R., et al., 2016. Scale-up on electrokinetic remediation: engineering and technological parameters. Journal of Hazardous Materials 315, 135–143.

Lotfy, S.M., Mostafa, A.Z., 2014. Phytoremediation of contaminated soil with cobalt and chromium. Journal of Geochemical Exploration 144, 367–373.

Lugli, F., Mahler, C.F., 2016. A soil-plant model applied to phytoremediation of metals. International Journal of Phytoremediation 18 (4), 295–307.

Mahmood, T., 2010. Phytoextraction of heavy metals - the process and scope for remediation of contaminated soils. Soil and Environment 29 (2), 91–109.

Mann, M.J., 1999. Full-scale and pilot-scale soil washing. Journal of Hazardous Materials 66 (1–2), 119–136.

Mao, X., et al., 2016. Electro-kinetic remediation coupled with phytoremediation to remove lead, arsenic and cesium from contaminated paddy soil. Ecotoxicology and Environmental Safety 125, 16–24.

Markert, B., 1994. Progress report on the element concentrations cadastre project (ECCP) of INTERCOL/IUBS. In: International Union of Biological Sciences. 25th General Assembly., Paris.

McGrath, S.P., Zhao, F.-J., 2003. Phytoextraction of metals and metalloids from contaminated soils. Current Opinion in Biotechnology 14 (3), 277–282.

Meegoda, J.N., et al., 2000. Remediation of chromium-contaminated soils— pilot-scale investigation. Practice Periodical of Hazardous, Toxic, and Radioactive Waste Management 4 (1), 7–15.

Miettinen, K., 1998. No-preference methods. In: Nonlinear Multiobjective Optimization. Springer, US: Boston, MA, pp. 67–76.

Miettinen, K., 1998. A Priori methods. In: Nonlinear Multiobjective Optimization. Springer, US: Boston, MA, pp. 115–129.

Miettinen, K., 1998. A Posteriori methods. In: Nonlinear Multiobjective Optimization. Springer, US: Boston, MA, pp. 77–113.

Miettinen, K., 1998. Interactive methods. In: Nonlinear Multiobjective Optimization. Springer, US: Boston, MA, pp. 131–213.

Mojiri, A., et al., 2013. Phytoremediation of heavy metals from urban waste leachate by Southern Cattail (Typha domingensis). International Journal of Scientific Research in Environmental Sciences 1 (4), 63–70.

Müller, H.R., 1969. Index of geoaccumulation in sediments of the rhine river. GeoJournal 2 (3), 108–118.

Ojuederie, O.B., Babalola, O.O., 2017. Microbial and plant-assisted bioremediation of heavy metal polluted environments: a review. International Journal of Environmental Research and Public Health 14 (12).

Papageorgiou, L.G., 2009. Supply chain optimisation for the process industries: advances and opportunities. Computers and Chemical Engineering 33 (12), 1931–1938.

Parton, W.J., et al., 1998. DAYCENT and its land surface submodel: description and testing. Global and Planetary Change 19 (1–4), 35–48.

Peltoniemi, M., et al., 2007. Models in Country Scale Carbon Accounting of Forest Soils, vol. 41.

Pérez, A.P., Eugenio, N.R., 2018. Status of Local Soil Contamination in Europe: Revision of the Indicator "Progress in the Management of Contaminated Sites in Europe", EUR 29124 EN. Publications Office of the European Union, Luxembourg.

Pollard, A.J., Reeves, R.D., Baker, A.J., 2014. Facultative hyperaccumulation of heavy metals and metalloids. Plant Science 217–218, 8–17.

Prasad, M.N.V., 2003. Phytoremediation of metal-polluted ecosystems: hype for commercialization. Russian Journal of Plant Physiology 50 (5), 686–700.

Qing, X., Yutong, Z., Shenggao, L., 2015. Assessment of heavy metal pollution and human health risk in urban soils of steel industrial city (Anshan), Liaoning, Northeast China. Ecotoxicology and Environmental Safety 120, 377–385.

Salt, D.E., et al., 1995. Phytoremediation: a novel strategy for the removal of toxic metals from the environment using plants. Biotechnology 13 (5), 468–474.

Salt, D.E., Smith, R.D., Raskin, I., 1998. Phytoremediation. Annual Review of Plant Physiology and Plant Molecular Biology 49, 643–668.

Sarwar, N., et al., 2017. Phytoremediation strategies for soils contaminated with heavy metals: modifications and future perspectives. Chemosphere 171, 710–721.

Sattelmacher, B., 2001. The apoplast and its significance for plant mineral nutrition. New Phytologist 149 (2), 167–192.

Schreurs, E., Voets, T., Thewys, T., 2011. GIS-based assessment of the biomass potential from phytoremediation of contaminated agricultural land in the Campine region in Belgium. Biomass and Bioenergy 35 (10), 4469–4480.

Selamat, S.N., et al., 2017. Optimization of lead (Pb) bioaccumulation in *Melastoma malabathricum* L. by response surface methodology (RSM). Rendiconti Lincei. Scienze Fisiche e Naturali 29 (1), 43–51.

Seuntjens, P., Nowack, B., Schulin, R., 2004. Root-zone modeling of heavy metal uptake and leaching in the presence of organic ligands. Plant and Soil 265 (1–2), 61–73.

Shah, N., 2004. Process industry supply chains: advances and challenges. In: BarbosaPovoa, A.P., Matos, H. (Eds.), European Symposium on Computer-Aided Process Engineering - 14. Elsevier Science, Amsterdam, pp. 123–138.

Sharma, B., et al., 2013. Biomass supply chain design and analysis: basis, overview, modeling, challenges, and future. Renewable and Sustainable Energy Reviews 24, 608–627.

Smith, J., et al., 2010. Estimating changes in Scottish soil carbon stocks using ECOSSE. I. Model description and uncertainties. Climate Research 45 (1), 179–192.

Smith, J., et al., 2010. Estimating changes in Scottish soil carbon stocks using ECOSSE. II. Application. Climate Research 45 (1), 193–205.

Sotenko, M., et al., 2017. Phytoremediation-biorefinery tandem for effective clean-up of metal contaminated soil and biomass valorisation. International Journal of Phytoremediation 19 (11), 965–975.

Stojanovic, M., et al., 2016. Biometric approach in selecting plants for phytoaccumulation of uranium. International Journal of Phytoremediation 18 (5), 527–533.

Sun, H., et al., 2018. The enhancement by arbuscular mycorrhizal fungi of the Cd remediation ability and bioenergy quality-related factors of five switchgrass cultivars in Cd-contaminated soil. PeerJ 6, 27.

Tang, Y.-T., et al., 2012. Designing cropping systems for metal-contaminated sites: A Review. Pedosphere 22, 470–488.

The Ministry of Environmental Protection, 2014. The Ministry of Land and Resources Report on the National General Survey of Soil Contamination.

Titah, H.S., et al., 2018. Statistical optimization of the phytoremediation of arsenic by Ludwiga octovalvisin a pilot reed bed using response surface methodology (RSM) versus an artificial neural network (ANN). International Journal of Phytoremediation 20 (7), 721–729.

Titah, H.S., et al., 2018. Statistical optimization of the phytoremediation of arsenic by Ludwigia octovalvis- in a pilot reed bed using response surface methodology (RSM) versus an artificial neural network (ANN). International Journal of Phytoremediation 20 (7), 721–729.

Tóth, G., Jones, A., Montanarella, L., 2013. LUCAS Topsoil Survey. Methodology, Data and Results. Publications Office of the European Union, Luxembourg, p. 141.

Toth, G., et al., 2016. Heavy metals in agricultural soils of the European Union with implications for food safety. Environment International 88, 299–309.

Tu, S., Ma, L.Q., 2003. Interactive effects of pH, arsenic and phosphorus on uptake of As and P and growth of the arsenic hyperaccumulator *Pteris vittata* L. under hydroponic conditions. Environmental and Experimental Botany 50 (3), 243–251.

Tunali Akar, S., et al., 2016. Removal of cadmium and manganese by an ecofriendly biomass. CLEAN – Soil, Air, Water 44 (2), 202–210.

UNEP/SETAC Life Cycle Initiative, 2011. Towards a Life Cycle Sustainability Assessment. UNEP.

Valentín, L., Nousiainen, A., Mikkonen, A., 2013. Introduction to organic contaminants in soil. Concepts and Risks 24, 1–29.

van der Ent, A., et al., 2012. Hyperaccumulators of metal and metalloid trace elements: facts and fiction. Plant and Soil 362 (1–2), 319–334.

Verma, P., et al., 2006. Modeling rhizofiltration: heavy-metal uptake by plant roots. Environmental Modeling and Assessment 11 (4), 387–394.

Virkutyte, J., Sillanpää, M., Latostenmaa, P., 2002. Electrokinetic soil remediation — critical overview. The Science of the Total Environment 289 (1–3), 97–121.

Voets, T., et al., 2013. GIS-BASED location optimization of a biomass conversion plant on contaminated willow in the Campine region (Belgium). Biomass and Bioenergy 55, 339–349.

von Caemmerer, S., Farquhar, G.D., 1981. Some relationships between the biochemistry of photosynthesis and the gas exchange of leaves. Planta 153 (4), 376–387.

Wang, J., et al., 2002. Mechanisms of arsenic Hyperaccumulation in *Pteris vittata.* Uptake kinetics, Interactions with phosphate, and arsenic speciation. Plant Physiology 130 (3), 1552–1561.

Wang, S., Wu, Z., Luo, J., 2018. Transfer mechanism, uptake kinetic process, and bioavailability of P, Cu, Cd, Pb, and Zn in macrophyte rhizosphere using diffusive gradients in thin films. Environmental Science and Technology 52 (3), 1096–1108.

Wasay, S.A., Barrington, S., Tokunaga, S., 2001. Organic acids for the in situ remediation of soils polluted by heavy metals: soil flushing in Columns. Water, Air, and Soil Pollution 127 (1/4), 301–314.

Wong, C.S.C., Li, X.D., Thornton, I., 2006. Urban environmental geochemistry of trace metals. Environmental Pollution 142 (1), 1–16.

Wuana, R.A., Okieimen, F.E., 2011. Heavy metals in contaminated soils: a review of sources, Chemistry, risks and best available strategies for remediation. ISRN Ecology 2011, 1–20.

Xiao, R., et al., 2017. Soil heavy metal contamination and health risks associated with artisanal gold mining in Tongguan, Shaanxi, China. Ecotoxicology and Environmental Safety 141, 17–24.

Yan, Y., et al., 2018. Application of iron-loaded activated carbon electrodes for electrokinetic remediation of chromium-contaminated soil in a three-dimensional electrode system. Scientific Reports 8 (1), 5753.

Yu, X.Z., Gu, J.D., 2007. Differences in Michaelis-Menten kinetics for different cultivars of maize during cyanide removal. Ecotoxicology and Environmental Safety 67 (2), 254–259.

Yue, D., You, F., Snyder, S.W., 2014. Biomass-to-bioenergy and biofuel supply chain optimization: overview, key issues and challenges. Computers and Chemical Engineering 66 (0), 36–56.

Zhang, Y., et al., 2002. A simulation model linking crop growth and soil biogeochemistry for sustainable agriculture. Ecological Modelling 151 (1), 75–108.

Zhang, L., et al., 2007. Heavy metal contamination in western Xiamen Bay sediments and its vicinity, China. Marine Pollution Bulletin 54 (7), 974–982.

Zhang, X., et al., 2015. Impact of soil heavy metal pollution on food safety in China. PLoS One 10 (8), e0135182.

Zhao, F.J., et al., 2015. Soil contamination in China: current status and mitigation strategies. Environmental Science and Technology 49 (2), 750–759.

Zhu, J., Raschke, K., Köhler, B., 2006. An electrogenic pump in the xylem parenchyma of barley roots. Physiologia Plantarum 129 (2), 397–406.

Zloch, M., et al., 2017. Modeling of phytoextraction efficiency of microbially stimulated *Salix dasyclados* L. in the soils with different speciation of heavy metals. International Journal of Phytoremediation 19 (12), 1150–1164.

Chapter 14

The sustainability of nanoremediation—two initial case studies from Europe

Brian Bone[1], Paul Bardos[2], Steve Edgar[3], Petr Kvapil[4]
[1]*Bone Environmental Consultant Ltd. and University of Coventry, Coventry, United Kingdom;* [2]*Environmental Technology Ltd., Reading, and University of Brighton, Brighton, United Kingdom;* [3]*FLI Global, Bristol, United Kingdom;* [4]*Photon Water Technology s.r.o, Prague, Czech Republic*

1. Introduction

Since the 1990s, a risk-based approach to the management of historically contaminated land has developed in Europe, and it has been adopted by key stakeholder groups including regulators, site owners, and practitioners (CLARINET & NICOLE, 1998). This approach is based on the prevention of unacceptable risks to human health and the environment to ensure that sites are fit for existing or future use. More recently, interest has been shown in including sustainability as a decision-making criterion. This enables the selection of a remediation approach that both meets remediation objectives and achieves a balanced net benefit when considering wider environmental, economic, and social impacts (NICOLE and Common Forum, 2013). Sustainable remediation has become an area of intense development across the world, with public and private sector organizations involved in a number of projects and networks, which aim to improve remediation practice and make it more sustainable (Bardos et al., 2016a; Rizzo et al., 2016).

Nanoremediation is an emerging remediation technology for the treatment of groundwater and soil. It is applied mainly in the saturated zone, primarily using iron-based engineered nanoparticles (that is, particles with at least one dimension less than 100 nm). To date, there have been relatively few commercial deployments of nanoremediation. Bardos et al. (2018) cite 100 examples of field-scale applications of nanoiron particles at pilot and full-scale.

The reasons for the slow market development of nanoremediation are linked to uncertainty, relatively high material costs, and perception that on balance, the benefits do not outweigh the risks in the context of sustainable risk management (Beddoes et al., 2015; Irving et al., 2006; Karn et al., 2009). A

Sustainable Remediation of Contaminated Soil and Groundwater. https://doi.org/10.1016/B978-0-12-817982-6.00014-8

number of national risk–benefit studies have been undertaken in North America and Europe (as cited in Bardos et al., 2011; Grieger et al., 2010), and NanoRem has conducted a Strength, Weakness, Opportunity, and Threat (SWOT) analysis (Bardos et al., 2016b).

In some regions, the potential risks of the deployment of nanoparticles for in situ remediation have been poorly understood, and this has resulted in precautionary regulatory positions. There is a special situation in the United Kingdom, which introduced a voluntary moratorium on the release of engineered nanoparticles in response to a Royal Society/Royal Academy of Engineering report (RS/RAE, 2004). This is described more in Annex 1.

The sustainability of all remediation approaches, including nanoremediation, will come under increasing scrutiny as part of an emerging worldwide discussion on sustainable remediation being advanced by developing global initiatives (Bardos et al., 2013) and international standards (e.g., American Society for Testing & Materials, 2013; International Organization for Standardization, 2017).

A European Commission FP7 project, NanoRem (taking nanotechnological remediation processes from lab scale to end-user applications for the restoration of a clean environment) (http://www.nanorem.eu/), ran from 2013 to early 2017. It addressed some of the uncertainties about remedial performance, fate and transport, and the impacts on ecosystems of the release of engineered nanoparticles for the remediation of soil and groundwater. One aspect of this project was a qualitative sustainability assessment of the nanoremediation used at one NanoRem pilot site, benchmarked against at least one possible alternative remediation strategy.

A small team of remediation specialists carried out a qualitative assessment for the Spolchemie site in the Czech Republic and a redevelopment site in the United Kingdom, referred to in this chapter as Site A. The latter site was not part of the NanoRem project. This chapter presents the findings from the two case studies and discusses them in the context of uncertainties associated with risks and benefits and the new evidence from the NanoRem project.

2. Approach to the sustainability assessments

A NanoRem workbook (Bardos et al., 2015) for sustainable remediation assessment was developed to be used in conjunction with risk-based land management approaches, based on recognized good practice from European and UK networks (see Section 14.2.2 in this chapter)

- The Network for Industrially Contaminated Land in Europe (NICOLE) “roadmap” to sustainable remediation to promote the incorporation of sustainability assessment into risk-based land management (NICOLE, 2010) and
- The UK Sustainable Remediation Forum (SuRF-UK) framework for assessing the sustainability of soil and groundwater remediation (CL:AIRE, 2010).

The NICOLE roadmap (NICOLE, 2010) describes a sustainable remediation project as "one that represents the best solution when considering environmental, social and economic factors—as agreed by the stakeholders" and sustainability assessment as "the process of gaining an understanding of possible outcomes across all three elements (environmental, social and economic) of sustainable development."

The roadmap lays down principles for the sustainability assessment of remediation projects, but does not provide a procedural approach. NanoRem therefore developed a workbook based on the SuRF-UK assessment framework (CL:AIRE, 2010), which is in line with the roadmap (and now also ISO 18504:2017).

2.1 Sites selected for sustainability assessment

A small team of remediation specialists carried out qualitative sustainability assessments for two sites. One site, a small area of the large Spolchemie site in the Czech Republic, was a NanoRem pilot test site for which treatability studies and a trial injection had already taken place. The sustainability assessment was therefore post hoc, because a remediation process had already been selected, and some pilot testing data were already available.

The second site, Site A (from the United Kingdom) was assessed in a fully ex ante (forward looking) sustainability assessment for a potential nanoremediation deployment. It was carried out at the remediation options appraisal stage (Stage B of the SuRF-UK framework). As a consequence, no site-specific information on the performance of nanoremediation was available to support the assessment. The two sites are considered collectively to determine whether there are any general sustainability drivers (positive or negative) that lead to common conclusions.

The two qualitative assessments, collectively labeled NanoRem, were planned and carried out by a small group of remediation professionals from AQUATEST, r3, and FLI Global, with an independent overview provided by CL:AIRE (Contaminated Land: Applications in Real Environments). This provided a blend of practical experience of remediation, sustainability assessment, and knowledge of the sites. At this initial stage, no wider stakeholders were involved, but the contractors at each site had discussions with other stakeholders, the site owners, and the regulators before taking part in the assessment process.

2.2 The NanoRem workbook for sustainability assessment

The NanoRem workbook (Bardos et al., 2015) involves three stages (Fig. 15.1).

1. Preparation—agreeing in advance on how the sustainability assessment will be reported, who will be involved, and how communication will take place with other stakeholders.

2. Definition—providing a clearly defined assessment procedure, which considers objectives, boundaries, scope, method, and uncertainty.
3. Execution—carrying out the assessment procedure with an appropriate level of dialogue and ensuring that the procedure, its findings, and its underlying assumptions are clearly communicated to all relevant parties.

Taken together, the preparation and definition stages provide the framework for sustainability assessment, for a qualitative, semiquantitative, or quantitative assessment. SuRF-UK refers to these two stages together as "framing." As Fig. 14.1 shows, each of these stages has been divided into individual steps, which were followed for each sustainability assessment and reported in a standard format spreadsheet. Sustainability assessment may be an iterative process, with successive iterations refining the sustainability assessment.

2.3 Preparation

Preparation is needed to specify the nature of the sustainability assessment to be undertaken. There are a number of steps that need to be completed to successfully plan for the assessment.

Step 1—identifies the required outcomes. The principal decisions of the NanoRem sustainability assessments are to provide a qualitative assessment of the sustainability of nanoremediation and to compare a nanoremediation option with a range of feasible remediation options for the treatment of groundwater and/or source contamination.

Step 2—project description. The project being considered is the use of nanoremediation, or feasible alternative remedial options, to attain a desired

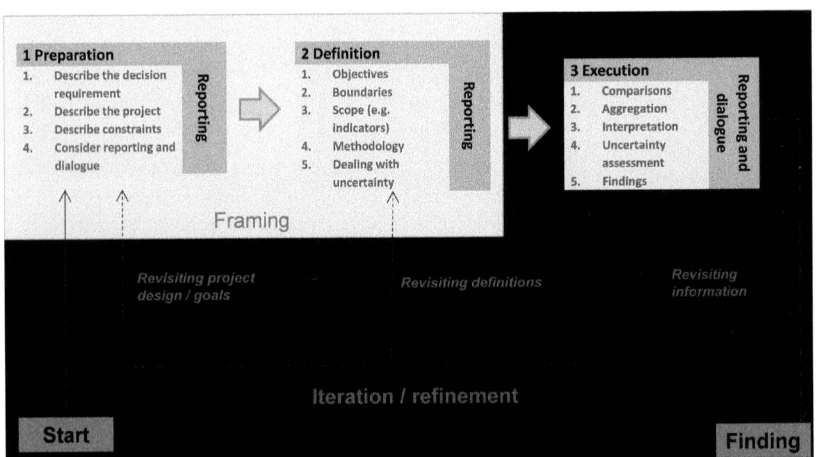

FIGURE 14.1 Approach to sustainability assessment (after SuRF-UK). *Credit: © CL:AIRE. Reproduced with permission.*

set of remediation or management objectives that may be actual or hypothetical. Both assessments compared nanoremediation with both in situ and ex situ technologies. A baseline option (maintain the status quo) was considered for one assessment.

Step 3—description of constraints. Constraints are limitations or restrictions and may be environmental (e.g., site conditions, emissions limits), economic (e.g., budgetary limits), or social (e.g., regulations and policy). The assessors identified the constraints for each site. However, this identification was provisional because the research was conducted at an early stage of the remediation, and it will be reviewed in any further iteration.

Step 4—reporting and dialogue. A reporting template and approach described in the NanoRem workbook were used for reporting the sustainability assessments discussed in this chapter.

2.4 Definition

The goal of this stage is to clearly specify the sustainability assessment approach that is to be used, so that it can be discussed with all relevant parties. It uses the outputs of the preparation stage as a starting point and involves five steps.

Step 1—objectives. Specific sustainability objectives are set from the information documented during the preparation stage to provide a concise description of what the sustainability assessment will consider, what its purpose is, what factors will affect it, who it will be discussed with, and how it will be reported. The objectives for both NanoRem projects were to carry out a qualitative assessment of the sustainability of nanoremediation technology in comparison to established technologies. As the NanoRem studies were initial assessments, the findings were not discussed with wider stakeholders identified during preparation, but the information is available for use in dialogue with stakeholders as remediation plans develop. The assessment reports were reviewed by the NanoRem Project Management Board and Advisory Group and the findings summarized in Bardos et al. (2016b).

Step 2—boundaries. Boundary conditions determine which effects will be considered within a sustainability assessment to ensure a fair, like-for-like comparison of options. A consistent boundary was set for the two assessments, limiting consideration of the life cycle to the operational stage only. The boundaries set are shown in Table 14.1.

Step 3—scope. The assessors decided to consider all the indicator sets developed to support the SuRF-UK framework (CL:AIRE, 2010; 2011). Five overarching indicators were identified for each of the environmental, economic, and social "pillars" or "elements" of sustainability (CL:AIRE, 2011), as shown in Table 14.2, with brief definitions as used for the NanoRem assessments. Each indicator category can be subdivided into a number of criteria as discussed in Section 14.4.

TABLE 14.1 Boundaries set for the NanoRem assessments.

Boundary aspect	Inclusions/exclusions
Time	For Spolchemie, the measure needs to remain effective (prevent off-site migration) until final remediation is completed in 2020. No timescale is available for site A.
Spatial	Local: impacts on-site and at adjacent sites Global: impact of reagent production and emissions of greenhouse gases.
System	All options must achieve risk management objectives.
Life cycle	Considers the operational stage only, i.e., the deterioration of plant/equipment that will be reused; the use of consumables, reagents, fuel, etc., during the works; and the production of reagents used, but not the manufacture of remediation equipment. No life-cycle data for the production of nanoparticles were available at the time of the assessment.

Step 4—methodology. This where individual comparisons across options are carried out and then aggregated into an overall sustainability assessment. For the NanoRem assessments, comparisons were made using a simple ranking system, with a value of 1 assigned to the best option. For the Spolchemie site, the scores ranged from 1 to 5 and from 1 to 3 for Site A, reflecting the number of options under consideration.

Step 5—uncertainties. Uncertainties may emerge during the assessment work because of limitations to the available information or as a result of the dialogue process because of differences in opinion. A convenient means of assessing the impact of uncertainty on the outcome is to use sensitivity analysis. Section 14.4 provides further discussion of uncertainty reported during the assessments.

2.5 Execution

In practice, a number of spreadsheets are produced, which provide a matrix of the appropriate sustainability indicators (identified earlier, in the Definition stage), against the remediation options to be considered.

Step 1—comparison across options. In this step, comparisons across all options are made for each sustainability category (e.g., emissions to air). Each category can be subdivided into a number of criterion.

The initial stages for both NanoRem assessments were undertaken by the same small project team for continuity across the assessments. The project

TABLE 14.2 Overarching indicator categories and their summary working descriptions.

Environmental	Social	Economic
ENV1 emissions to air: Emissions during operational stage that may impact climate or air quality.	SOC1 human health andand safety: Healthy andand safety of workers and public during operational stage.	ECON1 direct economic costs andand benefits: Direct financial costs of works and benefits to landowner, incl. site value.
ENV2 soil andand ground conditions: Changes in soil function (e.g., quality, filtration, structure, geotechnical), impacts on protected sites.	SOC2 ethics andand equity: Distribution of risks and benefits between stakeholders and generations.	ECON2 indirect economic costs andand benefits: Change in local property values, impact on corporate reputation, potential enforcement action.
ENV3 groundwater andand surface water: Changes in the release of contaminants and impact on water quality, impact on flood risk/prevention.	SOC3 neighborhoods andand locality: Impacts/benefits on neighborhood (noise, dust etc.) and local environment (e.g., conservation).	ECON3 employment andand employment capital: Job creation and increase in skill levels, opportunities for training.
ENV4 ecology: Impacts on flora, fauna, and food chains, not addressed in ENV 2 andand 3.	SOC4 communities andand community involvement: Communications plan and engagement in decision-making.	ECON4 induced economic costs andand benefits: Potential to create opportunity for inward investment.
ENV5 natural resources andand waste: Impacts on use and substitution of natural resources (e.g., aggregates, energy, andand fuel), water abstraction, waste management.	SOC5 uncertainty andand evidence: Quality of supporting evidence and robustness of appraisal of each option.	ECON5 project lifespan andand flexibility: Duration of measures and requirement for institutional controls, ability to adapt, and resilience in the face of climate change.

Credit: Based on Contaminated Land Applications in Real Environments (CL:AIRE). 2010. A Framework for Assessing the Sustainability of Soil and Groundwater Remediation. March 2010, CL:AIRE, London, UK. ISBN 978-1-905046-19-5. Available from: https://www.claire.co.uk/projects-and-initiatives/surf-uk.

team included the contractor responsible for delivering the remediation strategy for each site.

Step 2—aggregation. At its simplest, aggregation may be no more than tabulating individual comparisons in summary tables. This was the approach taken by the project team for the NanoRem assessments, with scores against each of the 15 headline indicator categories. Aggregation was carried out in two stages. Firstly, the five categories for each of the environmental, social, and economic "pillars" were aggregated, providing a combined ranking (CR) for each pillar with no weighting of scores. Secondly, the CRs were then summed to produce an overall combined ranking (OCR), the overall score of relative sustainability that can be compared for all remedial actions assessed.

Step 3—interpretation. The most obvious initial question is: Does the comparison yield a clear "winner" supported by all parties?

- If "yes," sufficient decision-making support has been provided, subject to any sensitivity analysis carried out.
- If "no," one of the following applies:
  - Two or more options are supported equally, so choosing between them may be based on operational convenience, cost, etc.
  - The assessment is inconclusive and requires further effort to improve its reliability.

The interpretation of the aggregated scores is simple, based on comparison tables. This was the approach taken for the NanoRem assessments. The intention was to use a simple communication tool to highlight clear similarities and differences between options and to determine whether there are any general sustainability drivers (positive or negative) affecting the deployment of nanoremediation on a site-specific basis that may be generically applied to the technology.

Step 4—uncertainties. Any uncertainties identified during the definition and execution phases should be collated and restated in a single section or table, which should be circulated to the participating stakeholders for their feedback. The easiest way to assess the implications of these uncertainties is to undertake a sensitivity analysis and assess the changes in the overall sustainability assessment resulting from altering assessment scores. The changes made to the input data for assessment against individual indicator categories should be agreed in advance and clearly documented.

Sensitivity analysis was carried out for the Spolchemie assessment (Section 14.4.1). An alternative approach was taken for the UK site, Site A (Section 14.4.2). As a voluntary moratorium on the release of engineered nanoparticles is in place in the United Kingdom, the uncertainties associated with their use were discussed against perceived risks and benefits documented from public dialogue exercises on nanotechnologies carried out in the United Kingdom in 2006 and 2015 (Irving et al., 2006, Beddoes et al., 2015). Evaluation of those perceptions was made against the research findings of the NanoRem project to establish whether the emerging evidence helps to address the concerns expressed during the public dialogue exercises (Annex 1 of this Chapter).

Step 5—conclusion. Where the uncertainties in the sustainability assessment mean that there is no clearly favored option, the further steps needed to improve assessment reliability will usually need to be agreed and clearly recorded. It is important to note that the NanoRem assessments were for research purposes to examine how nanoremediation compares against other technologies. This represents the first stage of an iterative process to determine the overall remediation strategies for both sites. It is anticipated that consultation with external stakeholders will be undertaken once the overall remediation strategies for the transition from research to full-scale implementation have been developed. As such, no external stakeholders were formally involved at this early stage in the remedial actions.

3. NanoRem site summary details

The Spolchemie site was chosen as one of the NanoRem pilot test sites, to assess the use of zerovalent iron nanoparticles (nZVI) for in situ cleanup of dissolved chlorinated hydrocarbons in groundwater. Spolchemie, located at Usti nad Labem in the Czech Republic, is one of the leading synthetic resin manufacturers in Europe. Over several decades of production and storage of raw materials and products, spillage and leakage have led to extensive contamination by chlorinated hydrocarbons, which have dispersed widely (in contaminant plumes) from the original contaminant source areas.

"Site A" is a former chemical works located in the east of England. The chemical works were operational from the 1940s until the 2000s. A chlorinated hydrocarbon-contaminated source zone is largely associated with historic waste water disposal trenches that were in use until the 1970s, after which they were backfilled. Much of the plant and associated site infrastructure remains on Site A, although large parts of the site comprise open space. It is proposed to redevelop the site for residential properties with the final land use already confirmed through the UK planning process.

FLI Global offered Site A for evaluating the sustainability of options that were being considered to remediate the site. Nanotechnology's potential to be an effective part of the remediation process was recognized by the remediation contractor, but it was also clear that constructive dialogue would be important to present evidence of remedial performance and wider environmental effects to UK stakeholders. It is appropriate that this dialogue is taking place while the voluntary moratorium on the release of engineered nanoparticles is in place.

3.1 Spolchemie site

A permeable reactive barrier (PRB) has been installed near the western boundary of the site to prevent the migration of contamination onto adjacent

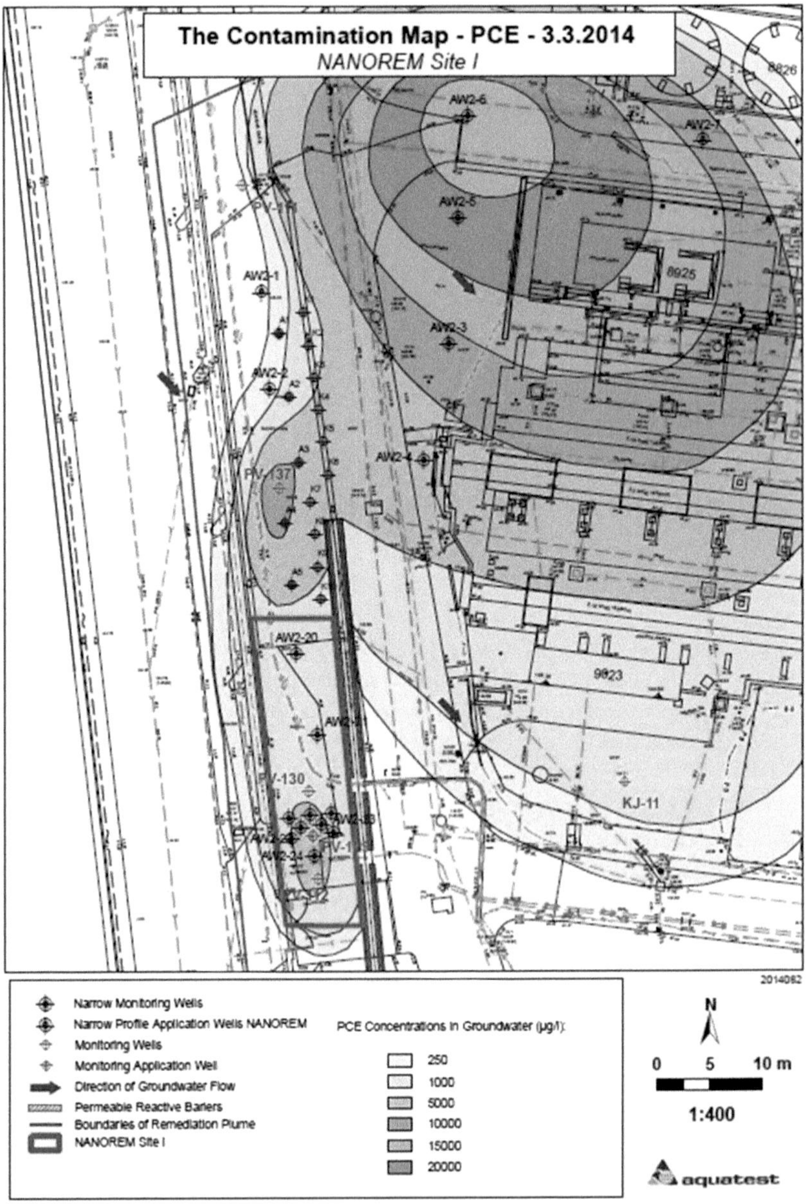

FIGURE 14.2 Spolchemie site layout plan. *Credit:© Photon Water Technology s.r.o. Reproduced with permission.*

land, but a contaminant plume of chlorinated hydrocarbons has been identified to the north, outside of the PRB. The pilot test area (Fig. 14.2) is located within this plume, and the pilot test results are being used to test the suitability of nZVI injection as a short-term remediation option before full remediation is completed in 2020. No change in land use is associated with the remediation plans for the site.

The contaminant plume is located within a 10 m thick unit of Quaternary sand and gravel underlain by a clay aquitard. The water level is about 4 m below ground level, with water flow to the southeast.

Remediation and management approaches were selected by AQUATEST as feasible options for managing the risks effectively, and a baseline condition was set to represent the current situation. The options assessed are summarized in Table 14.3. They were:

- Monitoring (baseline) (Mon)
- In situ chemical oxidation (ISCO)
- In situ bioremediation (ISB)
- In situ nanoremediation using nanoscale zero valent iron (nZVI)
- Pump and treat (P&T).

3.2 Site A

The contaminant source and associated groundwater plume are located within deposits of made ground and natural strata (interbedded sand and gravel and clay) up to 4 m thick, which overlie a clay aquitard. Groundwater level at the site is typically 1 m below ground level, and groundwater flows toward a local watercourse. A plume of dissolved chlorinated hydrocarbons has been identified, which represents a potential risk to the local watercourse, located approximately 200 m down the gradient of the source zone.

The overall project goal was to maximize contaminant mass removal from the source area and ultimately reduce dissolved concentrations of contaminants (by betterment) in the aquifer. Given that a change in land use at the site (to residential use) had already been confirmed through a planning process, this sustainability assessment was focused on the implementation of the remediation strategy (options appraisal) stage.

Remediation and management approaches were selected by FLI Global as being feasible options to manage the risks effectively. No baseline condition was set to represent the current situation as a change in land use is driving remediation. The options assessed are summarized in Table 14.4. They are:

- In situ chemical oxidation (ISCO)
- In situ integrated nanoremediation using a combination of nanoremediation, microscale iron, and direct current electrokinetics (nZVI + DC)
- Excavation and disposal (Ex&D).

4. Sustainability assessment findings

The assessors decided to initially consider all the indicator categories developed to support the SuRF-UK framework (CL:AIRE 2010): five overarching

TABLE 14.3 Remediation options for the Spolchemie site.

Option	How it works	Aim	Global track record
Baseline (monitoring, Mon)	Collection of groundwater samples from existing wells for chemical analysis.	Check for off-site migration of contaminants.	Established use, action may be triggered if migration occurs.
In situ chemical oxidation (ISCO)	Multiple injections (2/year) of strong oxidant using direct-push technology to produce less/nontoxic reaction products.	Contaminant destruction without toxic intermediate and daughter products.	Established use. Multiple injections to address contaminant rebound (return).
In situ bioremediation (ISB)	Multiple injections (2–3/year), into existing wells, of amendments and nutrients to enhance biological activity.	Contaminant degradation without toxic intermediate and daughter products.	Established use. Multiple injections to address toxic intermediate product formation.
In situ nanoremediation (nZVI)	Multiple injections (1.5/year) of nZVI using direct-push technology. nZVI to promote dechlorination. Rapid reaction, but toxic intermediate products may be formed.	Contaminant destruction without toxic intermediate and daughter products.	Emerging technology with a number of field projects, mainly in the United States. Multiple injections to address toxic intermediate product formation.
Pump and treat (P&T)	Hydraulic containment of plume by pumping groundwater from existing	Control of migration and removal of contaminants from groundwater.	Established use. May take time to achieve cleanup, rebound can occur,

TABLE 14.3 Remediation options for the Spolchemie site.—cont'd

Option	How it works	Aim	Global track record
	wells and above ground treatment of groundwater using air stripping andand granular activated carbon (GAC). Treated water to be reinjected into the aquifer.		potentially extending treatment timescale.

Based on Braun, J. (Ed.), 2017. Taking Nanotechnological Remediation Processes from Lab Scale to End User Applications for the Restoration of a Clean Environment. WP8: Up-Scaling, Risk and Sustainability. DL8.2: Final report on three large-scale experiments and generalized guideline for application. NanoRem Deliverable, European Union Seventh Framework Programme Project, Grant Agreement no 309517 (unpublished).

indicators for the environmental, economic, and social "pillars" or "elements" of sustainability (Table 14.2).

A simple ranked order was used, with 1 for the best option down to 5 for Spolchemie and 3 for Site A. The individual rankings were added to provide an OCR for each option, with the lowest value being the most sustainable one. Results are reported against each sustainability pillar and for the OCR for each site. The following abbreviations are used in the discussion to describe indicator categories:

ENV—Environmental
ECON—Economic
SOC—Social

4.1 Spolchemie

The OCR shows nanoremediation to be the most sustainable option with a score of 24, although there is little to separate the three in situ options, with OCRs of 26 and 27 for in situ bioremediation and ISCO (in situ chemical oxidation), respectively. The results are summarized in Table 14.5 and discussed for each of the environmental, economic, and social indicator categories.

Environmental indicator categories.

TABLE 14.4 Remediation options for Site A.

Option	How it works	Aim	Global track record
In situ chemical oxidation (ISCO)	Multiple injections of strong oxidant using existing boreholes, supplemented where appropriate with the installation of additional boreholes and/ or the use of direct-push technology.	Contaminant destruction without toxic intermediate and daughter products.	Established use. Multiple injections to address contaminant rebound.
In situ integrated nanoremediation with direct current (nZVI + DC)	Multiple injections of nZVI using direct-push technology. nZVI to promote dechlorination. DC electrokinetics will be applied across the remediation area to aid reactions. Potentially rapid reaction, but toxic intermediate products may be formed.	Contaminant destruction without toxic intermediate and daughter products.	The integrated nZVI and electrokinetic approach proposed has been developed and tested in the Czech republic (Czech patent no. 304152 CZ; Cernik et al. pending; Gomes et al., 2012).
Excavation and disposal (Ex&D)	Enabling works including dewatering in source zone followed by excavation of source zone, off-site disposal, and backfilling of excavations.	Removal of contaminanted soil from the source zone.	Established use in United Kingdom and elsewhere.

TABLE 14.5 Summary rankings from Spolchemie sustainability assessment.

Indicator category	ISCO	P and T	nZVI	Mon	ISB
Environmental					
ENV1: Air	3	5	3	1	2
ENV2: Soil	1	3	2	3	3
ENV3: Water	1	4	1	5	3
ENV4: Ecology	1	4	1	5	1
ENV5: Resources andand waste	3	5	3	1	2
ENV combined ranking	**9**	**21**	**10**	**15**	**11**
ECON1: Direct C&B	2	4	2	5	1
ECON2: Indirect C&B	1	4	1	5	1
ECON3: Employment	2	1	2	2	2
ECON4: Induced C&B	2	2	1	2	2
ECON5: Lifespan andand flexibility	1	4	1	5	1
ECON combined ranking	**8**	**15**	**7**	**19**	**7**
Social					
SOC1: Human H&S	4	5	2	1	2
SOC2: Ethics andand equity	1	1	1	1	1
SOC3: Neighborhood	3	5	3	1	2
SOC4: Community involvement	1	4	1	5	1
SOC5: Uncertainty andand evidence	2	2	1	2	2
SOC combined ranking	**10**	**17**	**7**	**10**	**8**
Overall combined ranking	**27**	**53**	**24**	**44**	**26**

Key: ISCO = in situ chemical oxidation; P&T = pump and treat; nZVI = nanoremediation; Mon - = monitoring; ISB = in situ bioremediation.
From Braun, J. (Ed.), 2017. Taking Nanotechnological Remediation Processes from Lab Scale to End User Applications for the Restoration of a Clean Environment. WP8: Up-Scaling, Risk and Sustainability. DL8.2: Final report on three large-scale experiments and generalized guideline for application. NanoRem Deliverable, European Union Seventh Framework Programme Project, Grant Agreement n° 309517 (unpublished).

The difference between the in situ remediation methods and the monitoring and pump and treat options were clearly revealed using environmental indicators, with little differentiation between the three in situ options. This is reflected in the CRs of 9–11 for the in situ options, and 15 and 21 for monitoring and pump and treat, respectively, as shown in the radar plot against

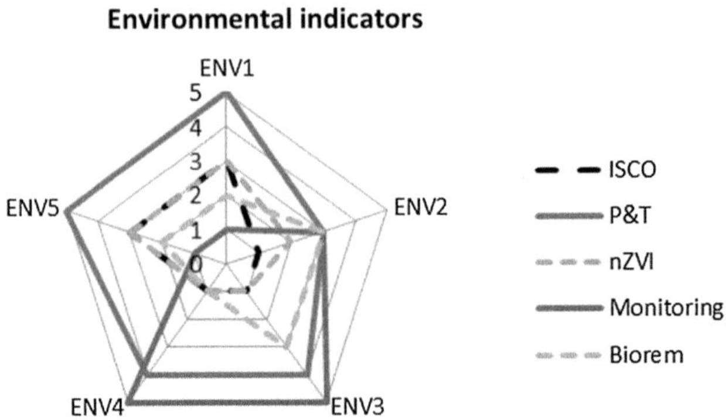

FIGURE 15.3 Radar plot: Environmental indicators. *From Braun, J. (Ed.), 2017. Taking Nanotechnological Remediation Processes from Lab Scale to End User Applications for the Restoration of a Clean Environment. WP8: Up-Scaling, Risk and Sustainability. DL8.2: Final report on three large-scale experiments and generalized guideline for application. NanoRem Deliverable, European Union Seventh Framework Programme Project, Grant Agreement n° 309517 (unpublished).*

each indicator (Fig. 14.3). Quantitative data would be required to separate the in situ options more clearly. Table 14.6 gives a summary of findings and uncertainties.

Monitoring has the highest (worst) scores against ENV3 and ENV4 as groundwater quality and ecology are not improved. In contrast, the lowest scores are evident for ENV1 and ENV5 due to the low emissions, energy use, and natural resource use. Pump and treat is clearly the worst performer against ENV1 and ENV5 due to the energy use and air emissions from an active pumping and treatment process. Even the use of sustainable energy resources will not have a significant impact on rankings.

Economic indicator categories.

There is little to differentiate between the three in situ options, with CRs of 78 in comparison to 15 and 19 for pump and treat and monitoring respectively. The CRs for all of the economic indicators are shown in the radar plot (Fig. 14.4). Quantitative data (for example, capital and operational costs) would be required to differentiate between the in situ options more clearly. However, there is unlikely to be any significant differentiation for ECON3 and ECON4. Table 14.7 gives a summary of findings and uncertainties.

Although monitoring is by far the cheapest option, there are no direct or indirect benefits and this option retains a potential ongoing liability for contamination impacting a neighboring property and substantial reparation costs. This resulted in much discussion by the participants and would likely produce a range of viewpoints in a wider stakeholder group.

TABLE 14.6 Summary of execution findings: Environment.

Indicator and criteria	Summary findings	Uncertainty
ENV1: Emissions to air Production of consumables, Transport, Operation	P&T has the highest emissions and VOC to atmosphere. Emissions from monitoring very low. Products for bioremediation are locally available and use of existing wells reduces emission of exhaust gases during operation in comparison to ISCO and nZVI.	Emissions differentiated for transport, fuel/power use, and production of reagents. Uncertainty over number of injections. No information available on production of reagents. For higher tier assessment, information can be found for plant andand equipment used and transport emissions estimated.
ENV2: Soil andand ground conditions Contaminants, Soil structure	Remediation to address dissolved phase only. Little/no sorbed phase present. Possible risk of clogging is highest with in situ technologies.	Aquifer heterogeneity—impact of effective delivery of reagents in situ and on removal efficiency using P&T. Relative timescale of impact on soil function (positive and negative) of the in situ technologies. Impact of clogging likely ISB > nZVI > ISCO (subjective).
ENV3: Groundwater andand surface water pH andand redox, Contaminants	ISCO and nZVI injection are comparable. They cause only temporary pH and redox changes and no or very low production of intermediates. ISB is slightly worse because of possible production of intermediates. P&T is slower at reducing contaminant concentrations. Mon has no impact on groundwater contamination.	Aquifer heterogeneity—impact of effective delivery of reagents in situ and on removal efficiency using P&T. Assessment based on results of pilot trial (nVZI) and generic information for other options.
ENV4: Ecology Aquifer, River	No off-sites impacts expected. Reduced contaminant concentration is a positive impact, mainly for the faster in situ technologies.	Timescale of impact on microbial community (positive and negative).

Continued

TABLE 14.6 Summary of execution findings: Environment.—cont'd

Indicator and criteria	Summary findings	Uncertainty
	ISB positive microbial community versus potential toxic effects of intermediates. No positive impact from Mon.	
ENV5: Natural resources andand waste Energy andand fossil fuel, Water use, Waste production	The best option is Mon due to few resources needed and little waste production. P&T poorest due to energy use and waste production. In situ techniques are far better, with ISB best of the three due to use of existing wells (although more injections).	None at this tier of assessment. Data for direct push versus existing wells per injection may better differentiate between the in situ options. P&T is clearly a higher producer of waste and a higher consumer of natural resources, even with water being reinjected.

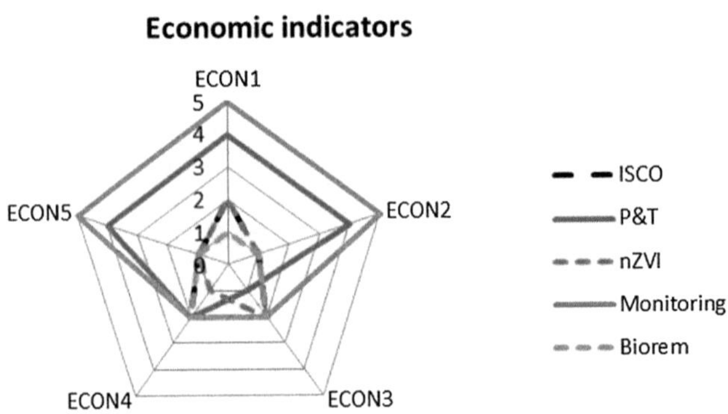

FIGURE 14.4 Radar plot: Economic indicators. *From Braun, J. (Ed.), 2017. Taking Nanotechnological Remediation Processes from Lab Scale to End User Applications for the Restoration of a Clean Environment. WP8: Up-Scaling, Risk and Sustainability. DL8.2: Final report on three large-scale experiments and generalized guideline for application. NanoRem Deliverable, European Union Seventh Framework Programme Project, Grant Agreement n° 309517 (unpublished).*

TABLE 14.7 Summary of findings and uncertainties: Economic.

Indicator and criteria	Summary findings	Uncertainty
ECON1: Direct economic costs andand benefits	ISB has the lowest price of the in situ options. ISCO and nZVI are more expensive. P&T is expensive to build and maintain, compared to all three in situ techniques. Mon is cheaper than the other methods, but has no direct benefits (e.g., uplift in land value, liability discharge).	Not all options benefit from laboratory or pilot tests. Actual configuration andand number of injections are dependent on reagent and site-specific performance.
ECON2: Indirect economic costs andand benefits	There are no significant differences between in situ techniques. A limitation of P&T is that it cannot be used at some sites due to space requirements. Mon retains an ongoing liability due to the consequences of potential off-site migration.	No significant uncertainties identified that would differentiate between the three in situ options.
ECON3: Employment andand employment capital	None of the options has a major influence. Neither in situ techniques nor monitoring generate any jobs in the site area. P&T generates one or two jobs.	None at this tier of assessment or scale of activity.
ECON4: Induced economic costs andand benefits	nZVI is the only option that has an induced economic benefit—ongoing development of a new remediation technique. The other options have no significant impacts.	None at this tier of assessment or scale of activity.

Continued

TABLE 14.7 Summary of findings and uncertainties: Economic.—cont'd

Indicator and criteria	Summary findings	Uncertainty
ECON5: Project lifespan andand flexibility	Short operational duration, long effect, and flexible approach of in situ techniques. P&T has longer operation and is less flexible without significant capital expenditure. Mon has no added benefits and requires a long timescale.	Uncertainty over flexibility if selected option does not deliver remediation objectives (additional injection points, change in reagent, etc.). In situ methods and PandT are considered equal in flexibility at this tier of assessment.

To test the sensitivity of direct and indirect costs and benefits scores, they were adjusted to reflect the low cost of monitoring. Assumptions made in this alternative scenario are that:

- Direct costs and benefits reflect only the financial cost of monitoring.
- There is no impact of monitoring on land values and no associated liability.
- Monitoring remains effective and no costs (punitive fines, emergency remediation measures, decreased land values, damage to corporate reputation) are associated with a failure to control off-site migration.

Table 14.8 shows the impact on OCR when the scores of both direct and indirect costs and benefits are adjusted.

The given table shows that adjustment of scores does not impact the outcome of the assessment, although the revised OCR for monitoring is closer to the in situ technologies.

Social indicator categories.

TABLE 15.8 Sensitivity analysis of costs and benefits of monitoring.

Indicator	ISCO	P and T	nZVI	Mon	ISB
ECON1: Direct costs andand benefits	3	5	3	1	2
ECON2: Indirect costs andand benefits	2	5	2	1	2
Original OCR	27	53	24	44	26
Revised OCR	29	56	26	36	28

Social indicators are difficult to evaluate for an action that is short-term and that impacts a relatively small part of the site. With no plans for regeneration of the site, wider social benefits are unlikely to be realized. Table 14.9 gives a summary of findings and uncertainties.

There is little to differentiate between the three in situ options and monitoring, with CRs of 7–10 in comparison to 17 for pump and treat. The scores against each indicator are shown in the radar plot (Fig. 14.5).

Of the in situ options, ISCO performs relatively poorly due to the health and safety risks perceived from handling the highly reactive reagent (Fenton's reagent). However, this assessment could change if the contractor's health and safety record with ISCO was taken into consideration. ISB is ranked more favorably than the other in situ techniques for SOC 3 as it utilizes existing boreholes, in contrast to the use of direct push for ISCO and nZVI.

For a remediation, as opposed to a regeneration project, it is difficult to differentiate between options in relation to ethics and equity and all options are ranked 1.

Overall outcomes.

The sustainability assessment for the Spolchemie pilot site gave total OCRs (low = best) of:

- 24 for in situ nanoremediation
- 26 for in situ bioremediation
- 27 for in situ chemical oxidation
- 44 for monitoring (the baseline condition)
- 53 for pump and treat.

None of the indicator categories were discounted and no weightings were applied to assign different levels of importance to different indicators. During the exercise it was clear that a number of indicators were considered to be priority and those indicators triggered the most discussion. In no particular order, the priority indicators and their OCRs are identified in Table 14.10. The OCRs are shown for the original assessment and the scores adjusted for costs and benefits are shown in parentheses (see Table 14.8).

It can be seen that the performances of the in situ options are similar, although the order has changed as in situ bioremediation now has the lowest (best) OCR when assessed against the priority indicator categories. When the priority indicators are compared against an alternative scenario that ranks monitoring as the best option in terms of both direct and indirect costs and benefits, the OCR for monitoring is comparable to the OCR for ISCO. In situ bioremediation retains the lowest OCR, followed by nanoremediation.

4.2 Site A

The OCRs show that integrated nanoremediation + direct current is the most sustainable option with a score of 22, although there is little to choose between

TABLE 14.9 Summary of findings and uncertainties: Social.

Indicator and criteria	Summary findings	Uncertainty
SOC1: Human health andand safety	Mon represents low health and safety risk. Injected chemical reagents for in situ techniques represent increased health risk during the applications, especially ISCO, that is far more reactive than nZVI or ISB. P&T method is worst option because of reagent and waste handling (long term), possibility of traffic accidents during waste transport or vandalism on the site.	Statistical data for the options (accidents, near misses) may be used to support or revise the assessment at the next stage.
SOC2: Ethics andand equity	There is nothing to discriminate between options, they are all equal.	This indicator is more relevant to regeneration schemes.
SOC3: Neighborhood andand locality Dust, Traffic, Noise	Mon has no impact on the neighborhoods and locality. For in situ injections, there is disturbance during the injection, with noise and potentially dust (pump and drilling). The highest impact is from P&T —increased traffic, longer duration, and noise potentially 24/7.	None at this tier of assessment or scale of activity.
SOC4: Communities andand community involvement	In situ techniques are only temporarily present on the site, so the duration of community involvement is short. P&T method is on the site long-term and can increase community involvement. Mon is ranked lowest due to	None at this tier of assessment or scale of activity. Communication strategy would be adjusted depending on option selected.

TABLE 14.9 Summary of findings and uncertainties: Social.—cont'd

Indicator and criteria	Summary findings	Uncertainty
	ongoing concerns about contamination that can increase community involvement.	
SOC5: Uncertainty andand evidence	There is uniform uncertainty for all five options except nZVI with the results of the pilot study informing the design.	None at this tier of assessment or scale of activity.

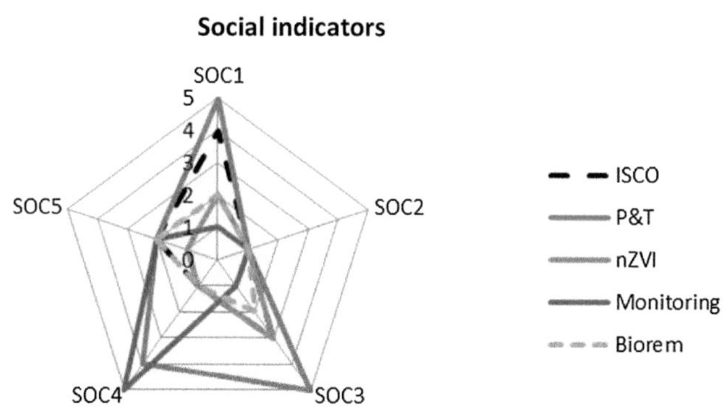

FIGURE 14.5 Radar plot: Social indicators. *From Braun, J. (Ed.), 2017. Taking Nanotechnological Remediation Processes from Lab Scale to End User Applications for the Restoration of a Clean Environment. WP8: Up-Scaling, Risk and Sustainability. DL8.2: Final report on three large-scale experiments and generalized guideline for application. NanoRem Deliverable, European Union Seventh Framework Programme Project, Grant Agreement n° 309517 (unpublished).*

nanoremediation and in situ chemical oxidation (OCR 23). However, the OCRs for both in situ options are significantly better than the OCR of 32 for the excavation and disposal option (Table 14.11).

As the OCRs of the two in situ methods are similar, a decision is most likely to be made following semiquantitative analysis of the three options based on a more detailed evidence base and/or considering the opinions of a wider range of stakeholders. The scoring reflects the uncertainty over the application of the nanoremediation option. Additional data from further field trials would reduce uncertainty and could further reduce the OCR for the nanoremediation option.

Environmental indicator categories.

TABLE 14.10 Assessment against priority headline indicators.

	Spolchemie OCR for priority indicators				
Indicator category	**ISCO**	**P and T**	**nZVI**	**Mon**	**ISB**
ENV1: Air	3	5	3	1	2
ENV3: Water	1	4	1	5	3
ENV5: Resources andand waste	3	5	3	1	2
ENV total CR	7	14	7	7	7
ECON1: Direct C&B	2 (3)	4 (5)	2 (3)	5 (1)	1 (2)
ECON2: Indirect C&B	1 (2)	4 (5)	1 (2)	5 (1)	1 (2)
ECON5: Lifespan andand flexibility	1	4	1	5	1
ECON total CR	4 (6)	12 (14)	4 (6)	15 (7)	3 (5)
SOC1: Human H&S	4	5	2	1	2
SOC3: Neighborhood	3	5	3	1	2
SOC4: Community involvement	1	4	1	5	1
SOC total CR	8	14	6	7	5
Overall CR	19 (21)	40 (42)	17 (19)	29 (21)	15 (17)

There is only a marginal difference between the two in situ methods assessed, with a CR of 6 for nZVI + DC and 7 for ISCO. The only difference between the two options was for groundwater and surface water (ENV3) where nZVI + DC is considered likely to have more positive longer-term impact on pH and redox potential in the wider groundwater plume. Table 14.12 gives a summary of findings and uncertainties.

The CR for the ExandD option is 13 compared to a worst-case score of 15. ExandD scored best for ENV 2 (soil and ground conditions) based on the certainty of removal that excavation provides. For all other environmental options, the excavation option scored 3, reflecting the far greater disturbance and use of natural resources compared to the in situ options (Fig. 14.6).

Economic indicator categories.

There was little difference between the three remedial options considered, with the nZVI + DC and ISCO both scoring a CR of 9 and ExandD scoring 8. Table 14.13 gives a summary of findings and uncertainties.

TABLE 14.11 Summary findings from sustainability assessments.

	Site A sustainability assessment		
Indicator category	**nZVI + DC**	**ISCO**	**Ex and D**
Environmental			
ENV1: Air	1	1	3
ENV2: Soil	2	2	1
ENV3: Water	1	2	3
ENV4: Ecology	1	1	3
ENV5: Resources andand waste	1	1	3
ENV combined ranking	6	7	13
Economic			
ECON1: Direct C&B	1	1	3
ECON2: Indirect C&B	2	2	1
ECON3: Employment	2	2	1
ECON4: Induced C&B	1	2	2
ECON5 lifespan andand flexibility	3	2	1
ECON combined ranking	9	9	8
Social			
SOC1: Human H&S	1	2	3
SOC2: Ethics andand equity	1	1	1
SOC3: Neighborhood	1	1	3
SOC4: Community involvement	1	1	3
SOC5: Uncertainty andand evidence	3	2	1
SOC combined ranking	7	7	11
Overall combined ranking	22	23	32

Key: nZVI + DC = nanoremediation andand direct current; ISCO = in situ chemical oxidation.Ex&D = excavation andand disposal.

Although ExandD scored 3 on ECON1 (direct economic costs and benefits), other rankings were similar to or better than the in situ options, with the increased certainty of contamination removal and likely remediation and aftercare timescales.

The nZVI + DC option was ranked worst for ECON5 (project lifespan and flexibility). This ranking reflects the uncertainty due to the lack of experience

TABLE 14.12 Summary of execution findings: Environment.

Indicator and criteria	Summary findings	Uncertainty
ENV1: Emissions to air Production of consumables, Transport, Operation	Emissions from Ex&D may be significant due to contaminant type and transport. Emissions from in situ options likely to be relatively low. Reuse of existing boreholes will minimize emissions.	Extent/nature of contaminants present relatively well defined but some inherent uncertainties in emissions during excavation. Uncertainty for in situ works in number of injection points required, number of injections, etc.
ENV2: Soil andand ground conditions Contaminants, Soil structure	ExandD will remove contamination and import of clean fill will improve soil structure. In situ methods will reduce contaminant mass in soils but may have negative impact (clogging of pores) on soil structure but this may not be significant in terms of site development.	Presence of previously unidentified contaminant sources, aquifer heterogeneity and impact on application/injection of reagents for in situ methods, impact on soil clogging, etc. Addition of EK with nZVI will partly mitigate issues associated with aquifer heterogeneity.
ENV3: Groundwater andand surface water pH andand redox, Contaminants	In situ methods largely comparable and will have the advantage of positively impacting the wider plume plus the source zones with only limited temporary changes to pH/redox. Excavation will remove more of the source and treat more groundwater in the source zone but may not have the benefits of the in situ techniques in the wider plume.	Aquifer heterogeneity and delivery of reagent, number of injection points, frequency of injection. Addition of direct current with nZVI will partly mitigate issues associated with aquifer heterogeneity. Potentially high uncertainty for in situ methods until further site-specific data/trial results are available.
ENV4: Ecology Aquifer, River	No off-site surface water impacts anticipated. Reduction in contaminant concentrations to have positive impact on aquifer microbial	Timescale of impact on aquifer microbial community (positive and negative).

TABLE 14.12 Summary of execution findings: Environment.—cont'd

Indicator and criteria	Summary findings	Uncertainty
	populations. ExandD will ultimately have positive impact.	
ENV5: Natural resources andand waste Energy andand fossil fuel, Water use Waste production	Both ISCO and nZVI options will require use of reagents and some materials in construction of injection points. However, waste will be minimal. ExandD will require significant fuel and generate significant waste for treatment/disposal.	Main uncertainty is for excavation/off-site treatment/disposal and the final volume of material to be taken off site. Some uncertainty on the number of injection points/frequency for in situ methods that can be partly mitigated by field trials.

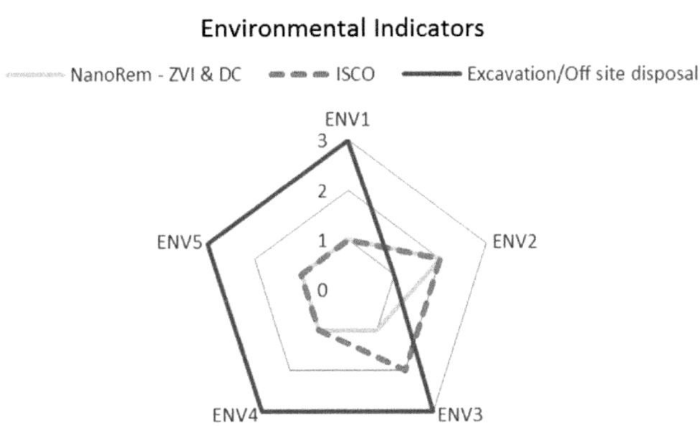

FIGURE 14.6 Radar plot: Environmental indicators.

with the method and the ranking has the potential to be reduced through undertaking field trials. The scores against each headline indicator are summarized in Fig. 14.7.

Social indicator categories.

As with the environmental and economic indicators, there is little to differentiate between the two in situ options considered, with both recording a CR of 7. nZVI + DC scores marginally better on health and safety because the risks associated with handling the zerovalent iron are lower than the risks

TABLE 14.13 Summary of findings and uncertainties: Economic.

Indicator and criteria	Summary findings	Uncertainty
ECON1: Direct economic costs andand benefits	In situ approaches are broadly similar in terms of cost—primary costs will be for the reagent, installation of injection points, etc. Ex&D will incur significant costs for both excavations and long-term reagent and disposal costs.	Uncertainty regarding the success of nZVI and ISCO techniques at the site (aquifer heterogeneity, number of injection points/injections, quantities of reagent, etc.) without site trials.
ECON2: Indirect economic costs andand benefits	Ex&D will provide more certainty of remedial success and as a result may offer better reduction of long-term liability.	In situ methods will have a higher uncertainty in success (possible rebound of contaminants) and therefore potential future liability.
ECON3: Employment andand employment capital	Excavation works will generate short-term local jobs. In situ methods unlikely to have any local impact.	Minimal uncertainty in potential for generation of additional employment opportunities. Excavation works will require plant operators, etc.
ECON4: Induced economic costs andand benefits	nZVI + DC is the only option that has an economic benefit via the ongoing technology development. There are no other induced benefits.	Only uncertainty is the extent of the benefit of the development of new technologies.
ECON5: Project lifespan andand flexibility	In situ methods have short operational duration but may require repeat injections. Both in situ methods can be flexible in terms of injection locations and number of injections, etc. Ex&D offers more flexibility in removing/chasing out previously unidentified contaminant mass.	Although the ex situ works may require a longer initial period, the key uncertainty is likely to be the overall period of treatment/ reinjection required to achieve the remediation goals using the in situ methods.

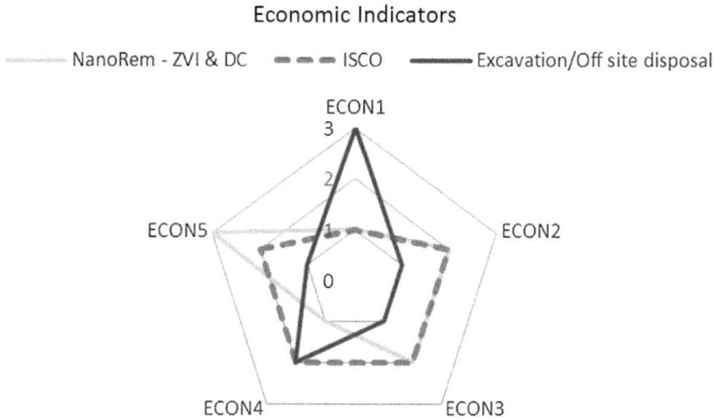

FIGURE 14.7 Radar plot: Economic indicators.

associated with the Fenton's reagent required for ISCO. However, ISCO is considered to have less uncertainty associated with its use. Table 14.14 gives a summary of findings and uncertainties.

The Ex&D option scored a CR of 11, with significant negative impacts considered likely on the neighborhood and local communities due to increases in dust/vapors/odors and increased traffic movements. The option was also considered to represent the greatest negative impact on health and safety due to the excavation and use of heavy plant.

For a remediation, as opposed to a regeneration project, it is difficult to differentiate based on ethics and equity and all options are ranked 1. The scores against each indicator are summarized in Fig. 14.8.

Overall outcomes.

The sustainability assessment gave total OCRs (low = best) of:

- 22 for in situ nanoremediation and direct current
- 23 for in situ chemical oxidation
- 32 for excavation and disposal.

The initial assessment shows that there is very little difference between the overall scores of the two in situ options, although both scored much better than the excavation and disposal option. Therefore, the in situ treatment of the source zone at Site A is considered to be the preferred remedial approach, and the next steps in the assessment should be to further refine the scoring of these options. These next steps may include, but are not limited to, the following:

- Undertake field trials of the nanoremediation + direct current option to reduce the current level of uncertainty regarding the operation and performance of the technology.
- Further assess and/or refine the scoring system so that any key differences between the in situ options can be more clearly distinguished.

TABLE 14.14 Summary of findings and uncertainties: Social.

Indicator and criteria	Summary findings	Uncertainty
SOC1: Human health andand safety	Excavations will present greatest risk to human health through exposure to contaminated soils/ groundwater, working with heavy plant, etc. In situ options have fewer health and safety implications but in ISCO handling Fenton's reagent may represent a higher risk than nZVI.	Potential human error likely to be key with respect to health and safety.
SOC2: Ethics andand equity	All options are considered to be equal.	None.
SOC3: Neighborhood andand locality Dust, Traffic, Noise	Given location of site (approx. 500m from closest residential receptors), in situ works unlikely to impact. Increased traffic movements, plant noise, and associated odors/dust from excavations likely to have significant negative impact.	Principal uncertainty is the generation of dust and odor from ex situ works. Based on experience on the adjacent site, potential for odor generation to impact on local residents during the works is high. Uncertainty of impact from in situ methods is considered to be low.
SOC4: Communities andand community involvement	Likely to be negligible community involvement for in situ methods. Excavation/disposal option issues are likely to result in significant community opposition (based on previous experience at the site).	

TABLE 14.14 Summary of findings and uncertainties: Social.—cont'd

Indicator and criteria	Summary findings	Uncertainty
SOC5: Uncertainty andand evidence	Excavation will minimize uncertainty of success. For in situ methods, there is inherent uncertainty about whether all contaminant sources have been treated. Further design work will be needed to optimize results.	Key uncertainty is in success of in situ methods and potential for contaminants remaining.

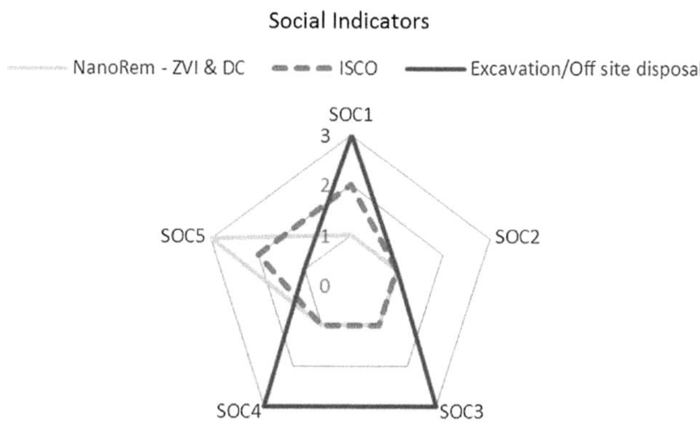

FIGURE 14.8 Radar plot: Social indicators.

5. Conclusions

The findings from both sustainability assessments indicate that nanoremediation compares favorably with other in situ options. This is an encouraging outcome, despite widely reported concerns over the release of nanoparticles and the emerging status of the technology.

One additional issue for Site A is that no UK-based field trial has been carried out for the nanoremediation with direct current option. This option is therefore at a disadvantage when it is compared to ISCO, which has a long track record of use (USEPA, 1998; ITRC, 2005; Siegrist et al., 2011). This is particularly the case in regard to environmental and uncertainty criteria. Although a significant body of published literature exists for nanoremediation at both the laboratory and field scales (Bardos et al., 2011; Karn et al., 2009), the primary focus is on remedial performance rather than the fate and toxicity of the nanoparticles (NPs).

Further differentiation of the in situ options may be refined by progressing to a more quantitative tier of assessment and/or engaging the opinions of a wider range of stakeholders. Both assessments were contractor-led and are therefore preliminary and, in practice, would be used to support further stakeholder engagement. This has not taken place at either site owing to site sensitivities and timing.

To consider wider stakeholder perceptions for Site A, the findings from two public dialogue projects on nanotechnology were assessed against the body of evidence obtained from the NanoRem project. This is discussed further in Annex 1, but we conclude that the findings from the NanoRem project provide evidence that indicates that extreme worst-case scenarios on the fate, transport, and toxicity of nanoparticles in groundwater are unfounded. The findings may have a significant role in future public engagement on nanoremediation projects in the United Kingdom, particularly if used in conjunction with site-specific treatability trials.

Annex References

Batka, V.N., Hofmann, T., 2016. Taking nanotechnological remediation processes from lab scale to end user applications for the restoration of a clean environment. WP4: Mobility and fate of nanoparticles. DL-4.2: Stability, mobility, delivery and fate of optimized NPs under field relevant conditions. NanoRem Deliverable, European Union Seventh Framework Program Project, Grant Agreement n° 309517. Available from: http://www.nanorem.eu/toolbox

Beddoes, D., McMillan, C., Peach, B., Litchfield, Z., Wild, M. Understanding public perceptions of specific applications of nanotechnologies. Report CB0440 for the Department of Environment, Food and Rural Affairs. Available from: http://sciencesearch.defra.gov.uk/Document.aspx?Document=13780_CB0486PublicDialogueofNanotechnologiesFinalReport-Appendices.pdf

Contaminated Land: Applications in Real Environments (CL:AIRE), 2017. NanoRem Bulletin NanoRem 7. NanoRem pilot site – Spolchemie 1, Czech Republic: Nanoscale zero valent iron remediation of chlorinated hydrocarbons. Available from: http://www.nanorem.eu/toolbox/project-deliverables.aspx

Coutris, C., Nguyen, N., Hjorth, R., 2015. DL-5.1: Dose response relationships, matrix effects on ecotox. NanoRem report. Available from: http://www.nanorem.eu/toolbox/project-deliverables.aspx

Coutris, C., Boothman, C., Hjorth, R., Ševců, A., 2016. DL-5.2: Influence of transformation and transport on ecotox. NanoRem report. Available from: http://www.nanorem.eu/toolbox/project-deliverables.aspx
HM Government, 2005. Response to the Royal Society and Royal Academy of Engineering report: "Nanoscience and nanotechnologies: Opportunities and uncertainties". Available from: https://webarchive.nationalarchives.gov.uk/20081107223413/http://www.dti.gov.uk/files/file14873.pdf
Irving, P., Bone, B., Hayes, E., Colvin, J., Irwin, J., Stilgoe, J., Jones, K., 2006. A people's inquiry on nanotechnology and the environment. Environment Agency Science Report SCHO0607BMUJ-E-P. Environment Agency, Bristol. ISBN 978-1-84432-782-9. Available from: https://www.gov.uk/government/publications/a-peoples-inquiry-on-nanotechnology-and-the-environment
Oughton, D., Auffan, M., Bleyl, S., Bosch, J. Filip, J., Klass, N., Lloyd, J., van der Kammer, F., 2015. DL 6.1: Feasibility and applicability of monitoring methods. NanoRem report. Available from: http://www.nanorem.eu/toolbox/project-deliverables.aspx
Royal Society and Royal Academy of Engineering, 2004. Nanoscience and nanotechnologies: Opportunities and uncertainties. Available from: http://www.nanotec.org.uk/finalreport.htm
Stilgoe, J., 2007. Nanodialogues. Experiments in public engagement with science. Demos. ISBN 978-1-84180-187-2. Available from: https://www.demos.co.uk/files/Nanodialogues%20-%20%20web.pdf

Annex 1: Public dialogue on nanoremediation

In the United Kingdom, a report by the Royal Society and the Royal Academy of Engineering (2004) reviewed the opportunities and risks from nanotechnologies and one of its conclusions was that "Until more is known about environmental impacts of nanoparticles and nanotubes, we recommend that the release of manufactured nanoparticles and nanotubes into the environment be avoided as far as possible" (Recommendation four in RS/RAE, 2004). The UK Government's response to this recommendation was that the use of manufactured nanoparticles in environmental applications such as remediation "be prohibited until appropriate research has been undertaken and it can be demonstrated that the potential benefits outweigh the potential risks" and that "It is the Government's current view that further information concerning the environmental fate and toxicity of nanoparticles is required before impact of such releases can be fully assessed" (HMG, 2005). This is in contrast to the position in the Czech Republic where nanoremediation is not considered to be a hazardous activity as the iron nanoparticles oxidize and readily agglomerate in the aquifer as naturally occurring minerals. The injection of nZVI into

TABLE 14.15 Summary of findings from the NanoRem project.

	Laboratory scale	Field scale	Uncertainty
Fate	Experiments on NPs optimized for mobility and stability. Transformation products are typically minerals commonly found in natural soils (carbonates, oxides, and hydroxides) (Batka & Hofmann, 2016).	Results from Spolchemie show rapid corrosion of NPs associated with reduction in dissolved contaminant concentration (CL:AIRE, 2017).	Fate will ultimately depend on NP properties, site conditions, and time. Findings provide reassurance that iron NPs corrode to naturally occurring minerals, even when optimized for performance.
Transport	Transport distance in columns and in large-scale tank tests is generally up to a few meters from the injection point (Batka & Hofmann, 2015).	Field data are difficult to gather, but numerical modeling supports the conclusion of limited transport and limited escape from the treatment zone.	Although transport is both site- and NP-specific, the evidence suggests that the worst-case scenario of significant escape of NPs beyond the treatment zone is highly unlikely.
Traceability (monitoring)	Proof of the efficacy of monitoring tools established in large laboratory tank (Oughton et al., 2015).	Monitoring of iron NP dispersion was carried out at pilot sites using a variety of techniques within the treatment area, but becomes more difficult andand expensive outside it (Oughton et al., 2015).	Monitoring is possible, and when combined with fate andand transport and toxicity evidence, the possibility of long-term harm is considered unlikely.
Toxicity	No significant toxic effects of seven candidate NPs on standard	Results variable, but positive impact on microbial communities at two sites (Coutris et al., 2016).	In the absence of evidence, a worst-case approach is

TABLE 14.15 Summary of findings from the NanoRem project.—cont'd

	Laboratory scale	Field scale	Uncertainty
	bacteria, soil and water organisms. All except one NP can be considered nontoxic (Coutris et al., 2015).		usually taken, e.g., toxicity of iron NPs may be high. Positive outcome from tests on soil and water samples from the pilot sites, but results are NP- and site-specific. Generally no significant toxic effects.

groundwater, in common with other chemical reagents, requires an authorization from the water authority.

The nanodialogues project (Stilgoe, 2007) tested the benefit of upstream engagement (early dialogue in technology development) for the safe development of new nanotechnologies. In 2006, nanoremediation was one of four technology areas addressed in a 3-day people's enquiry. The participants were 13 residents from East London (Irving et al., 2006).

In the inquiry, a number of experts shared their knowledge and responded to questions from the participants. A number of themes emerged from the discussion. They were related to regulation, governance, communication, and trust. Much discussion revolved around the scientific uncertainty regarding the fate and transport of nanoparticles once released, the potential long-term impact on humans and the environment, and the need for open discussion of uncertainties in a local context.

These points were echoed by the participants of the public dialogue project that took place in early 2015 (Beddoes et al., 2015). This project involved 40 members of the public in three separate workshops, each supported by experts. The objectives were to explore public attitudes to, and aspirations for, four nanotechnology applications, including remediation in order to support the UK government's thinking on regulation and governance. Concerning remediation, the participants saw benefits in returning contaminated land to use, but they expressed concerns about the traceability of the released nanoparticles and the potential long-term risks, for example, to drinking water and the food chain.

The paucity of evidence on the toxicity of nanoparticles released into the environment, their potential to enter drinking water and the food chain, the ultimate fate of the NPs ("Will they cause problems for our children?") and the difficulty of monitoring them in groundwater are seen as potential barriers to the public acceptance of nanoremediation in the United Kingdom. Evidence is therefore needed to mitigate against an extreme worst-case scenario being taken by regulators and wider stakeholders. Examples of such extreme views include:

- Iron NPs used for remediation are as toxic as silver NPs.
- Iron NPs may remain unreacted and escape from the treatment zone into the wider environment.

The NanoRem project addressed the issues of the fate, transport (traceability), and toxicity of nanoparticles at both laboratory and field scales. Table 14.15 summarizes the outputs.

Although the performance and environmental impacts of injected NPs are both particle- and site-condition-specific, there is no evidence to suggest that the worst-case conditions are likely to be widely realized. The findings from the NanoRem project may therefore have a significant role in public engagement, particularly if used in conjunction with site-specific treatability trials. This may reduce concerns over uncertainty and lead to a more favorable outcome when assessed against other in situ technologies.

References

American Society for Testing & Materials (ASTM), 2013. ASTM E2876-13. Standard Guide for Integrating Sustainable Objectives into Cleanup. ASTM International, West Conshohocken, PA, 2013. Available from: www.astm.org.

Bardos, P., Merly, C., Kvapil, P., Koschitzky, H.-P., 2018. Status of nanoremediation and its potential for future deployment: risk-benefit and benchmarking appraisals. Remediation Journal 28, 43–56. https://doi.org/10.1002/rem.21559.

Bardos, R.P., Bone, B.D., Boyle, R., Evans, F., Harries, N.D., Howard, T., Smith, J.W.N., 2016a. The rationale for simple approaches for sustainability assessment and management in contaminated land practice. The Science of the Total Environment 563–564, 755–768.

Bardos, P., Merly, M., Bardos, A., Bone, B., Bartke, S., Harries, N., Gillett, A., Nathanail, J., Nathanail, P., Gens, A., Koschitzky, H.-P., Braun, J., Klass, N., Limasset, E., Oughton, D., Tomkiv, Y., 2016b. Taking Nanotechnological Remediation Processes from Lab Scale to End User Applications for the Restoration of a Clean Environment WP9: Dissemination, Dialogue with Stakeholders and Exploitation DL9.2 Final Exploitation Strategy, Risk Benefit Analysis and Standardisation Status. NanoRem Deliverable. European Union Seventh Framework Programme Project, Grant Agreement no 309517. https://doi.org/10.13140/RG.2.2.21746.61126. Available from: www.nanorem.eu/toolbox.

Bardos, P., Bone, B., Daly, P., 2015. Sustainability Assessment Detailed Methodology and Workbook. https://doi.org/10.13140/RG.2.2.23369.60007. NanoRem Project. EU FP7 Project Contract 309517. April 2015. Available from: www.nanorem.eu/toolbox/nanoparticles_and_tools.aspx#TB1.

Bardos, P., Bakker, L., Darmendrail, D., Harries, N., Holland, K., Mackay, S., Pachon, C., Slenders, H., Smith, G., Smith, J., Wiltshire, L., 2013. ThS E3 Sustainable Use of the Subsurface Sustainable and Green Remediation – Global Update. Available from: http://www.

zerobrownfields.eu/HombreMainGallery/Docs/AquaConSoil%202013_Sustainable%20and%20Green%20Remediation_Global%20Update.pdf.

Bardos, P., Bone, B., Elliott, D.W., Hartog, N., Henstock, J.E., Nathanail, C.P., 2011. A risk/benefit approach to the application of iron nanoparticles for the remediation of contaminated sites in the environment. Report CB0486 for the Department of Environment, Food and Rural Affairs. Available from: http://randd.defra.gov.uk/Default.aspx?Menu=Menu&Module=More&Location=None&Completed=0&ProjectID=17502.

Beddoes, D., McMillan, C., Peach, B., Litchfield, Z. & Wild, M. 2015. Understanding public perceptions of specific applications of nanotechnologies. Report CB0440 for the Department of Environment, Food and Rural Affairs. Available from: http://sciencesearch.defra.gov.uk/Document.aspx?Document=13780_CB0486PublicDialogueofNanotechnologiesFinalReport-Appendices.pdf.

Braun, J. (Ed.), 2017. Taking Nanotechnological Remediation Processes from Lab Scale to End User Applications for the Restoration of a Clean Environment. WP8: Up-Scaling, Risk and Sustainability. DL8.2: Final report on three large-scale experiments and generalized guideline for application. NanoRem Deliverable, European Union Seventh Framework Programme Project, Grant Agreement no 309517. Unpublished.

CLARINET, NICOLE, 1998. Better Decision Making Now – the Use of Risk Assessment and Risk Management for Tackling the Problems of Contaminated land. CLARINET & NICOLE Joint Statement. Available from: http://www.nicole.org/uploadedfiles/1998-the-use-of-risk-assesment-and-risk-management.pdf.

Contaminated Land Applications in Real Environments (CL:AIRE), 2017. NanoRem Bulletin NanoRem 7. NanoRem pilot site – Spolchemie 1, Czech Republic: Nanoscale Zero Valent Iron Remediation of Chlorinated Hydrocarbons. Available from: http://www.nanorem.eu/toolbox/project-deliverables.aspx.

Contaminated Land Applications in Real Environments (CL:AIRE), 2011. SuRF-UK Framework Annex 1: The SuRF-UK Indicator Set for Sustainable Remediation Assessment, 2011. Available from: https://www.claire.co.uk/surfuk.

Contaminated Land Applications in Real Environments (CL:AIRE), March 2010. A Framework for Assessing the Sustainability of Soil and Groundwater Remediation. CL:AIRE, London, UK. Available from: https://www.claire.co.uk/projects-and-initiatives/surf-uk. ISBN 978-1-905046-19-5.

Gomes, H.I., Dias-Ferreira, C., Ribeiro, A.B., Pamukcu, S., 2012. Electrokinetic enhanced transport of zero valent iron nanoparticles for chromium (VI) reduction in soils. Chemical Engineering Transactions 28, 139–144.

Grieger, K.D., Fjordbøge, A., Hartmann, N.B., Eriksson, E., Bjerg, P.L., Baun, A., 2010. Environmental benefits and risks of zero-valent iron nanoparticles (nZVI) for in situ remediation: risk mitigation or trade-off? Journal of Contaminant Hydrology 118 (3), 165–183.

Interstate Technology and Regulatory Council (ITRC), 2005. Technical and Regulatory Guidance for the in Situ Chemical Oxidation of Contaminated Soil and Groundwater, second ed. Interstate Technology and Regulatory Council. Available from: https://clu-in.org/download/remed/chemox.pdf.

Irving, P., Bone, B., Hayes, E., Colvin, J., Irwin, J., Stilgoe, J., Jones, K., 2006. A People's Inquiry on Nanotechnology and the Environment. Environment Agency Science Report SCHO0607BMUJ-E-P. Environment Agency, Bristol, ISBN 978-1-84432-782-9. Available from: https://www.gov.uk/government/publications/a-peoples-inquiry-on-nanotechnology-and-the-environment.

International Organization Standards (ISO), 2017. Soil Quality – Sustainable Remediation. ISO 18504: 2017. http://iso.org.

Karn, B., Kuiken, T., Otto, M., 2009. Nanotechnology and in situ remediation: a review of the benefits and potential risks. Environmental Health Perspectives 117 (12), 1823–1831.

NICOLE, 2010. NICOLE road map for sustainable remediation. Network for Industrially Contaminated Land in Europe (NICOLE) Report. Available from: http://www.nicole.org/uploadedfiles/2010-wg-sustainable-remediation-roadmap.pdf.

NICOLE, Common Forum, 2013. Risk-informed and Sustainable Remediation. A Joint Position Statement by NICOLE & Common Forum. Available from: http://www.nicole.org/uploadedfiles/2013%20NICOLE-Common-Forum-Joint-Position-Sustainable-Remediation.pdf.

Rizzo, E., Bardos, P., Pizzol, L., Critto, A., Giubilato, E., Marcomini, A., Albano, C., Darmendrail, D., Döberl, G., Harclerode, M., Harries, N., Nathanail, P., Pachon, C., Rodriguez, A., Slenders, H., Smith, G., 2016. Comparison of international approaches to sustainable remediation. Journal of Environmental Management 184, 4–17. https://doi.org/10.1016/j.jenvman.2016.07.062.

Royal Society and Royal Academy of Engineering, 2004. Nanoscience and Nanotechnologies: Opportunities and Uncertainties. Available from: http://www.nanotec.org.uk/finalreport.htm.

Siegrist, R.L., Crimi, M., Simpkin, T.J., 2011. In Situ Chemical Oxidation for Groundwater Remediation. Springer-Verlag, New York. ISBN (Hardcover): 978-1-4419-7825-7.

Stilgoe, J., 2007. Nanodialogues. Experiments in Public Engagement with Science. Demos, ISBN 978-1-84180-187-2. Available from: https://www.demos.co.uk/files/Nanodialogues%20-%20%20web.pdf.

United States Environmental Protection Agency (USEPA), 1998. Field Applications of in Situ Remediation Technologies: Chemical Oxidation. USEPA Technology Innovation Office, Washington. Report EPA-542-R-98-008.

Chapter 15

Understanding the diverse norms and rules driving sustainable remediation: a study of positioning, aggregation, and scoping

Jason H. Prior
Institute for Sustainable Futures, University of Technology Sydney, Sydney, NSW, Australia

1. Introduction

The implementation of sustainable remediation approaches is augmenting the risk-based approach to managing the remediation of contaminated sites and transforming the norms and rules that guide the remediation of those sites. A broad cross-section of professional organizations have sought to state what some of these norms and rules might be through the development of sustainable remediation guidance (Sustainable Remediation Forum Australia et al., 2011a; Sustainable Remediation Forum United Kingdom, 2010). However, the ability to map the inventory of norms and rules that are used by participants in sustainable remediation practice, and in any remediation practice, has been restricted by the absence of systematic methodologies for carrying out such mappings. Addressing this challenge entails developing a systematic way of identifying and deconstructing the components (e.g., aims, performers, prescriptives, sanctions, and conditions) of the norms and rules being used every day by participants in sustainable remediation practice. This enables the emerging guidance, policy, and legislation to be informed by a clear understanding of what is considered appropriate practice.

This chapter presents a study that utilized the Institutional Grammar Tool (IGT) as a means to systematically understand the norms and rules that commonly guided participants' involvement in sustainable remediation processes at a series of Australian sites (McGinnis, 2011). In some contexts, sustainable remediation is limited to "greening" remediation processes so that

Sustainable Remediation of Contaminated Soil and Groundwater. https://doi.org/10.1016/B978-0-12-817982-6.00015-X

they maximize net environmental benefits (e.g., carbon mitigation) (Adamson et al., 2011; Simon, 2010). In other contexts, including the Australian context, which is examined in this study, a focus on sustainability also opens remediation processes to broader economic, cultural, and social accountability (Bardos et al., 2000; Dixon, 2006, 2007; Doick et al., 2009; Wernstedt et al., 2004). The grammar for the IGT was initially formulated by Crawford and Ostrom (Crawford, 2004; Crawford and Ostrom, 1995) and has since been revised and developed (Basurto et al., 2010; McGinnis, 2011; Schluter and Theesfeld, 2010; Siddiki et al., 2011). While the IGT has been applied in an increasingly broad range of contexts, it has not previously been applied to remediation until the study that is presented in this chapter (Feiock et al., 2014; Roditis et al., 2014).

The chapter firstly outlines the conceptualization of the institutional grammar for norms and rules in sustainable remediation that was used in the study. Secondly, the chapter presents the study's methods, which were used to identify the positioning, aggregation, and scoping rules and norms that guided participants' actions within the sustainable remediation processes at three Australian sites and the components of these norms and rules (e.g., aim, performer among other components). These norms and rules: define the position of participants within the sustainable remediation process; provide means for aggregating sustainability into established remediation approaches, such as the risk-based approach; and define the scope of outcomes from the sustainable remediation process. Thirdly, the chapter presents the study's findings. The study found a core set of 16 norms and 18 rules (sanctions) used by participants to implement sustainable remediation at the Australian sites and found strong interdependence between these norms and rules (sanctions) and the normative principles operating within the broader domain of environmental management and planning. While the chapter presents all 16 norms and 18 rules, the chapter, where possible, provides greater elaboration of the three norms and two rules involving neighbors and the broader community. Finally, the chapter concludes with a discussion of: the system of norms that the study found were operating within sustainable remediation (which far exceed those associated with Ecologically Sustainable Development (ESD)); and their link, through rules (sanctions) to contemporary styles of regulatory enforcement.

Understanding the study's inventory of norms and rules being used by participants within sustainable remediation processes can be extremely valuable for regulators and professional organizations in helping them to create effective guidance, policies, and legislation to support sustainable remediation (McGinnis, 2011). If these developed guidelines and frameworks are not effectively linked to actual practice, they become irrelevant to those directly responsible for implementing remediation initiatives. Moreover, they can mislead those who are seeking to understand what is appropriate, and they may contribute to the development of regulations and legislation that are difficult to comply with.

2. Conceptualizing the institutional grammar for norms and rules in sustainable remediation

This section outlines a conceptual understanding of the norms and rules in sustainable remediation using Crawford and Ostrom's IGT. Using the logic of their institutional analysis and development (IAD) framework (Ostrom, 2005, 2011), Crawford and Ostrom have sought to unify interpretations of norms and rules by developing a "grammar" for their contents (Crawford, 2004; Crawford and Ostrom, 1995). Their institutional grammar has since been revised and developed into the IGT (Basurto et al., 2010; McGinnis, 2011; Schluter and Theesfeld, 2010; Siddiki et al., 2011). The study presented within this chapter applies this IGT to identify the norms and rules that operate in sustainable remediation at three Australian sites. Each norm or rule is a "shared linguistic constraint or opportunity that prescribes, permits, or advises actions or outcomes for actors (both individual and corporate)" (Crawford and Ostrom, 1995, p. 583). With the study a norm or rule only existed in the sustainable remediation process at the three Australian sites if it was utilized by the diverse participants within the sustainable remediation processes (e.g., problem holder, local government, auditor, remediation service provider, regulatory authority, and neighbors).

Table 15.1 lists the components that each norm and rule in the institutional grammar is required to have. It also provides sample statements, which describe the components of particular norms or rules, components that are used in this study. Both norms and rules contain an aim component that outlines its goals, actions, or outcomes, and a prescriptive component that indicates what may, must, or must not be done. In both norms and rules, the aims are framed by three other components: performer of aim, receiver of aim, and condition (McGinnis, 2011; Ostrom, 2005, p. 149). While these shared components of norms and rules mean that they are conceptually related, rules are distinguished from norms by the presence of a tangible and formal sanction—what Crawford and Ostrom call the "or else" component (Crawford and Ostrom, 1995; McGinnis, 2011). For example, a sanction might be a monetary penalty, it might involve having to perform a particular task, or it might involve the withdrawal of a certain right (Crawford and Ostrom, 1995).

The implementation of norms and rules is dependent on their context. Norms and rules only exist within remediation decision-making if they have some collective authority among participants (Crawford and Ostrom, 1995). Furthermore, they may not be exclusive to such processes and may be shared across society (e.g., effective management of risks to human health is a broad aim across diverse sectors of society and is not just limited to remediating

TABLE 15.1 Sample coding of an institutional statement: Problem holder must accept liability for the contaminant's containment, remediation, avoidance, and abatement indefinitely, or else regulatory authority will impose responsibilities using orders, notices, and directions.

Component	Explanation of component	Norm	Rule	Coding of sample statement with relevant component underlined
Aim	Describes the particular goal or action that the prescriptive refers to.	✓	✓	Problem holder must accept liability for the contaminant remediation, avoidance, and abatement indefinitely.
Prescriptive.[a]	The prescriptive component stipulates whether the aim *may, must*, or *must not* be done.	✓	✓	Problem holder must accept liability for the contaminant remediation, avoidance, and abatement indefinitely.
Performer of aim.[b]	The individual(s), group(s) of individuals, or organization(s) that performs the aim.	✓	✓	Problem holder must accept liability for the contaminant remediation, avoidance, and abatement indefinitely
Receiver of aim.[c]	The receiver of the aim may be an object(s), individual(s) or group(s) of individuals, or organization(s)	✓	✓	Problem holder must accept liability for the contaminant remediation, avoidance, and abatement indefinitely
Condition	Qualifies the "when," "where," and "how" of the aim.	✓	✓	Problem holder must accept liability for the contaminant remediation, avoidance, and abatement indefinitely
Formal sanction.[d]	The sanction experienced by performer of aim if the prescriptive isn't adhered to.	X	✓	Problem holder must accept liability for the contaminant remediation, avoidance, and abatement indefinitely, or else regulatory authority will impose responsibilities using orders, notices, and directions.

[a] *Called the "Deontic" in the institutional grammar.*
[b] *Called the "Attribute" in the institutional grammar.*
[c] *Called the "Object" in the institutional grammar.*
[d] *Called the "Or Else" in the institutional grammar.*

contaminated land and groundwater, but is practiced by communities, markets, private associations, and governments at all scales) (Ostrom, 2005).

A number of studies have shown that the institutional grammar described here can be used to parse written legislation, regulations, and policies (Siddiki et al., 2011). The study presented in this chapter sought to use the institutional grammar to identify norms and rules through the analysis of interviews with respondents belonging to the participant types involved in sustainable remediation processes at the site level. Norms and rules can be spoken or tacitly understood; they do not need to be written (Crawford and Ostrom, 1995). In this study the titles of the components of the institutional grammar have been simplified for a general audience (see Table 15.1) (Basurto et al., 2010; Crawford and Ostrom, 1995; Schluter and Theesfeld, 2010).

Building on the aforementioned conceptual understanding of the components of norms and rules, the study presented within the chapter addressed two research questions to gain a better understanding of the norms and rules driving sustainable remediation:

RQ1. What are the norms that guide participants' involvement in sustainable remediation?

RQ2. To what extent do formal sanctions (rules) promote compliance with these norms?

In addressing research questions RQ1 and RQ2, the study focused on identifying three types of norms and rules using the IAD framework: positioning, aggregation, and scoping norms and rules (McGinnis, 2011). Position norms and rules specify a set of positions for participants within the remediation processes (Ostrom, 2005). Each position has a unique combination of resources, opportunities, preferences, and responsibilities within the remediation processes (Feiock et al., 2014). For participant types within the remediation processes, aggregation norms and rules clarify "who is to decide which action or set of activities is to be undertaken" in order to achieve intermediate or final outcomes (McGinnis, 2011, p. 174; Ostrom, 2005, p. 202). Ostrom acknowledges the great diversity of aggregation rules and norms (e.g., joint control over an action by different participant types, or designation of control to participant types). Scope norms and rules specify a set of outcomes for the remediation processes (McGinnis, 2011; Ostrom, 2005). While decisions about categorization were based on these three types of rules and norms, this does not mean that other rule types specified in the IAD framework were not present (e.g., boundary rules, information rules, payoff rules).

3. Methodology

To address research questions RQ1 and RQ2, the researcher collected data on the norms and rules used by participants within the sustainable remediation processes at three different Australian sites (Byrne, 2009). While the analysis in this study focuses on the implementation of sustainable remediation at particular sites, it aims to allow generalizability of findings beyond these unique instances (Byrne, 2009). It does so by using the IGT to elicit those positioning, aggregation, and scoping rules and norms that are commonly used by participants to implement sustainable remediation at all sites (Crawford and Ostrom, 1995). This study's methodology involved collecting and processing interview data about the positioning, aggregation, and scoping norms and rules used by participants in implementing sustainable remediation at these sites. The University of Technology Sydney Human Research Ethics Committee approved the research, and the research was conducted between 2011 and 2017.

At each site members of the same six participant types were interviewed (giving a total of 18 interviewees across all three sites). The participant types selected for the study were: problem holder (PH), regulatory authority (RA), remediation service provider (RSP), local government (LG), neighbors (N), and auditor (A). In this study the PH were all owners of the sites that were the points of origin of the environmental contamination (i.e., the source site). RA were representatives of state environmental protection authorities that had jurisdiction over the problem holders' sites. RSP were professionals of large engineering companies who specialized in remediation techniques. LG were local council planners who dealt with environmental remediation in their local areas. N were adjoining property owners affected by contamination from the source site. A were individuals appointed under a formal auditing scheme within each state to carry out reviews of site contamination investigations and remediation works. This study did not engage with other organizations involved at any of the sites, such as local or national environmental groups, or local or national media.

To protect the confidentiality of sites and participants, only generic information is provided on the processes at each site. One was located in Western Australia (WA), one in New South Wales (NSW), Australia, and the third in South Australia (SA). The WA site involved a small-scale soil and groundwater remediation project located in an industrial area of a city. The focus was on remediating contamination that had emanated from a single point source. This had resulted in a plume of contaminants in groundwater under adjacent industrial properties, which extended toward waterways. The NSW site was in an industrial area surrounded by residential neighborhoods. Contamination associated with this site included a groundwater plume, which extended under nearby residences and various areas of contaminated soil. The SA site was a single urban site surrounded on all sides by residential neighborhoods. Contamination associated with this site included various areas of contaminated soil.

3.1 Data collection

While archival documentation of policies, legislation, and other materials relating to remediation processes at the sites was collected at the outset of this project, the primary focus of the analysis was on semistructured in-depth interviews with representatives of the participant types engaged in the remediation processes at the three sites. These interviews, conducted between 2011 and 2014, were used to capture data on norms and rules that participants used to implement sustainable remediation at these sites.

At the time of our study the IGT had not previously been applied to interview data to identify norms and rules, but only to policy and legislation (Basurto et al., 2010). Each interview commenced with the interviewer explaining that the research project aimed to understand the norms and rules in use at the site, and outlining the key components of the institutional grammar. Questions asked during interviews were directly related to the interviewee's role as a participant type (e.g., auditor) and to their understanding of the roles of other participant types. Interviews were structured around the IGT components (see Table 15.1) in order to obtain data on the norms and rules that each participant type thought were relevant to the sustainable remediation process that was undertaken at their site. The interviewees were asked to address the following principal questions across the life of the sustainable remediation processes:

- As a [interviewee's participant type], what are the particular outcomes, goals, and actions that you must, may, or must not perform during the sustainable remediation process at [site name]?
- What are the particular outcomes, goals, and actions that [other participant types] must, may, or must not perform during the sustainable remediation process at [site name]?
- What sanctions can [interviewee's participant type] utilize during the sustainable remediation process at [site name] to ensure [other participant types] perform their role?
- Who is responsible for sanctioning [interviewee's participant type] if they do not perform their role during the sustainable remediation process?

When necessary, these questions were followed by a series of questions designed to elicit the when, where, and how of each answer. Finally, interviewees were asked to provide references to documentation that detailed their or other participant types' roles and responsibilities within the sustainable remediation processes (e.g., legislation, policies, or guidelines). The interview instrument was trialed and feedback was used to clarify the instrument. Interviews took between 2 and 4 hours each and were conducted by two interviewers. The interview questions sought to collect as complete and salient a set of norms and rules as possible. At the end of the interview, participants were asked if they were willing to be contacted later to provide additional information that might help the researchers to further understand the norms and rules discussed in the interviews. These additional communications occurred between 2014 and 2017, and in the context of this study provided the opportunity to collect further data that could be used to elaborate in greater detail the norms and rules involving neighbors.

3.2 Data coding and nested analysis

To address RQI and RQ2, the interview data was transcribed and coded using IGT components detailed in Table 15.1 using NVIVO qualitative analysis software (QSR International). Coding of norms and rules involved six steps:

- To maintain confidentiality, coding commenced by replacing the names of the participants with the six participant types listed earlier.
- Aims that were performed by the six participant types were then coded. In many cases more than one participant type performed the same aim.
- The coding of aims was restricted by omitting all aims that were not prefaced by a prescriptive: must, must not, or may, either explicit or implicit (e.g., the verb "required" or "shall" suggests a "must") (Basurto et al., 2010).
- Formal sanctions were then coded for each aim remaining from the previous step.
- Identified norms and rules were then separated out from the interview transcripts for further analysis. Norms or rules were reduced to their most fundamental forms (e.g., all synonyms for prescriptives were reduced to must, must not, or may). In addition, all norms and rules that were not common to the processes at all three sites were removed, and so were duplicate norms and rules.
- The final step involved nesting the norms and rules (formal sanctions) that emerged from the coding (see Fig. 15.1). Using the IAD framework, norms and rules were nested into specific categories (see Table 15.3): position, aggregation, and scope (McGinnis, 2011).

Two researchers coded the data. In order to maximize intracoder reliability, the second researcher verified the coding of the first. A remediation industry expert was engaged to adjudicate coding differences; in some cases the original interviewee was contacted to determine the accuracy of the researchers' interpretations. A high degree of agreement between the researchers was evident during the coding. NVIVO qualitative analysis software (QSR International) was used to code the interview data.

4. Results

4.1 Norms driving sustainable remediation

Table 15.2 shows the results of the basic frequency count of the total number of norms from our nested analysis. Frequency counts such as those displayed

Type of Norms and Rules: e.g. Positioning, Aggregation or Scoping

Norm No. X: [Performer of aim] [*Prescriptive*] [Aim] [Receiver of aim] [Condition]
Example: [Problem Holder] [*must*] [accept liability for] [the contaminant's containment, remediation, avoidance and abatement] [indefinitely]
Formal sanction: [*Or else* statement]
Example: [*Or else* Regulatory Authority will impose responsibilities using orders, notices and directions].

FIGURE 15.1 Structure of nested analysis of norms and rules (norm + formal sanction) using IGT components detailed in Table 15.1. *Credit: Created by the Author*

TABLE 15.2 Frequency counts of norms and formal sanctions and who performs them.

	Frequency of units			
	Position	Aggregation	Scope	Total
Norms identified within all three remediation processes				
Norms: Total	6	7	3	16
Norms: Must	5	7	3	15
Norms: May	1	–	–	1
Norms: Must not	–	–	–	–
Performers of norms				
Problem holder	4	7	3	14
Remediation service provider	–	7	3	10
Regulatory authority	1	–	–	1
Local government	1	–	–	1
Auditor	1	–	–	1
Neighbor	–	–	–	–
Sanctions associated with norms observed in all three remediation processes				
Sanction: formal	6	8	4	18
Performers of formal sanctions				
Problem holder	1	2	–	3
Remediation service provider	–	–	–	–
Regulatory authority	4	6	4	13
Local government	3	6	4	12
Auditor	–	4	–	4
Neighbor	1	–	–	1

in Table 15.2 are useful for identifying overall trends and tendencies. In our analysis, the frequency count revealed that all norms identified across the three sites, except one, were obligations (i.e., the prescriptive was "must"), and in most cases the problem holder was held to be responsible for performing them with support from the remediation service provider. The aims, receiver of aims, and conditions specified within the norms were diverse, so they could not be subjected to frequency analysis.

Table 15.3 details the 16 norms that emerged from the analysis. These norms are divided into three types: position norms, aggregation norms, and scoping norms. The first six norms (Norms 1 to 6) in Table 15.3 stipulate the position of each participant type within the sustainable remediation process. While the problem holder, regulatory authority, local government, and auditor all carried out their roles within the remediation process by fulfilling the particular, responsibilities prescribed to them through Norms 1 through 6, it was only possible to determine the positions of the neighbor and remediation service provider by analyzing their responses to the problem holder's fulfilment of Norms 2, 4, and 5. For example, in the context of Norm 5, it was only possible to assess the position of the neighbor within the remediation process by assessing their level of approval (where the problem holder's social license to operate (SLO) is related to the level of approval by neighbor) of the remediation and management plan for the contaminated site that was put forward to the neighbor by the problem holder (or their representative, the remediation service provider). The neighbor's approval may or may not be in writing, and differed from the type of approval that the problem holder needed to obtain from other participant types like government regulators or auditors. The neighbors' levels of approval ranged along a spectrum from complete approval, to finding a remediation and management plan tolerable, through to withholding approval for the remediation and management plan, and/or resisting the remediation and management plan being applied or protesting against it through the implementation of tangible sanctions (Formal Sanction 5). The types of issues of concern to neighbors in their approval of the remediation and management plan for the contaminated site were extensive and could not be included in Table 15.3; they included how the remediation and management plan addressed environmental impacts (e.g., minimizing risk to ecosystem services such as groundwater; conserving natural resources; minimizing waste generation, limitations on use of property); social impacts (e.g., reducing local health risks; reducing the intergenerational effect of the contamination; minimizing local-scale amenity impacts such as noise, odor, local air quality, traffic, and dust; minimizing contamination left behind; and changes in the services they can access); and economic impacts (e.g., property values). Furthermore, the neighbor's level of approval was guided by the perceived quality of the remediation and management plan, the perceived degree of transparency in which the problem holder had engaged with them

TABLE 15.3 Norms and rules (norm + formal sanction) guiding sustainable remediation.

PART 1: Norms and rules stipulating the position of participant types within the sustainable remediation process	
Norm 1: [Problem holder] [must] [accept liability and responsibility for] [the containment remediation, avoidance, or abatement] [indefinitely]. This norm is related to the polluter pays principle: Those who generate pollution or waste should bear the costs of containment cleanup, avoidance, or abatement.	All[n]
Formal sanction 1: [Or else] [Regulatory authority and/or other areas of government] [will apply a legislated hierarchy to identify the problem holder and impose responsibilities using orders, notices, and directions].[a] The hierarchy generally commences with the polluter. If it is not practicable to assign responsibility to the polluter, responsibility is allocated to others, including the owner who has acquired the title, followed usually by the relevant public authority.	
Norm 2: [Problem holder] [may] [contract] [remediation service providers to undertake the remediation processes for the contaminated site and engage with participants] [as needed].	
Formal sanction 2: [Or else] [the problem holder] [may sue remediation service provider for breach of agreed contract and possible associated penalties].	
Norm 3: [Regulatory authorities and local government] [must] [take responsibility for regulating] [the remediation processes for contaminated sites in their jurisdictions] [through reviews, licensing, notices, orders, and approvals in accordance with jurisdictional legislation]. This norm is related to the principle of share responsibility: Protection of the environment and society is a responsibility shared by all levels of government and industry, business, and communities.	All[n]
Formal sanction 3: [Or else] [Government regulator and/or other areas of government] [may issue penalties (e.g., monetary) or impose responsibilities using orders, notices, and directions][b]	
Norm 4: [Problem holder] [must] [legitimately engage with] [affected neighbors/surrounding population] [so they can understand and participate in decisions about any site assessment, remediation, and management planning that may affect them]. This norm is related to the principle of accountability: The local community has a legitimate right to understand and be engaged in decisions that may affect them about the restoration, protection, and enhancement of the environment.	All[n]

Continued

TABLE 15.3 Norms and rules (norm + formal sanction) guiding sustainable remediation.—cont'd

Formal sanction 4: [Or else] [regulatory authority and/or other areas of government] [may issue orders or directions to carry out local community engagement processes and demonstrate how the communities' input has been taken into consideration].[c]	
Norm 5: [Problem holder] [must] [gain approval from] [affected local community, particularly affected neighbors, for the remediation and management plan to be implemented, particularly in relation to the remediation technologies that are to be utilized.] [The approval may be unwritten.] This norm is related to the idea of Social License to Operate: the local community has a legitimate right to accept or reject a proposed remediation and management plan based on their concerns."	PH, RSP, N, LRC[n]
Formal sanction 5: [Or else] [neighbors and the broader affected community] [may impose a range of sanctions on the problem holder's plans, ranging from neighbors restricting the problem holder's access to their property through to collective public/political campaigns to oppose the plan's implementation].	
Norm 6: [Auditors] [must] [independently review and provide guidance on] [assessment and remediation work conducted by site consultants in undertaking contaminated site investigations and remediation] [to ensure that the work complies with the requirements of legislation (unless supported by clear reasoning), and that the remediation meets the standard applicable to the proposed land use.]	All[n]
Formal sanction 6: [Or else] [Regulatory authority and/or other areas of government] [may revoke the accreditation and/or impose penalties.][d]	
PART 2: Aggregation norms and rules 7 through 13 clarify who is to decide which action or set of activities is to be undertaken that leads to sustainability outcomes	
Norm 7: [Problem holder and remediation service provider] [must] [seek the trust and confidence] [of other participants, including the local community] [throughout the remediation and management of the contaminant(s)]	All[n]
Norm 8: [Problem holder and remediation service provider] [must] [effectively assess] [the remediation and management planning for contaminated sites] [using clear and transparent ecologically sustainable development (ESD) principles and tools with other participants]. This norm utilizes the principle of ESD: Integrated assessment of environmental, economic, and social impacts of development must	All[n]

TABLE 15.3 Norms and rules (norm + formal sanction) guiding sustainable remediation.—cont'd

meet the needs of the present generations without compromising the ability of future generations to meet their needs.	
Formal sanction 8a: [Or else] [regulatory authority, auditor, local government, and/or other areas of government in NSW and SA] [may issue notices, orders, or directives to the problem holder to ensure that contaminated sites are managed so as to maintain ESD].[e]	
Formal sanction 8b: [Or else] [the problem holder] [may take formal action against the remediation service provider for breach of agreed contract and seek associated penalties].	
Norm 9: [Problem holder and remediation service provider] [must] [use] [the precautionary principle] [to address the current site contamination and prevent further contamination at the site]. This norm utilizes the precautionary principle: Where there is a threat of serious or irreversible environmental degradation from a particular action, lack of scientific certainty about the environmental impacts of that action should not be used as a reason to postpone measures to prevent environmental degradation.	PH, RSP, A, LC, RA[n]
Formal sanction 9a: [Or else] [the regulatory authority, auditor, local government, and/or other areas of government] may issue orders, notices, or directions to ensure that the precautionary principle or precautionary approach is adopted by the problem holder].[f]	
Formal sanction 9b: [Or else] [the problem holder] [may sue the remediation service provider for breach of contract and possible associated penalties].	
Norm 10: [Problem holder and remediation service provider] [must] [clearly demonstrate that] [the remediation and management option selected for the contaminated site] [has an equivalent or higher level of performance than other options].	PH, RSP, A[n]
Formal sanction 10: [Or else] [regulatory authority, auditor and/or local government, and/or other areas as of government][will impose an options hierarchy to guide selection of the remediation and management option for the contaminated site].[g]	
Norm 11: [Problem holder and remediation service provider] [must] [ensure that] [sustainable remediation approaches] [are compatible with risk-based fit-for-purpose approaches when carrying out site assessment and developing the remediation and management plan for the contaminated site]. This norm utilizes the risk-based fit-for-purpose philosophy, which: Determines land-use scenarios for which risk-based health investigation levels and ecological investigation levels have been	**All**[n]

Continued

TABLE 15.3 Norms and rules (norm + formal sanction) guiding sustainable remediation.—cont'd

derived. The intended use of the site determines the level of contamination risk that may be permitted to remain on the site.	
Formal sanction 11: [Or else] [regulatory authorities, auditor, local government, and/or other areas of government] must issue instructions, orders, and notices requiring the problem holder to carry out site assessment and develop remediation and management plans that include a coordinated set of activities and methods to control risks so as to promote the likelihood that the site can be made suitable for the proposed use and provide adequate protection of human health, property, and the environment].[h]	
Norm 12: [Problem holder and remediation service provider] [must] [integrate] [contaminated site remediation and management decision-making] [with local, regional, and state planning and development processes] This norm utilizes the principle of localization: Locally tailored solutions are understood as key——site and local circumstances were understood to drive a remediation strategy that is sustainable.	PH, RSP,A, LRC, RA[n]
Formal sanction 12: [Or else] [regulatory authority, local government, and/or other areas of government] [will impose orders and directions such as management plans, cleanup notices, ongoing maintenance orders and enforcement, financial assurances, abatement notices, contaminated site registers, memorials or notations on land titles, and aligned planning and development permit approval processes. Should the land use of the site change, the process is reinstituted and the assessment, management, and remediation procedures are carried out in accordance with the new intended land use. All jurisdictions have postremediation/ management controls.]	
Norm 13: [Problem holder and remediation service provider] [must] [pursue site-specific remediation technologies and approaches] [using: • the most cost-effective option, by establishing incentive structures, including market mechanisms, that enable those best placed to maximize benefits or minimize costs to develop their own solutions and responses to environmental problems. • full life cycle of costs of providing services, including the use of natural resources and assets and the ultimate disposal of any waste, in a way that minimizes local, regional, and global impacts. Consideration needs to be given to management of energy, water, and other material resources. Taking into consideration the scarcity, minimization, depletion rates, regional availability, recycling and recovery of water, energy, and material resources, as well as hazardous and nonhazardous waste generated through remediation, residual contamination,	PH, RSP,A, LRC, RA[n]

TABLE 15.3 Norms and rules (norm + formal sanction) guiding sustainable remediation.—cont'd

rehabilitation needs (e.g., to restore soil and ground functions), and the fate of treated contaminants (destruction, vs. removal, vs. containment/stabilization, permanence of the solution)].This norm utilizes the principle of prevention: That steps need to be taken to minimize the creation of any additional contamination; to prevent further contamination of already contaminated sites by reducing risk to human health; and to prevent the degradation of the environment by using mechanisms that promote cleaner production, eliminate harmful wastes, reduce the use of materials, and promote the reuse, recovery, or recycling of materials. This norm also utilizes the principle of waste minimization: That wastes should be managed in accordance with the following order of preference: Avoidance, recycling, recovery of energy, containment, and only lastly disposal.	
Formal sanction 13: [Or else] [Regulatory authority, local government, and/or other part of government] [may impose restrictions, requirements, notices, or orders requiring that all decisions made to remediate and manage contamination at the site use the principles of prevention adopted in that jurisdiction].[i]	
PART 3: Norms and rules 14 through 16 scope the anticipated sustainability outcomes from the remediation processes	
Norm 14: [Problem holder and remediation service provider] [must] [ensure that] [remediation and management planning for contaminated sites] [is carried out to safeguard intragenerational and intergenerational equity]. This norm utilizes the ideas of intragenerational equity (the rights of people within the current generation to fair access to natural resources e.g., access to safe land and groundwater) and intergenerational equity (the responsibility of current generations to ensure that their resource use does not limit the resources available to future generations e.g., access to safe land and groundwater). Together they make up the "principle of equity."	PH, RSP, LC, RA, N[n]
Formal sanction 14a: [Or else] [regulatory authority, local government, and/or other area of government within NSW and SA] [may impose requirements upon the problem holder that ensure that the principle of intergenerational equity is incorporated into remediation decision-making processes].[j]	
Formal sanction 14b: [Or else] [regulatory authority, local government, and/or other area of government] [may impose requirements upon the problem holder that ensure that the principle of is incorporated into remediation decision-making processes]. While specific references to intragenerational equity were not found in NSW, SA, or WA remediation legislation, it was understood to be implicit within the "polluter pays" aspects of existing legislation, in	

Continued

TABLE 15.3 Norms and rules (norm + formal sanction) guiding sustainable remediation.—cont'd

which a person must be held responsible for the way in which they misuse current resources to the disadvantage of others within that generation.]	
Norm 15: [Problem holder and remediation service provider] [must] [protect] [essential ecological processes and life-support systems within the context of the principle of sustainable use] [during the remediation process by protecting: Human health and safety (anthroposphere) from impacts of contaminants and adverse effects resulting from remediation. Air (atmosphere) from the impacts of contaminants, the impacts of substances added during remediation such as chemical solvents, and the impacts of remediation by-products and emissions (toxic substances, greenhouse gases, stratosphere ozone-depleting gases such as CFCs). Ground and surface water (hydrosphere) by reducing the impacts of substances added during remediation (e.g., nutrients and fertilizers, chemical reagents) and the impacts of remediation by-products and emissions (e.g., toxic inorganic substances, substances that change pH), flora and fauna (biosphere) by reducing the impacts of organisms added during the remediation (e.g., bacteria, fungi, plants) on ecosystems (naturally occurring organisms), and the impacts on the "quality of the nature" (conservation of biological value and biodiversity), ecosystem soil function (lithosphere) by reducing the impacts of substances added during the remediation on soil systems (nutrients, fertilizers, surfactants), the impacts of organisms added to soil system (organisms), the impacts of process by-products and emissions on soil systems, and changes in soil function or the impacts on subsurface structure of remediation work.] This norm utilizes the principle of sustainable use: while a certain sovereign right exists over the exploitation of natural resources, this right is qualified by a duty to refrain from causing irreparable damage to the ecological system.[k]	All[n]
Formal sanction 15: [Or else] [Regulatory authority, local government, and other areas of government in NSW, WA, and SA] [may issue notices, orders, or directives ensuring that contaminated sites are managed so as to maintain ESD and broader legislation in WA, SA, NSW.].[l]	
Norm 16: [Problem holder and remediation service provider] [must] [minimize impacts] [on local and regional amenity] [during the remediation processes from: increased vehicular traffic; infrastructure changes to public rights of way for road, rail and other transport; noise, dust, litter and vibration; the visual impact of operations; restrictions to the physical use of space; and remediation by-products and emissions (light, heat, organisms)].	All[n]
Formal sanction 16: [Or else] [Regulatory authority, local government and/or other part of government] [may impose	

TABLE 15.3 Norms and rules (norm + formal sanction) guiding sustainable remediation.—cont'd

restrictions, requirements and penalties in accordance with a broad cross section of legislation and guidelines for: volatile odor and gas emissions, noise emissions, vibration etc.].[m]	

[a]*Western Australia Contaminated Site Act (2003); New South Wales Contaminated Land Management Act (1997); South Australia Environment Protection Act (1993); ANZECC position paper on Financial Liability for Contaminated Site Remediation (1994).*
[b]*New South Wales Contaminated Land Management Act (1997); South Australia Development Act 1993.*
[c]*New South Wales Environment Protection Act (1993); Western Australia Contaminated Sites Act (2003); Western Australia Contaminated Sites Management Series, Community Consultation (n.d); Australian National Environmental Protection (Assessment of Site contamination) Measures, Schedule B(8) (1999).*
[d]*New South Wales Contaminated Land Management Act (1997); Western Australia Contaminated Sites Act (2003).*
[e]*New South Wales Contaminated Land Management Act (1997); New South Wales Environment Protection Act (1993); South Australia Development Act (1993); and the Australian National Strategy of Ecologically Sustainable Development (1992) for references to ESD. ESD is not included in the Western Australia Contaminated Sites Act (2003) or Western Australia Environmental Protection Act 1993.*
[f]*The New South Wales Contaminated Land Management Act (1997) identifies the precautionary principle as one principle of ESD.*
[g]*See Western Australian Environment Protection Authority Guidance Statement for Remediation Hierarchy for Contaminated land No17. (n.d); South Australia Environment Protection Authority Guidelines for Site Contamination Audit System (EPA SA); New South Wales Department of Environment and Conservation Guidelines for the NSW Site Auditor Scheme (2006); and the Australian National Environment Protection (Assessment of Site Contamination) Measure (1999).*
[h]*See New South Wales Managing Land Contamination——Planning Guidelines SEPP 55 (1998); South Australian Environment Protection Act (1993); Western Australian Contaminated Sites Act (2003); AS/NZS ISO 31,000 Risk Management (2009); Australian National Environmental Protection (Assessment of Site contamination) Measures (1999).*
[i]*New South Wales Protection of the Environment Act (1997), South Australia Environment Protection Act (1993), New South Wales Contaminated Land Management Act (1997), and the Western Australia Contaminated Sites Act (2003). Central to this approach as stated in the New South Wales Contaminated Land Management Act (1997) is "improved valuation, pricing and incentive mechanisms—namely, that environmental factors should be included in the valuation of assets and services."*
[j]*See New South Wales Contaminated Land Management Act (1997), South Australia Environment Protection Act (1993). Intergenerational equity is not incorporated into the Western Australia Contaminated Site Act (2003).*
[k]*See Principle 21, Stockholm Declaration (1972).*
[l]*See New South Wales Contaminated Land Management Act (1997); New South Wales Protection of the Environment Operations Act (1997); South Australia Development Act (1993), and the Australian National Strategy of Ecologically Sustainable Development (1992); Australian National Environmental Protection (Assessment of Site contamination) Measures, Schedule B(9) Guideline on Protection of Health and the Environment During the Assessment of Site Contamination; and, specific guidelines for soil, surface, and ground water protection, fauna and flora (e.g., Development Acts, National Parks and Wildlife Act, Native Vegetation Acts, Environment Protection and Biodiversity Conservation Acts in each Australian state).*
[m]*For volatile odor and gas emissions see, for example: Australian National Environmental Protection (Assessment of Site contamination) Measures, Schedule B(9) Guideline on Protection of Health and the Environment During the Assessment of Site Contamination; Australian National Environment (Ambient Air Quality) Protection Measure (1998), and environmental health risk assessment guidelines in each state.*
[n]*Indicates the participant types that identified the associated Norm during the research: problem holder (PH), local government (LG), auditor (A), remediation service provider (RSP), regulatory authority (RA), and neighbors (N). "All" is used when all participant types identified the Norm.*

about the plan to seek their approval, and the degree to which the problem holder had engaged with the neighbor's concerns in developing the plan.

The problem holder, with the support of the remediation service provider, is the sole performer of the aggregation norms (see Norms 7–13, Table 15.3), which specify the means by which actions will achieve intermediate and final outcomes, and also the scoping norms (see Norms 14–16, Table 15.3), which specify a set of outcomes for the sustainable remediation process. The high number of aggregation norms (see Norms 7–13, Table 15.3) highlights the significant effort that is currently expended by participant types on developing a pathway for obtaining sustainable remediation outcomes that acknowledge, integrate, and build on well-established environmental planning and remediation principles (e.g., the precautionary principle, risk-based fit-for-purpose, ESD, waste minimization). The scoping norms (see Norms 14–16, Table 15.3) highlight that sustainable remediation as practiced at the study sites is not just limited to the selection of "greening" remedies that maximize net environmental benefits (e.g., carbon mitigation). Rather, sustainable remediation includes processes that involve a much broader consideration of environmental, economic, cultural (e.g., indigenous customs), and social (e.g., intergenerational and intragenerational equity) accountability.

Table 15.3 also highlights where norms guiding sustainable remediation were identified as comprising elements traced from, and interdependent with, a broader emerging context of normative practices (principles) associated with sustainability and environmental management. For example, the problem holder's responsibilities in Norm 1 are aligned with the polluter pays principle.

4.2 Rules (norm + formal sanction) encouraging compliance with norms

The frequency counts detailed in Table 15.2 highlight a total of 18 formal sanctions that in most cases the regulatory authorities were responsible for enforcing, supported by local government and other areas of government. These 18 formal sanctions were associated with 15 of the 16 norms detailed in Table 15.3. Within the institutional grammar a norm is transformed into a rule through the development of a formal sanction. Therefore, the study identified 18 rules operating in association with 15 of the norms in the studied sustainable remediation processes, with some of these norms having two associated rules (e.g., see Norm 8, Table 15.3). The nested analysis reveals a close alignment between the norms that emerge in the sustainable remediation processes and the rules that emerge in the same processes.

As one interviewee noted, this close relationship between norms and rules creates a certain economy of compliance within sustainable remediation processes because participant types are likely to act in accordance with norms that have been well socialized within the remediation industry. In many cases they will also be institutionalized as rules through the formation of a social sanction, but the sanction is rendered unnecessary by the socialization. This economy of compliance was further emphasized by the fact that all interviewees responsible for enforcing formal sanctions indicated that they preferred to deal with noncompliance issues on a case-by-case basis rather than by imposing blanket formal sanctions. Generally, they said they sought to be lenient when implementing sanctions, and said this largely reflected the desire to maintain harmonious and cooperative relations between regulating agencies and members of the industry. Of the six regulatory authorities and local government representatives interviewed, only one stated that they interpreted sanctions very literally and five stated that they interpreted the different sanctions with some degree of leniency. One of the latter interviewees noted that she only used sanctions when a problem holder or a member of another participant type was not actively seeking to comply with their obligations. She then stated that, oftentimes, noncompliance with norms had minor or negligible implications, and that it was not appropriate to enact sanctions in these cases. She further stated that conditions largely shaped her interpretation of the need to enact sanctions. For example, a "must" was really only a must in a norm under certain conditions, such as, in the case of extreme risk to human health, or where the problem holder neglects to respond to the neighbors' complaints during a site investigation or remediation and ongoing management (e.g., noise, dust, odor, traffic).

The participant types identified one sanction (i.e., Formal Sanction 5) that may be implemented by neighbors and broader community members against the problem holder if the problem holder didn't obtain an SLO a specific remediation technology from the affected community. The researchers decided to call this neighbor and broader communities-based sanction a formal sanction because it involved tangible acts such as:

- Inhibiting the movement of those carrying out the remediation, through e.g., restrictions on access to neighboring properties and communities near the site
- Lodging objections through formal regulatory processes
- Instigating economic sanctions or restrictions on specific actors carrying out the remediation process
- Protesting against the application of a technology at the site that they did not support
- Generating political sanctions e.g., through engaging with political figures
- "Spotlighting" the actions of those carrying out the remediation with support of the media.

These formal sanctions, if enacted by residents, can pose significant sociopolitical risk to those seeking to carry out remediation through the application of a technology that is not supported by residents. Some of these formal sanctions, such as "spotlighting" and instigating economic sanctions, are not unlike those that are often imposed by environmental courts. Environmental courts have demonstrated some willingness to use sanctions such as "spotlighting" instead of more traditional penalties such as monetary penalties and sentencing. Similarly, the Australian Institute of Criminology notes that orders that might be particularly effective are those that put a spotlight on the fact that a wrongdoing has been committed (Bricknell, 2010). The fact that an environmental wrongdoing has been committed can be highlighted by publicizing the offence in a medium available to the public and/or the offender's peer group. This targets the prestige, profit, and stability of larger corporations and may have a greater deterrent effect than traditional pecuniary penalties (Fisse and Braithwaite, 1988).

The overall mapping of the relationships between the norms and rules (sanctions) operating within the remediation processes at all three sites is detailed in Fig. 15.2.

5. Concluding discussion

The study presented in this chapter provided detailed insights into the complex arrangement of norms and rules (formal sanctions) that guide sustainable remediation. To do so the study used the recently developed IGT to identify the basic elements of the norms used by the participant types who engage in sustainable remediation, and to identify the rules (formal sanctions) that inform their compliance with those norms. The use of the IGT within this study addressed a key theoretical and practical challenge that has faced professionals, communities, and public policy scholars: the need to develop a systematic understanding of the institutional arrangements——the norms and rules——that guide practices such as remediation. While the norms and rules identified through the study are imperfect and will need modifications as research in this nascent area of study develops further, it is a first step toward developing insights that can be used to better navigate the complex landscape of norms and rules that guide the practice of sustainable remediation. These understandings, however nascent, are beneficial to the development of professional guidelines and public policies.

5.1 Normativity driving sustainable remediation

The study revealed the common norms shared by participants across three sustainable remediation processes. What the IGT helps to illustrate are the overall interrelationships between who is involved in fulfilling the norms that make up sustainable remediation processes, who is permitted to perform what actions, who is forbidden to perform them, and under what conditions. That is, the IGT helps to uncover the microelements of the norms at the same time as it reveals a macro view of the norms used in remediation, to help policy analysts and remediation professionals to understand who are the main participants (performers) involved, and where most of the decision-making power is

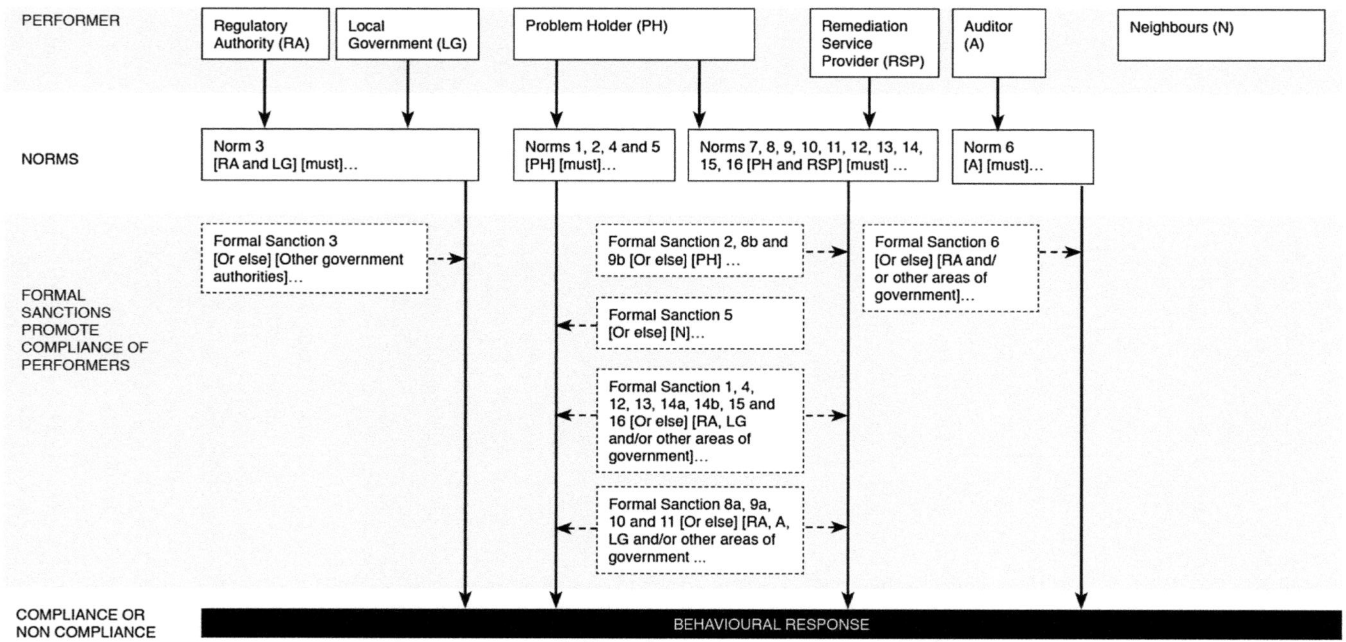

FIGURE 15.2 Mapping of relationships between participant types and norms and rules (norm + sanction) operating in sustainable remediation. *Credit: Created by the Author*

located, which in turn might determine the overall degree of enforceability of the set of norms being analyzed. Given that the formal sanctions provided for by specific remediation legislation and policy in Australia tend to be used only sparingly, and given that the emphasis is on negotiated and voluntary remediation processes, which is supported by the regulatory authority and initiated by the problem holder, it is important to understand and articulate the micro and macro structures of the norms that guide such negotiated and voluntary processes (Fowler, 2008). As Fowler (2008) notes:

Regulators have preferred … [a] negotiated approach because it avoids the possibility of legal appeals or other forms of litigation designed to contest liability to potentially responsible parties. The negotiated approach understandably has also been attractive to potentially responsible parties because it avoids the imposition of formal orders on them. This can be a significant consideration in terms of corporate reporting requirements (Fowler, 2008, n.p.).

Furthermore, it could be argued that the problem holder's willingness to initiate voluntary remediation is partly in response to the growing formalization of many of the norms (e.g., Norm 5, 14, 15) outlined by this study into corporate social responsibilities, which are often written in corporate social policies (e.g., the ethical reasonabilities of a company to do no harm to either humans or the environment). The concept of Corporate Social Responsibility is generally understood to mean that organizations have a degree of responsibility not only for the economic consequences of their activities but also for the social and environmental implications.

One of the clearest findings within the study is the key role that the problem holder plays in performing the vast majority of norms, and the imperative that they must perform these norms, usually supported by the remediation service provider. Notable by its absence in the majority of these norms is an expectation that affected neighbors, auditors, local councils, and regulatory authorities should contribute to the performance of the aims within the norms. Their responsibility, as has been discussed, is focused upon the implementation of sanctions to encourage compliance. Arguably, the focus of responsibility on the problem holder reflects the current acceptance and dominance of the polluter pays principle within Australian remediation practice, which does not just imply, as the phrase might suggest, that the problem holder monetarily support the remediation, but that they must perform the remediation. Furthermore, these identified norms can provide helpful guidance to the growing number of remediation policies and guidelines that call for the problem holder to engage with other participants within the remediation context to identify a mutually "acceptable resolution" (South Australia Environment Protection Authority, 2015).

The study revealed that the norms operating in sustainable remediation comprise elements traced from, and interdependent with, a broader emerging context of normative principles associated with sustainability and environmental management; that is, sustainable remediation can be thought of as being a component of a complex system of normative principles operating within society (see Table 15.3). As illustrated in Table 15.3, sustainable remediation has a normative core that extends beyond the principle of ESD to incorporate a broad range of other normative principles, which include, among others: the polluter pays principle, the principle of shared responsibility, the principle of accountability, the principle of localization, the precautionary principle, the risk-based fit-for-purpose principle, the principle of prevention, the principle of intergenerational and intragenerational equity, the principle of sustainable use, and the principle of waste minimization. Within the remediation processes at these sites participant types often failed to achieve the ideals expressed within each principle (as several of the participants noted during the data collection), and they often had to trade-off one principle against another in order to achieve the overall goal of sustainable remediation. The significance of sustainable remediation is that it allows the participant types to consider and engage a range of individual normative principles, but to gauge their importance collectively. That is, sustainable remediation is an exercise in interstitial normativity, pushing and pulling the boundaries of the set of sustainability and environmental normative principles that constitute it when they threaten to overlap or conflict with each other.

5.2 Formal sanctions promoting compliance

While the problem holders were obliged to perform most norms within the remediation processes (with the support of the remediation service providers), the IGT analysis highlights that the prime responsibility of other participant types——regulatory authority, local council, neighbor, auditor——across the three sites was to ensure that this performance complied with the expectations, needs, and interests of the broader public (e.g., see Norms 3 and 5). An interesting development in the Australian context in recent decades has been the development of a system of environmental auditors (see Norm 6), which represents a partial privatization of the administration of state governments' responsibilities for monitoring and supervising remediation in line with the public's interest (Fowler, 2008, p. 25).

The formal sanctions detailed in Table 15.3 were identified by participants as key means for ensuring that the public interest is protected, largely through deterrence. Broader research highlights that there is a lack of regulatory drivers for sustainable remediation in other national contexts, and that there is a need for a firmer mandate to carry it out (see Ellis and Hadley, 2009; Hou et al., 2014). However, this study identified a broad cross section of state and federal regulatory mechanisms that permit sustainable remediation to be enforced by government in Australia (see notes to Table 15.3). The degree to which regulatory authorities seek to enforce such regulations is reflective of the environmental regulatory hierarchy that has been in operation in the

Australian context (and in many other international jurisdictions) for some decades. Environmental misdemeanors are dealt with using a sequential regulatory pyramid, which commences at the base with attempts to promote voluntary cleanup by wrongdoers (via mechanisms such as corporate due diligence). If these attempts are not successful, the next steps are administrative actions such as notices that are designed to educate perpetrators and rectify wrongdoings. If these actions do not have the desired effect, the next step on the regulatory pyramid is to launch a prosecution that involves deterrents in the form of fines and custodial sentences (Abbot, 2005; Kagan and Scholz, 1984). The emphasis is on voluntary remediation by the industry and regulatory authorities, as discussed in the chapter's methodology. This reflects best practice deterrence and prevention models in which formal sanctions are a last resort. Both Scholz's tit-for-tat enforcement strategy (Scholz, 1984) and Ayers and Braithwaite's enforcement pyramid (Braithwaite, 1985; Grabosky and Gant, 2000) are based on the premise that best-practice regulation must involve a mix of persuasion and punishment, although they differ on how intricate or complex that mix needs to be (Bricknell, 2010). Punishment should be "in the background until there is no choice but to move it to the foreground" (Ayres and Braithwaite, 1992, p. 47). However, punishment must be perceived as unavoidable for those who do not cooperate and adjust their behavior following intervention at the lower levels of the pyramid. Best-practice models require environmental protection agencies to play dual roles as regulators and enforcers. A key consequence of this regulatory pyramid is that formal sanctions need to have an inbuilt flexibility so that authorities are able to use their enforcement discretion based on the problem holder's performance of remediation.

The study revealed one formal sanction that was implemented by the neighbor and broader community that was explicitly recognized by most participant types and the very tangible sociopolitical risks associated with this sanction. This sanction was associated with problem holders obtaining an SLO from neighbors (and broader surrounding communities) for the remediation option selected (see Norm 5). SLO refers to the (often tacit) "contract" with members of an affected community, which enables a problem holder to enter a community and implement a remediation solution (Nelsen, 2006). This contract is a complement to the provision of formal regulatory approval for the remediation solution, and the two are by no means identical. In the remediation processes studied in this research, the attainment of an SLO was closely linked to the selection of technologies and methods, and in particular, to the ability of experts engaged in the processes to incorporate external perspectives and social values into the design of those technologies and methods. Consequently, technology assessment was seen to exist not only within the "non-social" domain of technical expertise and the fields of science and engineering, but also within society more broadly. Ultimately, the selection of remediation technologies was seen as being based on a holistic examination of how they addressed the full spectrum of aims that participant types associated with sustainability.

5.3 Limitations and further research

Despite the study's significant new insights into the norms and rules that are used in sustainable remediation, this study is of course subject to many of the limitations that beset institutional studies. While in the theory behind the IGT the distinctions between norms and rules and their grammatical components are clear-cut, in practice, the borders between various components are seldom as clear. This is because norms and rules are social phenomena (Searle, 2005). Like the grammar of everyday language, any grammar created for norms and rules is subject to misinterpretation and human error. Moreover, in an empirical situation the identification of grammar components is often beset with problems. For example, in the present study people were at first reluctant to reveal the norms that drove their decisions. Nevertheless, the researcher believes that the use of the IGT has helped to increase analytic rigor and the ability to articulate what is happening within sustainable remediation processes. The great achievement of the IGT is that it allows differentiation and classification of the components of norms and rules. Such insights are rarely, if ever, available through other approaches.

Limitations are not restricted to the study's methodology; our findings may also be limited by the study's scope. The study was focused on identifying positioning norms and rules, aggregation norms/rules, and scope norms/rules (McGinnis, 2011; Ostrom, 2005); this does not mean that other rule types specified in the IAD framework were not present (e.g., boundary rules, information rules, payoff rules). Such norm and rule type limitations may only become apparent once further comparative studies have been conducted. Finally, it is worth noting that the study's findings are limited by its focus on Australian remediation processes.

This study of the norms and rules driving sustainable remediation opens up several new perspectives on how such processes operate. The evidence from this study certainly supports the argument that we need to pay greater attention to understanding how norms and rules are incorporated into remediation. The evidence from our analysis provides compelling support for the development of pragmatic tools, studies, and methods to encourage and enable participants within such processes to make explicit their diverse norms and rules and the associations between them.

Acknowledgments

This research has been assisted by the New South Wales Government through its Environmental Trust. Furthermore, this research has been funded by the Cooperative Research Centre for Contamination Assessment and Remediation of the Environment.

References

Abbot, C., 2005. The regulatory enforcement of pollution control laws: the Australian experience. Journal of Environmental Law 17, 161–180.

Adamson, D.T., Mcguire, T.M., Newell, C.J., Stroo, H., 2011. Sustained treatment: implications for treatment timescales associated with source-depletion technologies. Remediation Journal 21, 27–50.

Ayres, I., Braithwaite, J., 1992. Responsive Regulation: Transcending the Deregulation Debate. Oxford University Press, Oxford.

Bardos, P., Nathanail, C.P., Weenk, A., 2000. Assessing the Wider Environmental Value of Remediating Land Contamination: A Review. Environment Agency, Bristol.

Basurto, X., Kingsley, G., Mcqueen, K., Smith, M., Weible, C.M., 2010. A systematic approach to institutional analysis: applying Crawford and Ostrom grammar. Political Research Quarterly 63, 523–537.

Braithwaite, J., 1985. To Punish or Persuade: Enforcement of Coal Mine Safety. State University of New York Press, Albany.

Bricknell, S., 2010. Environmental crime in Australia. In: Australian Institute of Criminology Reports: Research and Public Policy Series 109. Australian Institute of Criminology, Canberra, ACT.

Byrne, D., 2009. Case-based methods: why we need them; what they are; how to do them. In: Bryne, D., Ragin, C.C. (Eds.), The SAGE Handbook of Case-Based Methods. SAGE Publications, London.

Crawford, S.E.S., 2004. An Institutional Grammar of Mores. Prepared for presentation at the Workshop on the Workshop, Bloomington.

Crawford, S.E.S., Ostrom, E., 1995. A grammar of institutions. The American Political Science Review 89, 582–600.

Dixon, T., 2006. Integrating sustainability into brownfield regeneration: rhetoric or reality?—an analysis of the UK development industry. Journal of Property Research 23, 237–267.

Dixon, T., 2007. The property development industry and sustainable urban brownfield regeneration in England: an analysis of case studies in Thames Gateway and Greater Manchester. Urban Studies 44, 2379–2400.

Doick, K., Sellers, G., Castan-Broto, V., Silverthorne, T., 2009. Understanding success in the context of brownfield greening projects: the requirement for out come evaluation in urban greenspace success assessment. Urban Forestry and Urban Greening 8, 163–178.

Ellis, D.E., Hadley, P.W., 2009. Sustainable remediation white paper—integrating sustainable principles, practices, and metrics into remediation projects. In: U.S. Sustainable Remediation Forum.

Feiock, R.C., Weible, C.M., Carter, D.P., Curley, C., Deslatte, A., Heikkila, T., 2014. Capturing structural and functional diversity through institutional analysis: the mayor position in city charters. Urban Affairs Review 52, 129–150.

Fisse, B., Braithwaite, J., 1988. The allocation of responsibility for corporate crime. Sydney Law Review 11, 469–513.

Fowler, R., 2008. The legal framework for management of contamination: international and Australian approaches compared. In: EcoForum Conference. Gold Coast, Queensland, 27–29 February 2008.

Grabosky, P., Gant, F., 2000. Improving environmental performance, preventing environmnetal crime. In: Australian Institute of Criminology Research and Public Policy Series No. 27. Australian Institute of Criminology, Canberra ACT.

Hou, D., O'Connor, D., Al-Tabbaa, A., 2014. Comparing the adoption of contaminated land remediation technologies in the United States, United Kingdom, and China. Remediation Journal 25, 33–51.

Kagan, R.A., Scholz, J.T., 1984. The "criminology of the corporation" and regulatory enforcement strategies. In: Hawkins, K., Thomas, J.M. (Eds.), Enforcing Regulation. Boston, Mass, U.S.A.

McGinnis, M.D., 2011. An introduction to IAD and the language of the Ostrom workshop: a simple guide to a complex framework. Policy Studies Journal 39, 169–183.

Nelsen, J.L., 2006. Social license to operate. International Journal of Mining, Reclamation and Environment 20, 161–162.

Ostrom, E., 2005. Understanding Institutional Diversity. Princeton University Press, Princeton.

Ostrom, E., 2011. Background on the institutional analysis and development framework. Policy Studies Journal 39, 7–27.

Roditis, M.L., Wang, D., Glantz, S.A., Fallin, A., 2014. Evaluating California campus tobacco policies using the American college health association guidelines and the institutional grammar tool. Journal of American College Health 63, 57–67.

Schluter, A., Theesfeld, I., 2010. The grammar of institutions: the challenge of distinguishing between strategies, norms, and rules. Rationality and Society 22, 445–475.

Scholz, J., 1984. Cooperation, deterrence and the ecology of regulatory enforcement. Law & Society Review 18, 179–224.

Searle, J.R., 2005. What is an Institution? Journal of Institutional Economics 1, 1–22.

Siddiki, S., Weible, C., Basurto, X., Calanni, J., 2011. Dissecting policy designs: an application of the institutional grammar tool. The Policy Studies Journal 39, 79–103.

Simon, J.A., 2010. Editor's perspective—green and sustainable remediation—fad or revolution? Remediation Journal 21, 1–8.

South Australia Environment Protection Authority, 2015. Stakeholder Engagement Delivery Framework (Site Contamination). South Australia Government, South Australia.

Sustainable Remediation Forum Australia, Cooperative Research Centre for Contamination Assessment and Remediation of the Environment & Australasian Land and Groundwater Association, 2011a. A Framework for Assessing the Sustainability of Soil and Groundwater Remediation. Sustainable Remediation Forum Australia.

Sustainable Remediation Forum United Kingdom, 2010. A Framework for Assessing the Sustainability of Soil and Groundwater Remediation: Final March 2010. Sustainable Remediation Forum UK, London.

Wernstedt, K., Alberini, A., Heberle, L., Meyer, P., 2004. The Brownfields Phenomenon: Much Ado About Something or the Timing of the Shrewd. Working Paper.

Chapter 16

Socioeconomic benefit of contaminated site remediation

Jingqi Dong[1,2], Hongzhen Zhang[2], Solvita Klapare[3], Jinnan Wang[2], Yongming Luo[4]
[1]*China University of Geosciences, Beijing, China;* [2]*Chinese Academy for Environmental Planning, Beijing, China;* [3]*The World Bank Group, Beijing, China;* [4]*Institute of Soil Science, Chinese Academy of Sciences, Nanjing, China*

1. Background

Over the past half century, China has undergone a rapid industrial and urbanization process. Large areas of farmlands were irrigated by sewage water in 1960–1970s. Municipal and industrial waste water was discharged without any efficient treatment into surface water and groundwater for decades. Soil environmental supervision and legislation are deficient for a long period, especially for heavy pollution industries such as nonferrous metals, chemical industries, petroleum chemistry, pesticides, steel and iron industries, etc. The historic issues in China caused large areas of farmlands polluted, hundreds of thousands of industrial contaminated sites left, and the ecological environment of thousands of mega mining areas and hundreds of river watersheds damaged. Rapid civilization and industrialization result in severe pollution problems that China is currently facing, with massive suspected industrial contaminated lands and polluted farmlands taking about 16% area of total agriculture lands in China. Unlike the integral soil and groundwater environment information bases and systematic risk-based contaminated sites management mechanism that developed countries had already built up, at present, China does not have the capabilities to normatively investigate soil and groundwater environment and identify contamination, overall inventory of contaminated sites, and rational risk management technologies for pollution, not even mention about the risk-based decision-making mechanism and framework, efficient redevelopment and planning system of contaminated lands, and land pollution information disclosure institution.

Under the social contexts of urgent expectations from central government for soil pollution investigation and remediation and huge investments from central soil special funds, following the current market trends of industrial land

Sustainable Remediation of Contaminated Soil and Groundwater. https://doi.org/10.1016/B978-0-12-817982-6.00016-1

remediation and redevelopment driven by urban real estates, backward industries elimination in east China and industrial parks demolition, soil environment pollution remediation industry in China is stepping into its fast growth early stage. However, the forming types of contaminated sites in China are far more than those mentioned earlier. A large amount of contaminated sites are distributed in downtown areas and suburbs of the cities all over the country, as well as rural areas with decentralized village factories.

2. A qualitative cost–benefit analysis case study of contaminated site remediation in China

2.1 Development of models for China

The sites that are classified as contaminated or potentially contaminated and require cleanup undergo analysis using the Early-Stage Planning Model (ESPlaM), which is designed as a Decision Support Tool (DST) to help prioritize the sites in terms of importance and urgency of remediation.

While such models are routinely used worldwide, prioritization of contaminated sites in China does not follow such methodologies as the Multi criteria decision analysis (MCDA) or other analytical tools, partly because such planning is time- and data-intensive, and most of the time such data are not easily available. Instead, the key driving factors for contaminated site remediation remain development potential and likely land price increase. Given these limitations, the study has developed ESPlaM that is largely based on qualitative assessments and could be readily applied in China. The planning model is intended to serve as a DST for the land use planning authorities and the environmental protection departments to come up with an informed decision of the redevelopment of an area of land. The model will help collect data and analyze in a more systematic way the potentially contaminated sites.

In a city or region with more than one contaminated site, all sites can be evaluated using ESPLaM by scoring the core elements, such as damage degrees, technical feasibility of remediation and funding sources, or a qualitative multicriteria method. The final scores of the two categories—"the feasibility of contaminated sites/soil remediation" and "the comprehensive benefits assessment of land redevelopment"—for each site are then compared indicating the priorities for remediation. The matrix can be used as a simple DST for policy, remediation, and land use planners.

Under the key pillars of brownfield redevelopment planning, the United States, EU, and other countries have developed methodologies for brownfield redevelopment planning (Nijkamp et al., 2002; Schädler et al., 2011; Chrysochoou et al., 2012; Chen and Ma, 2013; Chen et al., 2009; Stezar et al., 2013; GARÇÃO, 2015). The key elements considered in these methodologies and proposed criteria that will be applicable in the Chinese context are summarized in Table 16.1.

TABLE 16.1 Core elements considered in brownfield redevelopment planning.

Elements		The Netherlands	Sweden	Germany	The United States	Proposed in China
Indexes	Environment	**1.** Soil type/ composition **2.** Groundwater **3.** Proximity and exposure paths to environmentally sensitive receptors **4.** Land use/zoning	**1.** Soil **2.** Flora and fauna **3.** Groundwater **4.** Surface water **5.** Sediment **6.** Air **7.** Nonrenewable natural resources **8.** Nonrecyclable waste	**1.** Contaminant group **2.** Level of contamination **3.** Depth below grade **4.** Preservation of nature and the landscape **5.** Land use types	**1.** Past use **2.** Soil type (permeability) **3.** Zoning (redevelopment) **4.** Proximity to surface water body and aquifer protection area (drinking water well) **5.** Proximity to protected open space **6.** Proximity to natural diversity areas **7.** Proximity to protected habitats **8.** Proximity to state parks **9.** Characterization as floodplain or wetland **10.** Nature and extent of contamination **11.** Risk management strategy	**1.** Damage degree assess ments **(a)** Toxicities of Contamin ants of Concern (CoC) **(b)** Level of contamin ation **(c)** Predicted remediation soil volume **(d)** Environmentally sensitive receptors **(e)** Affected population **2.** Remedial technical feasibility **(a)** Feasible remediation technologies **(b)** Secondary pollution **3.** Health and environment benefits **(a)** Health risk reduction **(b)** Reducing possibilities of acute environmental accidents **(c)** Ecological risk reduction

Continued

TABLE 16.1 Core elements considered in brownfield redevelopment planning.—cont'd

Elements		The Netherlands	Sweden	Germany	The United States	Proposed in China
	Social	**1.** Situation **2.** Use before cleanup process **3.** Potential use after cleanup process **4.** Legal regulations **5.** Current owner is party concerned **6.** Current owner is causer **7.** Accountability of the current owner	**1.** Local environmental quality and amenity **2.** Cultural heritage **3.** Equity **4.** Health and safety **5.** Local participation **6.** Local acceptance	**1.** Mobility management that saves resources and reduces emissions **2.** High quality of environment for housing and living	**1.** Unemployment rate **2.** Population density **3.** Targeted development indicators **4.** Cultural heritage **5.** Local economic stimulus **6.** Local stakeholder (i.e., Brownfields recipient)	**1.** Comprehensive benefits of the urban sustainable development **(a)** Local economic improvements by site remediation **(b)** Main benefits of reme diation and land redevelopment for urban development **2.** Social impacts of reme diation technologies **(a)** Possibility of chan ging redevelopment land use types **(b)** Public acceptance **(c)** Time scales and urgency of remediation
	Economic	**1.** Costs **2.** Finances **3.** Instruments	**1.** Property value (after remediation) **2.** Health (monetized impaired or improved health)	**1.** Remediation costs for soil and groundwater within each land use classes	**1.** Property values **2.** Potential funding sources and financial guarantee for redevelopment/ restoration	**1.** Potential funding sources and financial guarantee **(a)** Remediation cost funded by polluters

			3. Provision of ecosystem services **4.** Remediation costs **5.** Property value (after redevelopment) **6.** Demolition and construction costs	**2.** Economical land management **3.** Strengthening of the local economy	**3.** Remediation cost funded by regulatory authority (state or federal) **4.** Strengthening of the local economy	**(b)** Remediation cost funded by land users **(c)** Possibility to receive central or provincial appropriation for remediation **2.** Direct and indirect economic benefits **(a)** Land price **(b)** Regional economic improvements
	Smart growth				**1.** Intersection density **2.** Utility service area **3.** Job housing balance **4.** Transportation connectivity improvement	
Method		Rough set analysis (meta-analytic methods)	MCDA-based tool (SCORE)	Multicriteria objective optimization; GIS-based spatial analysis	Multicriteria assessment; GIS-based spatial analysis	Multicriteria assessment; Scenario analysis
Scope		Multiple contaminated sites	One site in different remediation and redevelopment scenarios	One site in different remediation and redevelopment scenarios	Large numbers of brownfield sites in wide areas, each site containing several remedy and redevelopment scenarios	Large numbers of brownfield sites in wide areas, each site containing several remedy and redevelopment scenarios
Purpose		To identify important factors that influence the successes of urban brownfield redevelopment	To apply SCORE in the early planning stages of brownfield redevelopment	To establish a multicriteria assessment framework according to stakeholders' preferences to determine optimal land use configurations	To establish a decision support system for brownfield redevelopment prioritization in urban planning	To establish a decision support system for brownfield redevelopment prioritization in urban planning
References		Nijkamp et al., 2002	GARÇÃO, 2015	Schädler et al., 2011.	Chrysochoou et al., 2012.	

Data framework for the brownfield redevelopment planning model. The following data that fall into four categories are collected to serve as initial set of information: (a) historical data on urban industrial and mining activities, the inventory data of the types and distribution of pollutants; (b) information about site investigations, risk assessments, remediation plans, and construction designs; (c) management and technical documents and policies regarding land redevelopment and remediation; and (d) cities' development plans, especially the proposed land use at the contaminated sites or surrounding areas.

Based on the above-mentioned information, a map of areas of (potential) contaminated sites/an inventory of contaminated sites and the scenarios of land redevelopment are proposed, focusing on the key objects, areas, and categories.

Qualitative evaluation indexes for damages and remediation feasibility versus **comprehensive benefit assessment of land redevelopment.** Taking the concerned sites as evaluation objectives, and considering the core elements listed in Table 16.1, including pollution degree, technical feasibility of remediation, and funding sources, a qualitative multicriteria matrix is adopted for ranking the priority contaminated sites. The recommended criteria are summarized in Table 16.2.

Elements such as contaminated site location, redevelopment plan, or the implementation phasing need to be considered as well. As such, the methodology needs to consider the most significant key factors, including (a) positive health and environmental benefits of contaminated site remediation; (b) the direct and indirect benefits of contaminated site redevelopment; and (c) the comprehensive benefits of urban sustainable development. The qualitative scores of high, medium, and low are then assigned for each criterion.

Evaluation matrix. Based on local land redevelopment and site remediation plans, the areas considered for remediation could be divided into various zones. Each zone is evaluated by the planning model matrix considering feasibility of contaminated site remediation and comprehensive benefits of land redevelopment planning. After application of the evaluation matrix, each site in the zoning list or city/regional inventory is placed in one patch of the nine squares, where (a) the **green** patch indicates the highest redevelopment potential, (b) the **blue** patch indicates high redevelopment potential, (c) the **yellow** patch indicates medium redevelopment potential, (d) the **red** patch indicates low redevelopment potential, and (e) the **black** patch indicates no redevelopment possibility:

The final scores of the two categories (that is, two axes: "the feasibility of contaminated sits/soil remediation" and "the comprehensive benefits assessment of land redevelopment") for each site in the evaluation list are plotted as a point in Fig. 16.1, indicating the priorities for site remediation. The ranking results of the matrix can be evaluated by further available information and the order of the sites can be adjusted according to some external factors such as political preferences, financial support, public interests, and so on Fig. 16.2.

TABLE 16.2 Evaluation indexes for planning model.

Evaluation indexes for damages and remediation feasibility		Evaluation indexes for benefits of land redevelopment	
Damage degree assessment	Toxicity of contaminant of concern (COC) of the site (list of priority controlled pollutants)	Positive health and environmental benefits	Environmental health benefits of remediation, including the number of beneficiaries, level of risk reduction
	Contamination degree (times exceeding the screening level) and ranges (if larger than 1000 m^2)		Environmental benefits of remediation, including reducing environmental accidents; surface water safety; reduced risk of contaminating groundwater
	Environmentally sensitive receptors around the contaminated site (surface water, farmland, ecological protection zones) or groundwater affected		Ecological benefits of remediation, including the improved site's ecological and environmental conditions or the resorting conditions for nearby residents
	Population density within or around contaminated site (including presence of sensitive population kindergartens, schools, nursing homes, and so on)		—
Technical feasibility of remediation action	Feasibility and maturity of remediation technologies, facilities, and implementation practices (considering the fitness of technologies based on site characteristics)	Direct and indirect benefits of site redevelopment	Increase of land price of the site after remediation (housing availability)
	Management complexity and risk level of secondary pollution (or incidents) during remediation		Increase of land price around the site after remediation
	Comprehensive environmental benefits of remedial targets or effects (high, medium, and low)		Growth of the regional economy
	Acceptance degree of public/authorities of the remediation process and results		—
	Timing of remediation (months, years, or decades)		—

Continued

TABLE 16.2 Evaluation indexes for planning model.—cont'd

Evaluation indexes for damages and remediation feasibility		Evaluation indexes for benefits of land redevelopment	
Potential funding sources	Availability of fund sources (high, medium, and low)	Comprehensive benefits of the urban sustainable development of contaminated site redevelopment	Strengthening of the local/regional overall competitiveness as a result of site remediation (quality of life improvements, SMART Growth, etc.)
	Funding demand for remediation compared to local economies (more than average, average, less than average)		Strengthening of the local/regional scientific and technological base
	Possibility to receive central appropriation for remediation (high, medium, low)		Improving local/regional social culture (that is, as a landmark of the city)
	Cost of remediation compared to land value (more than average, average, less than average)		Strengthened local/regional education levels
	—		Growth of local/regional tourism industry (transportation, aesthetics, etc.)

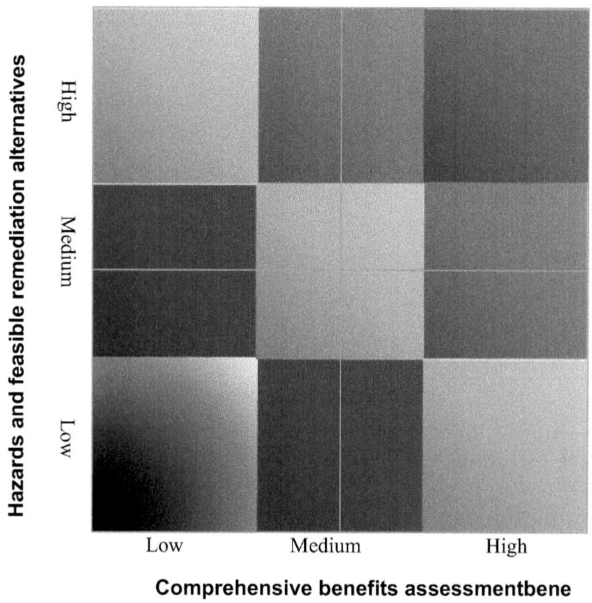

FIGURE 16.1 Comprehensive evaluation matrix of the planning model.

FIGURE 16.2 Sites map for the case study.

2.2 A planning model case at early redevelopment phase of contaminated site

To test the proposed ESPlaM, three typical contaminated sites in Chongqing were selected. The results of field visits, information collected from 170 respondents of the questionnaires, were aggregated in the model to conduct preliminary screening and decision analysis of the factors of feasible remediation technologies, remediation prioritization of the sites, and most suitable land redevelopment potential after remediation.

The questionnaire was in the topic of "land restoration and redevelopment." It mainly included the questions on: (1) environmental, social, and economic indices of the sites remediation and redevelopment decision-making, (2) contamianted sites cost−benefit analysis, and (3) remediation and redevelopment options for the case below. The survey was delivered by paper questionnaire and online questionnaire. About 170 effective respondents were collected from engineers and consultants who participate in the fields of sites remediation, administrators of environmental protection, public who care about contaminated sites, etc.

An area of 262 ha in a megacity houses three contaminated sites. All sites are located in the proximity of a large river and its tributaries. Based on site investigations, the key problem the sites face relates to soil contamination rather than groundwater(Tables. 16.3−16.5).

Site A is a former chromate production factory, located next to a significant tributary of large river. At present, there is existing equipment in place to collect chromium-polluted surface water, which can treat approximately 120,000 m^3 of polluted water annually. Additionally, at the downstream of the river (about 15 km from the site), there is a centralized drinking water source.

Site B is a former chlor-alkali production factory. The soil in the area is contaminated by various organic pollutants. The location of the site is 4 km from a megacity center.

Site C used to be a large iron and steel smelting factory. The area is huge and the pollution situation is complicated. The dominant pollutants include organic and heavy metals. The site is located in an important area of the city in terms of economic activity.

The results of the qualitative simulation of "the feasibility of contaminated sites/soil remediation" and "the comprehensive benefits assessment of land redevelopment" indicate that remediation of site C—the former iron and steel smelting factory—has the highest remediation and redevelopment priority among the three sites (Fig. 16.3). The selection of this site as a priority could be based on the relatively high risk for groundwater/surface water contamination, a largest population affected among the three sites, and the very high potential for redevelopment and finance availability from the sale of remediated land for urban development. These benefits largely surpass the fact that

TABLE 16.3 Contamination information.

Site	Current status	Area (ha)	Pollutant type	Area of soil to Be remediated (ha)	Quantity of the contaminated soil (m^3)	
A	Abandoned industrial land in suburban area	32	Cr^{6+}	30	3,000,000	
B	Abandoned industrial land in city	30	Chlorinated hydrocarbon, benzene, toluene, ethylbenzene, and xylene (BTEX)	3	30,000	
C	Abandoned industrial land in city center	200	PAHs, As, Ni, Pb, total petroleum hydrocarbons (TPHs), and so on	80	PAHs	500,000
				200	As, Ni, Pb, TPH	2,000,000

TABLE 16.4 Available remediation techniques.

Site	Feasible remediation technology		Technology readiness[a]	Expected remediation year	Estimated remediation cost (RMB, millions)	Financial availability[b]
A	Chemical reduction stabilization, in situ injection, pumping treatment		High	5–6	500–800	Low to medium
B	Cement kiln		Relatively high	3–5	160–200	Medium
C	PAHs	Cement kiln	Medium	10	900–1200	High
	As, Ni, Pb, TPHs, and so on.	Chemical reduction stabilization cement kiln				

[a]*This means the maturity of remediation technologies. Although there are limited choices of remediation technologies in chromium-contaminated sites, the technology is mature,*
[b]*Because most of the contaminated site remediation funding comes from the national funds, each provincial authority has to decide which sites require remediation most urgently ("financial availability").*

TABLE 16.5 Risk receptors.

Site	Affected population	Impact to groundwater/ surface water	Preliminary assessment of risk of environmental events in the event of emergency
A	<200	High	High
B	<1500	Medium	Relatively high
C	<80,000	Medium	Relatively high

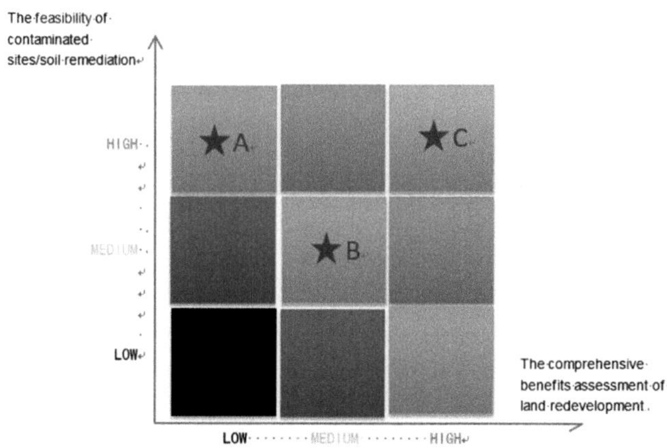

FIGURE 16.3 The evaluation results of the three sites.

remediation technologies are currently limited and may be expensive, and that remediation will be lengthy and complex.

The second ranked priority would be site A. Redeveloping the former chromate production factory site has the next highest priority because of its high potential for groundwater/surface water contamination, and because of its high redevelopment potential and finance availability based on land price increases and sales with limited potential in the area for alternative land uses that are less demanding on the level of cleanup. Also taken into consideration in this prioritization is the opportunity cost associated with not redeveloping this site. Leaving the site as abandoned/wasteland would require investment for fencing and monitoring, the continued cost of maintaining a chrome-contaminated water collection and treatment facility, the risk of negative impact on water quality downstream will remain high, and potential benefit of land value increases will be deferred.

The third ranked site would be Site B The remediation of this former chlor-alkali production factory site is considered a lower priority because of its lower potential for health or environmental risk and its lower potential for redevelopment and financing, compared to sites C and A However, the site is located close to an urban area and does have potential health and ecosystem hazards and should be considered for remediation. Its redevelopment financing potential would depend on the proposed land use and remediation technology choice. Self-financing would only be feasible if the site was developed into a residential or commercial area.

3. Preliminary social cost–benefit analysis of contaminated sites remediation in China at a national level

Based on the assessment of the remediation market by 2020, the study estimates the number of sites that could be remediated under the funding of RMB 150–200 billion using various redevelopment scenarios, ranging from strictest "multi-use" remediation standards to step-by-step risk-based remediation. The analysis suggests that use of smart risk-based approaches ("fit-for-purpose") in contaminated site management would allow financing of up to eight times more sites compared to the "multi-use" remediation.

A national averaged method is used to estimate the number of sites and the areas that could be remediated under the same remediation investment magnitude and the corresponding comprehensive benefits under different scenarios. For the purposes of this method, the following assumptions are made: (A) single site has an average area of 10,000 m^2 with remediation depth of 3 m; (B) the cost of remediation for Scenario 1 (remediation to the strictest standards) is RMB 5000 per m^3, the cost of remediation for Scenario 2 (remediation linked to the predetermined land use) is RMB 1500 per m^3, and the cost of remediation for Scenario 3 (phased risk-based remediation) is RMB 600 per m^3; (C) average land price is RMB 3826 per m^2 based on "2016 Chinese national land resources bulletin," which is the comprehensive land price for 105 main monitored cities all over China in the fourth quarter of 2016 (http://data.mlr.gov.cn/).

Table 16.6 summarizes the number of sites and net monitized benefits that could hypothetically be achieved in China comparing standard (Scenario 1), land-use-linked (Scenario 2), and phased risk-based remediation alternatives (Scenario 3).

TABLE 16.6 Institutional responsibilities for ssite remediation and redevelopment in China.

Scenarios	Description	Numbers of sites that can be remediation with 2016–20 budget of 150–200 RMB billion	Net monitized benefits at investment of 150–200 RMB billion
Scenario 1	The strictest remediation scenario at no risk/zero-risk level	1000–1300	39–52
Scenario 2	Risk-based remediation, with predefined land uses	3300–4400	129–172
Scenario 3	Adjusted future land use types and use of phased approach	8300–11,000	322–429

4. Conclusions

(1) Most of the cities in China are facing the tricky problems that there are insufficient contaminated sites information and it's difficult to complete overall investigations for all the industrial sites in a relative short time during the land redevelopment planning process of "industry-out and service-in" (which is a policy that most cities implemented to move out the secondary industries from the city and introduce the service sector in). It's a challenge for all the officers in Chinese local government, urban planning department, land resources department, and environmental protection department that how to make decisions based on the factors of soil pollution potential risk levels, remediation technology feasibilities, and land redevelopment demands under the premise of limited time and funding supports. To solve the problem, a comprehensive industrial contaminated lands remediation and redevelopment projects screening and optimization planning model is proposed, which is suitable at urban or regional scale and applicable to the early redevelopment planning stage with insufficient basic industrial land information.

(2) Three typical sites in Chongqing were selected to be applied in the model for screening and comparison in the case study. The results of field visits, information collection, and 170 respondents of the questionnaire are aggregated in the model to conduct preliminary screening and decision analysis for the factors of feasible remediation technologies, remediation prioritization of the sites, and best mode of land redevelopment after remediation.

(3) A preliminary social cost–benefit analysis of contaminated sites remediation in Chinese national level was conducted and the result suggests that under the estimated soil pollution prevention funding between RMB 150–200 billion in 2016–20, 1000–1300 sites could be remediated if the strictest remediation Scenario 1 is used, 3300–4400 sites in case of Scenario 2, and 8300–11,000 sites using Scenario 3. As such, using phased risk-based approaches could lead to ability to remediate up to eight times the number of sites compared with the business as usual. Analysis and prediction of Chinese soil remediation industry indicate that the release of *Soil Pollution Prevention and Control Action* will promote a booming growth to Chinese environmental remediation market. By applying the risk-based and phased approaches, **China will form an innovation initiative mechanism and practical situation based green and sustainable environmental remediation supervision, technology promotion, and public involvement system.**

References

Chen, I.C., Ma, H.W., 2013. Using risk maps to link land value damage and risk as basis of flexible risk management for brownfield redevelopment. Chemosphere 90, 2101–2108.

Chen, Y., Hipel, K.W., Kilgour, D.M., 2009. A strategic classification support system for brownfield redevelopment. Environmental Modelling & Software 24, 647–654.

Chrysochoou, M., Brown, K., Geeta Dahal, G., et al., 2012. A GIS and indexing scheme to screen brownfields for area-wide redevelopment planning. Landscape and Urban Planning 105, 187–198.

Garção, R., 2015. Assessment of Alternatives of Urban Brownfield Redevelopment- Application of the SCORE Tool in Early Planning Stages. Master's Thesis. Chalmers University of Technology, Sweden.

Stezar, I.C., Pizzol, L., Critto, A., et al., 2013. Comparison of risk-based decision-support systems for brownfield site rehabilitation: DESYRE and SADA applied to a Romanian case study. Journal of Environmental Management 131, 383–393.

Schädler, S., Morio, M., Bartke, S., Rohr-Zänker, R.R., Finkel, M., 2011. Designing sustainable and economically attractive brownfield revitalization options using an integrated assessment model. Journal of Environmental Management 92, 827–837.

Nijkamp, P., Rodenburg, C.A., Wagtendonk, A.J., 2002. Success factors for sustainable urban brownfield Development. a comparative case study approach to polluted sites. Ecological Economics 40, 235–252.

Further reading

Alam, K., 2008. Cost–benefit analysis of restoring Buriganga River, Bangladesh. International Journal of Water Resources Development 24 (4), 593–607.

Barbara Karn, B., Kuiken, T., Otto, M., 2009. Nanotechnology and in situ remediation: a review of the benefits and potential risks. Environmental Health Perspectives 117 (12), 1823–1831.

Chen, F.F., 2007. Recovering Urban Land: A Framework to Improve Brownfield Redevelopment Practices, Case of Shenzhen, China. Master thesis(the Netherlands).

CLARINET, 2002. Remediation of Contaminated Land Technology Implementation in Europe.

SADA. Cost Benefit Analysis, [EB/OL]. http://www.sadaproject.net/cost_benefit.html, 2007/2016.

Cost-benefit Analysis of Remediation, Livelihood and Income from the Environment Program, Lead Pollution Clean-Up in Qalyoubia, 2007.

De Sousa, C.A., 2002. Brownfield redevelopment in Toronto:an examination of past trends and future prospects. Land Use Policy 19, 297–309.

USEPA. Federal Agency Hazardous Waste Compliance Docket [EB/OL]. www.epa.gov/fedfac/previous-federal-agency-hazardous-waste-compliance-docket-updates,2016-3-8/2016-3-18.

Environment Agency, 2009. Dealing with Contaminated Land in England and Wales - A Review of Progress from 2000-2007 with Part 2A of the Environmental Protection Act [R]. Bristol: EA.

Faisal, K.I., Tahir, H., Ramzi, H., 2004. An overview and analysis of site remediation technologies. Journal of Environmental Management 71, 95–122.

Fisher BH. Brownfields Redevelopment and Gentrification: A Socio-Economic Evaluation of the EPA Brownfields.

Gamo, M., Oka, T., et al., 2003. Ranking the risks of 12 major environmental pollutants that occur in Japan. Chemosphere 53, 277–284.

Hardisty PE, Ozdemiroglu E. Costs and Benefits Associated with the Remediation of Contaminated Groundwater: Application and Example. R&D Technical Report P2-078/TR.

Kenney, M., White, M., 2007. Reclaiming the Land - Rethinking Superfund Institutions, Methods and Practices. Springer, pp. 169–196.

Lavee, D., Ash, T., Baniad, G., 2012. Cost-benefit analysis of soil remediation in Israeli industrial zones. Natural Resources Forum 36, 285–299.

Liedekerke, M., Prokop, G., Rabl-Berger, S., et al., 2014. Progress in the Management of Contaminated Sites in Europe. EUR 26376 EN.

Murray, C.J.L., Ezzati, M., Flaxman, A.D., et al., 2012a. GBD 2010: design, definitions, and metrics. The Lancet 380, 2063–2066.

A. Myrick Freeman III,J.A. Herriges,and C.L.Kling. The Measurement of Environmental and Resource Values, Theory and Methods. Resource for the Future.

USEPA. Road Map to Understanding Innovative Technology Options for Brownfields Investigation and Cleanup, third ed. EPA 542-B-01-001.

Office of Solid Waste and Emergency Response, 2000. Innovative Remediation Technologies: Field-Scale Demonstration Projects in North America, second ed. U.S. EPA, Washington, D.C.

Otsuka, N., Dixon, T., Abe, H., 2013. Stock measurement and regeneration policy approaches to 'hardcore' brownfield sites: England and Japan compared. Land Use Policy 33, 36–41.

Pediaditi, K., Doick, K.J., Moffat, A.J., 2010. Monitoring and evaluation practice for brownfield, regeneration to greenspace Initiatives, A meta-evaluation of assessment and monitoring tools. Landscape and Urban Planning 97, 22–36.

Pilot Program, 2011. Iowa State University, Master Thesis.

Postle, M., Fenn, T., Grosso, A., Steeds, J., 1999. Cost Benefit Analysis for Remediation of Land Contamination. Environment Agency, UK. R&D Technical Report P316.

Schilling, M., 2012. The Management of Contaminated Sites and the Soil Remediation Industry in Denmark. Danish Ministry of the Environment.

Sooderqvist, T., Petra Brinkhoff, P., Tommy Norberg, T., et al., 2015. Cost-benefit analysis as a part of sustainability assessment of remediation alternatives for contaminated land. Journal of Environmental Management 157, 267–278.

Sweden, E.P.B., 2008. Cost Benefit Analysis as a Tool for Prioritizing Remediation, Method Development and Application Examples.

The Office of the Under Secretary of Defense for Acquisition, Technology and Logistics, 2011. Defense Environmental Programs - FY10 Annual Report to Congress. Department of Defense, Washington D.C.

U.S. EPA, 2010. Guidelines for Preparing Economic Analyses.

USEPA, 2004. Cleaning up the Nation's Waste Sites: Markets and Technology Trends, 2004 Edition. EPA 542-R-04-015.

USEPA, 2011. Handbook on the Benefits, Costs, and Impacts of Land Cleanup and Reuse. EPA – 240 – R – 11 - 001.

USEPA. Economic Analysis of the Implementation of Permeable Reactive Barriers for Remediation of Contaminated Ground Water.

van Weze, A.P., Franken, R.O., Drissen, E., et al., 2007. Societal cost–benefit analysis for soil remediation in The Netherlands. Integrated Environmental Assessment and Management 4 (1), 61–74.

Waste, S., Emergency Response, 2007. Treatment Technologies for Site Cleanup: Annual Status Report, twelth ed. EPA, Washinton, D.C.

WHO, Metrics: Disability-Adjusted Life Year (DALY) [EB/OL]. http://www.who.int/healthinfo/global_burden_disease/metrics_daly/en/, 2016-6-12.

USEPA. Types of Brownfields Grant Funding, Brownfields Area-Wide Planning Program [EB/OL]. www.epa.gov/brownfields/types-brownfields-grant-funding#tab-5, 2016-05-09/2016-05-18.

Index

'*Note:* Page numbers followed by "f" indicate figures, "t" indicates tables.'

D

E

R

S